ISBN 978-3-662-42795-8   ISBN 978-3-662-43074-3 (eBook)
DOI 10.1007/978-3-662-43074-3

TABLE OF CONTENTS

	Page
I. INTRODUCTION	1
II. DATA	7
A. GERMANIUM	7
1. Tetraorganogermanes	7
Symmetrical Tetraorganogermanes	7
Asymmetrical Tetraorganogermanes	12
Non-Functional Tetraorganogermanes	12
Germanes Containing Olefinic Groups	14
Germanes Containing Halo- and Cyano-Substituted Groups	17
Germanes Containing Carbonyl and Carboxylic Groups	18
Germanes Containing Hydroxy Groups	22
Germanes Containing Functional Nitrogen	23
Germanes Containing Polyfunctional Groups	23
Tetraheterocyclylgermane	24
2. Alkali Metal Germanides	24
3. Organogermanium Hydrides	25
4. Organogermanium Halides and Homologs	31
Triorganogermanium Halides	31
Diorganogermanium Dihalides	38
Organogermanium Trihalides	42
Organogermanium Cyanides, Isocyanates, Thiocyanates, and Isothiocyanates	47
5. Organogermanium-Oxygen Compounds	52
Organogermanonic Acids and Acid Anhydrides	52
Organogermanium Oxides	53
Organogermanium Hydroxides and Peroxides	56
Organogermanium Alkoxides	59
Organogermanium Carboxylates	60
Organogermanium Sulfates and Chromates	66
6. Organogermanium-Sulfur Compounds	68
Organogermanium Sulfides	68
Triethylgermanium Mercaptides	71

	Page
7. Aminogermanes	72
8. Organogermanium Compounds Containing Silicon and Tin	73
9. Digermanes and Bisgermyl-Aklanes and -Acetylenes	75
10. Polymeric Organogermanium Compounds	78
11. Derivatives of Divalent Germanium	78

B. TIN — 79

1. Tetraorganotin Compounds — 79
    - Symmetrical, Non-Functional Tetraorganotin Compounds — 79
    - Symmetrical, Functionally Substituted Tetraorganotin Compounds — 90
    - Asymmetrical Tetraorganotin Compounds — 93
        - Non-Functional Tetraorganotin Compounds — 93
        - Halo-, Cyano-, and Thiocyano-Substituted Compounds — 101
        - Carboxy-, Carbalkoxy-, Carbamyl-, and Dicarboxyanhydro-Substituted Compounds — 105
        - Compounds Containing the Carbonyl Groups and Their Derivatives — 109
        - Compounds Containing the Hydroxy, Ether, and Ester Groups — 112
        - Amino- and Amido-Substituted Compounds — 115
        - Compounds Containing Olefinic Substituents — 117
        - Heterocycle-Substituted Compounds — 122
        - Compounds Containing Acetylenic Groups — 124
        - Hetero-Polyfunctionally Substituted Compounds — 124

2. Alkali Metal Organostannides — 126

3. Organotin Hydrides — 128

4. Organotin Halides — 132
    - Triorganotin Halides — 132
        - Homogeneous, Non-Functionally Substituted Triorganotin Halides — 132
        - Homogeneous, Functionally Substituted Triorganotin Halides — 143
        - Heterogeneous, Non-Functionally Substituted Triorganotin Halides — 146

		Page
	Heterogeneous, Functionally Substituted Triorganotin Halides	149
	Silicon-Containing Triorganotin Halides	154
	Diorganotin Dihalides	155
	Dialkyl- and Diaryltin Dihalides	155
	Functionally Substituted Diorganotin Dihalides	166
	Silyl- and Heterocycle- Substituted Diorganotin Dihalides	171
	Organotin Trihalides	172
	Alkyl- and Aryltin Trihalides	172
	Functionally Substituted Alkyl- and Aryltin Trihalides	175
5.	Organotin-Oxygen Compounds	177
	Organostannonic Acids and Their Derivatives	177
	Organotin Tricarboxylates	180
	Diorganotin Oxides and Their Complexes	180
	Diorganotin Bis(organoöxides)	186
	Dialkyl- and Diarylalkoxytin Carboxylates	189
	Diorganotin Dicarboxylates, Disulfonates, and Arsonates	190
	Polymeric Diorganotin Dialkoxides, Dicarboxylates, and Their Complexes	194
	Other Diorganotin Salts	197
	Triorganotin Hydroxides	198
	Bis(triorganotin) Oxides	200
	Triorganotin Alkoxides and Phenoxides	203
	Triorganotin Peroxy Derivatives	205
	Triorganotin Carboxylates	205
	Organotin Inner Salts	211
	Organotin Phosphorus Salts	211
	Other Triorganotin Salts	216
6.	Organotin-Sulfur and -Selenium Compounds	216
	Organotin Trisulfides and Triselenides	216
	Organotin Trimercaptides	217
	Diorganotin Sulfides	220
	Diorganotin Dimercaptides	221
	Organotin Thiocyanates, Xanthates, Dithiocarbamates, and Dithiophosphates	226
	Bis(triorganotin) Sulfides	227
	Triorganotin Mercaptides	227
7.	Organotin Grignard Compounds	231
8.	Organotin Derivatives of Boron and Aluminum	231

		Page
9.	Organotin Derivatives of Silicon and Germanium	232
10.	Organotin-Nitrogen Compounds	237
11.	Organotin-Phosphonium Complex Salts	238
12.	Organotin-Metal Carbonyls and -Metal Carbonyl Phosphine Complexes	239
13.	Organotin Heterocycles	241
14.	Hexaorganoditin Compounds and Tetraorganoditin Dihalides and Dinitrates	243
15.	Bridged Bis(triorganotin) Compounds	246
16.	Miscellaneous Organotin Compounds Including Polymers	249
17.	Organic Derivatives of Divalent Tin	251

C. LEAD   255

1. Tetraorganolead Compounds   255
   - Symmetrical, Non-Functional Tetraorganolead Compounds   255
   - Symmetrical, Functionally Substituted Tetraorganolead Compounds   266
   - Asymmetrical, Non-Functional Tetraorganolead Compounds   268
   - Asymmetrical, Homofunctional Tetraorganolead Compounds   273
   - Asymmetrical, Heterofunctional Tetraorganolead Compounds   279
   - Tetraheterocyclyllead Compound   281

2. Alkali Metal Triorganoplumbides   281

3. Organolead Hydrides   282

4. Organolead Halides   282
   - Homogeneous Triorganolead Halides   282
   - Heterogeneous Triorganolead Halides   287
   - Diorganolead Dihalides   288
   - Organolead Trihalides   291

		Page
5.	Organolead-Oxygen Compounds	292
	Organoplumbonic Acids	292
	Organolead Tricarboxylates	293
	Diorganolead Oxides	295
	Diorganolead Dicarboxylates	295
	Di- and Triorganolead Hydroxides	300
	Di- and Triorganolead Alkoxides and Aryloxides	301
	Di- and Triorganolead Sulfonates	302
	Triorganolead Carboxylates	304
6.	Organolead-Sulfur Compounds	309
	Organolead Sulfides, Mercaptides, and Thiocarboxylates	309
7.	Organolead-Amides and -Imides	310
8.	Organolead Cyanides, Cyanates, Thiocyanates, Selenocyanates, Nitrates, Sulfates, Sulfites, Selenites, and Phosphonate	312
9.	Hexaorganodilead Compounds	315
10.	Bridged Bis(triorganolead) Compounds	319
11.	Organolead-Metal Carbonyl Compounds	321
12.	Miscellaneous Organolead Compounds	322
13.	Organic Derivatives of Divalent Lead	323
III.	BIBLIOGRAPHY TO DATA SECTION	325
IV.	BIBLIOGRAPHY OF MONOGRAPHS AND REVIEW ARTICLES	361
V.	LIST OF COMPOUNDS WITH PAGE INDEX	363

I. INTRODUCTION

SCOPE OF SEARCH This volume comprises organic derivatives of germanium, tin, and lead which contain at least one covalent carbon-metal bond.

As stated in the Preface to the Series (RD-59-Lit. Survey No. 1250), the non-critical compendium of organometallic compounds reported in the period 1937-1958 was prepared for an easy reference to the methods for their preparation and to their physical and chemical properties.

The search is based upon Chemical Abstracts (CA), Volume 31 (1937) through Volume 51 (1957) and upon Current Chemical Papers published by the Chemical Society (London), 1958 through August 1959; thus this volume also comprises literature published in 1959.

Since the literature of organic derivatives of germanium, tin, and lead prior to 1937 is thoroughly covered by E. Krause and A. von Grosse in "Die Chemie der metallorganischen Verbindungen," Verlag von Gebruder Borntraeger, Berlin, 1937, CA 31 was taken as the starting point.

Volume III of the Series will cover organic derivatives of arsenic, antimony, and bismuth.

HIGHLIGHTS After 1936, new types of organic derivatives of the three metals were synthesized, numerous new compounds were added to the previously-existing classes, and previously known compounds were prepared by new or improved methods.

The progress made and amount of research work done in the field of organogermanium compounds in the last two decades are relatively small, as compared with the developments in the chemistry of organotin and organolead compounds. Attention of commercial research laboratories has been focused on organic derivatives of tin and lead because they are commercially useful.

GERMANIUM Organogermanium hydrides, carboxylates, peroxides, organogermyl-silicon and tin derivatives, and acetylene-bridged bis(triorganogermyl) compounds constitute the new classes of germanium derivatives.

The existence of organic derivatives of bivalent germanium as well as of "hexaphenylgermin" — a hexaphenyl - substituted six-membered germanium ring — has been reported. Their structure and composition, however, have not been proved.

Organogermanium hydrides, prepared by reducing organogermanium halides with lithium aluminum hydride or other suitable reducing agents, are very useful intermediates for the preparation of other organogermanium compounds. The hydrides combine with olefinic compounds to form the corresponding saturated organogermyl-substituted compounds. This reaction is used in the preparation of organogermanium trihalides from trihalogermanes and α-olefins. In reactions with diazoalkanes the organogermyl group and the hydride hydrogen combine with the carbon atom bearing the diazo group, and nitrogen is set free. The hydride hydrogen is replaceable by halogens or alkali metals. Atmospheric oxygen oxidizes organogermanium hydrides to the corresponding oxides or hydroxides.

Organolithium compounds have been extensively used for the preparation of organogermanium compounds from germanium tetrahalides.

Numerous haloalkylgermanium derivatives were prepared by treating germanium tetrahalides or organogermanium halides with diazoalkyl compounds. In this reaction, the carbon atom bearing the diazo group is inserted between the germanium and halogen atoms, and nitrogen is set free.

Arylgermanoic acid anhydrides were prepared by reacting germanium tetrahalides with aryldiazonium fluoborates in acetone in the presence of zinc powder.

Alkyl halides react with germanium powder in the presence of copper to yield organogermanium halides. The method, however, requires improvement.

Allyl-substituted germanes were polymerized and copolymerized with olefinic compounds at elevated temperatures under very high pressure in the presence of radical type catalysts.

No progress has been made in the extension of the "aliphatic" organogermane series. Octaphenyltrigermane, $Ph_3GeGePh_2GePh_3$, the longest chain of organo-substituted germanium atoms known so far, was reported in 1930.

Various methods for the preparation of the individual types of organogermanium compounds are summarized separately for each group of compounds in the Data Section.

TIN Organotin compounds have been extensively used as stabilizers for halogenated organic compounds, especially for halogen-containing polymers. Since some organotin compounds showed germicidal, fungicidal, and bactericidal properties, numerous new tin derivatives were prepared, and their activity and toxicity were biologically tested. Some organotin compounds found application as anthelmintics.

As shown in the Table of Contents, functionally-substituted, homogeneous and mixed tetraorganotin compounds were synthesized. Similar progress was made in the class of organotin halides.

Organotin hydrides have been used as versatile intermediates for the preparation of other organotin compounds. Their reactions, particularly or triphenyltin hydride, have been extensively investigated. Reactions with diazo compounds, such as diazomethane, diazoacetophenone, and diazoacetic acid esters proceed in the same manner as the corresponding reactions of organogermanium hydrides.

With alkenes or aralkenes, organotin hydrides yield the corresponding alkyl- and aralkyltin derivatives, respectively. If the olefinic compounds have another functional group, the course of the reaction, particularly with triphenyltin hydride, depends on the type of the other functional group. Thus, α,β-unsaturated aldehydes and ketones are reduced to α,β-unsaturated carbinols, while hexaphenylditin is formed; α,β-unsaturated halides exchange the halogen atoms for the hydride hydrogen; α,β-unsaturated mercaptans react with two moles of the hydride to form 1,1-bis(triphenylstannyl)-2-alkenes and hydrogen sulfide. α,β-Unsaturated carboxylic acids react with triphenyltin hydride to form a γ-lactone-like diphenyltin compound, $Ph_2SnCHRCH_2COO$, with elimination of one phenyl group. Similar reaction of tripropyltin hydride with acrylic acid yields tripropyltin acrylate.

Acetylenic compounds react with one and with two moles of triorganotin hydrides to yield alkenyltriorganotin and bis(triorganotin)alkanes, respectively.

Various haloalkyltin derivatives have been prepared from tin tetrahalides or organotin halides and diazoalkanes.

Numerous organotin mercaptides, carboxylates, alkoxides, and phenoxides found commercial application as stabilizers for halogenated vinyl polymers.

Reactions of organotin compounds with various organo-phosphorus compounds yielded only organotin phosphonates; no compounds with a tin-phosphorus bond were obtained.

Organotin derivatives of boron, aluminum, silicon, and germanium were synthesized; however, only germanium and silicon derivatives with the Si-Sn and Ge-Sn bonds are known. Organotin compounds containing a haloalkyl group form Grignard compounds, which are useful intermediates in reactions with boron and silicon halides.

Allyl-substituted organotin compounds were recently polymerized and copolymerized with other olefinic compounds in the presence of radical type catalysts at elevated temperatures under high pressure.

LEAD Organolead compounds, particularly tetraethyllead, which at the present time are the best antiknock fuel additives, became large volume commercial products. Commercial processes for the manufacture of tetraethyllead have been perfected. Toxicity of organolead compounds and their decomposition products, which contribute to air pollution in the form of exhaust gases from automobile engines, has been studied. Triorganolead salts, particularly tripropyllead salts which, irrespective of the acid radical, showed marked sternutatory properties, were prepared and biologically tested.

Magnetic and molecular weight data indicate that organic derivatives of "trivalent" lead are dimeric; thus pertinent compounds appear in this volume in the group of hexaorganodilead compounds.

Trimethyllead hydride, the only organolead hydride, was prepared by reducing trimethyllead chloride with potassium hydroborate. The reactivity of the very unstable compound, which decompose above its melting point ($-100°$) was not further investigated.

Alkali metal triorganoplumbides should be mentioned as a new class of organolead derivatives useful in the preparation of bridged diplumbyl derivatives and asymmetrical organolead compounds.

Organolead halides react with diazoalkanes in the same manner as do the corresponding germanium and tin derivatives.

New classes or organolead compounds include acetylene-bridged triorganolead derivatives; N-organoplumbyl-amides, -imides, and -sulfonamides; and organolead sulfonates.

NOMENCLATURE   An attempt was made to use the nomenclature of Chemical Abstracts. It should be pointed out that the CA nomenclature has varied in the course of years. Moreover, the nomenclature of organogermanium compounds differs considerably from that of the corresponding tin and lead compounds. For example, organogermanium hydrides, $R_{3-x}GeH_{1+x}$ are named mono-, di-, and tri-organogermanes, respectively, while the corresponding tin and lead compounds are named mono-, di-, and tri-organotin (or-lead) hydrides. Tetraorganogermanium compounds are named tetraorganogermanes, and the corresponding tin and lead compounds are named tetraorganotin and tetraorganolead. Hexaorganodigermanium compounds are named digermanes, but the corresponding tin and lead derivatives are listed as hexaorganoditin and hexaorganodilead, not distannanes and diplumbanes.

For the sake of clarity, formulas of all compounds are given in the report.

ABBREVIATIONS   The following abbreviations are used along with those used by CA: rxn. stands for reaction; r.t. for room temperature; m. for melting point; b. for boiling point; and prod(s). for product(s).

II. DATA

A. GERMANIUM

1. TETRAORGANOGERMANES

SYMMETRICAL TETRAORGANOGERMANES

The methods for the preparation of tetraorgano-substituted germanes, known before 1936, include the reactions of germanium tetrahalides with dialkyl-, diaralkyl-, or diaryl-zinc, with organomagnesium halides, or with aryl halides and sodium. Now these compounds are conveniently prepared by reacting germanium tetrahalides with organolithium compounds in the presence or absence of solvents.

$\quad$ TETRAMETHYLGERMANE $\quad$ Me_4Ge
Prepn.: By reacting $GeBr_4$ or $GeCl_4$ with MeMgI in Bu_2O or Et_2O (63, 320, 330).
By reacting $GeCl_4$ with Me_2Zn (1:2) at -70 to $0°$; theoret. yield (313).
Props.: b. $44.3°/740mm$; $n^{20}D$ 1.3882 (330, 579); b. $43.5°/760mm$; azeotropic mixt. with Et_2O, b. $34°$ (320); Mol. wt. measurements indicate monomeric structure (785); Data for the refractive index, density, and viscosity (100a). Magnetooptical data (320); Magnetic susceptibility (367); Polarity (464); Force consts. (774); Thermodynamic data (330); Mass spectrum (112); Raman spectrum (330, 510, 558, 579); IR spectrum (330, 379); Potential barrier value (136); Tetrahedral symmetry; Ge-C bond distance (60); Bond strength (511).
Rxns. with: $Br_2 \rightarrow Me_3GeBr$ and MeBr (63)
$\quad$ I_2 in the presence of $AlI_3 \rightarrow Me_3GeI$ and MeI (323)

TETRAETHYLGERMANE Et_4Ge

Prepn.: By reacting $GeBr_4$ with $EtMgBr$ in Et_2O; up to 80% yield along with some $(Et_3Ge)_2$ (187, 300).
By reacting $GeBr_4$ with $EtLi$; 12% yield along with 8.6% $(Et_3Ge)_2$ (187).

Props.: The compound crystallizes in two forms, one melting at 180.47°K and the other at 183.3°K (529); b. 163.5°/760 mm; magnetooptic data (320); Data for the refractive index, density, and viscosity (100a); Magnetic susceptibility (367, 407).

Rxns. with: Li or Na-K, the C-Ge bond remains intact (187)
  Na-K in $Et_2O \rightarrow$ unidentified prods. different from Et_4Ge (676)
  I_2 (1:1) in the presence of $AlI_3 \rightarrow Et_3GeI$ (323)

TETRAVINYLGERMANE $(CH_2:CH)_4Ge$

Prepn.: By reacting $GeCl_4$ with $CH_2:CHMgBr$ (1:4) in C_4H_8O under reflux; $[(CH_2:CH)_3Ge]_2$ is also formed (490).

Props.: b. 52-54°/27mm; $n^{25}D$ 1.4676; d_{20} 1.040 (490)

TETRAPROPYLGERMANE Pr_4Ge

Prepn.: By reacting Ge tetrahalides with $PrMgBr$ in Et_2O, and then refluxing the mixt. in a higher boiling solvent (320)

Props.: b. 225°/760 mm; magnetooptical data (320); Data for refractive index, density, and viscosity (100a); Magnetic susceptibility (367).

Rxn. with: Br_2 at 45° followed by excess 25% $NaOH \rightarrow (Pr_3Ge)_2O$ (10,11)
  I_2 in the presence of $AlI_3 \rightarrow Pr_3GeI$, PrI, and $n-C_6H_{14}$ (323)

TETRAALLYLGERMANE $(CH_2:CHCH_2)_4Ge$

Prepn.: By reacting $GeCl_4$ with $CH_2:CHCH_2MgBr$ in Et_2O (368)

Props.: b. 106-107°/15 mm; $n^{20}D$ 1.5030; d^{20}_4 1.0094 (368)

Rxns.: On heating in the presence of $Bz_2O_2 \rightarrow$ a solid, insol. polymer (368)
  with: Strong org. acids $\rightarrow$ cleavage of the allyl groups (368)

TETRABUTYLGERMANE Bu_4Ge

Prepn.: By reacting $GeBr_4$ with $BuMgBr$ in Et_2O; 80% yield (13,320)

Props.: b. 278°; b. 160-161°/17 mm; d^{20}_4 0.934; $n^{20}D$ 1.4571, M.R. 87.8 (13); b. 278°; magnetooptical data (320); Data for the refractive index, density, and viscosity (100a)

Rxns. with: $I_2 \rightarrow Bu_3GeI$ (13, 306)
  Br_2 in $(CH_2Br)_2$, followed by $KOH \rightarrow (Bu_3Ge)_2O$ (160)
  Li or K-Na, no cleavage of the C-Ge bond (187)
  Br_2 (1:1) in the presence of $AlBr_3 \rightarrow Bu_3GeBr$ (324)

TETRAISOBUTYLGERMANE $i\text{-}Bu_4Ge$
Props.: b. $135°/17$ mm; n^{20}_D 1.4594; d^{20}_4 0.9374 (323)
Rxn. with: I_2 (1:1) in the presence of $AlI_3 \rightarrow i\text{-}Bu_3GeI$ and $(i\text{-}Bu)_2$ (323)

TETRAPENTYLGERMANE $(n\text{-}C_5H_{11})_4Ge$
Prepn.: By reacting Ge tetrahalides with $n\text{-}C_5H_{11}MgBr$ in Et_2O and completing the rxn. in a higher boiling solvent (320).
Props.: b. $173°/4$ mm; magnetoöptical data (320); Data for the refractive index, density, and viscosity (100a); Magnetic susceptibility (367, 407)
Rxns. with: I_2 (1:1) in the presence of $AlI_3 \rightarrow (n\text{-}C_5H_{11})_3GeI$ and $C_{10}H_{22}$ (323)
Br_2 (1:1) in the presence of $AlBr_3 \rightarrow (n\text{-}C_5H_{11})_3GeBr$ (324)

TETRAHEXYLGERMANE $(n\text{-}C_6H_{13})_4Ge$
Prepn.: By reacting $GeCl_4$ with $n\text{-}C_6H_{13}Li$ at $0°$ (140)
By reacting Ge tetrahalides with $C_6H_{13}MgBr$ in Et_2O and completing the rxn. in a higher boiling solvent (320)
Props.: b. $158\text{-}161°/0.5$ mm; n^{27}_D 1.4567; d_{27} 0.908 (140)
b. $192°/3$ mm; magnetoöptical data (320); Data for the refractive index, density, and viscosity (100a); Magnetic susceptibility (367)

TETRAHEPTYLGERMANE $(n\text{-}C_7H_{15})_4Ge$
Prepn.: By reacting Ge tetrahalides with $n\text{-}C_7H_{15}MgBr$ in Et_2O and completing the rxn. in a higher boiling solvent (320)
Props.: b. $217°/3$ mm; n^{20}_D 1.4626; d^{20}_4 0.8985; magnetoöptical data (320); Magnetic susceptibility (367).

TETRAOCTYLGERMANE $(n\text{-}C_8H_{17})_4Ge$
Prepn.: By reacting $GeCl_4$ with $n\text{-}C_8H_{17}Li$; 21.3% yield (139).
By reacting Ge tetrahalides with $n\text{-}C_8H_{17}MgBr$ in Et_2O and completing the rxn. in a higher boiling solvent (320).
Props.: b. $210\text{-}213°/0.1$ mm; n^{28}_D 1.4530; d_{24} 0.891; IR spectrum (139); b. $236°/2$ mm; n^{20}_D 1.4540; d^{20}_4 0.8938; magnetoöptical data (320); Magnetic susceptibility (367).

TETRA(2-ETHYLHEXYL)GERMANE $(2\text{-}EtC_6H_{12})_4Ge$
Prepn.: By reacting $GeCl_4$ with $2\text{-}EtC_6H_{12}Li$ without solvent; 27% yield (139)
Props.: b. $193\text{-}196°/0.15$ mm; n^{24}_D 1.4688; d_{26} 0.909 (139).

TETRADECYLGERMANE $(n-C_{10}H_{21})_4Ge$
Prepn.: By reacting $GeCl_4$ with $n-C_{10}H_{21}Li$; 18% yield (139)
Props.: m. ~ -15°; b. 230-240° (bath temp.) /0.015 mm; $n^{25}D$ 1.4643; d_{31} 0.879 (139).

TETRADODECYLGERMANE $(n-C_{12}H_{25})_4Ge$
Prepn.: By reacting $GeCl_4$ with $n-C_{12}H_{25}MgBr$ in xylene; 47% yield (139).
Props.: b. 330-340° (bath temp.) /0.0005 mm; $n^{31}D$ 1.4654; d_{31} 0.879 (139).

TETRAKIS(TETRADECYL)GERMANE $(n-C_{14}H_{29})_4Ge$
Prepn.: By reacting $GeCl_4$ with $n-C_{14}H_{29}MgBr$ in xylene at 100°; 37.6% yield (139).
Props.: Yellow liquid, b. 360-370° (bath temp.) /0.0005 mm; $n^{31}D$ 1.4640; d_{31} 0.880 (139).

TETRAKIS(HEXADECYL)GERMANE $(n-C_{16}H_{33})_4Ge$
Prepn.: By reacting $GeCl_4$ with $n-C_{16}H_{33}MgBr$ in xylene; 36% yield (139).
Props.: m. 37-38° (139).

TETRAKIS(OCTADECYL)GERMANE $(n-C_{18}H_{37})_4Ge$
Prepn.: By reacting $GeCl_4$ with $n-C_{18}H_{37}MgBr$ in xylene; 24% yield (139).
Props.: m. 43-45° (139).

TETRAKIS(2-CYCLOHEXYLETHYL)GERMANE $(C_6H_{11}CH_2CH_2)_4Ge$
Prepn.: By reacting $GeCl_4$ with $C_6H_{11}CH_2CH_2Li$ in Et_2O and then in C_6H_6 at 80°; 15% yield (139).
Props.: m. 137.5-138.5°; IR spectrum (139).

TETRAPHENYLGERMANE Ph_4Ge
Prepn.: By refluxing $GeCl_4$ with excess PhLi in MePh; 90% yield (244)
By refluxing $GeCl_4$ with excess PhMgBr in MePh; up to 80% yield (211a, 245, 473, 573).
By reacting $GeCl_4$ with PhBr and Na-wire with cooling and completing the rxn. on a steam bath; 55.5% yield (476) or under reflux; 54% yield (573).
By refluxing Ph_3GeH with excess PhLi in Et_2O (246).
By reacting Ph_2Zn with $GeCl_4$ in C_6H_6 at 75-80° (551).
By prod. in the redn. of Ph_3GeBr with Zn-Hg in HCl (565).
By prod. in the rxn. of GeI_2 with PhLi (1:3) under reflux; the prod. contains mostly Ph_2Ge (?) along with a small amt. of Ph_4Ge (535).

Props.: White crystals, m. 235.7°; soly. data (211a);
m. 230-231° (244); m. 225-228° (245); m. 232° (476); Crystal structure (234); IR spectrum (473); Diffraction data (137); Magnetic susceptibility (407).
Rxns. with: Li in $(CH_2OMe)_2$ followed by $Ph_3GeCO_2Me \rightarrow (Ph_3Ge)_2$ (175).
Li in $(CH_2OMe)_2$ followed by 1-octadecene $\rightarrow$ n-$C_{18}H_{37}GePh_3$ and Ph_3GeOH (181).
Na-K followed by $Ph_3SnCl \rightarrow Ph_3GeSnPh_3$ (183).
Na in liq. $NH_3 \rightarrow Ph_3GeNa$ and Ph_2GeNa_2 (211a).
Br_2 in $CCl_4 \rightarrow Ph_3GeBr$ (211a, 244, 473).
Br_2 (1:2) in $C_2H_4Br_2 \rightarrow Ph_2GeBr_2$ and some Ph_3GeBr (245, 253, 551).
Br_2 (1:1) in $C_2H_4Br_2 \rightarrow Ph_3GeBr$ (254).
$GeCl_4$ (1:3) $\rightarrow PhGeCl_3$ (476).
Slow neutrons $\rightarrow$ radioactive Ph_4Ge (concentration method given) (722).

TETRA-p-TOLYLGERMANE $(p-MeC_6H_4)_4Ge$
Prepn.: By refluxing $GeCl_4$ with $p-MeC_6H_4MgBr$ in MePh; 70% yield (473).
Props.: IR spectrum (473).
Rxn. with: Br_2 in $CCl_4 \rightarrow (p-MeC_6H_4)_3GeBr$ (473).

TETRABENZYLGERMANE $(PhCH_2)_4Ge$
Rxns. with: Li, no cleavage of the C-Ge bond (187).

TETRAPHENETHYLGERMANE $(PhCH_2CH_2)_4Ge$
Prepn.: By reacting $GeCl_4$ with $PhCH_2CH_2MgBr$ in MePh; 21.6% yield (65, 139).
Props.: Intensively yellow, amorphous substance, sol. in alkali; m. 56-57° (65); m. 53-54° (139).
Rxn. with: Li, the C-Ge bonds remain intact (187).

TETRA(3-PHENYLPROPYL)GERMANE $(PhCH_2CH_2CH_2)_4Ge$
Prepn.: By reacting $GeCl_4$ with $Ph(CH_2)_3Li$ in toluene; 33% yield (139).
Props.: Liquid, b. 240-250° (bath temp.)/0.05 mm; n^{20}_D 1.5704; d_{25} 1.106 (139).

TETRA-1-NAPHTHYLGERMANE $(1-C_{10}H_7)_4Ge$
Prepn.: By reacting $(1-C_{10}H_7)_3GeBr$ with $1-C_{10}H_7Li$ in C_6H_6 under reflux; 38% yield (563).
Props.: Fine, colorless needles, m. 270°, sol. in warm $H_2SO_4-HNO_3$ mixt. (563).
Rxn. with: Alc. KOH, Br in CCl_4, $KMnO_4$, or concd. mineral acids $\rightarrow$ no rxn. (563).

ASYMMETRICAL TETRAORGANOGERMANES

NON-FUNCTIONAL TETRAORGANOGERMANES

The methods for the preparation of asymmetrical germanes containing non-functional substituents are exemplified by the following formulas:

I. $R_{3-n}GeX_{1+n} + R'MgX \rightarrow R_{3-n}GeR'$
II. $R_{3-n}GeX_{1+n} + R'M \rightarrow R_{3-n}GeR'_{1+n}$
III. $R_3GeH + CH_2:CHR'$ in the presence of Bz_2O_2 or in UV light $\rightarrow R_3GeCH_2CH_2R'$
IV. $R_3GeM + R'X \rightarrow R_3GeR'$
V. $R_3GeM + CH_2:CHR'$, followed by hydrolysis $\rightarrow R_3GeCH_2CH_2R'$

(R in one and the same compound are identical or different organic groups; X = halogen; n = 0, 1, or 2; M = alkali metal)

ASYMMETRICAL NON-FUNCTIONAL TETRAORGANOGERMANES

Compound	Prepared from	Yield %	Properties	Ref.
Me_3GeEt	$EtGeI_3$ + MeMg halides		b. 79-80°/760mm; n^{20}_D 1.4090; d^{20}_4 0.9843	323
Me_2GeEt_2	Et_2GeI_2 + MeMg halides		b. 108-109°/760mm; n^{20}_D 1.4221; d^{20}_4 0.9885	323
Et_2GeBu_2	Et_2GeI_2 + BuMg halides		b. 109-110°/14mm; n^{20}_D 1.4516; d^{20}_4 0.9547	323
Et_2GePh_2	Ph_2GeBr_2 + EtMgBr		b. 120-130°/mm (1)	551
Et_3GeMe	Et_3GeI + MeMg halides		b. 135°/760mm; n^{20}_D 1.4332; d^{20}_4 0.9906	323
Et_3GePr	Et_3GeI + PrMg halides		b. 73-74°/20mm; n^{20}_D 1.4460; d^{20}_4 0.9810	323
Et_3GeBu	Et_3GeI + BuMg halides		b. 91-92°/20mm; n^{20}_D 1.4483; d^{20}_4 0.9711	323
	Et_3GeCl + BuLi	51.1	b. 181-181.5°; n^{20}_D 1.4475	187
$Et_3Ge(n-C_5H_{11})$	Et_3GeI + $n-C_5H_{11}$-Mg halides		b. 104-105°/20mm; n^{20}_D 1.4483; d^{20}_4 0.9625	323
Et_3GeCH_2Ph	Et_3GeCl + $PhCH_2MgCl$	79	b. 78-81°/1mm; n^{20}_D 1.5178	187

(1) Rxn. with Br_2 in $C_2H_4Br_2$ → Et_2GeBr_2 (551).

ASYMMETRICAL NON-FUNCTIONAL TETRAORGANOGERMANES (Cont'd.)

Compound	Prepared from	Yield %	Properties	Ref.
$Et_3GeC_6H_4-2-Ph$	Et_3GeCl + $2-PhC_6H_4Li$	62.4	b. 150-152°/3.3mm; n^{20}_D 1.5697	187
$(n-C_5H_{11})_3GeEt$	$EtGeI_3$ + $n-C_5H_{11}-$ Mg halides		b. 148-149°/7mm; n^{20}_D 1.4565; d^{20}_4 0.9274	323
$(n-C_6H_{13})_3GePh$	$(n-C_6H_{13})_3GeBr$ + PhLi		b. 160-165°/0.5mm; n^{22}_D 1.4984; d_{25} 0.972 (2)	140
$(cyclo-C_6H_{11})_3GeMe$	$(cyclo-C_6H_{11})_3Ge-Br$ + MeMgBr	65	m. 48-48.5°	243
$(cyclo-C_6H_{11})_3GeEt$	$(cyclo-C_6H_{11})_3Ge-Br$ + EtMgBr	56	m. 38.5-39°	243
$(cyclo-C_6H_{11})_3GePr$	$(cyclo-C_6H_{11})_3Ge-Br$ + PrMgBr	70	m. 124-125°	243
$(cyclo-C_6H_{11})_3GeBu$	$(cyclo-C_6H_{11})_3Ge-Br$ + BuMgBr	61	M. 152.5-153.5°	243
$(cyclo-C_6H_{11})_3Ge(n-C_5H_{11})$	$(cyclo-C_6H_{11})_3Ge-Br$ + $n-C_5H_{11}MgBr$	71	m. 78-79°	243
$(cyclo-C_6H_{11})_3GePh$	$(cyclo-C_6H_{11})_3Ge-Br$ + PhLi	75	m. 210-211°	244
$(cyclo-C_6H_{11})_3GeCH_2Ph$	$(cyclo-C_6H_{11})_3Ge-Br$ + $PhCH_2MgBr$	69	m. 54-54.5°	243
$Ph_3Ge-(cyclo-C_6H_{11})$	Ph_3GeCl + $cyclo-C_6H_{11}MgBr$	14	m. 143-146°	138
	Ph_3GeH + cyclohexene (3)			138
$Ph_3Ge(n-C_6H_{13})$	Ph_3GeCl + $n-C_6H_{13}MgBr$		m. 54.3-55°	138
$Ph_3Ge(n-C_8H_{17})$	Ph_3GeH + $n-C_8H_{17}Li$		m. 72-73°; IR spectrum	138
	Ph_3GeH + 1-octene (3)	91		138
$Ph_3Ge-(n-C_{18}H_{37})$	Ph_3GeLi + 1-octadecene (4)		m. 79.5-80.5°	181
	Ph_3GeH + 1-octadecene (5)			181

(2) Rxn. with Li in $(CH_2OMe)_2$, the C-Ge bonds are preserved (187).
(3) The rxn. is carried out in the presence of Bz_2O_2 or in UV light (138).
(4) The rxn. of Li germanide with the olefinic compd. is followed by hydrolysis (181).
(5) The rxn. is carried out in the presence of Bz_2O_2 (181).

ASYMMETRICAL NON-FUNCTIONAL TETRAORGANOGERMANES (Cont'd.)

Compound	Prepared from	Yield %	Properties	Ref.
$Ph_3Ge-(n-C_{18}H_{37})$	$Ph_3GeCl + n-C_{18}H_{37}MgBr$			181
	$Ph_3GeLi + n-C_{18}H_{37}Br$	70	(6)	187
$Ph_3GeC_6H_4Me-m$	$Ph_3GeBr + m-MeC_6H_4Li$	89	m. 136-138° (7)	51
$Ph_3GeCH_2CH_2Ph$	$Ph_3GeLi + PhCH_2CH_2Br$	73	m. 147-149° (8)	187
$Ph_3GeCH_2CHPh_2$	$Ph_3GeK + Ph_2C:CH_2$ (4)		m. 96-97°	181
	$Ph_3GeLi + Ph_2C:CH_2$ (4)			181
	$Ph_3GeCH_2Cl + Ph_2CHNa$			181
$Ph_3GeC_6H_4-p-CHPh_2$	$Ph_3GeBr + p-(Ph_2CH)C_6H_4MgBr$	57	m. 208-210°	61
	By-prod. from $Ph_3C-Na + Ph_3GeBr$		(9)	61
Ph_3GeCPh_3	$Ph_3GeBr + Ph_3CNa$	56.5	m. 342-344° (10)	61
$Ph_2(PhCH_2CH_2)-Ge(n-C_{18}H_{37})$	$Ph_2(n-C_{18}H_{37})Ge-Li$ (11) + $PhCH_2CH_2Br$	17.6	m. 34.5-36°; IR spectrum	187
	$Ph_2(PhCH_2CH_2)Ge-Li$ (12) + $n-C_{18}H_{37}Br$	28.6		187

(6) Rxn. with Li → $Ph_2(n-C_{18}H_{37})GeLi$ (187).
(7) Rxn. with CrO_3 in $Ac_2O-AcOH$ → $Ph_3GeC_6H_4CO_2H-m$ (51).
(8) Rxn. with Li → $Ph_2(PhCH_2CH_2)GeLi$ (187).
(9) Rxn. with Na-K, followed by CO_2 → Ph_3GeCO_2H (61).
(10) Rxns. with Na-K, followed by CO_2 → $Ph_3GeCO_2H + Ph_3CCO_2H$ (61).
(11) Prepd. by reacting $Ph_3Ge(n-C_{18}H_{37})$ with Li in $(CH_2OMe)_2$ (187).
(12) Prepd. by reacting $Ph_3GeCH_2CH_2Ph$ with Li in $(CH_2OMe)_2$ (187).

GERMANES CONTAINING OLEFINIC GROUPS

Asymmetrical tetraorganogermanium compounds containing olefinic groups are prepared by reacting hydrocarbylgermanium halides with alkenylmagnesium halides or by reacting alkenylgermanium halides with hydrocarbyl halides.

ALLYLTRIMETHYLGERMANE $Me_3GeCH_2CH:CH_2$

Prepn.: By reacting $CH_2:CHCH_2GeCl_3$ with $MeMgCl$ under reflux; 68.1% yield (412).

Props.: b. 101°/764mm; n^{20}_D 1.4333; d_{20} 0.9952 (293, 412); Raman spectrum (121).

Rxns.: 1) Polymerization at 120°/6000 atm. in the presence of $(Me_3CO)_2$ → 34% liquid polymer, mol. wt. 560 (293, 688).
2) Copolymerization with $CH_2:CMeCO_2Me$ at 120°/6000 atm. in the presence of $(Me_3CO)_2$ → linear copolymer which on heating passes through a glassy to an elastic state (293, 688).
3) Copolymerization with styrene at 120°/6000 atm. in the presence of $(Me_3CO)_2$ → copolymers (688).

ALLYLTRIETHYLGERMANE $CH_2:CHCH_2GeEt_3$

Prepn.: By reacting Et_3Ge halides with $CH_2:CHCH_2MgBr$ in Et_2O (368).
By reacting $CH_2:CHCH_2GeCl_3$ with $EtMgBr$ under reflux; 52.3% yield (412).

Props.: b. 180°/732mm; n^{20}_D 1.4594; d_{20} 1.0004 (293, 412); b. 50°/6mm (412).

Rxns.: 1) Polymerization at 120°/6000 atm. in the presence of $(Me_3CO)_2$ → 64% oily polymer, mol. wt. 780 (293).
2) Copolymerization with $CH_2:CMeCO_2Me$ at 120°/6000 atm. in the presence of $(Me_3CO)_2$ → linear copolymer which on heating passes through a glassy to an elastic state (293).
3) I_2 → Et_3GeI + $CH_2:CHCH_2I$ (368).
4) $HSCH_2CH_2OH$ → $Et_3Ge(CH_2)_3SCH_2CH_2OH$ (368).
5) $HgCl_2$ → Et_3GeCl + $CH_2:CHCH_2HgCl$ (368).
6) Et_3GeH → $Et_3Ge(CH_2)_3GeEt_3$ (368).
7) 1% $KMnO_4$ in Me_2CO-H_2O → $Et_3GeCH_2CHOHCH_2OH$ (368).

ALLYLTRIBUTYLGERMANE $Bu_3GeCH_2CH:CH_2$

Prepn.: By reacting Bu_3GeI with $CH_2:CHCH_2MgBr$ in Et_2O (368).

Props.: b. 116-117°/2mm; n^{20}_D 1.4664; n^{26}_4 0.9629 (368).

Rxns. with: CCl_3CO_2H → $CCl_3CO_2GeBu_3$ (368).
HCO_2H → HCO_2GeBu_3 (368).
HBr → Bu_3GeBr + C_3H_6 (368).
$HgCl_2$ in $EtOH$ at 20° → Bu_3GeCl + $CH_2:CHCH_2HgCl$ (368).

ALLYLTRIPHENYLGERMANE $Ph_3GeCH_2CH:CH_2$

Prepn.: By reacting Ph_3GeCl with $CH_2:CHCH_2MgCl$ in Et_2O under reflux (181).

Props.: m. 90-91.5° (181).

TRIMETHYL(2-METHYLALLYL)GERMANE $CH_2{:}CMeCH_2GeMe_3$

Prepn.: By reacting $CH_2{:}CMeCH_2GeCl_3$ with MeMgBr under reflux; 42.5% yield (412).
Props.: b. 121°/733mm; n^{20}_D 1.4416; d_{20} 0.9908 (293, 412).

DIALLYLDIMETHYLGERMANE $(CH_2{:}CHCH_2)_2GeMe_2$

Prepn.: By reacting Me_2GeCl_2 with $CH_2{:}CHCH_2MgBr$ under reflux; 42% yield (412).
Props.: b. 146°/737mm; n^{20}_D 1.4645; d_{20} 1.0337 (293, 412).
Rxns.: Polymerization at 120°/6000 atm. in the presence of $(Me_3CO)_2 \rightarrow$ 100% colorless, flexible, apparently tridimensional polymer (293).
Copolymerization with $CH_2{:}CMeCO_2Me$ at 120°/6000 atm. in the presence of $(Me_3CO)_2 \rightarrow$ linear copolymers which, on heating, passed through a glassy to an elastic state (293).

TRIALLYLMETHYLGERMANE $(CH_2{:}CHCH_2)_3GeMe$

Prepn.: By reacting $MeGeCl_3$ with $CH_2{:}CHCH_2MgBr$ under reflux; 45% yield (412).
Props.: b. 190.5°/747mm; n^{20}_D 1.4867; d_{20} 1.0222 (293, 412).
Rxns.: Polymerization at 120°/6000 atm. in the presence of $(Me_3CO)_2 \rightarrow$ 100% transparent, glassy polymer; at atm. pressure $\rightarrow$ 38% viscous oily polymer (293).
Copolymer with $CH_2{:}CMeCO_2Me$ at 120°/6000 atm. in the presence of $(Me_3CO)_2 \rightarrow$ linear copolymers which, on heating, passed through a glassy to an elastic state (293).

TRIMETHYLVINYLGERMANE $Me_3GeCH{:}CH_2$

Rxns.: Polymerization at 120°/6000 atm. in the presence of $(Me_3CO)_2 \rightarrow$ liquid polymer (688).

TRIETHYLVINYLGERMANE $Et_3GeCH{:}CH_2$

Prepn.: By reacting Et_3GeCl with $CH_2{:}CHMgBr$ in C_4H_8O (368).
Props.: b. 61°/28mm; n^{20}_D 1.4501; d^{20}_4 1.0048 (368).
Rxns. with: Strong org. acids $\rightarrow$ Et_3Ge carborylates (368).
$HSCH_2CO_2H \rightarrow Et_3GeCH_2CH_2SCH_2CO_2H$ (368).
$Et_3GeH \rightarrow Et_3GeCH_2CH_2GeEt_3$ (368).
$BrCH_2CO_2Et \rightarrow Et_3GeCHBrCH_2CH_2CO_2Et$ (368).

TRIBUTYLVINYLGERMANE $Bu_3GeCH{:}CH_2$

Prepn.: By reacting Bu_3Ge halides with $CH_2{:}CHMgBr$ in C_4H_8O (368).
Props.: b. 108-109°/2mm; n^{20}_D 1.4598; d^{20}_4 0.9479 (368).
Rxns. with: Strong org. acids $\rightarrow$ Bu_3Ge carborylates (368).
$Br_2 \rightarrow Bu_3GeCHBrCH_2Br$ (368).
$HBr \rightarrow Bu_3GeCH_2CH_2Br$ (368).
$CHCl_3$ in the presence of $Bz_2O_2 \rightarrow Bu_3GeCH_2CH_2CCl_3$ (368).

DIETHYLDIVINYLGERMANE $Et_2Ge(CH:CH_2)_2$

Prepn.: By reacting Et_2GeCl_2 with $CH_2:CHMgBr$ in C_4H_8O on steam bath; 55% yield (413, 490).

Props.: b. 149.8°; $n^{20}D$ 1.4575; d_{20} 1.0193 (413).
b. 59-60°/28.5mm; $n^{25}D$ 1.4540 (490).

Rxns.: Polymerization at 120°/6000 atm. in the presence of $(Me_3CO)_2$ → tridimensional glassy polymer (688).

ETHYLTRIVINYLGERMANE $EtGe(CH:CH_2)_3$

Prepn.: By reacting $EtGeCl_3$ with excess $CH_2:CHMgBr$ in C_4H_8O on steam bath (413, 490).

Props.: b. 55-57°/28mm; $n^{25}D$ 1.4605 (490).

GERMANES CONTAINING HALO- AND CYANO-SUBSTITUTED GROUPS

Asymmetrical tetraorganogermanes containing halohydrocarbyl groups are prepared:

1. By reacting haloalkylgermanium halides with organomagnesium halides.
2. By reacting organogermanium halides with halohydrocarbyl-magnesium halides or halohydrocarbyllithium compounds.
3. By reacting alkenyltrihydrocarbylgermanes with halogens or halomethanes in the presence of Bz_2O_2.

Asymmetrical tetraorganogermanes containing cyano-substituted groups are prepared by reacting triorganogermanes, R_3GeH, with alkenyl cyanides.

HALO- AND CYANO-SUBSTITUTED GERMANES

Compound	Prepared from	Properties	Ref.
Me_3GeCH_2Cl	$ClCH_2GeCl_3 + MeMgI$ (80% yield)	b. $114°/762mm$ (1)	485
Me_3GeCH_2I	$Me(ClCH_2)GeCl_2 + MeMgI$	b. $154°/752mm$; d_{20} 1.1716 $n^{20}D$ 1.5112	419
$Bu_3GeCH_2CH_2CCl_3$	$Bu_3GeCH:CH_2 + CHCl_3$ in the presence of Bz_2O_2	b. $156°/0.5mm$; $d^{20}{_4}$ 1.474; $n^{20}D$ 1.4855	368
$Bu_3GeCHBr-CH_2Br$	$Bu_3GeCH:CH_2 + Br_2$	b. $147°/0.5mm$; $d^{20}{_4}$ 1.3619; $n^{20}D$ 1.5102	368
$Me_3GeC_6H_4-Br-p$	$Me_3GeBr + p-BrC_6H_4MgBr$	b. $118°/20mm$ (2)	90
$Et_3GeC_6H_4-Br-p$	$Et_3GeBr + p-BrC_6H_4MgBr$	b. $156°/14mm$ (2)	90
$Ph_3GeC_6H_4-Br-p$	$Ph_3GeBr + p-BrC_6H_4Li$	m. $159-160°$ (3)	51
Ph_3GeCH_2Cl	$ClCH_2GeCl_3 + PhMgBr$ (1:3)	m. $117.5-118.5°$ (4,5)	181
$Et_3GeCH_2CH_2CN$	$Et_3GeH + CH_2:CHCN$	b. $225-227°$	324
$Bu_3GeCH_2CH_2CN$	$Bu_3GeH + CH_2:CHCN$	b. $168-169°/11mm$	324
$(n-C_5H_{11})_3GeCH_2CH_2CN$	$(n-C_5H_{11})_3GeH + CH_2:CHCN$	b. $193-194°/11mm$	324

(1) Rxn. with Mg → Me_3GeCH_2MgCl; $Me_3GeCH_2MgCl + MeSi(OMe)_3$ → $Me_3GeCH_2SiMe(OMe)_2$ (484).
(2) Rxn. with Mg followed by CO_2 → $R_3GeC_6H_4CO_2H$ (90).
(3) Rxn. with Li followed by CO_2 → $Ph_3GeC_6H_4CO_2H$ (51).
(4) Rxn. with Mg followed by hydrolysis → Ph_3GeMe (181).
(5) Rxn. with Ph_2CHNa in Et_2O → $Ph_2CHCH_2GePh_3$ (181).

GERMANES CONTAINING CARBONYL AND CARBOXYLIC GROUPS

Asymmetrical tetraorganogermanes containing an aldehyde group are prepared by reacting triorganogermanes, R_3GeH, with unsaturated aldehydes, such as acrolein, in the presence of benzoyl peroxide or platinum catalyst.

Keto-substituted tetrorganogermanes are prepared:

1. By reacting triorganogermanes, R_3GeH, with diazoalkylketones such as N_2CH_2COMe and N_2CHBz.
2. By treating lithium triorganogermanides with unsaturated ketones and hydrolyzing the reaction product.

Triorganogermanecarboxylic acids are prepared by treating lithium triorganogermanides with carbon dioxide and hydrolyzing the reaction products. Their alkyl esters are obtained by treating the acids with diazoalkanes. Pyrolysis of triorganogermanecarboxylic acids yields triorganogermanium triorganogermanecarboxylates, $R_3GeCO_2GeR_3$. The latter are also obtained by reacting triorganogermanecarboxylic acid silver salts, R_3GeCO_2Ag, with triorganogermanium halides, R_3GeX.

Asymmetrical tetraorganogermanes containing carboxy groups are prepared:

1. By reacting haloaryltriorganogermanes with magnesium or lithium and treating the resulting products with carbon dioxide and hydrolyzing them.
2. By reacting triorganogermanes, R_3GeH, with α,β-unsaturated carboxylic acids or with diazoalkylcarboxylic acids. If esters of α,β-unsaturated carboxylic or diazoalkylcarboxylic acids are employed, the corresponding (triorganogermyl)-hydrocarbylcarboxylic acid esters are formed.
3. By oxidizing (hydroxymethylphenyl)- or (methylphenyl)-trihydrocarbylgermanes with potassium permanganate.
   For example:
   $Ph_3GeC_6H_4CH_2OH + KMnO_4 \rightarrow Ph_3GeC_6H_4CO_2H$
   $Ph_3GeC_6H_4Me + KMnO_4 \rightarrow Ph_3GeC_6H_4CO_2H$

TRIETHYL(2-FORMYLETHYL)GERMANE $Et_3GeCH_2CH_2CHO$
Prepn.: By refluxing Et_3GeH with acrolein in the presence of Pt and hydroquinone (324).
Props.: b. 129-131°/17 mm (324).

TRIBUTYL(2-FORMYLETHYL)GERMANE $Bu_3GeCH_2CH_2CHO$
Prepn.: By reacting Bu_3GeH with acrolein in the presence of Bz_2O_2 or platinum asbestos (324).
Props.: b. 125-128°/1 mm (324).

ACETONYLTRIETHYLGERMANE Et_3GeCH_2COMe
Prepn.: By reacting Et_3GeH with $N_2CHCOMe$ in Et_2O or C_6H_6 in the presence of Cu (324).
Props.: b. 111°/20 mm (324).

(BENZOYLMETHYL)TRIETHYLGERMANE Et_3GeCH_2Bz
Prepn.: By reacting Et_3GeH with N_2CHBz in Et_2O or C_6H_6 in the presence of Cu (324).
Props.: b. 135°/3 mm (324).

3-PHENYL-3-(TRIPHENYLGERMYL)PROPIOPHENONE $BzCH_2CHPhGePh_3$

Prepn.: By reacting Ph_3GeLi with $BzCH:CHPh$ in $(CH_2OMe)_2$ (181).
Props.: 116-118° (181).

TRIPHENYLGERMANECARBOXYLIC ACID Ph_3GeCO_2H

Prepn.: By carbonation of Ph_3GeLi; theoret. yield (179, 186).
By reacting Ph_3GeH with MeI in Et_2O under reflux and pouring
 the mixt. into Et_2O-CO_2 (179).
By cleaving $Ph_3GeSnPh_3$ with BuLi and carbonating the rxn.
 mixt.; 9% yield (183).
By cleaving Ph_3CGePh_3, $Ph_3GeC_6H_4CHPh_2$, $(Ph_3Ge)_2$, or $Ph_3GeSiPh_3$
 with Na-K alloy in Et_2O and carbonating rxn. mixt.;
 83% yield (61, 180).
Props.: m. 186° (179, 183); m. 190-195°; on heating loses CO
and H_2O and forms $Ph_3GeCO_2GePh_3$ (61); m. 184-186° (186).
Rxns. with: $CH_2N_2 \rightarrow Ph_3CO_2Me$ (61).
NaOH, followed by $Ph_3GeBr \rightarrow Ph_3GeCO_2GePh_3$ (61).
$Ph_3GeOH \rightarrow Ph_3GeCO_2GePh_3$ (62).

TRIPHENYLGERMANECARBOXYLIC ACID METHYL ESTER Ph_3GeCO_2Me

Prepn.: By heating Ph_3GeCO_2H with CH_2N_2; 87% yield (61).
Props.: m. 107-108° (61); on heating to 250° → Ph_3GeOMe (62).
Rxns. with: $Ph_3SiLi \rightarrow Ph_3SiGePh_3$ (175).
PhMgBr in $Et_2O \rightarrow Ph_3GeC(OH)Ph_2$ (177).
PhLi in $Et_2O \rightarrow Ph_3GeC(OH)Ph_2$ (177).

TRIPHENYLGERMANIUM TRIPHENYLGERMANECARBOXYLATE $Ph_3GeCO_2GePh_3$

Prepn.: By heating Ph_3GeCO_2H at 190-210°; 86% yield (61).
By converting Ph_3GeCO_2H to the Na salt and reacting it with
 Ph_3GeBr in C_6H_6 under reflux; 81% yield (61).
By converting Ph_3GeCO_2H to the Ag salt and reacting it with
 Ph_3GeBr in C_6H_6 under reflux; 53% yield (61).
By heating Ph_3GeCO_2H with Ph_3GeOH at 157° (62).
By-prod. in the rxn. of Ph_3GeLi with dibenzofuran followed
 by carbonation (186).
Props.: m. 163-164° (61); m. 165-166° (186); On heating to
250° decomp. → $(Ph_3Ge)_2$ (62).
Rxn. with: Alkali → Ph_3CO_2H and Ph_3GeOH (61).

p-TRIMETHYLGERMYLBENZOIC ACID $Me_3GeC_6H_4CO_2H$

Prepn.: By treating $Me_3GeC_6H_4Br$ with Mg in Et_2O and reacting
 the Grignard compd. with CO_2 (90).
Props.: m. 45-46° (90); Dissocn. const.: 1.07×10^{-6}; strength
of the $Ge-C_{ar}$ bond (90, 91, 641).

3-(TRIETHYLGERMYL)PROPIONIC ACID $\quad Et_3GeCH_2CH_2CO_2H$
Prepn.: By refluxing Et_3GeH with acrylic acid (324).
Props.: b. 158-160°/10 mm (324).

p-TRIETHYLGERMYLBENZOIC ACID $\quad Et_3GeC_6H_4CO_2H$
Prepn: By reacting $Et_3GeC_6H_4Br$ with Mg in Et_2O and treating the Grignard compd. with CO_2 (90).
Props.: m. 119.5-121° (90); Dissocn. const.: 1.08×10^{-6}; strength of the Ge-C_{ar} bond (90, 91, 641).

(2-CARBOMETHOXYETHYL)TRIPROPYLGERMANE $\quad Pr_3GeCH_2CH_2CO_2Me$
Prepn.: By refluxing Pr_3GeH with CH_2:$CHCO_2Me$ (324).
Props.: b. 150-154°/18 mm (324).

TRIBUTYL(CARBETHOXYMETHYL)GERMANE $\quad Bu_3GeCH_2CO_2Et$
Prepn.: By reacting Bu_3GeH with N_2CHCO_2Et in Et_2O in the presence of Cu (324).
Props.: b. 160-165°/10 mm (324).

TRIBUTYL(2-CARBETHOXYETHYL)GERMANE $\quad Bu_3GeCH_2CH_2CO_2Et$
Prepn.: By refluxing Bu_3GeH with CH_2:$CHCO_2Et$ (324).
Props.: b. 122-124°/0.5 mm (324).

m-TRIPHENYLGERMYLBENZOIC ACID $\quad Ph_3GeC_6H_4CO_2H$-m
Prepn.: By oxidation of m-$MeC_6H_4GePh_3$ with CrO_3 in $AcOH$-Ac_2O at 35-40°; 10% yield (51).
Props.: White crystals, m. 168-169° (51);
Derivs.: Amide, m. 162-163° (51).
Rxns. with: Ph_2CN_2, rate const. (51).

p-TRIPHENYLGERMYLBENZOIC ACID $\quad Ph_3GeC_6H_4CO_2H$-p
Prepn.: By reacting p-$BrC_6H_4GePh_3$ with Li in Et_2O-C_6H_6 and pouring the resulting Li deriv. onto crushed Dry-Ice in Et_2O (51).
By oxidation of $Ph_3GeC_6H_4CH_2OH$-p with $KMnO_4$; 52% yield (51).
Props.: m. 201-202° (51);
Derivs.: Acid chloride, m. 143-145° (51).
Amide, m. 171-172° (51).
Rxn. with: Ph_2CN_2, rate const. (51).

GERMANES CONTAINING HYDROXY GROUPS

Asymmetrical tetraorganogermanes containing alcoholic hydroxy groups are prepared:

1. By reacting lithium triorganogermanides with formaldehyde and hydrolyzing the reaction product.
2. By oxidation of tetraorganogermanes containing an olefinic double bond with potassium permanganate in aqueous acetone.
3. By reacting triorganogermanes, R_3GeH, with alcohols having an olefinic double bond, in the presence of benzoyl peroxide or platinum catalyst.
4. By reacting lithium triorganogermanides with benzophenone and hydrolyzing the reaction product.
5. By treating triorganogermanecarboxylic acid methyl ester with Grignard or organolithium compounds.
6. By reacting triorganogermanium halides with hydroxymethylphenyllithium.

GERMANES CONTAINING HYDROXY GROUPS

Compound	Prepared from	Properties	Ref.
$Et_3GeCH_2CHOH\text{-}CH_2OH$	$Et_3GeCH_2CH:CH_2$ + 1% $KMnO_4$ (1)	b. 128°/0.5mm; n^{20}_D 1.4837; d^{20}_4 1.1566	368
$Pr_3GeCH_2CH_2\text{-}CH_2OH$	$Pr_3GeH + CH_2:CHCH_2OH$ (2)	b. 175-180°/20mm	324
$Bu_3GeCH_2CH_2\text{-}CH_2OH$	$Bu_3GeH + CH_2:CHCH_2OH$ (2)	b. 153-155°/3mm	324
Ph_3GeCH_2OH	$Ph_3GeLi + CH_2O$ in $(CH_2OMe)_2$ (3)	m. 115-116°	177
$Ph_3GeC(OH)Ph_2$	$Ph_3GeLi + Ph_2CO$ in $(CH_2OMe)_2$ (3)	m. 153-155°	177
	Ph_3GeCO_2Me + PhMgBr or PhLi in Et_2O (3)		177
$Ph_3GeC_6H_4CH_2\text{-}OH\text{-}p$	$Ph_3GeBr + p\text{-}HOCH_2C_6H_4Li$	m. 108.5-110°	51

1. The rxn. is carried out in aqueous acetone
2. The rxn. is carried out in the presence of Bz_2O_2 or platinated asbestos.
3. And hydrolyzing the rxn. prod.

GERMANES CONTAINING FUNCTIONAL NITROGEN

Aminophenyltriphenylgermanes are prepared by reacting triphenylgermanium bromide with aminophenyllithium in a mixture of benzene and ether under reflux:

m-Me$_2$NC$_6$H$_4$GePh$_3$ obtained in 71% yield, m. 92-93° (51).
p-Me$_2$NC$_6$H$_4$GePh$_3$ obtained in 79% yield, m. 140-141° (51).

GERMANES CONTAINING POLYFUNCTIONAL GROUPS

Asymmetrical tetraorganogermanes containing hydroxyalkylthio ether and carboxyalkylthio ether groups are obtained by reacting tetraorganogermanes containing an alkenyl group with hydroxymercaptans and mercaptocarboxylic acids, respectively.

(Carbalkoxyhaloalkyl)triethylgermanes are prepared by reacting alkenyltriethylgermanes with haloalkylcarboxylic acid esters.

Hydroxyalkynyl-substituted tetraorganogermanes are prepared by reacting organogermanium halides with hydroxyalkynylmagnesium halides.

2-[3-(TRIETHYLGERMYL)PROPYLTHIO]ETHANOL   Et$_3$Ge(CH$_2$)$_3$SCH$_2$CH$_2$OH
Prepn.: By reacting Et$_3$GeCH$_2$CH:CH$_2$ with HSCH$_2$CH$_2$OH (368).
Props.: b. 150°/0.7 mm; n^{20}D 1.5027; d^{20}$_4$ 1.1158 (368).

3-(TRIETHYLGERMYL)PROPYLTHIOACETIC ACID   Et$_3$Ge(CH$_2$)$_3$SCH$_2$CO$_2$H
Prepn.: By reacting Et$_3$GeCH$_2$CH:CH$_2$ with HSCH$_2$CO$_2$H (368).
Props.: b. 168°/0.8 mm; n^{20}D 1.5005; d^{20}$_4$ 1.1664 (368).

2-(TRIETHYLGERMYL)ETHYLTHIOACETIC ACID   Et$_3$GeCH$_2$CH$_2$SCH$_2$CO$_2$H
Prepn.: By reacting Et$_3$GeCH:CH$_2$ with HSCH$_2$CO$_2$H (368).
Props.: b. 160°/0.3 mm; n^{20}D 1.5060; d^{20}$_4$ 1.1959 (368).

4-BROMO-4-(TRIETHYLGERMYL)BUTYRIC ACID ETHYL ESTER
Et$_3$GeCHBr(CH$_2$)$_2$CO$_2$Et
Prepn.: By reacting Et$_3$GeCH:CH$_2$ with BrCH$_2$CO$_2$Et (368).
Props.: b. 152°/2 mm; n^{20}D 1.4871; d^{20}$_4$ 1.2802 (368).

TRIS(3-HYDROXY-3-METHYL-1-BUTYNYL)METHYLGERMANE
MeGe[C:CC(OH)Me$_2$]$_3$
Prepn.: By adding MeGeBr$_3$ dissolved in Et$_2$O to Me$_2$C(OMgBr)C:CMgBr(1:3) and heating the mixt. on steam bath; 48% yield (773).
Props.: m. 208° (773).

(3-HYDROXY-3-METHYL-1-BUTYNYL)TRIMETHYLGERMANE $Me_3GeC\vdots CC(OH)Me_2$

Prepn.: By adding an Et_2O soln. of Me_3GeBr to $Me_2C(OMgBr)C\vdots CMgBr$ (1:1) at $-10°$ and heating the rxn. mixt. on steam bath; 45% yield (773).
Props.: m. $30°$; b. $72-73°/7$ mm; sol. in org. solvents (773).

BIS(3-ACETOXY-3-METHYL-1-BUTYNYL)DIMETHYLGERMANE
$Me_2Ge[C\vdots CC(OAc)Me_2]_2$

Prepn.: By reacting $Me_2Ge[C\vdots CC(OH)Me_2]_2$ with Ac_2O at $80°$; 57% yield (773).
Props.: m. $54°$; b. $142-143°/7$ mm (773).

BIS(3-HYDROXY-3-METHYL-1-BUTYNYL)DIMETHYLGERMANE
$Me_2Ge[C\vdots CC(OH)Me_2]_2$

Prepn.: By adding Me_2GeBr_2 dissolved in Et_2O to $Me_2C(OMgBr)C\vdots CMgBr$ (1:2) at $-10°$ and heating the mixt. on steam bath; 50-53% yield (773).
Props.: m. $97°$ (773).
Rxn. with: Ac_2O — $Me_2Ge[C\vdots CC(OAc)Me_2]_2$ (773).

TETRAHETEROCYCLYLGERMANE

TETRA-2-FURYLGERMANE $(C_4H_3O)_4Ge$

Prepn.: By refluxing $GeBr_4$ with $\overset{\frown}{OCH\vdots CHCH\vdots C}Li$ in C_6H_6; 32% yield (160, 306).
Props.: m. $99-100°$ (160, 306); b. $163°/1mm$ (306).

2. ALKALI METAL GERMANIDES

Alkali metal triorganogermanides are prepared by reacting tetraorganogermanes or hexaorganodigermanes with alkali metals or their alloys or by treating triorganogermanes with organoalkali metal compounds.

LITHIUM TRIETHYLGERMANIDE $LiGeEt_3$
Prepn.: By treating Et_3GeH with BuLi or PhLi in Et_2O; <10% yield (676).

LITHIUM TRIPHENYLGERMANIDE $LiGePh_3$

Prepn.: By treating $(Ph_3Ge)_2$ with Li in $(CH_2OMe)_2$ (177, 179, 181).
By treating Ph_4Ge with Li in $(CH_2OMe)_2$ (181).
Rxns. with: CH_2O, followed by hydrolysis → Ph_3GeCH_2OH (177).
Ph_2CO, followed by hydrolysis → $Ph_3GeC(OH)Ph_2$ (177).
Ph_3GeH in $(CH_2OMe)_2$, followed by H_2O → $(Ph_3Ge)_2$ and Ph_3GeOH (179).
$Ph_2C:CH_2$ → $Ph_2CHCH_2GePh_3$, $(Ph_3Ge)_2$, and $(Ph_3Ge)_2O$ (181).
1-Octadecene → $n-C_{18}H_{37}GePh_3$ and Ph_3GeOH (181).
trans-Stilbene → $(Ph_3Ge)_2O$ and unreacted stilbene (181).
1-Octene, followed by hydrolysis → Ph_3GeOH, Ph_3GeH, and $(Ph_3Ge)_2O$ (181).
Cyclohexene, followed by hydrolysis → Ph_3GeOH and Ph_3GeH (181).
$BzCH:CHPh$ → $Ph_3GeCHPhCH_2Bz$ (181).
Fluorene, followed by CO_2 → $(Ph_3Ge)_2$, $(Ph_3Ge)_2O$, and fluorene-9-carboxylic acid (186).
Dibenzofuran, followed by CO_2 → Ph_3GeCO_2H, $Ph_3GeCO_2GePh_3$, and $(Ph_3Ge)_2$ (186).
RCl → Ph_3GeR (187).
Br_2 at -25 to -20° in $(CH_2OMe)_2$ → $(Ph_3Ge)_2$ and Ph_3GeBr (186).

SODIUM TRIPHENYLGERMANIDE $NaGePh_3$

Rxns. with: Bu_2BCl in Et_2O → a prod. of unknown nature with a Ge-B bond (58).
$HSiCl_3$ → $(Ph_3Ge)_3SiH$ (375).
$SiCl_4$ → $(Ph_3Ge)_2Si:Si(GePh_3)_2(?)$ (375).
$Br(CH_2)_xBr$ in liq. NH_3 → $Ph_3Ge(CH_2)_xGePh_3$ and $(Ph_3Ge)_2O$ (519).
$(BrCH_2CH:)_2$, $CHCl_3$, $(ClCH:)_2$, or $(BrCH_2)_2$ in Et_2O → $(Ph_3Ge)_2$ (519).
BCl_3 in Et_2O → $NaCl$ and Ph_3Ge-B compd. (?) (519).
CCl_4 in Et_2O → $(Ph_3Ge)_2$ (519).

POTASSIUM TRIPHENYLGERMANIDE $KGePh_3$

Prepn.: By treating $(Ph_3Ge)_2$ with K-Na in Et_2O or Bu_2O (176, 181).
Props.: Solid insol. in Et_2O (176).

3. ORGANOGERMANIUM HYDRIDES

This class of compounds is relatively new. Organogermanium hydrides are not listed in "Die Chemie der Metallorganischen Verbindungen" by E. Krause and A. von Grosse.

Organogermanium hydrides, germanes, are prepared by reduction of organogermanium halides with lithium hydride, lithium aluminum hydride, or other reducing agents. Monoalkylgermanes are also prepared by reacting $NaGeH_3$ with alkyl halides.

METHYLGERMANE $MeGeH_3$
Prepn.: By reducing $MeGeCl_3$ or $MeGeBr_3$ with LiH (420).
By reacting $NaGeH_3$ with MeI in liq. NH_3 (547).
By reacting CH_2Br_2 with $NaGeH_3$ (1:2) in liq. NH_3 (547).
Props.: b. $-23.5°/745$ mm (420); m. $-158°$; b. $-23°$ (547);
Microwave spectra of 28 isotopic species, depole moment, and structural parameter (694).
Rxns. with: Na in $NH_3 \to NaGeH_2Me$ (190).
Li in $EtNH_2 \to$ 1 mole of H/1 at. Li was evolved (190).
Li in $EtNH_2$, followed by $i-C_5H_{11}Br \to Et(i-C_5H_{11})GeH_2$ (190).

ETHYLGERMANE $EtGeH_3$
Prepn.: By reducing $EtGeCl_3$ or $EtGeBr_3$ with LiH (420).
By reacting $NaGeH_3$ with EtBr in liq. NH_3 (547).
Props.: b. $11.5°/743.5$ mm; Raman spectrum (420); b. $9.2°$ (547).
Rxns. with: Na in $NH_3 \to EtGeH_2Na$. The main reaction with Na proceeds as follows: $Et_3GeH_3 + Na \to NaEtGeH_2 + \frac{1}{2}H_2$; a side rxn. takes place probably according to: $EtGeH_3 + NH_3 \xrightarrow{Na} EtGeH_2(NH_2) + H_2$. The total amt. of H evolved is greater than 1H/1Na (190).
Li in $EtNH_2 \to$ the metal and the germane react in almost equiv amts., evolving H_2 and C_2H_6 in a ration $\sim$4:1 (190).

DIISOPROPYLGERMANE $i-Pr_2GeH_2$
Prepn.: By refluxing $i-Pr_2GeCl_2$ with $LiAlH_4$ in Et_2O (23).
Props.: b. $110°$; d_{20} 0.982; $n^{20}D$ 1.432; MR_D 42.5 (23).
Rxns. with: Air $\to (i-Pr_2GeO)_3$ (23).
AgOAc $\to i-Pr_2Ge(OAc)_2$ (23).

ETHYLISOPENTYLGERMANE $Et(i-C_5H_{11})GeH_2$
Prepn.: By reacting $i-C_5H_{11}Br$ with a rxn. prod. obtd. from $EtGeH_3$ and Li(1:1) in $EtNH_2$; 64% yield (190).
Props.: Colorless liquid sol. in petr. ether and C_6H_6, insol. in H_2O; vapor pressure $\sim$ 1 cm at r.t. (190).
Rxns. with: Li(1:1) in $EtNH_2$ followed by EtI $\to$
$i-C_5H_{11}GeEt_2H$ (190).

DIPHENYLGERMANE Ph_2GeH_2
Prepn.: By reducing mixts. of phenylgermanium bromides, obtd. by bromination of Ph_4Ge or $(Ph_3Ge)_2$ in $C_2H_4Br_2$, with $LiAlH_4$ in Et_2O; 55 and 65% yield, respectively (245, 253).
Props.: Colorless liquid, b. $93°/1$ mm (245, 253); $n^{25}D$ 1.5921 (253).
Rxns. with: Br_2(1:2) in $CHCl_3 \to Ph_2GeBr_2$ (245,253).
I_2(1:2) in $CHCl_3 \to Ph_2GeI_2$ (245).

TRIETHYLGERMANE Et_3GeH

Prepn.: By reacting Et_3GeCl with $LiAlH_4$ in Et_2O; 79% yield (24).
Props.: b. 124°; n^{20}_D 1.4382; d^{20}_4 1.009 (24).
Rxns. with: Compds. of Pt, Pd, and Au → redn. to the metals;
  Hg(II) in excess → Hg(I); Hg(II) deficiency → Hg;
  Cu(II) → Cu(I); Ti(IV) → Ti(III) and Ti(II); and V(V) →
  V(IV) and V(III) (24).
Org. haloacids → the corresponding Et_3Ge esters (24).
CH_2:CHR (R= CN, CO_2H) under reflux → $Et_3GeCH_2CH_2R$ (324).
CH_2:CHCN at 140° in a sealed tube → a mixt. of di- and
  trinitriles (324).
N_2CHR (R= Ac, Bz) → Et_3GeCH_2R (324).
$Et_3GeCH:CH_2$ → $Et_3GeCH_2CH_2GeEt_3$ (368).
$Et_3GeCH_2CH:CH_2$ → $Et_3GeCH_2CH_2CH_2GeEt_3$ (368).
BuLi or PhLi in Et_2O → Et_3GeLi (676).

TRIPROPYLGERMANE Pr_3GeH

Prepn.: By reacting Pr_3GeCl with $LiAlH_4$ in $(Me_2CH)_2O$ (247, 255).
Props.: b. 65°/20 mm; b. 183°/742 (247).
Rxns. with: CH_2:CHR (R= CO_2Me, CH_2OH) → $Pr_3GeCH_2CH_2R$ (324).

TRIBUTYLGERMANE Bu_3GeH

Prepn.: By reacting Bu_4Ge with Br_2(1:1) in the presence of
  $AlBr_3$ and reducing the resulting Bu_3GeBr with $LiAlH_4$ in
  Et_2O (324).
Props.: b. 123°/20 mm; n^{20}_D 1.4508; d^{20}_4 0.9490 (324).
Rxns. with: CH_2:CHR (R= CN, CO_2Et, CH_2OH, CHO) →
  $Bu_3GeCH_2CH_2R$ (324).
N_2CHCO_2Et → $Bu_3GeCH_2CO_2Et$ (324).

TRICYCLOHEXYLGERMANE $(C_6H_{11})_3GeH$

Prepn.: By the reduction of $(C_6H_{11})_3GeCl$ or $(C_6H_{11})_3GeOH$
  with $LiAlH_4$ in Et_2O under reflux; 87% yield (244).
Props.: m. 24-25° (244).
Rxns. with: Br_2 in EtBr → $(C_6H_{11})_3GeBr$ (244).
I_2 in CCl_4 → $(C_6H_{11})_3GeI$ (244).
Air in CCl_4 — $(C_6H_{11})_3GeOH$ (244).

TRIPHENYLGERMANE Ph_3GeH

Prepn.: By reducing Ph_3GeBr or $(Ph_3Ge)_2$ with $LiAlH_4$ in Et_2O; up to 79% yield (244, 246, 254).
By reducing Ph_3GeBr with Zn-Hg in HCl (565).
By-prod. in the rxn. of Ph_3GeLi with 1-octene in $(CH_2OMe)_2$ (181).

Props.: The compound occurs in two forms, a higher-melting α-form and a lower-melting β-form. α-Form: b. 125-130°/0.03mm; m. 42.5-45° (181); m. 41-41.5° (246, 254). β-Form: m. 27° (244, 565). Soly. data (254). Phrolysis at 300° → Ph_4Ge, Ph_3GeH, and Ph_2GeH_2 (246).

Rxns. with: $n-C_8H_{17}Li$ (1:4) → $Ph_3GeC_8H_{17}$ (138).
MeLi, then poured onto CO_2 and washed with H_2O → $(Ph_3Ge)_2$ and $(Ph_3Ge)_2O$ (179).
BuLi, then poured onto CO_2 → Ph_3GeCO_2H (179).
PhLi (excess) → Ph_4Ge and some $(Ph_3Ge)_2$; if excess of Ph_3GeH is employed, $(Ph_3Ge)_2$ is the main prod. (246, 254).
MeI, then poured onto CO_2 → Ph_3GeCO_2H and a small amt. of $(Ph_3Ge)_2$ (179).
I_2 in CCl_4 → Ph_3GeI (246).
BzCl → Ph_3GeCl (246).
$Ph_3GeCH_2CH:CH_2$ in the presence of Bz_2O_2 → $(Ph_3GeCH_2)_2CH_2$ (181).
1-Octene in the presence of Bz_2O_2 or in UV light → $n-C_8H_{17}GePh_3$ (138).
1-Octadecene in the presence of Bz_2O_2 → $n-C_{18}H_{37}GePh_3$ and $(Ph_3Ge)_2O$ (181).
Cyclohexene in the presence of Bz_2O_2 or in UV light → $C_6H_{11}GePh_3$ (138).
$Ph_2C:CH_2$ in the presence of Bz_2O_2 → oily olefinic polymer (138).

OTHER ORGANOGERMANES

$R_{4-x}GeH_x$	Prepared from	Yield %	Properties	Ref.
MeGeD$_3$	MeGeCl$_3$ or MeGeBr + LiD		b. -23.5°/752mm;	420
EtGeD$_3$	EtGeCl$_3$ or EtGeBr + LiD		b. 11.3°/748.5mm; Raman spectrum	420
CH$_2$:CHGeH$_3$	CH$_2$:CHGeCl$_3$ + LiAlH$_4$		b. -3.5°(calcd.); IR spectrum; instable, in daylight polymerizes	626a
PrGeH$_3$	PrGeCl$_3$ + LiAlH$_4$ in i-Pr$_2$O NaGeH$_3$ + PrBr in liq. NH$_3$	85	b. 30° (1)	247, 255 547
Me$_2$GeH$_2$	Me$_2$GeCl$_2$ or Me$_2$GeBr$_2$ + LiH Me$_2$GeS ∓ Zn-Hg in HCl		b. 6.5°/744mm	420 565
Me$_2$GeD$_2$	Me$_2$GeCl$_2$ or Me$_2$GeBr$_2$ + LiD		b. 6.5°/745mm	420
Et$_2$GeH$_2$	Et$_2$GeCl$_2$ or Et$_2$GeBr$_2$ + LiH		b. 72.5°/740.5mm; d$_{20}$ 1.0378; n$^{20}_D$ 1.4208; Raman spectrum	420
Me$_3$GeH	Me$_3$GeCl or Me$_3$GeBr + LiH		b. 26°/755.5mm; d$_{20}$ 1.0128; n$^{20}_D$ 1.3890	420
Me$_3$GeD	Me$_3$GeCl or Me$_3$GeBr + LiD		b. 26°/758mm; d$_{20}$ 1.0207; n$^{20}_D$ 1.3893	420
Me$_2$EtGeH	Me$_2$EtGeCl or Me$_2$EtGeBr + LiH		b. 62°/755.5mm; d$_{20}$ 1.10158; n$^{20}_D$ 1.4090; Raman spectrum	420
Me$_2$EtGeD	Me$_2$EtGeCl or Me$_2$EtGeBr + LiD		b. 60°/737.4mm; d$_{20}$ 1.0262; n$^{20}_D$ 1.4083; Raman spectrum	420

(1) Rxn. with Br$_2$ in (CH$_2$Br)$_2$ → PrGeBr$_3$ (247).

OTHER ORGANOGERMANES (CONT'D.)

$R_{4-x}GeH_x$	Prepared from	Yield %	Properties	Ref.
$(n-C_5H_{11})_3GeH$	$(n-C_5H_{11})_3GeBr$ + $LiAlH_4$		b. 150°/11mm; n^{20}_D 1.4542; d^{20}_4 0.9310 (2)	324
$(i-C_5H_{11})_3GeH$	$(i-C_5H_{11})_3GeBr$ + $LiAlH_4$		b. 140°/17mm; n^{20}_D 1.4517; d^{20}_4 0.9238	324
$Et_2(i-C_5H_{11})GeH$	$Et(i-C_5H_{11})_2GeH_2$ + Li (1:1) in $EtNH_2$ and treating the resulting $Et(i-C_5H_{11})_2GeLi$ with EtI	87.2	Colorless liquid, v.p. ~ 2mm at r.t.	190
$(n-C_6H_{13})_3GeH$	$(n-C_6H_{13})_3GeBr$ + $LiAlH_4$		b. 122-125°/0.5mm; n^{21}_D 1.4565; d_{25} 0.917; IR spectrum	140
$(1-C_{10}H_7)_3GeH$	$(1-C_{10}H_7)_3GeBr$ + $LiAlH_4$	82	Colorless needles, m. 249-250° (3,4)	563

(2) Rxn. with $CH_2:CHCN \rightarrow (n-C_5H_{11})_3GeCH_2CH_2CN$ (324).
(3) Rxn. with KOH in moist $C_5H_5N \rightarrow$ evolution of H (563).
(4) Rxn. with Br_2 in $CCl_4 \rightarrow$ evolution of HBr (563).

4. ORGANOGERMANIUM HALIDES AND OTHER SALTS

TRIORGANOGERMANIUM HALIDES

Triorganogermanium halides are prepared:

1. By treating diorganogermanium dihalides or diorgano-(dialkoxy)germane with one mole of a Grignard compound.
2. By treating triorganogermanium oxides, hydroxides, or carboxylates with hydrogen halides, silver fluoride, or acyl halides.
3. By reacting germanium tetrahalides or tetraalkoxides with Grignard or organolithium compounds in a molar ratio of 1:3.
4. By treating tetraorganogermanes with halogens; if iodine is used, the presence of aluminum iodide catalyst is required.
5. By cleaving hexaorganodigermanes with halogens.
6. By reacting organogermanium trihalides with two moles of azoalkanes.
7. By reacting organic halides with powdered germanium at elevated temperatures in the presence of copper; the product contains a mixture of organogermanium halides.
8. By treating triorganogermanes, R_3GeH, or lithium triorganogermanides with halogens.

Triorganogermanium chlorides and bromides exchange the halogens for fluorine in reactions with antimony fluoride.

TRIMETHYLGERMANIUM FLUORIDE Me_3GeF
Prepn: By reacting Me_3Ge chloride or bromide with SbF_3 (419).
Props: b. 76°/746mm; d_{20} 1.2300; $n^{20}D$ 1.3863 (419).

TRIMETHYLGERMANIUM CHLORIDE Me_3GeCl
Prepn.: By reacting Me_2GeCl_2 with MeMgBr; 12% yield along with 25% Me_2GeBr_2 (419); with MeMgCl the yield was 25% (447).
Props.: b. 98°/736mm; d_{20} 1.2493; $n^{20}D$ 1.4337 (419); b. 115°; m-13°; $n^{29}D$ 1.4314 (447).
Rxns. with: $MgBrCl \rightarrow Me_3GeBr$ (419).
$SbF_3 \rightarrow Me_3GeF$ (419).
LiH or LiD $\rightarrow Me_3GeH$ and $MeGeD_3$, respectively (420).

TRIMETHYLGERMANIUM BROMIDE   Me$_3$GeBr

Prepn.: By treating Me$_4$Ge with Br$_2$ (63).
By refluxing Me$_3$GeCl with MgClBr in Et$_2$O; 22% yield (419).
By-prod. in the rxn. of MeBr with a mixt. of Ge and Cu (1:1) at 340-350° (446).
Props.: b. 115°/755mm; n$_{20}$ 1.5486; n^{20}D 1.4660 (419).
Rxns. with: Molten K → (Me$_3$Ge)$_2$ (63).
p-BrC$_6$H$_4$MgBr → Me$_3$GeC$_6$H$_4$Br (90).
SbF$_3$ → Me$_3$GeF (419).
LiH or LiD → Me$_3$GeH and Me$_3$GeD, respectively (420).

TRIMETHYLGERMANIUM IODIDE   Me$_3$GeI

Prepn.: By treating Me$_4$Ge with I$_2$ (1:1) in the presence of AlI$_3$ (323).
By reacting Me$_2$Ge(OMe)$_2$ with MeMgI in Et$_2$O under N (566).
Props.: b. 136° (566).
Rxn. with: NaOR → Me$_3$GeOR (566).

BIS(CHLOROMETHYL)METHYLGERMANIUM CHLORIDE   Me(CH$_2$Cl)$_2$GeCl

Prepn.: By-prod. in the rxn. of MeGeCl$_3$ with CH$_2$N$_2$(1:>2) in Et$_2$O at -60°; 20.5% yield (485).
Props.: b. 95-97°/30mm (485).

ETHYLDIMETHYLGERMANIUM CHLORIDE   Me$_2$EtGeCl

Prepn.: By reacting Me$_2$GeCl$_2$ with EtMgBr; Me$_2$EtGeBr and Me$_2$GeBr$_2$ are also formed (420).
Rxn. with: LiH or LiD → Me$_2$EtGeH and Me$_2$EtGeD, respectively (420).

ETHYLDIMETHYLGERMANIUM BROMIDE   Me$_2$EtGeBr

Prepn.: By reacting Me$_2$GeCl$_2$ with EtMgBr; Me$_2$EtGeCl and Me$_2$GeBr$_2$ are also formed (420).

TRIETHYLGERMANIUM FLUORIDE   Et$_3$GeF

Prepn.: By refluxing (Et$_3$Ge)$_2$O with AgF; 60% yield (23).
By refluxing Et$_3$GeOAc with AgF; 80% yield (23).
Props.: b. 149.4° (23).

TRIETHYLGERMANIUM CHLORIDE   Et$_3$GeCl

Props.: n^{20}D 1.4643; d^{20}$_4$ 1.175 (24).
Rxns. with: AgNCO → Et$_3$GeNCO (7).
RLi → Et$_3$GeR (187).
RMg halide → Et$_3$GeR (187, 368).
Li in Et$_2$O → (Et$_3$Ge)$_2$ (676).

TRIETHYLGERMANIUM BROMIDE Et_3GeBr
Props.: $n^{20}D$ 1.4892; d^{20}_4 1.412 (24).
Rxns. with: $Ag_2CO_3 \rightarrow (Et_3Ge)_2O$ (10).
$AgNO_3 \rightarrow$ instable Et_3GeNO_3 (?) (10).
$AgOCOR \rightarrow Et_3GeOCOR$ (21, 22).
$p\text{-}BrC_6H_4MgBr \rightarrow Et_3GeC_6H_4Br$ (90).
$(:CMgBr)_2 \rightarrow Et_3GeC:CGeEt_3$ (213).

TRIETHYLGERMANIUM IODIDE Et_3GeI
Prepn.: By reacting Et_4Ge with I_2 (1:1) in the presence of AlI_3 (323).
Props.: $n^{20}D$ 1.528; d^{20}_4 1.608 (24).
Rxns. with: I_2(1:1) in the presence of $AlI_3 \rightarrow Et_2GeI_2$, EtI, and C_4H_{10} (323).
RMg halides $\rightarrow Et_3GeR$ (323).

TRIVINYLGERMANIUM BROMIDE $(CH_2:CH)_3GeBr$
Prepn.: By treating $[(CH_2:CH)_3Ge]_2$ with Br_2 in $CHCl_3$ under reflux (490).
Props.: b. 58°/10 mm; $n^{25}D$ 1.5057; lacrimator (490).

TRIVINYLGERMANIUM IODIDE $(CH_2:CH)_3GeI$
Prepn.: By treating $[(CH_2:CH)_3Ge]_2$ with I_2 in $CHCl_3$ under reflux (490).
Props.: b. 71-74°/12 mm; extremely sensitive to light (490).

TRIPROPYLGERMANIUM FLUORIDE Pr_3GeF
Prepn.: By treating $(Pr_3Ge)_2O$ with 48% HF (11).
Props.: Colorless liquid sol. in org. solvents, b. 203°; m. -27.5°; d^{20}_4 1.074; $n^{20}D$ 1.4340; mol. R 53.6 (11).
Rxns. with: Fe + Br on steam bath, followed by hydrolysis and treatment with HF $\rightarrow Pr_2GeF_2$ (15).
$(Et_3Sn)_2O \rightarrow (Pr_3Ge)_2O$ and Et_3SnF (19).

TRIPROPYLGERMANIUM CHLORIDE Pr_3GeCl
Prepn.: By shaking $(Pr_3Ge)_2O$ with 12N HCl; theoret. yield (11).
Props.: Colorless liquid sol. in org. solvents, b. 227°; b. 98-99°/11 mm; m. -70°; d^{20}_4 1.100; $n^{20}D$ 1.4641; mol. R 59.5 (11).
Rxns. with: H_2O_2 or alkyl hydroperoxides in the presence of tert. amines $\rightarrow$ peroxides of Pr_3Ge (111).

TRIPROPYLGERMANIUM BROMIDE Pr_3GeBr
Prepn.: By shaking $(Pr_3Ge)_2O$ with 48% HBr; 93% yield (11).
Props.: Colorless liquid, sol. in org. solvents, b. 242°; b. 112-113°/12 mm; m. -47°; d^{20}_4 1.282; $n^{20}D$ 1.4832; mol. R. 62.7 (11).

TRIPROPYLGERMANIUM IODIDE Pr_3GeI
Prepn.: By shaking $(Pr_3Ge)_2O$ with 45% HI; 85% yield (11)
Props.: Colorless liquid sol. in org. solvents, b. 259°; b. 122-123°/10mm; m. -38°; d^{20}_4 1.443; $n^{20}D$ 1.5144; mol. R 68.7 (11).

TRIISOPROPYLGERMANIUM FLUORIDE $i-Pr_3GeF$
Prepn.: By heating $i-Pr_3GeBr$ with SbF_3 (17).
By shaking $i-Pr_3GeOH$ with concd. HF (17)
Props.: Colorless liquid, b. 198°; m. -65°; d^{20}_4 1.069; $n^{20}D$ 1.440 (17).

TRIISOPROPYLGERMANIUM CHLORIDE $i-Pr_3GeCl$
Prepn.: By shaking $i-Pr_3GeOH$ with concd. aq. HCl (17).
Props.: Colorless liquid, b. 222°; d^{20}_4 1.110; $n^{20}D$ 1.472 (17).
Rxns. with: $AgO_2CR \rightarrow i-Pr_3GeO_2CR$ (20).
$AgNCO \rightarrow i-Pr_3Ge(NCO)$ (20).
$AgNCS \rightarrow i-Pr_3Ge(NCS) + AgCl$ (23).

TRIISOPROPYLGERMANIUM BROMIDE $i-Pr_3GeBr$
Prepn.: By reacting $i-PrMgBr$ with $GeBr_4$ in Et_2O or C_6H_6 (17).
Props.: Colorless liquid with a penetrating, nauseating odor, b. 234°; m. -45°; d^{20}_4 1.231; $n^{20}D$ 1.4852 (17).
Rxns. with: $SbF_3 \rightarrow i-Pr_3GeF$ (17).
6 M aq. $NAOH \rightarrow i-Pr_3GeOH$ (17).
$AgCN \rightarrow i-Pr_3Ge(CN)$ (23).
$AgNCS \rightarrow i-Pr_3Ge(NCS)$ (23).

TRIISOPROPYLGERMANIUM IODIDE $i-Pr_3GeI$
Prepn.: By shaking $i-Pr_3GeOH$ with concd. aq. HI (17).
Props: Colorless liquid, b. 254°; m. 4°; d^{20}_4 1.446; $n^{20}D$ 1.524 (17).
Rxn. with: Ag_2CO_3 in $C_6H_6 \rightarrow (i-Pr_3Ge)_2O$ (17).

TRIBUTYLGERMANIUM FLUORIDE Bu_3GeF
Prepn.: By stirring $(Bu_3Ge)_2O$ with 48% HF or heating with 60% aq. KHF_2; 80% yield (13).
Props.: Colorless liquid, b. 245°; b. 128-130°/14 mm; d^{20}_4 1.038; $n^{20}D$ 1.4419; M.R. 67.0 (13).

TRIBUTYLGERMANIUM CHLORIDE Bu_3GeCl
Prepn.: By refluxing $(Bu_3Ge)_2O$ with 12N HCl; theoret. yield (13)
Props.: Colorless liquid; b. 271°; b. 139-140°/13 mm; d^{20}_4 1.054; $n^{20}D$ 1.4652; MR 73.3 (13)
Rxns. with: $AgNCO \rightarrow Bu_3Ge(NCO)$ (13)
$RMgBr \rightarrow Bu_3GeR$ (368).

TRIBUTYLGERMANIUM BROMIDE Bu_3GeBr
Prepn.: By refluxing $(Bu_3Ge)_2O$ with 48% HBr; theoret. yield (13).
Props.: Colorless liquid; b. 279°; b. 143-144°/10 mm; d^{20}_4 1.195; $n^{20}D$ 1.4809; MR 77.1 (13).

TRIBUTYLGERMANIUM IODIDE Bu_3GeI
Prepn.: By reacting Bu_4Ge with I_2 under reflux; 69% yield (13, 306).
By reacting $(Bu_3Ge)_2O$ with concd. HI; 49% yield (13, 160)
Props.: Colorless liquid, b. 297°; 163-164°/15 mm; d^{20}_4 1.340; $n^{20}D$ 1.508; MR 82.6 (13); b. 125-127°/4 mm; (160); b. 126-128°/4 mm (306).
Rxns. with: Aq. $NaOH \rightarrow (Bu_3Ge)_2O$ (13).
RMg halide $\rightarrow Bu_3GeR$ (368).

TRIISOBUTYLGERMANIUM IODIDE $i-Bu_3GeI$
Prepn.: By reacting $i-Bu_4Ge$ with I_2 (1:1) in the presence of AlI_3 (323).
Props.: b. 137-138°/17 mm; $n^{20}D$ 1.5098; d^{20}_4 1.3543 (323).

TRIPENTYLGERMANIUM IODIDE $(n-C_5H_{11})_3GeI$
Prepn.: By reacting $(n-C_5H_{11})_4Ge$ with I_2 (1:1) in the presence of AlI_3 (323).
Props.: b. 184-185°/11 mm; $n^{20}D$ 1.5023; d^{20}_4 1.2686 (323).

TRIHEXYLGERMANIUM CHLORIDE $(C_6H_{13})_3GeCl$
Prepn.: By treating $[(C_6H_{13})_3Ge]_2O$ with concd. HCl (140).
Props.: b. 138-139°/0.5 mm; $n^{20}D$ 1.4661; d_{20} 0.989 (140).
Rxns. with: NaI in $Me_2CO \rightarrow (C_6H_{13})_3GeI$ (140).

TRIHEXYLGERMANIUM BROMIDE $(C_6H_{13})_3GeBr$
Prepn.: By reacting $GeCl_4$ with $C_6H_{13}MgBr$ (1:3) (140).
Props.: b. 143-145°/0.5 mm; $n^{27}D$ 1.4763; d_{26} 1.117 (140).
Rxns. with: $PhLi \rightarrow (C_6H_{13})_3GePh$ (140).
10% $KOH \rightarrow [(C_6H_{13})_3Ge]_2O$ (140).
$LiAlH_4 \rightarrow (C_6H_{13})_3GeH$ (140).

TRIHEXYLGERMANIUM IODIDE $(C_6H_{13})_3GeI$
Prepn.: By treating $(C_6H_{13})_3GeCl$ with NaI in Me_2CO (140).
Props.: b. 154-155°/0.5 mm; n^{26}_D 1.4935; d_{26} 1.188; turns brown on standing for several days (140).

TRICYCLOHEXYLGERMANIUM CHLORIDE $(C_6H_{11})_3GeCl$
Prepn.: By reacting $GeCl_4$ with excess $C_6H_{11}Li$ in petr. ether; 70% yield (244).
By refluxing $(C_6H_{11})_3GeOH$ with AcCl (244).
Props.: m. 101° (244).
Rxns. with: 5% alc. KOH → $(C_6H_{11})_3GeOH$ (244).
$LiAlH_4$ → $(C_6H_{11})_3GeH$ (244).

TRICYCLOHEXYLGERMANIUM BROMIDE $(C_6H_{11})_3GeBr$
Prepn.: By reacting $Ge(OEt)_4$ with $C_6H_{11}MgBr$ in MePh on steam bath; 50% yield (244).
By reacting $(C_6H_{11})_3GeH$ with excess Br_2 in EtBr; theoret. yield (244).
Props.: m. 109° (244).
Rxns. with: $(:CMgBr)_2$ → $(C_6H_{11})_3GeC:CGe(C_6H_{11})_3$ (213).
RMgBr → $(C_6H_{11})_3GeR$ (243).
PhLi → $(C_6H_{11})_3GePh$ (244).

TRICYCLOHEXYLGERMANIUM IODIDE $(C_6H_{11})_3GeI$
Prepn.: By treating $(C_6H_{11})_3GeH$ with I_2 in CCl_4 (244).
Props.: m. 99° (244).

TRIPHENYLGERMANIUM FLUORIDE Ph_3GeF
Prepn.: By slurrying $(Ph_3Ge)_2O$ in 47% HF; 71% yield (473).
Props.: IR spectrum (473).
Rxns. with: H_2O, rate of hydrolysis and rxn. mechanism (250, 473).
Slow neutrons → radioactive Ph_3GeF (method of concentration) (722).

TRIPHENYLGERMANIUM CHLORIDE Ph_3GeCl
Prepn.: By refluxing Ph_3GeH with BzCl (246).
By reacting $(Ph_3Ge)_2O$ with concd. HCl in hot EtOH; 92% yield (473).
Props.: m. 114-115° (246); Dipole moment (397); Conductance in non-aqueous solvents (788, 789); IR spectrum (473).
Rxns. with: RMgCl → Ph_3GeR (138, 181).
Ph_3SiK → $Ph_3SiGePh_3$ and $(Ph_3Ge)_2$ (180).
H_2O, rate of hydrolysis and rxn. mechanism (250, 473).

TRIPHENYLGERMANIUM BROMIDE Ph_3GeBr

Prepn.: By brominating Ph_4Ge in CCl_4 or $C_2H_4Br_2$; up to 82% yield (61, 244, 246, 254, 473).
By-prod. in the rxn. of Ph_3GeLi with Br_2 in $(CH_2OMe)_2$ at -25 to -20° (186).
Props.: m. 137-138° (61, 246); m. 138-140° (186); Conductance in dioxan-H_2O (250, 473); IR spectrum (473); Dipole moment (397, 520, 521).
Rxns. with: $RLi \rightarrow Ph_3GeR$ (51).
$Ph_3CNa \rightarrow Ph_3CGePh_3$ (61).
$MeONa \rightarrow Ph_3GeOMe$ (62).
$ROONa \rightarrow Ph_3GeOOR$ (753).
$H_2O_2 \rightarrow (Ph_3GeO)_2$ (753).
$LiAlH_4 \rightarrow Ph_3GeH$ (244, 246, 254).
H_2O, rate of hydrolysis and rxn. mechanism (250, 473).
H_2O or aq. alkali $\rightarrow (Ph_3Ge)_2O$ (254).
Aq.-alc. alkali $\rightarrow Ph_3GeOH$ (61).
Alc. KOH $\rightarrow (Ph_3Ge)_2O$ (244).
$Ph_3SiK \rightarrow Ph_3SiGePh_3$ (180).
$NH_3 \rightarrow (Ph_3Ge)_3N$ (254).
Dry NH_3 in $Et_2O \rightarrow Ph_3GeNH_2$ (753).
$Na_2S \rightarrow (Ph_3Ge)_2S$ (65).
$Ph_2CHC_6H_4MgBr \rightarrow Ph_3GeC_6H_4CHPh_2$ (61).
$(:CMgBr)_2 \rightarrow Ph_3GeC:CGePh_3$ (213).
Zn + HCl $\rightarrow Ph_3GeH$, Ph_4Ge, and Ph_3GeOH (565).
$AgNO_3$ in EtOH $\rightarrow (Ph_3Ge)_2O$ (473).

TRIPHENYLGERMANIUM IODIDE Ph_3GeI

Prepn.: By reacting Ph_3GeH with I_2(1:1) in CCl_4 under reflux (246).
Props.: White crystals, m. 156-157° (246).

TRI-p-TOLYLGERMANIUM FLUORIDE $(p-MeC_6H_4)_3GeF$

Prepn.: By treating $[(p-MeC_6H_4)_3Ge]_2O$ with 47% HF (250, 473).
Props.: IR spectrum (473).
Rxn. with: H_2O, rate of hydrolysis and rxn. mechanism (250, 473).

TRI-1-NAPHTHYLGERMANIUM BROMIDE $(1-C_{10}H_7)_3GeBr$

Prepn.: By refluxing $1-C_{10}H_7MgBr$ with $GeBr_4$(4:1) in C_6H_6; 38% yield (563).
Props.: Colorless, irregular prisms showing low birefringence and parallel extinction; m. 242-243.5° (563).
Rxns. with: Dild. alc. NaOH in $Me_2CO \rightarrow (1-C_{10}H_7)_3GeOH$ (563).
$1-C_{10}H_7Li$ in $C_6H_6 \rightarrow (1-C_{10}H_7)_4Ge$ (563).
$LiAlH_4 \rightarrow (1-C_{10}H_7)_3GeH$ (563).

DIORGANOGERMANIUM DIHALIDES

Diorganogermanium dihalides are prepared:

1. By reacting organic halides with powdered germanium at elevated temperatures in the presence of copper or other metal catalysts.
2. By treating tetraorganogermanes with halogens in a molar ratio of 1:2.
3. By treating triorganogermanium halides with halogen in a molar ratio of 1:1.
4. By treating diorganogermanium oxides or sulfates with hydrogen halides.
5. By treating diorganogermanium sulfides with silver halides.
6. By reacting germanium tetrahalides or organogermanium trihalides with diazoalkanes in a molar ratio of 1:2 and 1:1, respectively.
7. By reacting germanium tetrahalides with organolithium compounds in a molar ratio of 1:2.
8. By treating diorganogermanes, R_2GeH_2, with halogens in a molar ratio of 1:2.

Diorganogermanium fluorides are obtained by treating diorganogermanium dichlorides or dibromides with antimony trifluoride.

DIMETHYLGERMANIUM DIFLUORIDE Me_2GeF_2
Prepn.: By treating Me_2GeCl_2 or Me_2GeBr_2 with SbF_3 (419).
Props.: b. 112°/750 mm; d_{20} 1.5726; $n^{20}D$ 1.3743 (419).

DIMETHYLGERMANIUM DICHLORIDE Me_2GeCl_2
Prepn.: By reacting MeCl with Ge at elevated temp. in the presence of a Cu catalyst; the prod. contains a mixt. of related organo-Ge chlorides (252).
By reacting MeCl with Ge in the presence of absence of a catalyst (Cu, Ag, Ni, Sn, Sb, Mn, Ti) at 200-500° (600).
By-prod. in the rxn. of MeCl with Ge-Cu mixt. at 550°; 26% yield (412); if the ratio Ge:Cu= 4:1 and the rxn. is effected at 320-360°, the prod. contains 56% Me_2GeCl_2 along with some $MeGeCl_3$ (445, 446, 449); in the absence of Cu the rxn. occurs at 460° (446).
Props.: b. 122°/727 mm; $n^{20}D$ 1.4600; d_{20} 1.5053 (412); b. 124°; m. -22°; d_{26} 1.488; d_{20} 1.492; $n^{29}D$ 1.4552 (445, 449). In H_2O the compd. dissociates completely and behaves as a strong acid (457).

Rxns. with: RMgX (X= Cl or Br) → Me_2GeR_2 and Me_2GeRCl, depending on the rxn. conditions (412, 419, 420, 447).
H_2O ⇌ $Me_2Ge(OH)_2$ (448, 457, 600).
H_2S in 6N H_2SO_4 or in H_2O → Me_2GeS (448, 450, 601).
Cl_2 under reflux in UV light → $Me(ClCH_2)GeCl_2$ (419).
SbF_3 → Me_2GeF_2 (419).
CrO_4^{--} → Me_2GeCrO_4 (457).
CNS^- → $Me_2Ge(CNS)_2$ (457).
NaOR(1:2) – $Me_2Ge(OR)_2$ (566).
LiH or LiD – Me_2GeH_2 and Me_2GeD_2, respectively (420).

DIMETHYLGERMANIUM DIBROMIDE Me_2GeBr_2

Prepn.: By passing MeBr at 340-450° over a powder or alloy of Ge and Cu(4:1-1:1); the prod. contains $MeGeBr_3$ and Me_2GeBr_2 (419, 445, 445).
By reacting Me_2GeCl_2 with MeMgBr; 25% yield along with 12% Me_3GeCl (419) or with EtMgBr, Me_2GeBr_2, along with $Me_2EtGeCl$ and $Me_2EtGeBr$, is formed (420).
Props.: b. 153°/746 mm; d_{20} 2.1163; $d^{20}D$ 1.5268 (419); b. 153°/755 mm (420).
Rxns. with: SbF_3 → Me_2GeF_2 (419).
LiH or LiD → Me_2GeH_2 and Me_2GeD_2, respectively (420).

(CHLOROMETHYL)METHYLGERMANIUM DICHLORIDE $Me(CH_2Cl)GeCl_2$

Prepn.: By reacting $MeGeCl_3$ with CH_2N_2(1:1) in Et_2O at -60°; 78% yield. If the molar ratio of the reactants is 1:>2, 33.5% yield of $Me(CHCl)GeCl_2$ along with 20.5% $Me(CH_2Cl)_2GeCl$ resulted (485).
By passing Cl_2 into boiling Me_2GeCl_2 in UV light (419).
Props.: b. 71-74°/40 mm (485); b. 155°/750 mm; d_{20} 1.6694; $n^{20}D$ 1.4920 (419).
Rxn. with: MeMgI → Me_3GeCH_2I (419).

BIS(CHLOROMETHYL)GERMANIUM DICHLORIDE $(ClCH_2)_2GeCl_2$

Prepn.: By-prod. in the rxn. of $GeCl_4$ with CH_2N_2(12:12.8) in Et_2O at -60° in the presence of Cu (485).
Props.: b. 90°/21 mm (485).

DIETHYLGERMANIUM DIBROMIDE Et_2GeBr_2

Prepn.: By heating Ph_2GeEt_2 with Br_2 in $C_2H_4Br_2$ at 90-100° (551).
Props.: b. 90-95°/22 mm. (551).
Rxn. with: 10% aq. NaOH → $(Et_2GeO)_x$ (551).

DIETHYLGERMANIUM DICHLORIDE Et_2GeCl_2

Prepn.: By reacting $[Et_2GeSO_4]_2$ with HCl gas in C_6H_6 in the presence of powdered $(NH_4)_2SO_4$ (8).
By reacting EtCl with powdered Ge in the presence of Cu(1:1-1:6) at >317°; $EtGeCl_3$ is also formed (445, 446, 451).
Props.: b. 172° (8); b. 172.8° (451); Raman spectrum (331).
Rxns. with: AgNCO — $Et_2Ge(NCO)_2$ (7).
$CH_2:CHMgBr \rightarrow Et_2Ge(CH:CH_2)_2$ (413, 490).
$H_2O(1:2) \rightarrow$ colorless oil which condenses to $(Et_2GeO)_x$ (451).

DIETHYLGERMANIUM DIIODIDE Et_2GeI_2

Prepn.: By reacting Et_3GeI with I_2(1:1) in the presence of AlI_3 (323).
Rxns. with: I_2(1:1) in the presence of $AlI_3 \rightarrow EtGeI_3 + C_4H_{10}$ (323).
RMg halide $\rightarrow Et_2GeR_2$ (323).

DIPROPYLGERMANIUM DIFLUORIDE Pr_2GeF_2

Prepn.: By treating Pr_3GeF with Fe and Br on steam bath, hydrolyzing the rxn. mixt. with alkali, and converting the resulting $(Pr_2GeO)_3$ to fluoride (15).
Props.: Colorless liquid, b. 182.8°; m. 0.5°; d^{20}_4 1.248; n^{20}_D 1.4128 (15).
Rxn. with: 20% NaOH $\rightarrow (Pr_2GeO)_3$ (15).

DIPROPYLGERMANIUM DICHLORIDE Pr_2GeCl_2

Prepn.: By reacting $(Pr_2GeO)_3$ with concd. HCl (15).
By reacting $GeCl_4$ with PrLi(1:1) in petr. ether with CO_2-$CHCl_3$ cooling; the prod. contains 20% Pr_2GeCl_2, 16% $PrGeCl_3$ and some Pr_3GeCl; with 2 equivs. $GeCl_4$, the prod. contains 54% Pr_2GeCl_2 and 31.6% $PrGeCl_3$ (247).
Props.: Colorless liquid, b. 209.5°; m. -45°; d^{20}_4 1.275; n^{20}_D 1.4725 (15); b. 60-63°/5 mm (247).
Rxn. with: Peroxides $\rightarrow Pr_3Ge(OOR)_2$ (111).

DIPROPYLGERMANIUM DIBROMIDE Pr_2GeBr_2

Prepn.: By reacting $(Pr_2GeO)_3$ with concd. HBr (15).
Props.: Colorless liquid, b. 240.5°; m. -52.0°; d^{20}_4 1.689; n^{20}_D 1.5173 (15).
Rxn. with: NaOH in $H_2O \rightarrow (Pr_2GeO)_n$ (15).

DIPROPYLGERMANIUM DIIODIDE Pr_2GeI_2

Prepn.: By reacting $(Pr_2GeO)_3$ with 48% HI at 70°; 96% yield (15).
Props.: Colorless liquid, b. 276.5°; m. -53.5°; d^{20}_4 2.024 (15).

DIISOPROPYLGERMANIUM DIFLUORIDE   i-Pr$_2$GeF$_2$
Prepn.: By heating i-Pr$_2$GeBr$_2$ with SbF$_3$ (17).
Props.: Liquid with a repulsive, musty odor, b. 174°; m -24°;
d$^{20}_4$ 1.222; n^{20}D 1.4146 (17).
Rxn. with: NaOH → (i-Pr$_2$GeO)$_3$ (17).

DIISOPROPYLGERMANIUM DICHLORIDE   i-Pr$_2$GeCl$_2$
Prepn.: By heating (i-Pr$_2$GeO)$_3$ with concd. aq. HCl at 85°;
theoret. yield (17).
Props.: b. 203°; m. -52°; d$^{20}_4$ 1.268; n^{20}D 1.4738 (17).
Rxns. with: AgOCOR → i-Pr$_2$Ge(OCOR)$_2$ (23).
AgOCN → i-Pr$_2$Ge(NCO)$_2$ (23).
LiAlH$_4$ → i-Pr$_2$GeH$_2$ (23).

DIISOPROPYLGERMANIUM DIBROMIDE   i-Pr$_2$GeBr$_2$
Prepn.: By heating (i-Pr$_2$GeO)$_3$ with concd. aq. HBr at 85°;
theoret. yield (17).
By refluxing (i-Pr$_2$GeS)$_2$ with AgBr (23).
Props.: Colorless liquid, b. 234°; m. -22°; d$^{20}_4$ 1.670;
n^{20}D 1.519 (17).
Rxns. with: SbF$_3$ → i-Pr$_2$GeF$_2$ (17).

DIISOPROPYLGERMANIUM DIIODIDE   i-Pr$_2$GeI$_2$
Prepn.: By heating (i-Pr$_2$GeO)$_3$ with concd. aq. HI at 85°;
theoret. yield (17).
Props.: b. 268°; m. -9°; d$^{20}_4$ 2.008; n^{20}D 1.597 (17).
Rxn. with: Ag$_2$S — (i-Pr$_2$GeS)$_2$ (23).

DIPHENYLGERMANIUM DICHLORIDE   Ph$_2$GeCl$_2$
Prepn.: By reacting PhCl with a finely powdered mixture or an
alloy of Ge and Ag(4:1) at 440° (445, 446).
Rxn. with: H$_2$O → Ph$_2$Sn(OH)$_2$ (446).

DIPHENYLGERMANIUM DIBROMIDE   Ph$_2$GeBr$_2$
Prepn.: By bromination of Ph$_4$Ge in C$_2$H$_4$Br$_2$ under reflux (245,
253) or in a mixt. of C$_2$H$_4$Br$_2$ and CCl$_4$ at 80-90° (551).
By brominaton of Ph$_2$GeH$_2$ in CHCl$_3$ (245).
Rxns. with: LiAlH$_4$ in Et$_2$O → Ph$_2$GeH$_2$ (245, 253).
NH$_4$OH in EtOH → Ph$_2$GeO (245).
RMgBr(1:2) → Ph$_2$GeR$_2$ (551).

DIPHENYLGERMANIUM DIIODIDE   Ph$_2$GeI$_2$
Prepn.: By hydrolyzing Ph$_2$GeBr$_2$ with NH$_4$OH and treating the
resulting prod. (Ph$_2$GeO (?)) with HI (245).
By reacting Ph$_2$GeH$_2$ with I$_2$ in CHCl$_3$ at 70° (245).
Props.: m. 70-71° (245).

ORGANOGERMANIUM TRIHALIDES

Organogermanium trihalides are prepared:

1. By reacting organic halides with germanium powder at elevated temperatures in the presence of copper.
2. By reacting germanium tetrahalides with diazoalkanes in a molar ratio of 1:1 in the presence of copper.
3. By reacting germanium tetrahalides with Grignard, organolithium, or diorganomercury compounds in a molar ratio of 1:1.
4. By reacting organogermanes, $RGeH_3$, with halogens in a molar ratio of 1:3.
5. By reacting trihalogermanes with α-olefins.
6. By heating organic halides with germanium dihalides in a sealed tube.
7. Be cleaving one organic radical of diorganogermanium dihalides with one mole of halogen in the presence of aluminum chloride or in ultraviolet light.
8. By heating a mixture of tetraorganogermanes with germanium tetrahalides in a ratio of 1:3.
9. By reacting organic halides with cesium trihalogermanides.

Organogermanium trifluorides are prepared from other organogermanium trihalides by treating them with antimony trifluoride.

METHYLGERMANIUM TRIFLUORIDE $MeGeF_3$
Prepn.: By treating $MeGeCl_3$ or $MeGeBr_3$ with SbF_3 (419).
Props.: b. 96.5°/751 mm; m. 38° (419).

METHYLGERMANIUM TRICHLORIDE $MeGeCl_3$
Prepn.: By passing MeCl over a powder mixt. of Ge and Cu at 550°; 35% yield along with Me_2GeCl_2 (412).
By-prod. in the rxn. of MeCl with Ge-Cu alloy or with finely divided Ge at 320-360° in the presence of metallic Cu (445, 446, 449).
Props.: b. 110°/727 mm; $n^{20}D$ 1.4685; d_{20} 1.7053 (412); b. 111°; $d_{24.5}$ 1.72 (445, 449).
Rxns. with: H_2O followed by condensation → methylgermanium oxide polymer (305).
NaOR(1:3) → $MeGe(OMe)_3$ (566).
LiH or LiD → $MeGeH_3$ and $MeGeD_3$, respectively (420).
RMgBr — $MeGeR_3$ (412).
CH_2N_2(1:1) → $Me(CH_2Cl)GeCl_2$ and $Me(CH_2Cl)_2GeCl$ (485).
Cl_2 under reflux in UV light → Cl_3CGeCl_3 (419).
SbF_3 → $MeGeF_3$ (419).

METHYLGERMANIUM TRIBROMIDE   MeGeBr$_3$

Prepn.: By passing MeBr at 340-450° over a powder or alloy of Ge and Cu(4:1-1:1); the prod. consists of a mixt. of MeGeBr$_3$, Me$_2$GeBr$_2$ and Me$_3$GeBr (419, 445, 446).

Props.: b. 168°/750mm; d$_{20}$ 2.6337; n^{20}D 1.5770 (419).

Rxns. with: SbF$_3$ → MeGeF$_3$ (419).
LiH or LiD → MeGeH$_3$ or MeGeD$_3$ (420).

METHYLGERMANIUM TRIIODIDE   MeGeI$_3$

Prepn.: By reacting GeI$_2$ with MeI in a sealed tube at 110°; theoret. yield (131).

Props.: Yellow solid, m. 48-50°; b. 237°/752 mm (131).

CHLOROMETHYLGERMANIUM TRICHLORIDE   ClCH$_2$GeCl$_3$

Prepn.: By passing Cl$_2$ into refluxing Me$_2$GeCl$_2$ under UV irradiation 18.5% yield (419).
By reacting GeCl$_4$ with CH$_2$N$_2$(1:1) in Et$_2$O in the presence of Cu powder at -60°; 93.7% yield (485).

Props: b. 149°/759 mm; d$_{20}$ 1.8415; n^{20}D 1.5003 (419); b. 65°/35 mm (485).

Rxn. with: MeMgBr → Me$_3$GeCH$_2$Cl (485).

TRICHLOROMETHYLGERMANIUM TRICHLORIDE   Cl$_3$CGeCl$_3$

Prepn: By passing Cl$_2$ into refluxing MeGeCl$_3$ in UV light (419).

Props: b. 130°/200 mm; m. 106-107° (419).

ETHYLGERMANIUM TRICHLORIDE   EtGeCl$_3$

Prepn.: By passing EtCl over a mixt. of powdered Ge and Cu at 550°; 65% yield (412); if the reaction is carried out >317° with a mixt. contg. Ge and Cu in a ratio of 1:6 to 1:1, EtGeCl$_3$ and Et$_2$GeCl$_2$ are formed (445, 451, 734).
By reacting HGeCl$_3$ with C$_2$H$_4$ at 40 atm. in i-PrOH in the presence of H$_2$PtCl$_6$·6H$_2$O (420).
By reacting EtI with Cs(GeCl$_3$) in a sealed tube at 110°; 60% yield (543, 545).

Props.: b. 141°/760 mm; n^{20}D 1.4750; d$_{20}$ 1.6041 (412); b. 139°/763.5 mm; d$_{20}$ 1.6006; n^{20}D 1.4730 (420); b. 140°/763 mm (451, 734); b. 60°/18 mm (545); Raman spectrum (331); Dielec. const. and dipole moment (734).

Rxns. with: H$_2$O → EtGeO$_2$H (545).
H$_2$O(1:3) → a colorless oil and some EtGe(OH)$_3$ which condenses to (EtGeO$_{1.5}$)$_x$ (451).
LiH or LiD → EtGeH$_3$ and EtGeD$_3$, respectively (420).
RMgBr → EtGeR$_3$ (413, 490).
AgNCO → EtGe(NCO)$_3$ (7).
RCO$_2$Ag → EtGe(OCOR)$_3$ (16).
SO$_2$Cl$_2$ in the presence of Bz$_2$O$_2$ → ClCH$_2$CH$_2$GeCl$_3$ and MeCHClGeCl$_3$ (412).

ETHYLGERMANIUM TRIIODIDE   EtGeI$_3$
Prepn.: By reacting Et$_2$GeI$_2$ with I$_2$(1:1) in the presence of AlI$_3$ (323).
Rxns. with: AgCN in C$_6$H$_6$ → EtGe(CN)$_3$ (?) (10).
AgNCS in C$_6$H$_6$ → liquid, b. 130-132°/1 mm (10).
(HCO$_2$)$_2$Pb → EtGe(OOCH)$_3$ (16).
RMg halides → EtGeR$_3$ (323).

1-CHLOROETHYLGERMANIUM TRICHLORIDE   MeCHClGeCl$_3$
Prepn.: By-prod. in the rxn. of EtGeCl$_3$ with SO$_2$Cl$_2$ in the presence of Bz$_2$O$_2$ (412).
Props.: b. 167°/746.5 mm; n^{20}D 1.4948; d$_{20}$ 1.6975 (412).

2-CHLOROETHYLGERMANIUM TRICHLORIDE   ClCH$_2$CH$_2$GeCl$_3$
Prepn.: By heating EtGeCl$_3$ with SO$_2$Cl$_2$ in the presence of Bz$_2$O$_2$; (MeCHClGeCl$_3$ is also formed) (412).
Props.: b. 188°/746.5 mm; n^{20}D 1.5094; d$_{20}$ 1.7587 (412).
Rxn. with: Quinoline at up to 200° → CH$_2$:CHGeCl$_3$ (412).

VINYLGERMANIUM TRICHLORIDE   CH$_2$:CHGeCl$_3$
Prepn.: By heating ClCH$_2$CH$_2$GeCl$_3$ with quinoline up to 200°; 26.6% yield (412).
Props.: b. 127.5°/745.5 mm; n^{20}D 1.4816; d$_{20}$ 1.6520 (412).
Rxn. with: LiAlH$_4$ → CH$_2$:CHGeH$_3$ (626a).

PROPYLGERMANIUM TRICHLORIDE   PrGeCl$_3$
Prepn.: By reacting GeCl$_4$ with PrLi(1:1) in petr. ether with CO$_2$-CHCl$_3$ cooling; the prod. contains 16% PrGeCl$_3$, 20% Pr$_2$GeCl$_2$ and some Pr$_3$GeCl; with 2 equivs. GeCl$_4$ 31.6% PrGeCl$_3$, and 54% Pr$_2$GeCl$_2$ were obtd. (247, 255).
By reacting PrCl with a Ge-Cu mixt. (11.5:2.5) at 320° (453).
By reacting GeCl$_4$ with PrMgCl(1:1.5) in Et$_2$O (453).
Props.: Colorless liquid fuming in moist air with formation of HCl, b. 48-50°/20 mm; d^{25}$_4$ 1.51; decomp. 150° (247).
b. 167°/763 mm; b. 43-45°/12 mm; d$_{20}$ 1.513; n^{20}D 1.4779; n^{25}D 1.4720; IR spectrum (453).
Rxns. with: NaOH in H$_2$O → (PrGeO)$_2$O (via intermediate PrGeOOH) (247, 255).
H$_2$O → (PrGeO)$_2$O, amorphous ppt. (247).
LiAlH$_4$ in (Me$_2$CH)$_2$O → PrGeH$_3$ (247, 255).
LiAlH$_4$ in dioxan under reflux → GeH$_4$ (247).

PROPYLGERMANIUM TRIBROMIDE   PrGeBr$_3$
Prepn.: By bromination of PrGeH$_3$ in (CH$_2$Br)$_2$ (247).
Props.: b. 50-55°/5 mm (247).

ISOPROPYLGERMANIUM TRICHLORIDE   i-PrGeCl$_3$
Prepn.: By reacting GeCl$_4$ with i-PrMgCl(1:1) in Et$_2$O (453).
Props.: b. 164.5°/767 mm; n^{20}D 1.4760; n^{25}D 1.4700;
IR spectrum (453).

ISOPROPYLGERMANIUM TRIBROMIDE   i-PrGeBr$_3$
Prepn.: By reacting (i-PrGe)$_x$ with Br in CCl$_4$ (17).

ALLYLGERMANIUM TRICHLORIDE   CH$_2$:CHCH$_2$GeCl$_3$
Prepn.: By passing CH$_2$:CHCH$_2$Cl at 340° over a mixt. of 25g Ge and 10g Cu powder (412).
Props.: b. 153.8°/743.5 mm; n^{20}D 1.4928; d$_{20}$ 1.5274 (412).
Rxn. with: RMgCl → CH$_2$:CHCH$_2$GeR$_3$ (412).

2-METHYLALLYLGERMANIUM TRICHLORIDE   CH$_2$:CMeCH$_2$GeCl$_3$
Prepn.: By passing CH$_2$:CMeCH$_2$Cl at 340° over a powder mixt. of Ge and Cu; 24% yield (412).
Props.: b. 42°/7 mm (412).
Rxns. with: RMgBr → CH$_2$:CMeCH$_2$GeR$_3$ (412).

ALKYLGERMANIUM TRICHLORIDES PREPARED FROM TRICHLOROGERMANE AND OLEFINS

The following compounds were prepared by reacting $HGeCl_3$ with α-alkenes at a moderate temperature (up to 85°) in the presence of a peroxide (Bz_2O_2).

$RGeCl_3$	Alkene	Yield %	Properties	Ref.
n-$C_5H_{11}GeCl_3$	1-Pentene	21	Colorless oil, b. 101-103°/40mm	443
$MeCHCl(CH_2)_3GeCl_3$	4-Chloro-1-pentene	10	Colorless liquid, b. 79-82°/0.1mm	443
n-$C_6H_{13}GeCl_3$	1-Hexene	24	Colorless liquid, b. 93-95°/13mm; b. 97°/14mm; n_D^{25} 1.4719	443, 127
n-$C_6H_{12}ClGeCl_3$	Chloro-1-hexene	16	Colorless liquid, b. 89-91°/0.2mm	443
n-$C_6H_{11}F_2GeCl_3$	Difluoro-1-hexene	22	Colorless liquid, b. 75-76°/2mm	443
n-$C_7H_{15}GeCl_3$	1-Heptene	18	Colorless liquid, b. 101-103°/10mm	443
$Me_2CH(CH_2)_5GeCl_3$	Isoctene	10	Colorless liquid, b. 82-85°/0.9mm	443
n-$C_8H_{17}GeCl_3$	1-Octene	11	Colorless liquid, b. 78-82°/0.6mm	443
n-$C_{10}H_{21}GeCl_3$	1-Decene	9	Colorless liquid, b. 87-90°/0.1mm	443

PHENYLGERMANIUM TRICHLORIDE $PhGeCl_3$

Prepn.: By reacting Ph_4Ge with $GeCl_4$ (1:3) in an autoclave at $350°$; 75% yield (476).
By reacting PhI with $Cs(GeCl_3)$ in a sealed tube at $250°$; 80% yield (543, 545).
Props.:   b. $115°/19$ mm (545).
Rxn. with: K in xylene in O-free CO_2 under reflux → $(PhGe)_x$ (476).

METHYLENEBIS(GERMANIUM TRICHLORIDE) $CH_2(GeCl_3)_2$

Prepn.: By heating CH_2I_2 with $Cs(GeCl_3)$ in a sealed tube at $200°$; 80% yield (543, 545).
Props.:   b. $110°/18$ mm (543, 545).
Rxn. with: NH_4OH → $CH_2(GeO_2H)_2$ (543, 545).

ORGANOGERMANIUM CYANIDES, ISOCYANATES, THIOCYANATES, and ISOTHIOCYANATES

Organogermanium cyanides are prepared:

1. By reacting organogermanium halides or sulfides with silver cyanide.
2. By reacting organogermanium oxides with hydrogen cyanide.

Organogermanium isocyanates are prepared by reacting organogermanium halides or isothiocyanates with silver isocyanates.

Organogermanium isothiocyanates are prepared:

1. By reacting organogermanium halides or cyanides with hydrogen or silver isothiocyanate.
2. By reacting organogermanium oxides or hydroxides with hydrogen isothiocyanate.

ORGANOGERMANIUM CYANIDES

Compound	Prepared from	Properties	Ref.
$EtGe(CN)_3$ (?)	$EtGeI_3$ + AgCN	White solid, m. 127°, b. 255°, polymerizes during distn. to a black solid	10, 23
$(Me_2CH)_2Ge(CN)_2$	$[(Me_2CH)_2GeS]_2$ + AgCN		
Et_3GeCN	$(Et_3Ge)_2O$ + HCN	Colorless liquid, b. 113°, m. 18°, n^{20}_D 1.4509; d^{20}_4 1.111	10
Pr_3GeCN	$(Pr_3Ge)_2O$ + HCN	Colorless liquid, b. 253°, b. 115-117°/10mm; m. -13°; n^{20}_D 1.4544; d^{20}_4 1.041	10, 23
$(Me_2CH)_3GeCN$	$(Me_2CH)_3GeBr$ + AgCN		

ETHYLGERMANIUM TRIISOCYANATE   EtGe(NCO)$_3$
Prepn.: By reacting a 20% soln. of EtGeCl$_3$ in C$_6$H$_6$ with
   AgNCO; 90% yield (7).
Props.: Colorless liquid, b. 138.4-139.2°/52 mm; b. 225.4°;
m. -31°; d$^{20}_4$ 1.5344; n^{20}D 1.4739; log P= 8.4195- 2760/T;
thermodynamic data (7).
Rxns. with: H$_2$O → rapid hydrolysis (7).
Alcs. (ROH) → EtGe(OR)$_3$ (7).

DIETHYLGERMANIUM DIISOCYANATE   Et$_2$Ge(NCO)$_2$
Prepn.: By reacting 20% soln. of Et$_2$GeCl$_2$ in C$_6$H$_6$ with
   AgNCO; 90% yield (7).
Props.: Colorless liquid, b. 226.0°; b. 134.5-135.1°/52mm;
m. -32°; d$^{20}_4$ 1.330; n^{20}D 1.4619; log P= 8.0861- 2597/T;
thermodynamic data (7).
Rxns. with: H$_2$O → rapid hydrolysis (7).
Alcs.(ROH) → Et$_2$Ge(OR)$_2$ (7).

DIISOPROPYLGERMANIUM DIISOCYANATE   i-Pr$_2$Ge(NCO)$_2$
Prepn.: By refluxing i-Pr$_2$GeCl$_2$ with AgNCO at 72mm pressure
   (23).
Props.: b. 239°; b. 72-73°/1 mm; d$_{20}$ 1.225; n^{20}D 1.464;
MR$_D$ 54.7 (23).
Rxns. with: AgOCOR → i-Pr$_2$Ge(OCOR)$_2$ (23).

TRIETHYLGERMANIUM ISOCYANATE   Et$_3$Ge(NCO)
Prepn.: By reacting 20% soln. of Et$_3$GeCl in C$_6$H$_6$ with AgNCO;
   90% yield (7).
By refluxing Et$_3$Ge(NCS) with AgNCO without any solvent;
   90% yield (10).
Props.: Colorless liquid with a camphor-like odor, b. 200.4°;
b. 109.4-110.0°/48 mm; m. -26.4°; d$^{20}_4$ 1.1514; n^{20}D 1.4519;
log P= 7.9424- 2396/T; thermodynamic data (7); b. 200°;
n^{20}D 1.454 (10).
Rxns. with: H$_2$O at 25° slow, at 80° violent hydrolysis (7).
Alcs.(ROH) → Et$_3$GeOR (7).

TRIPROPYLGERMANIUM ISOCYANATE   Pr$_3$Ge(NCO)
Prepn.: By reacting Pr$_3$GeCl with AgNCO in C$_6$H$_6$ under reflux (11).
Props.: Colorless liquid, sol. in org. solvents, b. 247°;
b. 114°/10 mm; m. -19°; d$^{20}_4$ 1.055; n^{20}D 1.4575; MR$_D$ 61.4 (11).

TRIISOPROPYLGERMANIUM ISOCYANATE   i-Pr$_3$Ge(NCO)
Prepn.: By refluxing i-Pr$_3$GeCl with AgNCO; 96% yield (20).
Props.: b. 65-66°/1 mm; b. 238°; d$_{20}$ 1.097; n^{20}D 1.4602 (20).

TRIBUTYLGERMANIUM ISOCYANATE $Bu_3Ge(NCO)$
Prepn.: By refluxing Bu_3GeCl with AgNCO in C_6H_6; theoret. yield (73).
Props.: Colorless liquid; b. $283°$; b. $109-110°/2$ mm; d^{20}_4 1.044; n^{20}_D 1.4595; MR_D 75.0 (13).

DIMETHYLGERMANIUM THIOCYANATE $Me_2Ge(CNS)_2$
Prepn.: By adding Me_2GeCl_2 to an aq. soln. contg. CNS^- ions (457).
Props.: White needles, m. $45.5-47.0°$; b. $266-268°$; sol. in EtOH and C_6H_6 (457).

ORGANOGERMANIUM ISOTHIOCYANATE

Compound	Prepared from	Properties	Ref.
$Et_2Ge(NCS)_2$	$(Et_2GeO)_4$ + HNCS in Et_2O	b. 298°; b. 113-116°/1mm; m. 16°; d^{20}_4 1.356	10
	Et_3GeCN + AgNCS		10
$(Me_2CH)_2Ge(NCS)_2$	$(Me_2CH)_2GeO)_3$ + HNCS in Et_2O	Viscous, yellowish liquid, b. 296°; b. 105-107°/1mm; d_{20} 1.234; n^{20}_D 1.558 (1)	23
Et_3GeNCS	$(Et_3Ge)_2O$ + HNCS in Et_2O	b. 252°; b. 113-114°/8mm; m. -46°; n^{20}_D 1.517; d^{20}_4 1.184; MR_D 55.7 (2)	10
Pr_3GeNCS	$(Pr_3Ge)_2O$ + HNCS in Et_2O	b. 287°; b. 143-144°/9mm; m. -56°; n^{20}_D 1.5063; d^{20}_4 1.105; MR_D 69.8	10
$(Me_2CH)_3GeNCS$	$(Me_2CH)_3GeBr$ + AgNCS	b. 277°	10
	$(Me_2CH)_3GeOH$ + HNCS in Et_2O	b. 277°, m. 18°; d^{30}_4 1.10; n^{20}_D 1.512	23
Bu_3GeNCS	$(Bu_3Ge)_2O$ + HNCS in Et_2O	b. 319°; b. 135-136°/2mm; n^{20}_D 1.5039; d^{20}_4 1.071; MR_D 84.3	17
			10

(1) Rxn. with AgOAc → $(Me_2CH)_2Ge(OAc)_2$ (23).
(2) Rxn. with AgNCO → Et_3GeNCO (10).

5. ORGANOGERMANIUM-OXYGEN COMPOUNDS

ORGANOGERMANONIC ACIDS AND ANHYDRIDES

Organogermanonic acids are prepared by hydrolyzing organogermanium trihalides with aqueous ammonia or alkali solutions.

Arenegermanonic anhydrides are prepared by reacting aryldiazonium fluoroborates with germanium tetrahalides in acetone in the presence of zinc and hydrolyzing the reaction products.

METHANEDIGERMANONIC ACID $CH_2(GeO_2H)_2$
Prepn.: By hydrolyzing $CH_2(GeCl_3)_2$ with NH_4OH (543, 545).

PROPANEGERMANONIC ANHYDRIDE $(PrGeO)_2O$
Prepn.: By hydrolysis of $PrGeCl_3$ with NaOH in H_2O (247).
Props.: Decomp. at 285-290° (247).
Rxns. with: HCl — $PrGeCl_3$ (247).

ARENEGERMANONIC ANHYDRIDES $(RGeO)_2O$
The following arenegermanonic anhydrides, $(RGeO)_2O$, were prepared by reacting aryldiazonium fluoborates, RN_2BF_4, with $GeCl_4$ in Me_2CO in the presence of Zn at -8 to 5° and hydrolyzing the reaction products with sodium hydroxide.

$(RGeO)_2O$	RN_2BF_4	Yield %	Ref.
$(PhGeO)_2O$	PhN_2BF_4	28	398
$(p-MeOC_6H_4GeO)_2O$	$p-MeOC_6H_4N_2BF_4$		398
$(p-EtOC_6H_4GeO)_2O$	$p-EtOC_6H_4N_2BF_4$		398
$(p-ClC_6H_4GeO)_2O$	$p-ClC_6H_4N_2BF_4$		398
$(p-BrC_6H_4GeO)_2O$	$p-ClC_6H_4N_2BF_4$		398

$(p-HO_2CC_6H_4GeO)_2O$, amorphous solid, was prepared by oxidation of $(p-MeC_6H_4GeO)_2O$ with $KMnO_4$ (65).
$(3-O_2N-4-Me_2NC_6H_3GeO)_2O$ was prepared by reacting $(p-Me_2NC_6H_4GeO)_2O$ with $HNO_3-H_2SO_4$ (65).

ORGANOGERMANIUM OXIDES

Methylgermanium oxide polymer was prepared by hydrolyzing methylgermanium trichloride and condensing the intermediate trihydroxide.

Diorganogermanium oxides are prepared by hydrolyzing diorganogermanium dihalides or sulfides.

Triorganogermanium oxides, $(R_3Ge)_2O$, are prepared:

1. By treating triorganogermanium halides with silver carbonate in benzene or with silver nitrate in ethanol under reflux.
2. By hydrolyzing triorganogermanium halides with aqueous alkali and dehydrating the hydrolysis product.
3. By hydrolyzing alkali metal triorganogermanides.
4. By refluxing triorganogermanium hydroxides or fluorides with bis(triorganotin)oxides.
5. By pyrolysis of triorganogermanecarboxylic acid esters.

METHYLGERMANIUM OXIDE POLYMER $(MeGeO)_n$

Prepn.: By hydrolyzing $MeGeCl_3$ and condensing the intermediate $MeGe(OH)_3$ (305).
Props.: Colorless crystals, insol. in H_2O and common org. solvents; stable in air up to $480°$, above that decompn. occurs (305).

DIMETHYLGERMANIUM OXIDE TETRAMER $(Me_2GeO)_4$

Prepn.: By hydrolyzing Me_2GeS (448, 450, 601)
Props.: Crystalline substance, m. $133.4°$, remelting at $125°$; b. $211°$ (448, 601); Toxicity (452).
Rxn. with: $Ac_2O \rightarrow Me_2Ge(OAc)_2$ (16).

DIETHYLGERMANIUM OXIDE Et_2GeO

Prepn.: By hydrolyzing Et_2GeCl_2 or Et_2GeBr_2 with aq. NaOH (8) and heating the hydrolysis prod. to $175°$ (551).
Props.: The compound, colorless liquid, exists in the trimeric or tetrameric form. The trimer: m. $\sim 19°$ (551). The tetramer: m. $27.1°$; b. 128.5-$129.5°/3$ mm; b. $291°/762$ mm; mol. wt. determination (8).
Rxns. with: 100% $H_2SO_4 \rightarrow (Et_2GeSO_4)_2$ (7).
Aliphatic org. acids or acid anhydrides $\rightarrow Et_2Ge(OCOR)_2$ (9).
$HNCS \rightarrow Et_2Ge(NCS)_2$ (10).
Use: Antifoam agent for lubricating oils (551).

DIPROPYLGERMANIUM OXIDE Pr_2GeO

Prepn.: By hydrolyzing Pr_2GeF_2 or Pr_2GeCl_2 with aq. NaOH (15, 247).

Props.: The compound is reported to exist in the trimeric and tetrameric form. The trimer: colorless liquid, b. $320°$; b. $148-149°/1mm$; m. $5.8°$; slowly changes (reversibly) into another form m. $\sim 153°$ (15). The tetrameric form was established by the cryoscopic method (247).

Rxns. with: Hydrogen halides → Pr_2Ge dihalides (15).
Ac_2O → $Pr_2Ge(OAc)_2$ (15).

DIISOPROPYLGERMANIUM OXIDE $i-Pr_2GeO$

Prepn.: By reacting $i-PrMgBr$ or $i-PrMgCl$ with $GeCl_4$ or $GeBr_4$ in Et_2O or C_6H_6 and hydrolyzing the prod.; the rxn. yields the oxide along with $i-Pr_3GeOH$ and $(i-PrGe)_n$ (17, 23).
By hydrolyzing pure $i-Pr_2GeF_2$ with aq. NaOH; theoret. yield (17).

Props.: The oxide exists in the trimeric form; m. $44°$; b. $321°$ (decompn.) (17).

Rxns. with: Concd. hydrogen halides → $i-Pr_2Ge$ dihalides (17).
HNCS → $i-Pr_2Ge(NCS)_2$ (23).

DIPHENYLGERMANIUM OXIDE Ph_2GeO

Prepn.: By heating Ph_2GeBr_2 with concd. NH_4OH in EtOH (245).

Props.: White crystals of undetermined mol. wt.; sinters at $145°$; m. $180-210°$ (245).

Rxn. with: 47% HI → Ph_2GeI_2 (245).

BIS(TRIETHYLGERMANIUM) OXIDE $(Et_3Ge)_2O$

Prepn.: By reacting Et_3GeBr with Ag_2CO_3 in C_6H_6 (10).

Rxns. with: 95% H_2SO_4 → $(Et_3Ge)_2SO_4$ (8).
Aliphatic org. acids or acid anhydrides → $Et_3GeOOCR$ (9, 21).
HNCS → Et_3GeNCS (10).
HCN → Et_3GeCN (10).
PhSH → Et_3GeSPh (22).
AgF → Et_3GeF (23).
AgOAc → Et_3GeOAc (23).

BIS(TRIPROPYLGERMANIUM) OXIDE $(Pr_3Ge)_2O$

Prepn.: By cleaving Pr_4Ge with Br_2 and treating the resulting bromide with excess alkali; 66% yield (10, 11).
By refluxing Pr_3GeF with $(Et_3Sn)_2O$ (19)

Props.: Colorless liquid sol. in org. solvents; b. $305°$; b. $175°/14mm$; m. $-55°$; d^{20}_4 1.068; n^{20}_D 1.4648; RM_D 108.1 (11).

Rxns. with: HNCS → Pr_3GeNCS (10).
HCN → Pr_3GeCN (10).
Hydrogen halides → Pr_3Ge halides (11).
Org. acids → Pr_3Ge carboxylates (12).

BIS(TRIISOPROPYLGERMANIUM) OXIDE   (i-Pr$_3$Ge)$_2$O
Prepn.: By reacting i-Pr$_3$GeI with Ag$_2$CO$_3$ in C$_6$H$_6$; 87% yield (17).
By refluxing i-Pr$_3$GeOH with (Et$_3$Sn)$_2$O (19).
Props.: Colorless, viscous liquid, b. 315° (decompn.);
d$^{20}_4$ 1.112; n^{20}D 1.4836; at the b. temp. turns red; below 0°
forms a glass without a true m. (17).

BIS(TRIBUTYLGERMANIUM) OXIDE   (Bu$_3$Ge)$_2$O
Prepn.: By reacting Bu$_4$Ge with Br$_2$ and treating the rxn. mixt.
  with excess alkali (10).
Rxn. with: HNCS → Bu$_3$GeNCS (10).

BIS(TRIBUTYLGERMANIUM) OXIDE   (Bu$_3$Ge)$_2$O
Prepn.: By hydrolyzing Bu$_3$Ge halides (13, 160).
Props.: Colorless liquid, b. 353°; b. 173-174°/1 mm; d$^{20}_4$ 1.027;
n^{20}D 1.4652; RM$_D$ 135.7 (13).
Rxns. with: Hydrogen halides → Bu$_3$Ge halides (13, 160).
Organic acids → Bu$_3$Ge carboxylates (13).

BIS(TRIHEXYLGERMANIUM) OXIDE   [(C$_6$H$_{13}$)$_3$Ge]$_2$O
Prepn.: By hydrolyzing (C$_6$H$_{13}$)$_3$GeBr with 10% KOH (140).
Props.: b. 210-211°/0.04 mm; n^{25}D 1.4645; d$_{25}$ 0.963 (140).
Rxn. with: HCl → (C$_6$H$_{13}$)$_3$GeCl (140).

BIS(TRIPHENYLGERMANIUM) OXIDE   (Ph$_3$Ge)$_2$O
Prepn.: By pyrolyzing Ph$_3$GeCO$_2$Me at 250° (62).
By refluxing (Ph$_3$Ge)$_2$ with I$_2$ in xylene in the presence of
  quinoline (176).
By cleaving (Ph$_3$Ge)$_2$ with Na-K in (CH$_2$OMe)$_2$ and hydrolyzing
  the prod.; 24.7% yield (176).
By hydrolyzing Ph$_3$GeBr in 5% KOH in EtOH under reflux
  (244, 254).
By refluxing Ph$_3$GeBr with AgNO$_3$ in EtOH; 83% yield (473).
By-prod. in the rxn. of Ph$_3$GeNa with (:CHCH$_2$Br)$_2$ in liq. NH$_3$;
  82.1% yield (519).
By-prod. in the rxn. of Ph$_3$GeLi with HCHO or with Ph$_2$CO (177).
By-prod. in the rxn. of Ph$_3$GeH with MeLi followed by hydrolysis
  (179).
By-prod. in the rxn. of Ph$_3$GeK with Ph$_2$C:CH$_2$ or with stilbene
  (181).
By-prod. in the rxn. of Ph$_3$GeLi with fluorene (186).
Props.: m. 179-181° (62); m. 182° (244); m. 183-185° (176);
m. 185-187° (177); IR spectrum (473).
Rxns. with: LiAlH$_4$ → Ph$_3$GeH (244).
Concd. hydrogen halides → Ph$_3$Ge halides (473).

BIS(TRI-p-TOLYLGERMANIUM)OXIDE $[(p-MeC_6H_4)_3Ge]_2O$
Prepn.: By reacting $(p-MeC_6H_4)_3GeBr$ with $AgNO_3$ in EtOH under reflux; 60% yield (473).
Rxn. with: 47% HF → $(p-MeC_6H_4)_3GeF$ (473).

ORGANOGERMANIUM HYDROXIDES AND PEROXIDES

Triorganogermanium hydroxides are prepared by hydrolyzing triorganogermanium halides.

Organogermanium peroxides are prepared by reacting organogermanium halides with hydrogen peroxide or hydrocarbyl hydroperoxides, or by treating aminogermanes with hydrogen peroxide.

TRIISOPROPYLGERMANIUM HYDROXIDE $i-Pr_3GeOH$
Prepn.: By hydrolyzing $i-Pr_3GeBr$ with aq. NaOH; theoret. yield (17).
By-prod. in the rxn. of isopropylmagnesium halides with germanium tetrahalides followed by hydrolysis (17, 23).
Props.: Colorless liquid with a penetrating odor, b. 216°; m. -15°; d^{20}_4 1.077; n^{20}_D 1.472; at low temp. may be dimeric (17).
Rxns. with: Concd. hydrogen halides → the corresponding $i-Pr_3Ge$ halides (17).
HNCS → $i-Pr_3GeNCS$ (17).
$(Et_3Sn)_2O$ → $(i-Pr_3Ge)_2O$ and Et_3SnOH (19).
RCO_2H → $i-Pr_3GeOCOR$ (20).
H_2SO_4 → $(i-Pr_3Ge)_2SO_4$ (20).

TRICYCLOHEXYLGERMANIUM HYDROXIDE $(C_6H_{11})_3GeOH$
Prepn.: By hydrolysis of $(C_6H_{11})_3GeCl$ with 5% alc. KOH under reflux; theoret. yield (244).
By passing air through $(C_6H_{11})_3GeH$ refluxed in CCl_4 (244).
Props.: m. 175-176° (244).
Rxns. with: $LiAlH_4$ → $(C_6H_{11})_3GeH$ (244).
AcCl under reflux → $(C_6H_{11})_3GeCl$ (244).
Ac_2O under reflux → $(C_6H_{11})_3GeOAc$ (244).

TRIPHENYLGERMANIUM HYDROXIDE Ph_3GeOH
Prepn.: By hydrolyzing Ph_3GeBr or $Ph_3GeCO_2GePh_3$ with dil. alkali; 84% yield (61).
By-prod. in the rxn. of Ph_3GeH with Ph_3GeLi in $(CH_2OMe)_2$ or Et_2O followed by hydrolysis (179).
By-prod. in the rxn. of Ph_3GeH with $BuMgBr$ in Et_2O followed by carbonation and hydrolysis (179).
By-prod. in the rxn. of Ph_3GeLi with 1-octadecene followed by hydrolysis (181).
By-prod. in the reduction of Ph_3GeBr with Zn-Hg in HCl (?) (565).
Props.: m. $130°$ (565); m. $130-132°$ (181); m. $132-133°$ (179); m. $132-134°$ (61).
Rxn. with: $Ph_3GeCO_2H \rightarrow Ph_3GeCO_2GePh_3$ (62).

TRI-1-NAPHTHYLGERMANIUM HYDROXIDE $(1-C_{10}H_7)_3GeOH$
Prepn.: By reacting $(1-C_{10}H_7)_3GeBr$ dissolved in Me_2CO with alc. NaOH; 95% yield (563).
Props.: Colorless, triclinic needles, effloresces at $130-140°$, losing H_2O, to give a powder, m. $206-208°$ (563).
Rxn. with: HCO_2H under reflux → polymeric 1-naphthylgermanium oxide (563).

BIS(TRIPROPYLGERMYL) PEROXIDE Pr_3GeO_2
Prepn.: By reacting Pr_3GeCl with H_2O_2 in an inert solvent in the presence of a tertiary amine (111).
Props.: Stable at r.t., b. $65°/0.1-0.05$ mm; n^{26}_D 1.4608; the compound catalyzes polymerization of α-olefins (111).

BIS(TRIPHENYLGERMYL) PEROXIDE $(Ph_3GeO)_2$
Prepn.: By treating Ph_3GeNH_2 with H_2O_2 (2:1) (753).
Props.: m. $146-148°$ (753).
Rxn. with: HCl (1:2) → Ph_3GeCl and H_2O_2 (753).

OTHER ORGANOGERMANIUM PEROXIDES

The following organogermanium peroxides were prepared by reacting organogermanium halides with hydrocarbyl hydroperoxides in an inert solvent in the presence of a tertiary amine or with hydrocarbylperoxysodium.

Organogermanium Peroxide	Prepared from	Properties	Ref.
$Ph_2Ge(OOC_{10}H_{17})_2$ (a)	$Pr_2GeCl_2 + 9\text{-}C_{10}H_{17}OOH$	Stable at r.t., b. >90°/ 0.01 mm; (b)	111
$(CH_2)_5Ge(OOCMe_3)_2$	$(CH_2)_5GeCl_2 + Me_3COOH$	Stable at r.t.; b. 65°/ 0.2mm; m. -35°; n_D^{26} 1.4553; (b)	111
$Pr_3GeOOCMe_3$	$Pr_3GeCl + Me_3COOH$	Stable at r.t., b. 35° 0.001; n_D^{26} 1.4383; (b)	111
$Pr_3GeOOC_{10}H_{17}$ (a)	$Pr_3GeCl + 9\text{-}C_{10}H_{17}OOH$	Stable at r.t., b. 70°/ 0.01mm; n_D^{26} 1.4779; (b)	111
$Ph_3GeOOCMe_3$	$Bu_3GeBr + Me_3COONa$	m. 55-57°; with HCl(1:1) → Ph_3GeCl and Me_3COOH	753
$Ph_3GeOOCPh_3$	$Ph_3GeBr + Ph_3COONa$	m. 188-195°; with HCl(1:1) → Ph_3GeCl and Ph_3COOH	753
$Ph_3GeOOCMe_2Ph$	$Ph_3GeBr + PhMe_2COONa$	m. 104-106°; with HCl(1:1) → Ph_3GeCl and $PhMe_2COOH$	753

(a) Decahydro-9-naphthyl group
(b) Catalyzes polymerization of α-olefins.

ORGANOGERMANIUM ALKOXIDES

The compounds are prepared by reacting alkyl- or arylgermanium halides with alkali metal alkoxides in suitable alcohols under reflux.

$R_{3-n}Ge(OR')_{1+n}$	Prepared from	Yield %	Properties	Ref.
$MeGe(OMe)_3$	$MeGeCl_3$ + NaOMe	66	b. 136.5–138°/760mm; n_D^{25} 1.4053; d_{25} 1.264	566
$Me_2Ge(OMe)_2$	Me_2GeCl_2 + NaOMe	77	b. 118–118.5°/763mm; n_D^{25} 1.4093; d_{25} 1.207; (a)	566
Me_3GeOMe	Me_3GeCl + NaOMe	56	b. 87–88°/753mm; n_D^{25} 1.401; d_{25} 1.075	566, 62
Ph_3GeOMe (c)	Ph_3GeBr + NaONa (b)		m. 66–67° (d)	

(a) $Me_2Ge(OMe)_2$ + MeMgI in Et_2O under N — Me_3GeI (566).
(b) The reaction occurs at room temp.
(c) The compound was also prepared by heating Ph_3GeCO_2Me at 250° (62).
(d) The compound reacts with atm. moisture or with abs. EtOH to form $(Ph_3Ge)_2$ (62).

ORGANOGERMANIUM CARBOXYLATES

The compounds are prepared:

1. By reacting organogermanium halides or cyanides with metal carboxylates.
2. By reacting organogermanium oxides with organic acids or their anhydrides.
3. By transesterification of organogermanium carboxylates with carboxylic acids.
4. By heating organogermanium hydrides with carboxylic acids.
5. By refluxing organogermanium hydroxides with carboxylic acids.

TRIETHYLGERMANIUM (TRIETHYLGERMYLTHIO)ACETATE $Et_3GeSCH_2CO_2GeEt_3$
Prepn.: By heating $(Et_3Ge)_2O$ with $HSCH_2CO_2H$ at 100°; 56% yield (9).
Props.: Light-yellow liquid, b. 326°; b. 158.8-159.5°/4mm; d^{20}_4 1.2224; $n^{20}D$ 1.4993 (9).
Rxn. with: $H_2SO_4 \rightarrow (Et_3Ge)_2SO_4$ (12).
$HgCl_2 \rightarrow Et_3GeCl + HgSCH_2COO$ (17a).

TRIPHENYLGERMANIUM TRIPHENYLGERMANECARBOXYLATE $Ph_3GeCO_2GePh_3$
Prepn.: By heating Ph_3GeCO_2H at 190-210°; 86% yield (61).
By heating gently Ph_3GeCO_2H with NaOH, removing the H_2O in vacuo, and treating the Na salt with Ph_3GeBr in C_6H_6 under reflux; 81% yield (61).
By converting Ph_3GeCO_2H to its Ag salt and refluxing it in C_6H_6 with Ph_3GeBr; 53% yield (61).
By heating Ph_3GeCO_2H with Ph_3GeOH to 157° (62).
By-prod. in the rxn. of Ph_3GeLi with dibenzofuran followed by carbonation (186).
Props.: m. 163-164° (61); on heating slowly to 250° → $(Ph_3Ge)_2$ (62); m. 165-166° (186).
Rxn. with: Alkali → Ph_3GeCO_2H and Ph_3GeOH (61).

OTHER ORGANOGERMANIUM CARBOXYLATES $R_{3-n}Ge(OCOR')_{1+n}$

$R_{3-n}Ge(OCOR')_{1+n}$	Prepared from	Yield %	Properties	Ref.
$EtGe(OOCH)_3$	$EtGeI_3 + Pb(OOCH)_2$		$b_9 118°/9mm; d_4^{20} 1.617;$ $n_D^{20} 1.452$	16
$EtGe(OAc)_3$	$EtGeCl_3 + AgOAc$	95	$b. 99-101°/mm; b. 249°; d_4^{20}$ (1) $1.393; n_D^{20} 1.444; Rxn.$	16
$EtGe(OCOEt)_3$	$EtGeCl_3 + AgOCOEt$	95	$b. 114-116°/2mm; b. 256°;$ $d_4^{20} 1.271; n_D^{20} 1.4434$	16
$EtGe(OCOPr)_3$	$EtGeCl_3 + AgOCOPr$	95	$b. 136-137°/2mm; b. 271°$ (dec.)(2) $d_4^{20} 1.186; n_D^{20} 1.4432; Rxn.$	16
$EtGe(OCOBu)_3$	$EtGeCl_3 + AgOCOBu$		$b. 157-159°/2mm; b. 305°; d_4^{20}$ (3) $1.136; n_D^{20} 1.4456$	16
$Me_2Ge(OAc)_2$	$(Me_2GeO)_4 + Ac_2O$		$b. 94-95°/25mm; b. 188°; Rxn.$	16
$Et_2Ge(OOCH)_2$	$(Et_2GeO)_4 + HCOOH$	30	$b. 210°; b. 59-61°/3mm; d_4^{20}$ $1.366; n_D^{20} 1.4454$	16
$Et_2Ge(OAc)_2$	$(Et_2GeO)_4 + Ac_2O$	90	$b. 217°; d_4^{20} 1.268; n_D^{20} 1.4404$	9
$Pr_2Ge(OAc)_2$	$(Pr_2GeO)_3 + Ac_2O$	90	$b. 244.6°; m. 35.6°; Rxn.$ (4)	9
$Pr_2Ge(OCOCH_2Cl)_2$	$Pr_2Ge(OAc)_2 + CH_2ClCO_2H$		$b. 296°$ (dec.)(5) $d_4^{20} 1.374; n_D^{20}$ $1.4793; Rxn.; distils in$ $vacuo without decompn.$	15
$1-Pr_2Ge(OAc)_2$	$1-Pr_2GeCl_2 + AgOAc$		$b. 244°; b. 79-80°/1mm; d_4^{20}$ $1.193;(6) n_D^{20} 1.445; RM_D 61.8;$ $Rxn.$	5
	$1-Pr_2Ge(NCS)_2 + AgOAc$			23
	$1-Pr_2GeH_2 + AgOAc$			23
	$(1-Pr_2Ge)_2 + AgOAc$			23
				23

Rxns:
(1) $EtGe(OAc)_3 + Ph_2SiCl_2 (2:3) \rightarrow 2EtGeCl_3 + 3Ph_2Si(OAc)_2$ (16).
(2) $EtGe(OCOPr)_3 + Ph_2SiCl_2 (2:3) \rightarrow 2EtGeCl_3 + 3Ph_2Si(OCOPr)_2$ (16).
(3) $Me_2Ge(OAc)_2 + H_2SO_4 \rightarrow (Me_2GeSO_4)_2$ (16).
(4) $Pr_2Ge(OAc)_2 + CH_2ClCO_2H (1:2) \rightarrow Pr_2Ge(OCOCH_2Cl)_2$ (15).
(5) $Pr_2Ge(OCOCH_2Cl)_2 + H_2SO_4 \rightarrow Pr_2GeSO_4$ (15).
(6) $1-Pr_2Ge(OAc)_2 + ClCH_2CO_2H (1:2) \rightarrow 1-Pr_2Ge(OCOCH_2Cl)_2$ (23).

OTHER ORGANOGERMANIUM CARBOXYLATES $R_{3-n}Ge(OCOR')_{1+n}$ (Cont'd.)

$R_{3-n}Ge(OCOR')_{1+n}$	Prepared from	Yield %	Properties	Ref.
$1-Pr_2Ge(OCOCH_2Cl)_2$	$1-Pr_2Ge(OAc)_2$ + CH_2ClCO_2H		b. 298°; b. 136-138°/1mm; d^{20}_4 1.372; n^{20}_D 1.480; RM_D 71.6	23
$1-Pr_2Ge(OCOEt)_2$	$1-Pr_2GeCl_2$ + $AgOCOEt$		b. 255°; d^{20}_4 1.146; n^{20}_D 1.447; MR_D 71.0; Rxn. (7)	23
$1-Pr_2Ge(OCOCH_2-CH_2Cl)_2$	$1-Pr_2Ge(OCOEt)_2$ + $CH_2ClCH_2CO_2H$		b. 140-142°/1mm; d_{20} 1.294; n^{20}_D 1.474; MR_D 81.2	23
$1-Pr_2Ge(OCOPr)_2$	$1-Pr_2GeCl_2$ or $1-Pr_2Ge(NCO)_2$ + $AgOCOPr$		b. 274°; b. 106-108°/1mm; d^{20}_4 1.112; n^{20}_D 1.452; RM_D 80.6; Rxn. (8)	23
$1-Pr_2Ge(OCOPh)_2$	$1-Pr_2GeCl_2$ + $AgOCOPh$		b. 385°; b. 186-188°/1mm; m. 54°	23
$Et_3GeOOCH$	$(Et_3Ge)_2O$ + $HCOOH$	90	b. 185.7°; d^{20}_4 1.1672; n^{20}_D 1.4436; colorless with camphor-like odor	9
Et_3GeOAc	$(Et_3Ge)_2O$ + Ac_2O	90	b. 190.5°; d^{20}_4 1.1299; n^{20}_D 1.4413	9
	$(Et_3Ge)_2O$ + $AgOAc$		Rxn. (9,10)	23
$Et_3GeOCOCH_2Cl$	$(Et_3Ge)_2O$ + CH_2ClCO_2H		b. 234°; b. 113-114°/6mm; d^{20}_4 1.243; n^{20}_D 1.4672; (a)	21
$Et_3GeOCOCHCl_2$	$(Et_3Ge)_2O$ + $CHCl_2CO_2H$		b. 248°; b. 105-106°/1mm; d^{20}_4 1.304; n^{20}_D 1.4672; (a)	21
$Et_3GeOCOCCl_3$	$(Et_3Ge)_2O$ + CCl_3CO_2H		b. 125-126°/3mm; d^{20}_4 1.368; n^{20}_D 1.4790; (a)	21
$Et_3GeOCOCH_2Br$	$(Et_3Ge)_2O$ + CH_3BrCO_2H		b. 244°; b. 74-75°/1mm; d^{20}_4 1.423; n^{20}_D 1.4842; (a)	21
$Et_3GeOCOCF_3$	Et_3GeH + CF_3CO_2H		b. 190°	24
$Et_3GeOCOCH_2I$	$(Et_3Ge)_2O$ + CH_2ICO_2H		b. 254°; b. 85-86°/1mm; d^{20}_4 1.593; n^{20}_D 1.5112	21
$Et_3GeOCOCH_2CN$	Et_3GeOAc + $CH_2(CN)CO_2H$	90	b. 98-99°/1mm; d^{20}_4 1.194; n^{20}_D 1.464	21
$Et_3GeOCOEt$	$(Et_3Ge)_2O$ + $EtCO_2H$		b. 205°; b. 119-120°/50mm; d^{20}_4 1.109; n^{20}_D 1.4481	21
$Et_3GeOCOCHClMe$	Et_3GeBr + $AgOCOCHClMe$	~100	b. 235°; b. 73-75°/1mm; d^{20}_4 1.201; n^{20}_D 1.4619; (a)	21

Et$_3$GeOCOCH$_2$CH$_2$Cl	Et$_3$GeCl + AgOCOCH$_2$CH$_2$Cl	~100	b. 236°; b. 74-75°/1mm; d$^{20}_4$ 1.210; n$^{20}_D$ 1.4672; (a)	21
Et$_3$GeOCOCHBrMe	Et$_3$GeOAc + MeCHBrCO$_2$H		b. 74-75°(a); d$^{20}_4$ 1.370; n$^{20}_D$ 1.4725;	21
Et$_3$GeOCOCH$_2$CH$_2$Br	Et$_3$GeOAc + CH$_2$BrCH$_2$CO$_2$H	90	b. 85-86°(a); d$^{20}_4$ 1.376; n$^{20}_D$ 1.4832;	21
Et$_3$GeOCOCH$_2$CH$_2$I	Et$_3$GeOAc + CH$_2$ICH$_2$CO$_2$H	95	b. 94-96°(a); d$^{20}_4$ 1.544; n$^{20}_D$ 1.5002;	21
Et$_3$GeOCOC$_2$F$_5$	Et$_3$GeH + CF$_3$CF$_2$CO$_2$H		b. 190°	21
Et$_3$GeOCOPr	(Et$_3$Ge)$_2$O + PrCO$_2$H		b. 221°; b. 133-134°/47mm; d$^{20}_4$ 1.084; n$^{20}_D$ 1.4481	24
Et$_3$GeOCOCF$_2$CF$_2$CF$_3$	Et$_3$GeH + C$_3$F$_7$CO$_2$H		b. 2010	21
Et$_3$GeOCOBu	Et$_3$GeBr + AgOCOBu	90	b. 230°; b. 62-63°/1mm; d$^{20}_4$ 1.062; n$^{20}_D$ 1.4467	24
Et$_3$GeOCOPh	Et$_3$GeBr + AgOCOPh	90	b. 290°; b. 105-107°/1mm; d$^{20}_4$ 1.172; n$^{20}_D$ 1.513	22
Et$_3$GeOCOC$_6$H$_4$NH$_2$-o	Et$_3$GeBr + AgOCOC$_6$H$_4$NH$_2$-o		b. 331°; b. 174-176°/1mm; d$^{20}_4$ 1.215; n$^{20}_D$ 1.544	22
Pr$_3$GeOOCH	(Pr$_3$Ge)$_2$O + HCOOH		b. 233°; b. 108-109°/12mm; d$^{20}_4$ 1.094; n$^{20}_D$ 1.4505; RM$_D$ 60.7 Rxn. (11,12,13)	22
Pr$_3$GeOAc	(Pr$_3$Ge)$_2$O + AcOH	~100	b. 236°; b. 112-113°/13mm; d$^{20}_4$ 1.091; n$^{20}_D$ 1.4464; RM$_D$ 65.0 Rxn. (14)	12

{7} 1-Pr$_2$Ge(OCOEt)$_2$ + CH$_2$ClCH$_2$CO$_2$H (1:2) → 1-Pr$_2$Ge(OCOCH$_2$CH$_2$Cl)$_2$ (23).
{8} 1-Pr$_2$Ge(OCOPr)$_2$ + H$_2$SO$_4$ → (1-Pr$_2$GeSO$_4$)$_2$ (23).
(9) Et$_3$GeOAc + RCO$_2$H → Et$_3$GeOCOR (21).
(10) Et$_3$GeOAc + RSH → Et$_3$GeSR (22).
(11) Pr$_3$GeOOCH + H$_2$SO$_4$ → (Pr$_3$Ge)$_2$SO$_4$ (12).
{12} Pr$_3$GeOOCH + CHCl$_2$CO$_2$H → Pr$_3$GeOCOCHCl$_2$ {12}.
{13} Pr$_3$GeOOCH + CH$_2$ClCO$_2$H → Pr$_3$GeOCOCH$_2$Cl {12}.
(14) Pr$_3$GeOAc + H$_2$SO$_4$ → (Pr$_3$Ge)$_2$SO$_4$ (12).

(a) On heating under reflux at atm. pressure, hydrolyzable halogen is formed (21).

OTHER ORGANOGERMANIUM CARBOXYLATES $R_{3-n}Ge(OCOR')_{1+n}$ (Cont'd.)

$R_{3-n}Ge(OCOR')_{1+n}$	Prepared from	Yield %	Properties	Ref.
$Pr_3GeOCOCF_3$	$(Pr_3Ge)_2O + F_3CCO_2H$	~100	b. 226°; d^{20}_4 1.189; n^{20}_D 1.4103; RM_D 65.7	12
$Pr_3GeOCOCH_2Cl$	$(Pr_3Ge)_2O + CH_2ClCO_2H$		b. 265° (dec.); b. 143°/10mm; d^{20}_4 1.166; n^{20}_D 1.4640; RM_D 69.9; (a)	12
$Pr_3GeOCOCHCl_2$	$(Pr_3Ge)_2O + CHCl_2CO_2H$		b. 266°(dec.); b. 150°/10mm; d^{20}_4 1.226; n^{20}_D 1.4708; RM_D 75.2; Rxn.(15); (a)	12
$Pr_3GeOCOCH_2Br$	$(Pr_3Ge)_2O + CH_2BrCO_2H$	~100	b. 267°(dec.); b. 149°/10mm; d^{20}_4 1.325; n^{20}_D 1.4778; RM_D 73.2; (a)	12
$Pr_3GeOCOEt$	$(Pr_3Ge)_2O + EtCO_2H$	~100	b. 246°; d^{20}_4 1.054; n^{20}_D 1.4473; RM_D 69.7	12
1-Pr_3GeOAc	$Pr_3GeCN + AgOCOEt$ 1-$Pr_3GeCl + AgOAc$	80 90	b. 220°; b. 152-153°/33mm; d^{20}_4 1.070; n^{20}_D 1.4532	12 23 20
1-$Pr_3GeOCOCF_3$	1-$Pr_3GeOH + CF_3CO_2H$	96	b. 220°; d^{20}_4 1.178; n^{20}_D 1.4200	20
1-$Pr_3GeOCOCH_2Cl$	1-$Pr_3GeCl + AgOCOCH_2Cl$	89	b. 289°; b. 101-103°/1mm; d^{20}_4 1.182; n^{20}_D 1.4722	20
1-$Pr_3GeOCOCHCl_2$	1-$Pr_3GeOH + CHCl_2CO_2H$	~100	b. 275°; b. 92-93°/1mm; d^{20}_4 1.236; n^{20}_D 1.4772	20
1-$Pr_3GeOCOCCl_3$	1-$Pr_3GeOH + CCl_3CO_2H$	~100	b. 107-109°/1mm; d^{20}_4 1.291; n^{20}_D 1.4850	20
1-$Pr_3GeOCOCH_2Br$	1-$Pr_3GeOH + CH_2BrCO_2H$	~100	b. 271°(dec.); b. 98-99°/1mm; d^{20}_4 1.331; n^{20}_D 1.4872	20
1-$Pr_3GeOCOCH_2I$	1-$Pr_3GeOH + CH_2ICO_2H$	~100	b. 281°; b. 107-108°/1mm; d^{20}_4 1.465; n^{20}_D 1.5096	20
1-$Pr_3GeOCOEt$	1-$Pr_3GeCl + AgOCOEt$	90	b. 242°; b. 164-166°/90mm; d^{20}_4 1.059; n^{20}_D 1.4536	20
1-$Pr_3GeOCOCHClMe$	1-$Pr_3GeCl + AgOCOCHClMe$	81	b. 269°; b. 95-97°/1mm; d^{20}_4 1.144; n^{20}_D 1.4672	20
1-$Pr_3GeOCOCH_2CH_2Cl$	1-$Pr_3GeCl + AgOCOCH_2CH_2Cl$		b. 97-99°/1mm; d^{20}_4 1.154; n^{20}_D 1.4717; (b)	20
1-$Pr_3GeOCOPr$	1-$Pr_3GeCl + AgOCOPr$	~100	b. 256°; b. 180-182°/96mm; d^{20}_4 1.045; n^{20}_D 1.5486	20

i-Pr$_3$GeOCOBu	i-Pr$_3$GeCl + AgOCOBu	93	b. 266°; b. 82-83°/1mm; d^{20}_4 1.034; n^{20}_D 1.4566	20
i-Pr$_3$GeOCOPh	i-Pr$_3$GeCl + AgOCOPh		b. 317°; b. 130-132°/1mm; d^{20}_4 1.132; n^{20}_D 1.5136	20
Bu$_3$GeOOCH	(Bu$_3$Ge)$_2$O + HCOOH		b. 267°; b. 149-151°/16mm; d^{20}_4 1.051; n^{20}_D 1.4538	13
Bu$_3$GeOAc	Bu$_3$GeCl + AgOAc		b. 272°; b. 147-148°/14mm; d^{20}_4 1.027; n^{20}_D 1.4514	13
Bu$_3$GeOCOCF$_3$	(Bu$_3$Ge)$_2$O + CF$_3$CO$_2$H	~100	b. 262°; b. 147-149°/19mm; d^{20}_4 1.144; n^{20}_D 1.419	13
Bu$_3$GeOCOC$_6$H$_{13}$	Bu$_3$GeOAc + n-C$_{16}$H$_{13}$CO$_2$H		b. 324°(dec.); b. 147-148°/1mm; d^{20}_4 0.988; n^{20}_D 1.4539	13
(cyclo-C$_6$H$_{11}$)$_3$GeOAc	(C$_6$H$_{11}$)$_3$GeOH + Ac$_2$O		b. 82-83°	244

(15) Pr$_3$GeOCOCHCl$_2$ + H$_2$SO$_4$ → (Pr$_3$Ge)$_2$SO$_4$ (12).
(a) On heating under reflux at atm. pressure, hydrolyzable halogen is formed (21).
(b) Under reflux at atm. pressure decomposes into i-Pr$_3$GeCl and CH$_2$:CHCO$_2$H (20).

ORGANOGERMANIUM SULFATES AND CHROMATES

Organogermanium sulfates are prepared by treating organogermanium oxides, hydroxides, and carboxylates with sulfuric acid.

Organogermanium chromates are prepared by treating organogermanium halides with solutions containing chromate ions.

The sulfates are compiled in the table on the following page.

ORGANOGERMANIUM SULFATES

Compound	Prepared from	Properties	Ref.
Me_2GeSO_4	$Me_2Ge(OAc)_2 + H_2SO_4$	Exists in a dimeric form	16
Et_2GeSO_4	Et_2GeO_3 or $_4 + H_2SO_4$	Fluffy, white crystalline solid, m. 116° (1) exists in a dimeric form (1)	7
Pr_2GeSO_4	$Pr_2Ge(OCOCH_2Cl)_2 + H_2SO_4$	Colorless or white needles, m. 129°; exists in a dimeric form	15
$(Me_2CH)_2GeSO_4$	$(Me_2CH)_2Ge(OCOPr)_2 + H_2SO_4$	Solid, exists in a dimeric form	23
$(Et_3Ge)_2SO_4$	$(Et_3Ge)_2O + H_2SO_4$	Colorless liquid, b. 165°/3mm; b. 342° (2), m. -4°; soly. data	8
$(Pr_3Ge)_2SO_4$	$Pr_3GeOOCH$ or $Pr_3GeOAc + H_2SO_4$	Colorless liquid, b_{20} 370°; b. 180-182°/1mm; d^{20}_4 1.186 (3)	12
$[(Me_2CH)_3Ge]_2SO_4$	$Me_2CH_3GeOH + H_2SO_4$	b. 380°; b. 182-184°/1mm; d_{20} 1.217; n^{20}_D 1.482	20

1. Rxn. with HCl gas in the presence of $(NH_4)_2SO_4$ in $C_6H_6 \rightarrow Et_2GeCl_2$ (8).
2. Rxn. with $HgCl_2 \rightarrow Et_3GeCl + HgSO_4$ (17a).
3. Rxn. with $HgBr_2 \rightarrow Pr_3GeBr + HgSO_4$ (17a).

DIMETHYLGERMANIUM CHROMATE Me_2GeCrO_4
Prepn.: By adding Me_2GeCl_2 to an aq. soln. contg. CrO_4^{--} (457).
Props.: Orange crystals, decomp. in light or on heating to
158°; insol. in Me_2CO, $CHCl_3$, C_6H_6, EtOH (457).

6. ORGANOGERMANIUM-SULFUR COMPOUNDS

ORGANOGERMANIUM SULFIDES

Organogermanium sulfides are prepared by reacting organogermanium halides with hydrogen sulfide, sodium sulfide, or silver sulfide.

DIMETHYLGERMANIUM SULFIDE Me_2GeS
Prepn.: By reacting Me_2GeCl_2 with H_2S in $6N\ H_2SO_4$ or in H_2O; 94.5% yield (448, 450, 601).
Props.: m. 55.5°; b. 302° (448, 601).
Rxns. with: H_2O, dil. acids, or H_2O_2 → H_2S and $(Me_2GeO)_4$ (448, 450, 601).
Zn-Hg in HCl → Me_2GeH_2 (565).

DIISOPROPYLGERMANIUM SULFIDE DIMER $i-Pr_2GeS_2$
Prepn.: By refluxing $i-Pr_2GeI_2$ with Ag_2S (73).
Props.: b. 312; b. 117-121°/1mm; d_{20} 1.327; $n^{20}D$ 1.551 (23).
Rxns. with: AgOAc → $i-Pr_2Ge(OAc)_2$ (23).
AgCN → $i-Pr_2Ge(CN)_2$ (23).
AgBr → $i-Pr_2GeBr_2$ (23).

TRIHYDROCARBYLGERMANIUM SULFIDES ($R_3Ge)_2S$

The following compounds were prepared by fusing trihydrocarbylgermanium halides, R_3GeX, with Na_2S or by reacting the compounds in ethanol.

$(R_3Ge)_2S$	Prepared from	Properties	Ref.
$(Et_3Ge)_2S$	Et_3Ge halides + Na_2S	Colorless oil, b. 148-150°/12mm	65
$(Ph_3Ge)_2S$	Ph_3Ge halides + Na_2S	Crystals, m. 138°, insol. in alk., sol. in org. solvents	65
$[(PhCH_2)_3Ge]_2S$	$(PhCH_2)_3Ge$ halides + Na_2S	Crystals, m. 124°, insol. in alk., sol. in org. solvents	65
$[(p-MeC_6H_4)_3Ge]_2S$	$(p-MeC_6H_4)_3Ge$ halides + Na_2S	Crystals, m. 156-157°; insol. in alk., sol. in org. solvents	65
$[(PhC_6H_4)_3Ge]_2S$	$(PhC_6H_4)_3Ge$ halides + Na_2S	Crystals, m. 238°; insol. in alk., sol. in org. solvents	65

TRICYCLOHEXYLGERMANIUM DISULFIDE $(C_6H_{11})_3Ge_2S_2$

Prepn.: By reacting $(C_6H_{11})_3GeBr$ with Na_2S in alc. under reflux (65).

Props.: m. 87-88° (65).

TRIETHYLGERMANIUM MERCAPTIDES

Compounds of this group were prepared by refluxing triethylgermanium acetate or bis(triethylgermanium)oxide with thiols.

Et_3GeSR	Prepared from	Yield %	Properties	Ref.
$Et_3GeSC_6H_{13}$-n	Et_3GeOAc + n-$C_6H_{13}SH$		b. 277°; b_{20} 108-109°/1mm; d_{20} 1.029; n_D^{20} 1.488	22
$Et_3GeSC_7H_{15}$-n	Et_3GeOAc + n-$C_7H_{15}SH$		b. 288°; b_{20} 117-118°/1mm; d_{20} 1.019; n_D^{20} 1.489	22
$Et_3GeSC_{12}H_{25}$-n	Et_3GeOAc + n-$C_{12}H_{25}SH$		b. 357°; b_{20} 184-186°/1mm; d_{20} 0.975; n_D^{20} 1.481	22
$Et_3GeSC:CHCH:CHO$	Et_3GeOAc + furfuryl thiol	90	b. 276°; b_{20} 130-132°/1mm; d_{20} 1.177; n_D^{20} 1.522	22
Et_3GeSPh	Et_3GeOAc + PhSH	90	b. 286°; b_{20} 112-113°/1mm; d_{20} 1.153; n_D^{20} 1.553; (a)	22
	$(Et_3Ge)_2O$ + PhSH	60		22
$Et_3GeSC_6H_4Me$-m	Et_3GeOAc + m-MeC_6H_4SH	90	b. 300°; b_{20} 143-145°/1mm; d_{20} 1.131; n_D^{20} 1.550	22
$Et_3GeSC_6H_4Me$-o	Et_3GeOAc + MeC_6H_4SH		b. 298°; b_{20} 123-124°/1mm; d_{20} 1.141; n_D^{20} 1.552	22
$Et_3GeSC_6H_4NH_2$-o	Et_3GeOAc + o-$H_2NC_6H_4SH$	90	b. 326°; b_{20} 163-164°/1mm; d_{20} 1.197; n_D^{20} 1.583	22
Et_3GeSCH_2Ph	Et_3GeOAc + $PhCH_2SH$	90	b. 305°; b_{20} 130-131°/1mm; d_{20} 1.139; n_D^{20} 1.549	22
$Et_3GeSC_{10}H_7$-β	Et_3GeOAc + β-$C_{10}H_7SH$	90	b. 367°; b_{20} 195-197°/1mm; d_{20} 1.184; n_D^{20} 1.613; (b)	22
	Et_3GeSPh + β-$C_{10}H_7SH$			22
	$(Et_3Ge)_2O$ + HS-$C_{10}H_7$			22
$Et_3GeSCH_2CO_2GeEt_3$	CH_2CO_2H at 100°	56	b. 326°; b_{20} 158.8-159.5°/4mm; d_{20} 1.2224; n_D^{20} 1.4993; (c,d)	9

(a) $Et_3GeSPh + β-C_{10}H_7SH → Et_3GeSC_{10}H_7$ (22).
(b) $Et_3GeSC_{10}H_7-β + Et_2SiCl_2 → Et_3GeCl$ (22).
(c) $Et_3GeSCH_2CO_2GeEt_3 + H_2SO_4 → (Et_3Ge)_2SO_4$ (12).
(d) $Et_3GeSCH_2CO_2GeEt_3 + HgCl_2 → Et_3GeCl + HgSCH_2COO$ (17a).

7. AMINOGERMANES

Organogermaneamines are prepared by reacting organogermanium halides with amines or ammonia.

Octaalkylgermanetetramines, $(R_2N)_4Ge$, are prepared by reacting germanium tetrahalides with secondary amines.

ETHYL-N,N',N"-HEXAMETHYLGERMANETRIAMINE $EtGe(NMe_2)_3$
Prepn.: By reacting $EtGeCl_3$ with Me_2NH in cyclohexane at $-50°$ (14).
Props.: Colorless liquid, b. 191°; b. 105-107°/34mm; m. -46°; d^{22}_4 1.049 (14).

N,N',N",1-HEPTAETHYLGERMANETRIAMINE $EtGe(NEt_2)_3$
Prepn.: By reacting $EtGeCl_3$ with Et_2NH in C_6H_6; 90% yield (14).
Props.: Colorless liquid, b. 249°; b. 117-118°/12mm; d^{22}_4 1.108 (14).
Rxn. with: HI → $EtGeI_3$ (14).

TRIPHENYLGERMYLAMINE Ph_3GeNH_2
Prepn.: By reacting Ph_3GeBr with dry NH_3 in an inert solvent (Et_2O) (753).
Rxn. with: H_2O_2 (2:1) → $(Ph_3GeO)_2$ (753).

TRIS(TRIPHENYLGERMYL)AMINE $(Ph_3Ge)_3N$
Prepn.: By dissolving Ph_3GeBr in NH_3 (254).

OCTAMETHYLGERMANETETRAMINE $(Me_2N)_4Ge$
Prepn.: By reacting Me_2NH with $GeCl_4$ at -60 to 25°; 80% yield (14).
Props.: Colorless liquid, b. 203°; b. 87-89°/15mm; m. 14°; d^{22}_4 1.069; decomp. when heated above 203° (14).
Rxn. with: BzCl → $GeCl_4$ (14).

OCTAETHYLGERMANETETRAMINE $(Et_2N)_4Ge$
Prepn.: By reacting Et_2NH with $GeBr_4$ in cyclohexane at 25°; 45% yield (14).
Props.: Colorless liquid, b. 266° (decompn.); b. 108-110°/2mm; d^{22}_4 1.215 (14).

8. ORGANOGERMANIUM COMPOUNDS CONTAINING SILICON AND TIN

TRIPHENYL(TRIPHENYLSILYL)GERMANE $Ph_3SiGePh_3$
Prepn.: By reacting Ph_3GeCO_2Me with Ph_3SiLi in $(CH_2OMe)_2$ (175).
By reacting Ph_3SiK with Ph_3GeBr or Ph_3GeCl in Et_2O (180).
Props.: m. 357-359° (175); m. 354-355°; resistant to oxidation by atm. oxygen and by iodine in xylene (180).
Rxn. with: Na-K alloy followed by carbonation → Ph_3GeCO_2H (180).

TRIS(TRIPHENYLGERMYL)SILANE $(Ph_3Ge)_3SiH$
Prepn.: By reacting $NaGePh_3$ with $HSiCl_3$ in Et_2O; theoret. yield (375).
Props.: α-form, m. 187.5-188.5° (375); β-form, m. 170-171° (375).
Rxns. with: Li(1:1) in $EtNH_2$, followed by EtBr → $(Ph_3Ge)_3SiEt$ (375).
Br_2(1:1) → $(Ph_3Ge)_3SiBr$ + HBr (375).

TRIS(TRIPHENYLGERMYL)SILYL BROMIDE $(Ph_3Ge)_3SiBr$
Prepn.: By reacting $(Ph_3Ge)_3SiH$ with Br_2 in $EtBr$ (375).
Props.: m. 241.5-242.5°; crystallizes from C_6H_6 with 3 moles of the solvent (375).
Rxn. with: NH_3 in moist C_6H_6 → $(Ph_3Ge)_3SiOH$ (375).
NH_3 → $(Ph_3Ge)_3SiNH_2$ (375).
Alkali under reflux → $(Ph_3Ge)_2$ (375).
Addn. compds.: $(Ph_3Ge)_3SiBr \cdot 2ClCH_2CH_2Cl$, crystals stable at 110°; the solvent is removed at 170° in vacuo.

TRIS(TRIPHENYLGERMYL)SILOL $(Ph_3Ge)_3SiOH$
Prepn.: By treating $(Ph_3Ge)_3SiBr$ with NH_3 gas in moist C_6H_6 (375).
Props.: m. 196-15-197.5°, sol. in most org. solvents, excepting EtOH (375).

TRIS(TRIPHENYLGERMYL)SILYL CHLORIDE $(Ph_3Ge)_3SiCl$
Prepn.: By treating $(Ph_3Ge)_3SiNH_2$ with HCl gas in Et_2O (375).
Props.: Crystals, m. 230-231°, does not react with H_2O (375).

ETHYLTRIS(TRIPHENYLGERMYL)SILANE $(Ph_3Ge)_3SiEt$
Prepn.: By reacting $(Ph_3Ge)_3SiLi$ with EtBr in NH_3 (375).
Props.: m. 283-284.5°, sol. in common org. solvents (375).

TRIS(TRIPHENYLGERMYL)SILYLAMINE $(Ph_3Ge)_3SiNH_2$
Prepn.: By reacting $(Ph_3Ge)_3SiBr$ with liq. NH_3 (375).
Props.: m. 206-206.5° (375).
Rxn. with: HCl gas in Et_2O → $(Ph_3Ge)_3SiCl$ (375).

(TRIETHYLSILYL)TRIPHENYLGERMANE $Et_3SiGePh_3$
Prepn.: By treating $(Ph_3Ge)_2$ with Na-K alloy in Et_2O in the presence of C_4H_8O and reacting the resulting Ph_3GeK with Et_3SiCl (176).
By treating $(Ph_3Ge)_2$ with Li in $(CH_2OMe)_2$ and reacting the resulting Ph_3GeLi with Et_3SiCl (176).
Props.: m. 97-98° (176).

TETRAKIS(TRIPHENYLGERMYL)DISILENE (?) $(Ph_3Ge)_2Si:Si(GePh_3)_2$
Prepn.: By reacting $SiCl_4$ with Ph_3GeNa (375).
Props.: Amorphous compd.; analytical data suggest the above constitution (375).

TETRAKIS[p-(TRIMETHYLSILYL)PHENYL]GERMANE $(Me_3SiC_6H_4)_4Ge$
Prepn.: By reacting p-$(Me_3Si)C_6H_4Li$ with $GeCl_4$ (5.3:1) in MePh under reflux; 49% yield (370).
Props.: m. 351-354° (370).

3-METHOXY-3,5,5-TRIMETHYL-2-OXA-3-SILA-5-GERMAHEXANE
$Me_3GeCH_2SiMe(OMe)_2$
Prepn.: By reacting Me_3GeCH_2MgCl with $MeSi(OMe)_3$ in Et_2O; 75% yield (484).
Props.: b. 65-68°/18mm; $n^{25}D$ 1.4241; d_{25} 1.056 (484).
Rxns. with: H_2O (acidified H_2O-Et_2O mixt.) → $[Me_3GeCH_2Si(Me)O]_n$ (494).

(TRIMETHYLGERMYLMETHYL)METHYLSILOXANE POLYMER $[(Me_3GeCH_2)MeSiO]_x$
Prepn.: By hydrolyzing $(Me_3GeCH_2)MeSi(OMe)_2$ in a mixt. of H_2O and Et_2O contg. a small amt. HCl (494).
Props.: b. 90-95°/0.45-0.2mm; $n^{25}D$ 1.4640; d_{25} 1.194 (494).

(TRIPHENYLGERMYL)TRIPHENYLTIN $Ph_3GeSnPh_3$
Prepn.: By reacting Ph_3GeK with Ph_3SnCl; 60% yield (183).
Props.: m. 284-286° (183).
Rxns. with: I_2 → Ph_3GeI (183).
PhLi → mixt. of prods. (183).
BuLi, followed by CO_2 and acidification → Ph_3GeCO_2H and $(Ph_3Ge)_2$ (183).

9. DIGERMANES AND DIGERMYL-ALKANES AND -ACETYLENES

Hexaorganodigermanes are prepared:

1. By treating triorganogermanium halides with alkali metals in a ratio of one mole to one gram atom.
2. By reacting alkali metal triorganogermanides with halogens in $(CH_2OMe)_2$ at low temperatures, or with di- or polyhalogenated alkanes in ether, or with triorgano-(alkoxy)germanes.
3. By reacting triorganogermanes with lithium triorganogermanides or with organolithium compounds.
4. By dissolving triorgano(alkoxy)germanes in alcohol and evaporating the solns. to dryness or by exposing powdered triorgano(alkoxy)germanes to air.
5. By treating germanium tetrahalides with Grignard or organolithium compounds; digermanes are formed as by-products along with tetraorganogermanes.

Tetraorganodigermanium dihalides are obtained by reacting germanium dihalides with diorganomercury compounds.

Bis(triorganogermyl)alkanes are prepared:

1. By reacting alkenyltrihydrocarbylgermanes with triorganogermanes.
2. By reacting alkali metal triorganogermanides with α,ω-dihaloalkanes in liquid ammonia.

Bis(triorganogermyl)acetylenes are prepared by reacting triorganogermanium halides with ethynylenebis(magnesium bromide).

HEXAMETHYLDIGERMANE $(Me_3Ge)_2$
Prepn.: By reducing Me_3GeBr with molten K in the presence or absence of an inert solvent (63).
Props.: m. $-40°$; b. $138°/750mm$; n^{20}_D 1.4564 (63).

HEXAETHYLDIGERMANE $(Et_3Ge)_2$
Prepn.: By reacting Et_3GeCl with Li in Et_2O or C_4H_8O; 60 and 18% yield, respectively (676).
By-prod. in the rxn. of $GeBr_4$ with EtMgBr or EtLi in Et_2O (187).
Props.: b. $61-62°/0.007mm$; n^{20}_D 1.4960 (187).

HEXAVINYLDIGERMANE $(CH_2:CH)_3Ge_2$
Prepn.: By-prod. in the rxn. of $CH_2:CHMgBr$ with $GeCl_4$(1:4) in C_4H_8O under reflux (490).
Props.: b. 55°/0.35-0.25mm; $n^{25}D$ 1.5217; d_{25} 1.171; moderately stable in air, gradually turning yellow (490).
Rxns. with: I_2 in $CHCl_3$ → $(CH_2:CH)_3GeI$ (490).
Br_2 in $CHCl_3$ → $(CH_2:CH)_3GeBr$ (490).

HEXACYCLOHEXYLDIGERMANE $(C_6H_{11})_3Ge_2$
Prepn.: By refluxing $(C_6H_{11})_3GeBr$ with excess Na in MePh; 85% yield (243).
Props.: m. 316° (243).

HEXAPHENYLDIGERMANE $(Ph_3Ge)_2$
Prepn.: By dissolving Ph_3GeOMe in abs. EtOH and evapg. the soln. to dryness; theoret. yield (62).
By exposing finely powdered Ph_3GeOMe to air for 3 hrs. and recrystallizing the prod. from ligroin (62).
By reacting Ph_4Ge with Li in $(CH_2OMe)_2$ and treating the resulting Ph_3GeLi with Ph_3GeOMe or with Et_2CO_3 (175).
By reacting Ph_3GeH with Ph_3GeLi (179, 186).
By reacting Ph_3GeLi with Br_2 in $(CH_2OMe)_2$ at -25 to -20° (186).
By refluxing $GeCl_4$ with excess PhMgBr in Et_2O-MePh; 62% yield (Ph_4Ge is also formed) (211a, 245).
By reacting Ph_3GeH (large excess) with PhLi in Et_2O (along with some Ph_4Ge) (246).
By reacting Ph_3GeNa with CCl_4, $CHCl_3$, $(ClCH:)_2$, $(BrCH_2)_2$, or $(BrCH_2CH:)_2$ in Et_2O; up to 91% yield (519).
By-prod. in the rxn. of Ph_3GeH with MeLi (179).
By-prod. in the rxn. of Ph_3GeCl with Ph_3SiK (180).
By-prod. in the rxn. of Ph_3GeLi with fluorene or dibenzofuran (186).
Props.: m. 340-342° (175); m. 345-347° (186); m. 330-331° (211a, 245, 246); Inert to oxygen (176); Sublimes at 275°/1mm (211a); Magnetic measurements set upper limits of dissocn. at 1% in the solid state and 20% in a nearly satd. C_6H_6 soln. at 25° (478).
Rxns. with: Na-K in Et_2O-C_4H_8O → Ph_3GeK (176, 181); if the K deriv. is reacted with CO_2 and hydrolyzed, Ph_3GeCO_2H is formed (61).
Na-K in Et_2O, followed by $Ph_2C:CH_2$ and hydrolysis → $Ph_3GeCH_2CH_2Ph_2$, $(Ph_3Ge)_2O$, and some unreacted starting matl. (181).
Na-K in Et_2O, followed by stilbene → $(Ph_3Ge)_2O$ and unreacted stilbene (181).
Na-K in $(CH_2OMe)_2$ → $(Ph_3Ge)_2O$ and unreacted $(Ph_3Ge)_2$ (181).
Na in liq. NH_3 → Ph_3GeNa (211a).
Li in $(CH_2OMe)_2$ → Ph_3GeLi (176, 177, 179, 181).
Br_2 in CCl_4 → Ph_3GeBr (211a).
Br_2(1:2.6) in $C_2H_4Br_2$ → Ph_2GeBr_2 (245).
I_2 in boiling xylene → Ph_3GeI (176).
I_2 in boiling xylene in the presence of quinoline → $(Ph_3Ge)_2O$ (176).

TETRABUTYLDIGERMANIUM DIIODIDE (?) $(Bu_2IGe)_2$
Prepn.: By reacting GeI_2 with Bu_2Hg in Me_2CO in an inert atm. (327).
Props.: Oily prod. (237).

1,2-BIS(TRIETHYLGERMYL)ETHANE $(Et_3GeCH_2-)_2$
Prepn.: By reacting $Et_3GeCH:CH_2$ with Et_3GeH (368).
Props.: b. $126°/1.5mm$; $n^{20}D$ 1.4773; $d^{20}4$ 1.0935 (368).

1,3-BIS(TRIETHYLGERMYL)PROPANE $Et_3Ge(CH_2)_3GeEt_3$
Prepn.: By reacting $Et_3GeCH_2CH:CH_2$ with Et_3GeH (368).
Props.: b. $128-129°/1.4mm$; $n^{20}D$ 1.4759; $d^{20}4$ 1.0807 (368).

1,3-BIS(TRIPHENYLGERMYL)PROPANE $(Ph_3GeCH_2)_2CH_2$
Prepn.: By reacting Ph_3GeH with $Ph_3GeCH_2CH:CH_2$ in the presence of Bz_2O_2 in hexane under reflux (181)
By reacting Ph_3GeNa with $Br(CH_2)_3Br$ in liq. NH_3; 60.5% yield (519).
Props.: m. $134-136°$ (181); m. $132.5-133°$ (519).

1,5-BIS(TRIPHENYLGERMYL)PENTANE $Ph_3Ge(CH_2)_5GePh_3$
Prepn.: By reacting Ph_3GeNa with $Br(CH_2)_5Br$ in liq. NH_3; 59.5% yield (519).

BIS(TRIETHYLGERMYL)ACETYLENE $Et_3GeC:CGeEt_3$
Prepn.: By reacting Et_3GeBr with $(CMgBr)_2$ in $CHCl_3$ (213).
Props.: b. $50°/14\ mm$ (213).
Rxns. with: Alkali → Et_3GeOH and C_2H_2 (213).

BIS(TRICYCLOHEXYLGERMYL)ACETYLENE $(C_6H_{11})_3GeC:CGe(C_6H_{11})_3$
Prepn.: By reacting $(C_6H_{11})_3GeBr$ with $(CMgBr)_2$ in $CHCl_3$ (213).
Props.: m. $158°$ (213).
Rxns. with: Alkali → $(C_6H_{11})_3GeOH$ and C_2H_2 (213).

BIS(TRIPHENYLGERMYL)ACETYLENE $Ph_3GeC:CGePh_3$
Prepn.: By reacting Ph_3GeBr with $(CMgBr)_2$ in $CHCl_3$ (213).
Props.: m. $127°$ (213).
Rxns. with: Alkali → Ph_3GeOH and C_2H_2 (213).

10. POLYMERIC ORGANOGERMANIUM COMPOUNDS

HEXAPHENYLHEXAGERMANIN (?) $(PhGe)_6$
Prepn.: By refluxing $PhGeCl_3$ with K in xylene in O-free CO_2 (476).
Props.: Colorless substance, the constitution of which has not yet been satisfactorily elucidated (476).
Rxns. with: Br or I → absorption of 8 atoms of halogen (476).
O → absorption of 3-4 atoms O and polymerization to a compd. with 18 Ge atoms (?) (476).

ISOPROPYLGERMANIUM POLYMER $(i-PrGe)_x$
Prepn.: By-prod. in the rxn. of $i-PrMgBr$ or $i-PrMgCl$ with $GeBr_4$ or $GeCl_4$ in Et_2O or C_6H_6 (17)
Props.: Colorless matl. sol. in CCl_4 (17).
Rxn. with: Br in CCl_4 → $i-PrGeBr_3$ (17).

PROPYLGERMANIUM POLYMER $(PrGe)_x$ (?)
Formation: By-prod. in the rxn. of $GeCl_4$ with PrLi at 85-90° (247).
Props.: Colorless liquid insol. in H_2O, inert to dil. alkali and dil. acids, decomp. in boiling 10N NaOH and fuming HNO_3 (247).

11. DERIVATIVE OF BIVALET GERMANIUM

DIPHENYLGERMANIUM Ph_2Ge
Prepn.: By reacting PhLi with GeI_2, Ph_2Ge is formed along with Ph_4Ge (535).

B. TIN

1. TETRAORGANOTIN COMPOUNDS

SYMMETRICAL, NON-FUNCTIONAL TETRAORGANOTIN COMPOUNDS

The compounds are prepared:

1. By reacting tin tetrahalides with Grignard compounds in a ratio of 1:4.
2. By treating a mixture of tin tetrahalides and hydrocarbyl halides (1:4) with alkali metals.
3. By reacting tin tetrahalides with trialkylaluminum compounds.
4. By reacting tin dichloride or tin tetrahalides with organolithium compounds and treating the resulting lithium triorganostannides with halides of the same hydrocarbons.
5. By reacting tin powder or tin alloys, such as Na-Sn, Na-Sn-Zn, and Mg_2Sn, with organohalides or organomercury halides at elevated temperatures in the presence or absence of catalysts. The reactions yield tetraorganotin compounds along with triorgano-, diorgano-, and monoörganotin halides.
6. By reacting tin powder or tin alloys with diorganomercury compounds.
7. By reacting tin amalgam with organic halides in the presence of lithium shavings.

TETRAMETHYLTIN Me_4Sn

<u>Prepn.</u>: By reacting Sn tetrahalides with MeMgI in Bu_2O or with MeMgBr in Et_2O; up to 91% yield was obtd. in Bu_2O (120, 292, 330, 633).

<u>Props.</u>: m. 218.18°K (529); b. 78.3°/740mm; n^{21}_D 1.4393 (330); b. 76.6°/748mm (120); d_{25} 1.2905; d_{30} 1.2817; n^{20}_D 1.4415; n^{25}_D 1.4386; vapor pressure: at 0°= 32.3mm, at 19.5°= 86.5mm (292); Raman and IR spectra (800); Thermodynamic data (44, 113, 330, 529); Viscosity data (677); Force consts. (511, 774); Mol. structure (60, 511); C-Sn bond length (60); Polarity (464); Potential barrier value (136); Molar magnetic rotation (555); Magnetic susceptibility and parachor (258, 261); Self-diffusion coeff. (618); Mass spectrum (112); Raman spectrum (330, 510, 558, 579, 650); IR spectrum (510, 558, 579); Pyrolysis → CH_4 and C_2H_6; mechanism of decompn. proposed (335, 768); Kinetics of pyrolytic decompn. (768).

TETRAMETHYLTIN Me_4Sn (Cont'd.)

Rxns. with: $Et_4Sn \to$ 24.6% $Me_4Sn + Me_3EtSn$, 38.4% Me_2Et_2Sn, and 37% $MeEt_3Sn + Et_4Sn$ (75).
$SnBr_4$ (3:1) $\to Me_3SnBr$ (268).
$I_2 \to Me_3SnI$ and Me_2SnI_2, stepwise, and SnI_4 simultaneously (357).
Br_2 at 25°, heat of rxn. (411).
Br_2 at -40° $\to Me_3SnBr$ (487).
HCl in $CHCl_3 \to Me_3SnCl$ (361).
$HgCl_2$ in EtOH $\to Me_3SnCl$ and Me_2SnCl_2 (361).
RCO_2H in the presence of silica gel $\to$ cleavage of the Me group, the extent of which depends upon the strength of the acid and upon basicity of the solvent (316, 318).
Phenols contg. electroneg. groups (NO_2, Br) $\to$ cleavage of the Me group (318).
X-ray radiation $\to$ formation of $(C_2H_4)^+$ radical (?) and Me_2SnH_2 (662).
$Al(BH_4)_3 \to MeAl(BH_4)_2$ with intermediate formation of $Me_2Sn(BH_4)_2$ (?) (675).
Biol. props.: Toxicity and symptoms (586, 638).
Use: Filler for counter tubes (619).

TETRAETHYLTIN Et_4Sn

Prepn.: By reacting EtMgBr with $SnCl_4$ (5:1) in Et_2O under reflux; 85% yield (596).
By reacting EtBr (60% excess) with a Sn-Na-Zn alloy (34, 195, 203).
By reacting $SnCl_4$ with EtMgBr in a mixt. of Et_2O and MePh or in Et_2O; up to 96% yield (34, 264, 633); in C_6H_6 the yield is 22% (467).
By reacting $SnCl_2$ with EtLi in Et_2O at -10° and treating the resulting Et_3SnLi with EtBr (172).
By reacting Mg_2Sn with EtBr or EtCl in cyclohexane at 160° in a sealed tube; 61% yield along with some Et_3SnBr and Et_2SnBr_2. If C_6H_6 is employed as solvent the prod. contains about equal amts. of Et_4Sn, Et_3SnBr, and Et_2SnBr_2. The rxn. is catalyzed by $HgCl_2$ (266, 678). If Mg_2Sn is mixed with a small quant. of Hg and then refluxed with EtBr in cyclohexane the prod. contains only Et_4Sn (678). The rxn. is carried out in xylene (741).
By reacting NaSn with EtBr (4:2); the activation energy and kinetics were detd. (384a, 724).
Props.: b. 178.5°; n^{20}_D 1.4691 (172); b. 181°/760mm (195); b. 63-65°/12mm (596); n^{25}_D 1.4693-1.4699; d^{25}_4 1.1916; d_{25} 1.1916 (264); d_{30} 1.1857; n^{25}_D 1.4693; n^{30}_D 1.4668 (292); b. 181-182° (467); b. 136-144°/5.8-6.4mm (741); At low temps. the prod. crystallizes in several forms, m. (extrapolated) 142.15°K (529); At least 10 melting points were observed between 137.45 and 147.05°K due to polymorphism (527, 528); Viscosity data (677); Mol. wt. in various solvents: monomer (785); Thermodynamic data (44, 133, 528, 529); Dipole moment, $\mu = \sim 0$ (352); Molar magnetic rotation (555); UV spectrum (439); IR spectrum (261a).

TETRAETHYLTIN Et_4Sn (Cont'd.)

Rxns. with: $SnCl_4 \to EtSnCl_3$, Et_2SnCl_2, and Et_3SnCl (113, 256).
$Me_4Sn \to 24.6\%$ Me_4Sn + Me_3SnEt, 38.4% Me_2SnEt_2, and 37% $MeSnEt_3$ + Et_4Sn (75).
H_2 at 200°/60 atm. $\to Sn + C_2H_4$ (143).
$Br_2 \to Et_3SnBr$ (157, 487) and Et_2SnBr_2 (195).
$I_2 \to Et_3SnI$ (195) and Et_2SnI_2 + EtI (354, 357).
$Na \to Et_3SnNa$ (195).
i-Bu_4Sn (1:1) in the presence of $AlCl_3 \to$ i-Bu_2SnEt_2, i-Bu_3SnEt, and i-$BuSnEt_3$ (241).
RCO_2H in the presence of SiO_2 gel $\to$ cleavage of the Et group (316, 318).
RCO_2H in excess in the absence of catalyst $\to Et_3SnO_2CR$ (467).
$Na \to Et_3SnNa$ (195).
Phenols contg. electroneg. groups (NO_2 or Br) $\to$ cleavage of the Et group (318).
$HgCl_2$ in EtOH or $HCCl_3$ under reflux $\to Et_3SnCl$ and Et_2SnCl_2 (361).
$AlCl_3$ in $HCCl_3 \to Et_2SnCl_2$ (362).
$FeCl_3$ in $HCCl_3 \to Et_3SnCl$ (362).
$BiCl_3$ in $HCCl_3 \to Et_3SnCl$ and some Et_2SnCl_2 (363).
$RSH \to Et_3SnSR$ (467).
PhOH (1:1) $\to Et_3SnOPh$ (467).
RCOCl in the presence of $AlCl_3 \to EtCOR + Et_2SnCl_2$ (515).
PBr_3 at 150° $\to Et_3SnBr$ and $EtPBr_2$ (764).
P and I_2 in $CS_2 \to Et_3SnI$, P_2I_4, and unreacted Et_4Sn (464).
Biol. props.: Toxicity and symptoms (532, 586, 695); Metabalic conversion to Et_3Sn derivs. (108).
Distribution in animal body (635).
Effect on leucocytes (638).

TETRAPROPYLTIN Pr_4Sn

Prepn.: By heating granular Sn-Zn-Na alloy with PrBr (211); with PrCl at 140-145° under N in autoclave (590).
By refluxing Mg_2Sn with PrBr in xylene; Pr_3SnBr is also formed (741).
By reacting $SnCl_4$ with PrMgBr (1:4) in C_4H_8O; 81% yield (463); in C_6H_6, 24% yield (468).
Props.: m. 163.9°K (529); b. 110-111°/10mm (468); b. 113°/10mm; n_D^{20} 1.4748; d_4^{20} 1.1070 (634); b. 221-226° (741); Viscosity data (677); Magnetic susceptibility (261); Molar magnetic rotation (555).
Rxns. with: Br_2 or $I_2 \to Pr_3SnBr$, Pr_2SnBr_2 and Pr_3SnI and Pr_2SnI_2, respectively (211, 357).
RCO_2H in the presence of silica gel $\to$ cleavage of the Pr group, the extent of which depends on the strength of the acid and the basicity of the solvent employed (316, 318).
RCO_2H without any catalyst $\to Pr_3SnOCOR$ (468).
$SnCl_4$ (3:1) at 200° $\to Pr_3SnCl$ (463).
RSH (1:1) $\to Pr_3SnSr$ (468).
Biol. props.: Toxicity (634, 637); Effect on leucocytes (638); Intravenous lethal dose for dogs (87).
Use: Stabilizer for vinyl resins (578).

TETRAISOPROPYLTIN   i-Pr$_4$Sn

Prepn.: By reacting SnCl$_4$ with i-PrMgCl in boiling xylene; i-Pr$_3$SnCl, i-Pr$_2$SnCl$_2$, and (i-Pr$_3$Sn)$_2$ are also formed (87, 743); in C$_6$H$_6$, 22.6% yield of the title prod. (468).
By-prod. in the rxn. of SnCl$_4$ with i-PrLi (1:3) (743).
Props.: b. 89°/4mm; decomp. 200°; $n^{20.2}$D 1.4851; d_{20} 1.1237; MR 74.25 (87); b. 103-104°/10mm (468); b. 100.6-101.2°/8mm (743).
Rxns. with: SnBr$_4$ (3:1) → i-Pr$_3$SnBr (268, 743).
RSH (1:1) → i-Pr$_3$SnSR (468).
RCO$_2$H (1:1) → i-Pr$_2$Sn(OCOR)$_2$ (468).
Biol. props.: Toxicity (637); Intravenous lethal dose for dogs (87).

TETRABUTYLTIN   Bu$_4$Sn

Prepn.: By reacting SnCl$_4$ with BuCl and Na (1:4:8) in light petr. ether; 64% yield (265, 268).
By reacting Bu$_2$SnCl$_2$ with BuCl and Na (1:2:4) in light petr. ether; 88% yield along with a small quant. of (Bu$_3$Sn)$_2$ (265).
By reacting powdered Na-Sn-Zn alloy (10-20% Na and 10-25% Zn) with BuBr in an A atm. on steam bath; 63.5% yield. If the alloy contains a small amt. of Fe, 70-62% yield is obtd. (417).
By reacting Bu$_3$SnLi with BuI in Et$_2$O under reflux (172).
By refluxing BuCl with SnCl$_4$ and Mg (4:1:4) in MePh contg. Et$_2$O; 90% yield of a mixt. of Bu$_4$Sn and Bu$_3$SnCl along with a small amt. of Bu$_2$SnCl$_2$ (427).
By reacting SnCl$_4$ with BuMgCl in MePh >100°; 94.6% yield along with small amts. of Bu$_3$SnCl and Bu$_2$SnCl$_2$ (428).
By-prod. in the rxn. of Bu$_3$SnLi with PhLi (172).
By reacting a mixt. of Mg (treated with BuCl in Et$_2$O) and Bu$_2$SnCl$_2$ in MePh with BuCl at 95°; the prod. contains about equal amts. of Bu$_4$Sn and Bu$_3$SnCl along with a very small amt. of Bu$_2$SnCl$_2$ (430).
By heating BuCl with NaSn at 140-155°, or with NaSnZn (2% Zn) at 160-170°, or with NaKSn at 160-165° in an autoclave under N; the prod. contains Bu$_4$Sn and Bu$_3$SnCl (590).
By reacting Mg$_2$Sn with BuBr in xylene under reflux (741).
Props.: b. 127°/1.7 mm; n^{20}D 1.4727 (172); b. 125-130°/2-3mm (417); b. 107-112°/1 mm (741); b. 109-112°/0.3 mm; n^{20}D 1.4736 (417); b. 145°/11mm; n^{19}D 1.4739; d^{20}_4 1.0541 (634); m. 176.1°K (529); Viscosity data (677); Magnetic susceptibility and parachor (258, 261); Thermodynamic data (113); Molar magnetic rotation (555).

TETRABUTYLTIN Bu_4Sn (Cont'd.)

Rxns. with: $SnCl_4$ (1:1) → Bu_2SnCl_2, $BuSnCl_3$, and Bu_3SnCl (240, 241, 248, 256).
$SnCl_4$ (1:1) at 240° → Bu_2SnCl_2 (265, 417).
$SnCl_4$ (1:1) at 0-20° → $BuSnCl_3$ and Bu_3SnCl (609).
$SnCl_4$ (Bu_4Sn in excess) → Bu_3SnCl (256, 270).
$AlCl_3$ in $CHCl_3$ → Bu_3SnCl and Bu_2SnCl_2 (362).
$BiCl_3$ at 120° → Bu_2SnCl_2 (362).
I_2 → Bu_3SnI (357).
R_4Sn in the presence of $AlCl_3$ → Bu_3SnR, Bu_2SnR_2, and $BuSnR_3$ (240, 241).
Br_2 (1:1) at 25° → Bu_3SnBr and $BuBr$; heat of rxn. determined (740).
RCO_2H in the presence of silica gel → cleavage of the Bu group in an extent depending on the strength of the acid and the basicity of the solvent employed (316, 318).
Biol. props.: Toxicity (634, 637).
Use: Anticrazing agent for poly(dichlorostyrene) (641a).
Fuel additive in a combination with Et_4Pb and halogenated hydrocarbons (232).
Stabilizer for vinyl halide polymers (904).
Stabilizer for chlorinated esters of fatty acids (856).
Stabilizer for polymeric organosilicones (759).

TETRAISOBUTYLTIN $i-Bu_4Sn$

Prepn.: By reacting $SnCl_4$ with $i-Bu_3Al$ in $n-C_7H_{16}$ at 40-48° and treating the rxn. mixt. with NaOH; 53.6% yield along with $(i-Bu_3Sn)_2O$ (581).
Props.: b. 128-129°; $n^{19.5}D$ 1.4751; d_{20} 1.0517 (581).
Rxn. with: Et_4Sn (1:1) in the presence of $AlCl_3$ → $Et_2Sn-i-Bu_2$, $EtSn-i-Bu_3$, and $Et_3Sn-i-Bu$ (241).
Biol. props.: Toxicity (637); Anthelmintic props. (847).

TETRAPENTYLTIN $(n-C_5H_{11})_4Sn$

Prepn.: By heating $C_5H_{11}Cl$ with NaSnZn (2%Zn) at 162° under N in an autoclave; the prod. contains $(C_5H_{11})_4Sn$ along with $(C_5H_{11})_3SnCl$ (590).
Props.: b. 182°/10mm; $n^{17}D$ 1.4738; d^{20}_4 1.0159 (634); Viscosity data (677); Magnetic susceptibility (261); Molar magnetic rotation (555).
Rxn. with: Bu_4Sn (1:1) in the presence of $AlCl_3$ → $C_5H_{11}SnBu_3$, $(C_5H_{11})_2SnBu_2$ and $(C_5H_{11})_3SnBu$ (241).
Biol. props.: Toxicity (634, 637); Distribution in animal body (636).

TETRAISOPENTYLTIN $(i\text{-}C_5H_{11})_4Sn$

Prepn.: By reacting powdered Na-Sn-Zn alloy (10-20% Na and 10-25% Zn) with $i\text{-}C_5H_{11}Br$ under argon on a steam bath; 31% yield. If the alloy contained a small amt. of Fe the yield was 70% (417).
By reacting isopentyl bromide with Mg_2Sn in xylene under reflux (741).
Props.: b. 145-150°/2.4mm (741).
Rxn. with: $I_2 \rightarrow (i\text{-}C_5H_{11})_3SnI$ (357).
Biol. props.: Toxicity (637).

TETRACYCLOPENTADIENYLTIN $(C_5H_5)_4Sn$

Prepn.: By reacting $SnCl_4$ with C_5H_5MgBr in Et_2O; 41.5% yield (182.)
Props.: m. 71-73° (182).

TETRACYCLOHEXENYLTIN $(C_6H_9)_4Sn$

Props.: Magnetic susceptibility and parachor (258).

TETRAHEXYLTIN $(n\text{-}C_6H_{13})_4Sn$

Prepn.: By treating a mixt. of $SnCl_4$ and $n\text{-}C_6H_{13}Cl$ with Na in light petroleum under reflux; 51% yield (273).
Props.: b. 187-190°/1.5mm (273); b. 193°/3mm; n^{16}_D 1.4756; n^{20}_4 0.9936 (634); Thermodynamic data (113); Magnetic susceptibility (261); Molar magnetic rotation (555).
Rxn. with: $SnCl_4$ (1:1) at 220-225° $\rightarrow (C_6H_{13})_2SnCl_2$ (273).
Biol. props.: Toxicity (634); Effect on leucocytes (638).
Use: Catalyst for the polymerization of vinyl compounds (690).

TETRAHEPTYLTIN $(n\text{-}C_7H_{15})_4Sn$

Prepn.: By reacting $SnCl_4$ with $n\text{-}C_7H_{15}MgBr$ in Et_2O at r.t.; 89% yield (360).
Props.: b. 219-221°/2mm; n^{20}_D 1.4702; d^{20}_4 0.9746 (360); b. 218°/3mm; n^{19}_D 1.4729; d^{20}_4 0.9743 (634); Viscosity data (677); Magnetic susceptibility (261); Molar magnetic rotation (555).
Rxn. with: I_2 in boiling MePh or xylene $\rightarrow (n\text{-}C_7H_{15})_3SnI$; no rxn. occurred in Et_2O, and in C_6H_6 the rxn. was very slow (360).
Biol. props.: Toxicity (634).

TETRAOCTYLTIN $(n-C_8H_{17})_4Sn$

Prepn.: By treating $SnCl_4$ and $n-C_8H_{17}Cl$ with Na in light petroleum under reflux; 65-80% yield contaminated with $(C_8H_{17})_3Sn]_2$ and octyltin chlorides (273).
By reacting $SnCl_4$ with $n-C_8H_{17}MgBr$ in Et_2O at r.t.; 59% yield (360).
Props.: Yellow oil (273); b_0 250-255°/5-6mm; $n^{20}D$ 1.4681 (360); b. 224°/1mm; $n^{18}D$ 1.4730; d^{20}_4 0.9609 (634); Magnetic susceptibility (261); Molar magnetic rotation (555).
Rxns. with: $SnCl_4$ (1:1) at 110° then at 230-240° → $(n-C_8H_{17})_2SnCl_2$ (273).
I_2 in MePh or xylene → $(n-C_8H_{17})_3SnI$; no rxn. in Et_2O and in C_6H_6 the rxn. was very slow (360).
$HgCl_2$ in EtOH → $n-C_8H_{17}HgCl$ and $(n-C_8H_{17})_{4-x}SnCl_x$ (361).
Biol. props.: Toxicity (634); Effect on leucocytes (638).

TETRAKIS(2-ETHYL-n-HEXYL)TIN $(BuCHEtCH_2)_4Sn$

Prepn.: By refluxing $BuCHEtCH_2MgBr$ with $SnCl_4$ (>4:1) in Et_2O; 40% yield (273).
Rxn. with: $SnCl_4$ (1:1) at 225-230° → $(BuCHEtCH_2)_2SnCl_2$ (273).

TETRAKIS(3,5,5-TRIMETHYL-n-HEXYL)TIN $(Me_3CCH_2CHMeCH_2CH_2)_4Sn$

Prepn.: By refluxing $Me_3CCH_2CHMeCH_2CH_2MgBr$ with $SnCl_4$ (4:1) in Et_2O; 66.5% yield (273).
Rxn. with: $SnCl_4$ (1:1) at 225-230° → $(Me_3CCH_2CHMeCH_2CH_2)_2SnCl_2$ (273).

TETRADODECYLTIN $(n-C_{12}H_{25})_4Sn$

Prepn.: By reacting $SnCl_4$ with $n-C_{12}H_{25}MgBr$ (1:3) in C_6H_6 under reflux; 45% yield (369).
Props.: m. 15-16°, $n^{30}D$ 1.4692, $n^{20}D$ 1.4736 (369).
Rxn. with: Dry HCl in Et_2O → $(n-C_{12}H_{25})_3SnCl$ (369).
Use: Stabilizer for chlorinated esters of fatty acids (856).

TETRAKISTETRADECYLTIN $(n-C_{14}H_{29})_4Sn$

Prepn.: By reacting $SnCl_4$ with excess $n-C_{14}H_{29}MgBr$ in Et_2O under reflux; 66% yield (369).
Props.: m. 33-34° (369).
Rxn. with: Dry HCl → $(n-C_{14}H_{29})_3SnCl$ (369).

TETRAKISHEXADECYLTIN $(n-C_{16}H_{33})_4Sn$

Prepn.: By reacting $SnCl_4$ with $n-C_{16}H_{33}MgBr$ in C_6H_6; 76% yield (369).
Props.: m. 36-41° (369).
Rxn. with: Dry HCl in Et_2O → $(n-C_{16}H_{33})_3SnCl$ (369).

TETRAKISOCTADECYLTIN $(n\text{-}C_{18}H_{37})_4Sn$

<u>Prepn.</u>: By reacting $SnCl_4$ with excess $n\text{-}C_{18}H_{37}MgBr$; 56.8% yield (369).

<u>Props.</u>: m. 47° (369).

TETRAPHENYLTIN Ph_4Sn

<u>Prepn.</u>: By reacting $SnCl_4$ with PhLi at -10° in Et_2O; 14.6% yield (170).

By reacting $SnCl_4$ with PhCl or PhBr and Na in an inert solvent; 66.5% yield (212).

By reacting $SnCl_4$ with PhNa (1:4) in MePh at 45°; 50.5% yield (239).

By adding dropwise $SnCl_4 \cdot 2Et_2O$ to PhMgBr; 70% yield (592).

By reacting $SnBr_4$ with PhLi in $Et_2O\text{-}C_6H_6$; 80% yield (786, 787).

By reacting Sn or Na-Sn with Ph_2Hg; up to 28.6% yield (382).

By reacting Sn powder with PhLi in Et_2O; 46% yield (537, 538).

By refluxing PhLi with Sn-Hg in Et_2O; 67.7% yield (538).

By reacting PhBr with a mixt. of Sn-Hg and Li shavings in Et_2O; 69% yield (538).

By reacting powdered Na-Sn-Zn alloy (10-20% Na and 10-25% Zn) with PhBr on steam bath in an A atm.; 65% yield. Somewhat higher yield results if the alloy contains a small amt. Fe (417).

By reacting a double amalgam (Pb-Sn-Hg) with PhLi in Et_2O under reflux; 46.5% yield (538).

By reacting PhBr with Mg_2Sn in xylene under reflux (741).

By refluxing Ph_3SnLi with PhLi, PhBr, or PhCl in Et_2O; up to 82.2% yield (163, 167).

By reacting Ph_3SnLi with I_2 in Et_2O (170).

By reacting Ph_3SnLi with Ph_2CO at -20°; 9.5% yield (170).

By passing CO_2 over Ph_3SnLi in Et_2O at -10°; 23.6% yield (170).

By treating Ph_3SnCl with PhLi (154).

By reacting Ph_3SnCl with Na-Sn; 37% yield (382).

By refluxing Na_2Sn with PhHgCl (1:2) in xylene; 50.1% yield (382).

By cleaving $p\text{-}XC_6H_4SnPh_3$ (X= F, Cl, or Br) with PhLi in Et_2O under reflux; 95% yield (584).

<u>Props.</u>: m. 228-230° (154); m. 224-225° (163, 382, 787); m. 226-227° (537); m. 225° (538, 592, 741); d. 1.521 (144); Soly. in C_6H_6 (723); Crystal structure (234, 235, 587, 588); Lattice consts. and symmetry (144, 587, 588); Interatomic distances (235); UV spectrum (373, 663); Diffraction data (137); Magnetic susceptibility (261); Diamagnetic anomaly (406); Analytical method (100); Cocrystallization with Ph_4Pb (723).

TETRAPHENYLTIN Ph_4Sn (Cont'd.)

Rxns. with: $SnCl_4$ (1:3) → $PhSnCl_3$ (185, 594).
$SnCl_4$ (3:1) → Ph_3SnCl (154, 270).
$SnBr_4$ (3:1) → Ph_3SnBr (740).
$SnCl_4$ (1:1) → Ph_2SnCl_2 (185, 240, 417, 435, 492, 594).
$SnBr_4$ (1:1) → Ph_2SnBr_2 (740).
Br_2 (1:1 or 1:2) in CCl_4 → Ph_2SnBr_2 and PhBr; heat of rxn. determined (740).
I_2 in $CHCl_3$ → Ph_3SnI (157).
HCl in $CHCl_3$ → Ph_3SnCl (36).
H_2 at 200°/60 atm. → Sn and C_6H_6 (143).
Bu_4Sn (1:1) → Bu_2SnPh_2, Bu_3SnPh, and $BuSnPh_3$ (240, 241).
RLi → displacement of the Ph group (150).
$AgNO_3$ → 80.6% C_6H_6 and 5.2% Ph_2 (152).
AcCl at 50° → Ph_2SnCl_2; at 100-130° → $SnCl_4$ and AcPh (297).
RCOCl in the presence of $AlCl_3$ → Ph_3SnCl and PhCOR (355, 515).
HCO_2H at 100° in a sealed tube → C_6H_6 and a solid, probably $(HO)_3SnO_2CH$ (297).
AcOH at 125° in a sealed tube → C_6H_6 and $AcOSnO_2H$ (29).
$ClSO_3H$ → cleavage of the C-Sn bond (34, 157).
$PhSO_2Cl$ in the presence of $AlCl_3$ → Ph_2SO_2 and Ph_2SnCl_2 (515).
o-$HSC_6H_4CO_2H$ in Et_2O at 130° → $C_6H_4SSn(O)OCO$ (689).
CCl_4 in the presence of $AlCl_3$, followed by H_2O → Ph_2SnCl_2 and Ph_2CO (515).
CCl_4 under reflux in the presence of Bz_2O_2 → 84.1% $SnCl_4$ (399); the rxn. does not occur in the presence of Ac_2O_2 (400).
$AlCl_3$ in $CHCl_3$ → $SnCl_4$ and C_6H_6 (362).
$FeCl_3$ or $NiCl_3$ → $SnCl_4$ (433).
$CoCl_2$ → traces of C_6H_6 (433).
$CuCl_2$ → PhCl (433).
Slow neutrons → radioactive Ph_4Sn (780).
N_2O_3 (1:8) → $4PhN_2NO_3$ (350a).
p-$HOC_6H_4NO_2$ → $HOC_6H_4NO_2 \cdot SnO_2 \cdot 2H_2O$ (689a).
p-MeC_6H_4OH → $MeC_6H_4OH \cdot SnO_2 \cdot 3H_2O$ (689a).
$2,4,6$-$Cl_3C_6H_2OH$ → $SnCl_4$ (689a).
PhSH at 150° → $(PhS)_3SnOH$ and $(PhS)_4Sn$ (298a).
p-MeC_6H_4SH at 150° → $(p$-$MeC_6H_4S)_2Sn(OH)_2$ (298a).
Use: Polymerization catalyst for vinyl monomers (690).
Stabilizer for poly(vinyl halide) resins (578).
Anticrazing agent for poly(dichlorostyrene) (641a).
Stabilizer for polymeric organosilicon compositions (759).
"HCl chaser" for synthetic, non-flammable liquids used as elec. insulators (833).
Additive to fuels: lowering effect on the burning rate of mixts. of $C(NO_2)_4$ and a fuel of the general formula C_nH_{2n-6} (194).

TETRA-m-TOLYLTIN $(m\text{-}MeC_6H_4)_4Sn$

Prepn.: By reacting $SnCl_4$ with $m\text{-}MeC_6H_4Li$ (1:3) in Et_2O at $-10°$ and treating the resulting $(m\text{-}MeC_6H_4)_3SnLi$ with $m\text{-}MeC_6H_4Br$ in Et_2O under reflux; 80.2% yield (170).

Props.: m. 122-123° (170); Crystal structure and lattice consts. (356, 548).

Rxn. with: BzCl in the presence of $AlCl_3 \rightarrow m\text{-}MeC_6H_4Bz + SnCl_4$ (317).

TETRA-o-TOLYLTIN $(o\text{-}MeC_6H_4)_4Sn$

Prepn.: By reacting $SnCl_4$ with $o\text{-}MeC_6H_4MgBr$ in Et_2O under reflux (380).

By-prod. in the rxn. of $SnCl_4$ with $o\text{-}MeC_6H_4Li$ (1:3) in Et_2O at $-10°$ followed by refluxing with $o\text{-}MeC_6H_4I$ (170).

Props.: m. 214-215° (170).

Rxns. with: $SnCl_4$ (1:1) $\rightarrow (o\text{-}MeC_6H_4)_2SnCl_2$ (280).
BzCl in the presence of $AlCl_3 \rightarrow o\text{-}MeC_6H_4Bz$ and $SnCl_4$ (317).
I_2 in $CHCl_3$ followed by HBr $\rightarrow (o\text{-}MeC_6H_4)_3SnBr$ (380).

TETRA-p-TOLYLTIN $(p\text{-}MeC_6H_4)_4Sn$

Prepn.: By reacting $SnCl_4$ with $p\text{-}MeC_6H_4Li$ (1:3) in Et_2O at $-10°$ and treating the resulting $(p\text{-}MeC_6H_4)_3SnLi$ with $p\text{-}MeC_6H_4I$; 74.2% yield (170).

By reacting $SnCl_4$ with $p\text{-}MeC_6H_4Cl$ and Na (1:4:8) in boiling MePh; 50% yield (315).

By reacting Na_2Sn with $p\text{-}MeC_6H_4Cl$ in xylene; 41% yield (382).

By reacting $p\text{-}MeC_6H_4Br$ with a mixt. of Sn-Hg and Li shavings; 39% yield (538).

Props.: m. 233-235° (170); Crystal structure and interat. distance data (233); m. 238° (315); m. 236-237° (382); m. 238° (538); Crystal structure and lattice consts. (587, 588); Analytical method (100).

Rxns. with: $SnCl_4$ (3:1) $\rightarrow (p\text{-}MeC_6H_4)_3SnCl$ (63).
$SnCl_4$ (1:1) $\rightarrow (p\text{-}MeC_6H_4)_2SnCl_2$ (280).
$SnCl_4$ (1:3) $\rightarrow p\text{-}MeC_6H_4SnCl_3$ (280).
AcCl or BzCl in the presence of $AlCl_3 \rightarrow p\text{-}MeC_6H_4Ac$ and $p\text{-}MeC_6H_4Bz$, resp., and $SnCl_4$; without $AlCl_3$ only 5% of $p\text{-}MeC_6H_6Bz$ is formed (317).

TETRAMESITYLTIN $(Me_3C_6H_2)_4Sn$

Prepn.: By reacting $2,4,6\text{-}Me_3C_6H_2Li$ with $SnCl_4$ in Et_2O-MePh; 15.3% yield (40).

By reacting $2,4,6\text{-}Me_3C_6H_2MgBr$ with $SnCl_4$ (317).

Props.: White, crystalline powder, decomp. >320° (40).

Rxn. with: AcCl in the presence of $AlCl_3 \rightarrow AcC_6H_2Me_3 + SnCl_4$ (317).

TETRABENZYLTIN $(PhCH_2)_4Sn$
Prepn.: By reacting $SnCl_4$ with $PhCH_2MgBr$ (317).
Rxns. with: AcCl in the presence of $AlCl_3 \to AcCH_2Ph + SnCl_4$ (317).
BzCl in the presence of $AlCl_3 \to BzCH_2Ph + SnCl_4$ (317).
$PhCH_2Cl$ in the presence of $AlCl_3 \to$ complete cleavage of the
  stannane, while alkyl halides leave the Sn in org. compds.,
  such as R_3SnCl, R_2SnCl_2, and $RSnCl_3$ (317).

TETRA-1-INDENYLTIN $(C_9H_7)_4Sn$
Prepn.: By reacting $SnCl_4$ with C_9H_7Li (1:4) in Et_2O; 48.4%
  yield; with C_9H_7Na the yield was 41.4% (594).
Props.: m. 215° (593, 594).
Rxn. with: Concd. HCl or HCl in $Et_2O \to$ no rxn. (594).

TETRA(1-NAPHTHYL)TIN $(1-C_{10}H_7)_4Sn$
Prepn.: By reacting $1-C_{10}H_7Li$ with $SnCl_4$ in $Et_2O-C_6H_6$;
  21% yield (40).
Props.: Colorless prisms decomp. at 310-320° (40).
Rxn. with: AcCl in the presence of $AlCl_3 \to 1-C_{10}H_7Ac +$
  $SnCl_4$ (317).

TETRA(2-BIPHENYLYL)TIN $(2-PhC_6H_4)_4Sn$
Prepn.: By reacting $2-PhC_6H_4Li$ with $SnCl_4$ in $Et_2O-C_6H_6$;
  71.2% yield (39).
Props.: m. 300-301°; insol. in Et_2O and petr. ether, slightly
  sol. in hot AcOEt, EtOH, and C_4H_8O (39).
Rxns. with: $HgCl_2 \to 2-PhC_6H_4HgCl$ (39).
HCl $\to (2-PhC_6H_4)_2SnCl_2$ (39).

TETRA(3-BIPHENYLYL)TIN $(3-PhC_6H_4)_4Sn$
Prepn.: By reacting $3-PhC_6H_4Li$ with $SnCl_4$ in $Et_2O-C_6H_6$ (39).
Props.: m. 145.5-145.8°; slightly sol. in Et_2O; sol. in C_6H_6,
  C_5H_5N, $CHCl_3$, hot EtOAc and hot ligroin (39).

TETRA(4-BIPHENYLYL)TIN $(4-PhC_6H_4)_4Sn$
Prepn.: By reacting $p-PhC_6H_4Li$ with $SnCl_4$ in $Et_2O-C_6H_6$;
  77% yield (39); 68.6% yield (538).
Props.: m. 268.5°; soly. data (39); m. 260° (538).
Rxns. with: $SnCl_4$ (1:1) $\to (PhC_6H_4)_2SnCl_2$ (539).
$SnBr_4$ (1:1) $\to (PhC_6H_4)_2SnBr_2$ (539).

TETRA-9-FLUORENYLTIN $(C_{13}H_9)_4Sn$
Prepn.: By reacting $SnCl_4$ with 9-fluorenyllithium in Et_2O;
  41% yield (593, 594).
Props.: m. 290°, decomp. >310°; stable in boiling HCl
  (593, 594).

TETRA(9-PHENANTHRYL)TIN $(C_{14}H_9)_4Sn$
Prepn.: By reacting 9-phenanthryllithium with $SnCl_4$ in $Et_2O-C_6H_6$; 55% yield (40).
Props.: White, crystalline powder, m. 360-370° (decompn.) (40).
Rxn. with: HCl → $(C_{14}H_9)_2SnCl_2$ (40).

SYMMETRICAL, FUNCTIONALLY SUBSTITUTED TETRAORGANOTIN COMPOUNDS

Tetrakis(haloalkyl)tin compounds are prepared by reacting tin tetrahalides or haloalkyltin halides with diazoalkanes at low temperature.

Tetrakis(haloaryl)tin compounds are formed as by-products in the reactions of Na_2Sn with haloarylmercury halides.

Hydroxy-, alkoxy-, amino-, and organosilyl-substituted tetra-aryltin compounds are prepared by reacting tin tetrahalides with the correspondingly substituted aryllithium compounds.

Tetraalkenyl- and tetra(arylalkynyl)-tin compounds are prepared by reacting tin tetrahalides with alkenylmagnesium halides and arylalkynylmagnesium halides, respectively.

Tetraisoalkenyltin compounds are prepared by reacting tin dihalides with diisoalkenylmercury compounds.

SYMMETRICAL, FUNCTIONALLY SUBSTITUTED TETRAORGANOTIN COMPOUNDS

R_4Sn	Prepared from	Yield %	Properties	Ref.
$(ClCH_2)_4Sn$	$ClCH_2SnCl_3 + CH_2N_2$ (1:6)	52	m. 49-49.5°; b. 148.5°/5mm	574, 575
	$SnCl_4 + CH_2N_2$ (1:6)	57		575
$(BrCH_2)_4Sn$	$(BrCH_2)_2SnBr_2 + CH_2N_2$ (1:3)		m. 57°	574, 575
$(MeCHCl)_4Sn$	$(MeCHCl)_2SnCl_2$ or		b. 142°/2mm; d^{20}_{20} 1.568;	574, 575
	$(MeCHCl)_3SnCl + MeCHN_2$		n^{20}_D 1.5363 (1)	
$(p-ClC_6H_4)_4Sn$	$Na_2Sn + p-ClC_6H_4HgCl$	by-prod.	m. 197-198° (1)	382
$(m-HOC_6H_4)_4Sn$	$SnCl_4 + m-HOC_6H_4Li$		m. 180-184°(from $CHCl_3$-ligroin) (2)	185
			m. 134.8° (3)	538
$(p-MeOC_6H_4)_4Sn$	$SnCl_4 + p-MeOC_6H_4Li$	55	Crystal structure and lattice consts.	587, 588
			m. 104° (4)	538
$(p-EtOC_6H_4)_4Sn$	$SnCl_4 + p-EtOC_6H_4Li$	77.5	Crystal structure and lattice consts. (5,6)	587, 588
$(p-Me_2NC_6H_4)_4Sn$	$SnCl_4 + p-Me_2NC_6H_4Li$	58	m. 343-345°	174
$(p-Me_3SiC_6H_4)_4Sn$	$SnCl_4 + p-Me_3SiC_6H_4Li$	74	b. 160-163°/766mm	370
$(CH_2:CH)_4Sn$	$SnCl_4 + CH_2:CHMgBr$ in C_4H_8O		b. 67-70°/28mm; d_{25} 1.246; n^{25}_D 1.4914	487

(1) ClC_6H_4 groups are displaceable by alkyl- and aryllithium compounds (150).
(2) The substance recrystallized from $EtOH$-C_6H_6 decomp. at 170-173° (185).
(3) Rxns. with $SnCl_4$ or $SnBr_4$ → $(p-MeOC_6H_4)_2SnCl_2$ and $(p-MeOC_6H_4)_2SnBr_2$, respectively (539).
(4) Rxn. with $SnCl_4$ → $(p-EtOC_6H_4)_2SnCl_2$ (539).
(5) $Sn(C_6H_4NMe_3I)_4$ forms colorless needles, m. 190° (decompn.), sol. in hot H_2O and slightly sol. in hot MeOH (174).
(6) $Sn(C_6H_4NMe_2 \cdot Me_2SO_4)_4$, pale reddish-tan solid, decomp. >180°, sol. in H_2O and hot MeOH; insol. in most org. solvents (174).

SYMMETRICAL FUNCTIONALLY SUBSTITUTED TETRAORGANOTIN COMPOUNDS (Cont'd.)

R_4Sn	Prepared from	Yield %	Properties	Ref.
$(CH_2:CH)_4Sn$	$SnCl_4$ + $CH_2:CHMgCl$ in C_4H_8O	82	b. 55-57°/17mm; $n^{25}D$ 1.4993; d_{25} 1.267 (7,8,9) IR spectrum	459, 261a
$(CH_2:CMe)_4Sn$	$SnBr_2$ + $(CH_2:CMe)_2Hg$	by-prod.	b. 66-67°/8mm; $n^{20}D$ 1.5110; d^{20}_4 1.3153; MR 64.43	728, 729
$(CH_2:CHCH_2)_4Sn$	$SnCl_4$ + $CH_2:CHCH_2MgBr$ in C_6H_6	80	b. 69-70°/1.5mm; b. 87-88°/4mm; $n^{32}D$ 1.5324; d^{30}_5 1.243; decomp. >170° → infusible, insol. mat'l.(10,11)($CH_2:CHCH_2SnO_2H$ (?))	298, 553
$(PhC:C)_4Sn$	$SnCl_4$ + $PhC:CMgBr$ in $C_4H_8O-C_6H_6$		m. 174° (decompn.); reacts with moist air	664

(7) Rxns. with $SnCl_4$ (1:3, 1:1, and 3:1) → $CH_2:CHSnCl_3$, $(CH_2:CH)_2SnCl_2$, and $(CH_2:CH)_3SnCl$, respectively (460, 487).
(8) Rxn. with I_2 (1:1) in Et_2O → $(CH_2:CH)_3SnI$ (488).
(9) Rxn. with $HgCl_2$ in Et_2O → no rxn. (489).
(10) On heating at 160° in the presence of $[Me_2C(CN)N]_2$ a white infusible substance is formed (298).
(11) Rxns. with EtOH or EtOH-HCl → C_3H_6 and SnO_2 (298). Rxn. with HCOOH → an infusible solid (298). Rxn. with $SnBr_4$ (1:1) → $(CH_2:CHCH_2)_2SnBr_2$ (553). Br_2 and I_2 → mixtures of halogenated prods. (553).

TETRA-α-THIENYLTIN $(\alpha\text{-}C_4H_3S)_4Sn$
Use: Corrosion inhibitor in dielectrics (122).

ASYMMETRICAL TETRAORGANOTIN COMPOUNDS

NON-FUNCTIONAL TETRAORGANOTIN COMPOUNDS

Tetraorganotin compounds containing two different pairs of organic groups, $R_2SnR'_2$ are prepared:

1. By reacting diorganotin dihalides, R_2SnX_2, with organomagnesium halides, $R'MgX$, in a molar ratio of 1:2.
2. By reacting diorganotin dihalides, R_2SnX_2, with organolithium compounds, $R'Li$, in a molar ratio of 1:2.
3. By metathetic reaction of a homogeneous tetraorganotin compound, R_4Sn, with a different, homogeneous tetraorganotin compound, R'_4Sn, at a molar ratio of 1:1, at elevated temperatures under reduced pressure in the presence of aluminum chloride.
4. By reacting diorganotin dihydrides with olefins or with diazoalkanes in the presence or absence of copper.
5. By reacting homogeneous diorgano compounds of bivalent tin with diorganomercury derivatives.

Compounds of this group are compiled in Table I.

Asymmetrical tetraorganotin compounds containing three identical groups, R_3SnR', are prepared:

1. By reacting triorganotin halides, R_3SnX, with organomagnesium halides, $R'MgX$, or with organolithium or organosodium compounds, $R'Li$ or $R'Na$.
2. By reacting lithium triorganostannides, R_3SnLi, with organic halides, $R'X$.
3. By refluxing lithium triorganostannides, R_3SnLi, with organolithium compounds, $R'Li$, in Et_2O.
4. By treating organotin trihalides, $RSnX_3$, with organomagnesium halides, $R'MgX$, organolithium, $R'Li$, or organosodium, $R'Na$, compounds.
5. By reacting triorganotin hydrides with olefins or with diazoalkanes in the presence of copper.
6. By treating hexaorganoditin compounds, $(R_3Sn)_2$, with organomercury halides, $R'HgCl$, or diorganomercury compounds, R'_2Hg, at elevated temperature.
7. By reacting triorganotin hydrides, R_3SnH, with organolithium compounds, $R'Li$.
8. As by-products in the reactions of two different, symmetrical tetraorganotin compounds, R_4Sn and R'_4Sn.

Compounds of this group are compiled in Table II.

TABLE I

$R_2SnR'_2$	Prepared from	Yield %	Properties	Ref.
Me_2SnEt_2	$Et_2SnH_2 + CH_2N_2$ (Cu)		b. 144–147°; d^{19}_{19} 1.2319	321
	$Et_2SnCl_2 + MeMgI$	63	b. 131–132°; n^{19}_D 1.4650	361
			Thermodynamic data (1)	113
$Me_2Sn(CHMe_2)_2$	$Me_2SnCl_2 + Me_2CHMgBr$ (3:4:10) in C_4H_8O	76.6	b. 68°/29mm; n^{25}_D 1.4621; d^{25} 1.161 (2)	492
Me_2SnBu_2	$Bu_2SnCl_2 + MeMgBr$ (3.12:10) in C_4H_8O	90	b. 70°/44mm; n^{25}_D 1.4640; d^{25} 1.124	492
$Me_2Sn(CHMeEt)_2$	$Me_2SnCl_2 + EtMeCHMg-Br$(3.4:10) in C_4H_8O	86.5	b. 68°/5.5 mm; n^{25}_D 1.4738; d^{25} 1.143 (2)	492
$Me_2Sn(CMe_3)_2$	$Me_2SnCl_2 + Me_3CMgCl$ in Et_2O		b. 84.5–85°/40mm; n^{25}_D 1.4662; d^{25} 1.1043; MR 4.87	564
$Me_2Sn(cyclo-C_5H_9)_2$	$Me_2SnCl_2 + $ cyclo-C_5H_9MgBr	51.2	b. 76–77°/6.3mm; n^{25}_D 1.5109; d^{25} 1.231 (2)	492
$Me_2Sn(CMe_2Et)_2$	$Me_2SnCl_2 + EtMe_2CMgCl$		b. 119.5–120°/29mm; n^{25}_D 1.4870; d^{25} 1.29; MR 4.52	564
$Me_2Sn(cyclo-C_6H_{11})_2$	$Me_2SnCl_2 + C_6H_{11}MgCl$	82	b. 100–102°/0.45–0.6mm; (2)	492
	$Me_2SnCl_2 + C_6H_{11}MgBr$	69	n^{25}_D 1.5184; d^{25} 1.298 (2)	492
$Me_2Sn(n-C_8H_{17})_2$	$Me_2SnCl_2 + n-C_8H_{17}MgCl$	72	b. 121–122°/0.2mm; n^{25}_D 1.4659	460a
$Me_2Sn(CH_2CHEtBu)_2$	$Me_2SnCl_2 + BuEtCHCH_2MgCl$	73	b. 101–102°/0.2mm; n^{25}_D 1.4715	460a
Et_2SnBu_2	$Et_2SnCl_2 + BuMgBr$		b. 112°/10mm; n^{20}_D 1.4734; d^{20} 1.1035; thermodynamic data	113
$Et_2Sn(CH_2CHMe_2)_2$	$Et_4Sn + (Me_2CHCH_2)_4Sn$	90		241
$Et_2Sn(n-C_5H_{11})_2$	$Et_2SnCl_2 + n-C_5H_{11}MgBr$	75	b. 139–141°/14mm (3)	269
Et_2SnPh_2	$Et_2Sn + Ph_2Hg$ at 150°	60	b. 154–156°/4mm	283, 387
$Pr_2Sn(CH_2CH_2Ph)_2$	$Ph_2SnCl_2 + EtMgBr$	31	b. 135.5–137°/1mm (4,5,6)	319
	$Pr_2SnH_2 + PhCH:CH_2$ at 80°	82	b. 172°/0.015mm	273a
Pr_2SnPh_2	$Ph_2SnI_2 + PrMgBr$		b. 160–161°/3mm (4)	319

i-Pr$_2$SnBu$_2$	Bu$_2$SnCl$_2$ + i-PrMgBr (3.12:10) in C$_4$H$_8$O	88.1	b. 102°/2.9mm; n^{25}D 1.4756; d$_{25}$ 1.074	492
Bu$_2$Sn(CMe$_3$)$_2$	Bu$_2$SnCl$_2$ + Me$_3$CMgCl		b. 123-125°/40mm; n^{25}D 1.4809; d$_{25}$ 1.0527; MR 4.89	564
Bu$_2$Sn(cyclo-C$_5$H$_9$)$_2$	Bu$_2$SnCl$_2$ + cyclo-C$_5$H$_9$MgBr	83.5	b. 128°/0.3mm; n^{25}D 1.5067; d$_{25}$ 1.127	492
Bu$_2$Sn(CEtMe$_2$)$_2$	Bu$_2$SnCl$_2$ + EtMe$_2$CMgCl		Waxy, white solid	564
Bu$_2$Sn(cyclo-C$_6$H$_{11}$)$_2$	Bu$_2$SnCl$_2$ + cyclo-C$_6$H$_{11}$MgBr Bu$_4$Sn + Ph$_4$Sn at 190°/2.5mm	71.5	b. 143°/0.45mm; n^{25}D 1.5126; d$_{25}$ 1.119 (8) (9,10)	492 240, 241
Bu$_2$SnPh$_2$	Bu$_2$SnCl$_2$ + PhMgBr Bu$_2$Sn(OMe)$_2$ + indene	74	b. 137°/0.2mm; n^{23}D 1.5605 Liquid (11)	248 349

(1) Rxns.: with HCl gas → EtMeSnCl$_2$ (321); with HgCl$_2$ in EtOH → Et$_2$SnCl (361); with Br$_2$ at -40° → MeEt$_2$SnBr (487).
(2) Rxn. with I$_2$ (1:1) in C$_6$H$_6$ → R$_2$R'SnI and RR'$_2$SnI (492).
(3) Rxn. with I$_2$ in Et$_2$O → Et(n-C$_5$H$_{11}$)$_2$SnI (269).
(4) Rxn. with AgNO$_3$ in EtOH → (PhAg)$_2$·AgNO$_3$, yellow ppt. (319).
(5) Rxn. with CCl$_4$ or CHCl$_3$ under irradiation (Hg lamp) → cleavage of the Ph group (434).
(6) Rxn. with MeOH the Hg light → Sn, CH$_2$O, Et$_4$Sn, Ph$_2$SnEt$_2$, and Et$_2$SnO (?) (434).
(7) Rxn. with I$_2$ (1:1) in C$_6$H$_6$ → Bu$_2$(cyclo-C$_5$H$_9$)SnI and an unidentified mixt. (492).
(8) Rxn. with I$_2$ (1:1) → Bu$_2$(cyclo-C$_6$H$_{11}$)SnI and (cyclo-C$_6$H$_{11}$)$_2$BuSnI (492).
(9) Rxn. with Br$_2$ (1:1) in CCl$_4$ at 0° → Bu$_2$PhSnBr (460a).
(10) Use: Catalyst for the prepn. of polyesters (67) and for the ester interchange between α,ω-glycols and alkylurethanes of aromatic m-or p-diamines (629).
Stabilizer: for chlorinated paraffins (832), for chlorinated esters of fatty acids (856), for vinyl halide polymers (752, 867), for halogenated org. compds. (855), and for cured rubber (851).
(11) Use: Stabilizer for vinyl halide polymers (349).

TABLE I (Cont'd.)

$R_2SnR'_2$	Prepared from	Yield %	Properties	Ref.
$(cyclo-C_5H_9)_2SnPh_2$	Ph_2SnCl_2 + $cyclo-C_5H_9MgBr$	81.7	m. 49-50.2° (12)	492
$(cyclo-C_5H_5)_2SnPh_2$	Ph_2SnCl_2 + C_5H_5MgCl	70	m. 105-106° (12)	182
$(cyclo-C_6H_{11})_2SnPh_2$	Ph_2SnCl_2 + $C_5H_{11}MgCl$	82	m. 118-120° (13)	492
	Ph_2SnCl_2 + $C_6H_{11}MgBr$	50.1	m. 119-120°	57
$Ph_2Sn(CH_2Ph)_2$	Ph_2SnCl_2 + $PhCH_2MgCl$	92	Yellow oil, d_{15}^{15} 1.271 (14,15,16,17)	434
$Ph_2Sn(1\text{-indenyl})_2$	Ph_2SnCl_2 + C_9H_7MgBr	29.9	m. 108-110°	182
	Ph_2SnCl_2 + C_9H_7Li	40	m. 116-117°	594
$Ph_2Sn(1\text{-naphthyl})_2$	$(1-C_{10}H_7)_2SnCl_2$ + $PhMgBr$	88	m. 209-210° (17)	57
$Ph_2Sn(9\text{-fluorenyl})_2$	Ph_2SnCl_2 + $C_{13}H_9Li$	60	Colorless rods, m. 179° (18)	594

(12) Rxn. with I_2 (1:1) in C_6H_6 → PhI and a mixt. of organotin compds. (492).
(13) Rxn. with HCl → $(C_6H_{11})_2SnCl_2$ (57).
(14) Rxns. with $CHCl_3$, CCl_4, or alc. HCl in UV light → $(PhCH_2)_2SnCl_2$ + C_6H_6 (434).
(15) Rxn. with $(CH_2CO)_2NH$ at 170° → mixt. of C_6H_6 and MePh (434).
(16) Rxn. with $(CH_2CO)_2NBr$ in $CHCl_3$ → PhBr and a residue which, after treatment with alc. HCl, yields $(PhCH_2)_2SnCl_2$ (434).
(17) Rxn. with HCl → Ph_2SnCl_2 (57).
(18) Rxn. with HCl in Et_2O → $(C_{13}H_9)_2PhSnCl$ (594).

TABLE II

R₃SnR'	Prepared from	Yield %	Properties	Ref.
Me₃SnEt	Me₃SnI + EtMgI	90.9	b. 106°/746mm; n^{20}_D 1.4527 Thermodynamic data (1,2,3)	359 113
Me₃SnBu	Me₃SnCl + BuMgBr	71	b. 150-158°/764mm	269
	Me₃SnI + BuMgBr	87	b. 149-150°/724mm; d^{20}_4 1.183; n^{20}_D 1.4560 (1)	358
Me₃SnCH₂CHMe₂	Me₃SnI + Me₂CHCH₂MgI		b. 140-141°/726mm; $d^{21.5}_4$ 1.1804; $n^{21.5}_D$ 1.4544 (1)	358
Me₃Sn(n-C₅H₁₁)	Me₃SnI + n-C₅H₁₁MgBr		b. 171-172°/721mm; d^{15}_4 1.1586; n^{20}_D 1.4559 (1)	358
Me₃Sn(1-C₅H₁₁)	Me₃SnI + 1-C₅H₁₁MgI		b. 162-164°/725mm; d^{21}_4 1.1306; n^{20}_D 1.4470	358
Me₃Sn(n-C₈H₁₇)	Me₃SnI + n-C₈H₁₇MgBr	98	b. 102-118°/12mm	269
	Me₃SnCl + n-C₈H₁₇MgCl	74	b. 56-58°/0.03mm; n^{25}_D 1.4587; d^{25}_4 1.0802 (1)	460a
Me₃Sn(n-C₁₀H₂₁)	Me₃SnCl + n-C₁₀H₂₁MgCl	73	b. 67°/0.05mm; n^{25}_D 1.4602 (4) d^{25}_4 1.0487	460a
Me₃Sn(n-C₁₂H₂₅)	Me₃SnCl + n-C₁₂H₂₅MgBr	15.5	b. 158-160°/14mm	269
	Me₃SnCl + n-C₁₂H₂₅MgCl	73	b. 93-98°/0.15mm; n^{25}_D 1.4610 (1,4) d^{25}_4 1.0285 (5)	460a
Me₃SnPh	Me₃SnBr + PhMgBr			740
Me₃SnCH₂Ph	Me₃SnBr + PhCH₂MgBr		b. 90°/9mm (6)	740

(1) In the rxns, with one mole iodine in EtOH Me₃SnR' lose one methyl group and form Me₂R'SnI (269, 358, 359).
(2) HCl in CHCl₃ cleaves one methyl group, and Me₂EtSnCl is formed (361).
(3) HgCl₂ in EtOH cleaves one methyl group, and Me₂EtSnCl is formed (361).
(4) Me₃SnR' compds. lose one methyl group in reactions with one mole Br in CCl₄ at 40° and yield Me₂R'SnBr (460a).
(5) One Mole of Br cleaves the phenyl group, and Me₃SnBr is formed (740).
(6) One mole of Br cleaves the benzyl group, and Me₃SnBr and PhCH₂Br are formed; the heat of rxn. was detd. (740).

TABLE II (Cont'd.)

R_3SnR'	Prepared from	Yield %	Properties	Ref.
Et_3SnMe	$Et_3SnI + MeMgI$		b. 159°/745mm; n_D^{20} 1.4656 (7) d_4^{20} 1.2160;	359
			Thermodynamic data (8)	113
Et_3SnPr	$Et_3SnI + PrMgBr$		b. 193°/748mm; n_D^{20} 1.4726 (8)	359
$Et_3SnCHMe_2$	$Et_3SnI + Me_2CHMgBr$		b. 192°-194°/719mm; d_4^{12} 1.1733; n_D^{19} 1.4772 (9)	360
Et_3SnBu	Et_3SnCl or Et_3SnI + $BuMgBr$	>78 76	b. 73-75°/4mm b. 99-101°/15mm; d_4^{20} 1.1457; n_D^{20} 1.4736 (10)	360 269
$Et_3Sn(n-C_6H_{13})$	$Et_3SnCl + n-C_6H_{13}MgBr$	63	b. 102-104°/0.75mm (10)	358 269
$Et_3Sn(n-C_8H_{17})$	$Et_3SnCl + n-C_8H_{17}MgBr$	70	b. 135-138° 3-4mm (10)	269
$Et_3Sn(n-C_{12}H_{25})$	$Et_3SnCl + n-C_{12}H_{25}MgBr$	33	b. 158-164° 1.1mm (10)	269
Et_3SnPh	$(Et_3Sn)_2 + PhHgCl$	30	b. 113-114°/6mm	283, 387
	$(Et_3Sn)_2 + Ph_2Hg$	40		283, 387
Pr_3SnMe	$PhSnCl_3 + EtMgBr$	25	b. 128.5°/12mm (11,12)	319
	$Pr_3SnH + CH_2N_2$ (Cu)		b. 94-96°/11mm; d_{23}^{23} 1.125	321
	$Pr_3SnCH_2CH_2CN + MeMgI$		b. 49-53°/1.1mm	681
Pr_3SnBu	$Pr_3SnI + BuMgBr$		b. 137-138°/37mm; d_4^{20} 1.0917; n_D^{20} 1.4741 (13,14)	359
Pr_3SnPh	$PhSnCl_3 + PrMgBr$	55	b. 150°/15mm (12)	319
$Pr_3SnCH_2CH_2Ph$	$Pr_3SnH + PhCH:CH_2$	86	b. 118-121°/0.007mm (13)	273a
Bu_3SnMe	$Bu_3SnH + CH_2N_2$ (Cu)		b. 122-124°/11mm; d_{20}^{20} 1.0901	321
$Bu_3SnCHMeEt$	$Bu_3SnI + EtMeCHMgBr$	78	b. 143-144°/3mm; d_4^{15} 0.9485; n_D^{15} 1.4796 (15)	360
$Bu_3Sn(i-C_5H_{11})$	$Bu_3SnI + i-C_5H_{11}MgBr$		b. 177-178°; d_4^{20} 1.0409; n_D^{20} 1.4715 (16)	359
$Bu_3Sn(n-C_6H_{13})$	$Bu_3SnI + n-C_6H_{13}MgBr$	78	b. 165°/7mm; d_{17}^{17} 1.035; n_D^{17} 1.4762 (17)	359
Bu_3SnPh	$Bu_3SnI + PhI$ $Bu_4Sn + Ph_4Sn$ at 190°/2.5mm	by-prod.	b. 139°/0.6mm; (n_D^{20})(18,19) 1.5155 b. 145°/2.5mm	360 172 240 241

Compound	Reaction	Yield (%)	Properties	Refs.
Bu₃Sn(indenyl)	Bu₃SnCl + 1-indenyl-Na		Oily prod. (20)	341, 706, 708
(cyclo-C₅H₅)₃SnPh	PhSnCl₃ + cyclopentadienyl-MgCl	40	m. 64-65°, decomp. within 4 hrs. forming a compd. m.₁₅ 200-220°	182
(1-C₅H₁₁)₃SnMe	(1-C₅H₁₁)₃SnI + MeMgI		b. 138-140°/4mm (21); d¹⁵₄ 1.0519; n¹⁵D 1.4700	359
(1-C₅H₁₁)₃Sn-(n-C₇H₁₅)	(1-C₅H₁₁)₃SnI + n-C₇H₁₅MgBr	78	b. 158-160°/3mm (22); d²⁰₄ 1.003; n²⁰D 1.4696	360
Ph₃SnMe	Ph₃SnH + MeLi	15	m. 60-61°	173
	Ph₃SnF + MeMgI	74	m. 60° (23)	409
	MeSnCl₃ + PhLi	82		409
Ph₃SnEt	Ph₃SnI + EtI	68.5	m. 56-58°	163, 167
	Ph₃SnCl + EtMgBr	70	m. 56-57° (24,25)	319
Ph₃SnPr	Ph₃SnCl + PrMgBr			319
Ph₃SnBu	BuSnCl₃ + PhNa	65	m. 74-74.5° (25,26)	239

(7) One mole of I splits off the methyl group, and Et₃SnI is formed (359).
(8) One mole of I splits off one ethyl group, and Et₂PrSnI is formed (359).
(9) Rxn. with one mole of I in xylene yields Et₃SnI and Et₂(Me₂CH)SnI (360).
(10) One mole of iodine splits off one ethyl group to form Et₂R'SnI (269, 358).
(11) Dipole moment (352).
(12) Rxn. with AgNO₃ in EtOH at 25° → (PhAg)₂AgNO₃, yellow ppt. (319).
(13) One mole of I or Br splits off one propyl group (359, 682).
(14) Rxn. with BiCl₃ at 120° → PrBuSnCl₂ (363).
(15) Rxn. with I₂ in xylene → Bu₃SnI (360).
(16) Rxn. with I₂ in Et₂O → Bu₂(1-C₅H₁₁)SnI (359).
(17) Rxn. with I₂ in xylene → Bu₂(n-C₆H₁₃)SnI (360).
(18) Rxn. with BuSnPh₃ (1:1) → Bu₂SnPh₂ (240).
(19) Useful as stabilizer for chlorinated paraffins (382).
(20) Useful as stabilizer for halogen-contg. resins (341, 706, 708).
(21) Rxn. with I₂ in xylene → (1-C₅H₁₁)₃SnI (359).
(22) Rxn. with I₂ in xylene → (1-C₅H₁₁)₂(n-C₇H₁₅)SnI (360).
(23) Rxn. with SnCl₄ (1:3) at 185-190° → MeSnCl₃ + PhSnCl₃ (409).
(24) Forms addition compd. with HBr, Ph₃SnEt.HBr; dipole moment (352).
(25) Rxn. with AgNO₃ in EtOH → deposition of Ag (319).
(26) Useful as stabilizer for vinyl resins (578).

TABLE II (Cont'd.)

R_3SnR'	Prepared from	Yield %	Properties	Ref.
Ph_3SnBu	$Bu_4Sn + Ph_4Sn$ at $190°/2.5mm$ in the presence of $AlCl_3$	by-prod.	(27, 28)	240, 241
Ph_3Sn (cyclopentadienyl)	$Ph_3SnCl + C_5H_5MgCl$	71.5	m. 130-131° (29)	182 460a
$Ph_3Sn(n-C_6H_{13})$	$Ph_3SnCl + n-C_6H_{13}MgCl$	83	m. 54° (30)	271, 276
$Ph_3Sn(n-C_8H_{17})$	$Ph_3SnH + 1$-octene	72	m. 54-55°	167
$Ph_3Sn(C_6H_4Me$-$o)$	$Ph_3SnLi + o$-MeC_6H_4I	25.3	(31)	584
$Ph_3Sn(C_6H_4Me$-$p)$				167
$Ph_3Sn(C_6H_3$-$2,4$-$Me_2)$	$Ph_3SnLi + 2,4$-$Me_2C_6H_3I$	42.7	m. 113-115°	167
	$Ph_3SnLi + 2,4$-$Me_2C_6H_3Li$	71.2		167
$Ph_3Sn(C_6H_3$-$2,5$-$Me_2)$	$Ph_3SnLi + 2,5$-$Me_2C_6H_3I$	43	m. 97-99°	167
	$Ph_3SnLi + 2,5$-$Me_2C_6H_3Li$			167
$Ph_3Sn(C_6H_3$-$2,6$-$Me_2)$	$Ph_3SnLi + 2,6$-$Me_2C_6H_3I$	34.1	m. 118-119°	167
	$Ph_3SnLi + 2,6$-$Me_2C_6H_3Li$			167
$Ph_3Sn(C_6H_2$-$2,4,6$-$Me_3)$	$Ph_3SnLi + 2,4,6$-$Me_3C_6H_2Br$	38.8	m. 157-158°	167
Ph_3SnCH_2Ph	$Ph_3SnLi + PhCH_2Cl$	72	m. 90-91°	163, 167
$Ph_3SnCH_2CH_2Ph$	$Ph_3SnH + PhCH:CH_2$ at 80°		m. 127-127.5° (32)	271, 274, 276
Ph_3SnCPh_3			m. 272-273° (decompn.) (33)	43
$Ph_3Sn(1$-indenyl)	$Ph_3SnCl + 1$-C_9H_7MgBr	45.5	m. 129-130°	182
	$Ph_3SnCl + 1$-C_9H_7Li	60	m. 128-129°	593, 594
$Ph_3Sn(1$-naphthyl)	1-$C_{10}H_7SnCl_3 + PhMgBr$	73.9	m. 125-125.5°	414
Ph_3Sn-(9-fluorenyl)	$Ph_3SnCl + 9$-$C_{13}H_9Li$	67.7	m. 129-130° (33)	593, 594
$(o$-$MeC_6H_4)_3SnCH_2Ph$	$(o$-$MeC_6H_4)_3SnLi + PhCH_2Cl$	2.3	m. 108-109°	170
	$(o$-$MeC_6H_4)_3SnCl + PhCH_2MgCl$	58.6		170
$(9$-Fluorenyl$)_3$-$SnPh$	$PhSnCl_3 + 9$-$C_{13}H_9Li$	50	m. 262° (decompn.)	594

(27) Rxn. with $BuSnPh_3$ (1:1) → Bu_2SnPh_2 (240, 241).
(28) Rxn. with Br_2 (1:1) in CCl_4 at 0° → $BuPh_2SnBr$ (460a).
(29) Rxns: with 95% EtOH → $(Ph_3Sn)_2O$; with $BuLi$ → Ph_4Sn; with Br_2 in CCl_4 → Ph_3SnBr; with (:CHCO)$_2O$ → 7-(Ph_3Sn)bicyclo[2,2,1]hept-5-ene-2,3-dicarboxylic anhydride; with (:CHCO$_2$Et)$_2$ → 7-(Ph_3Sn)bicyclo[2,2,1]hept-5-ene-2,3-dicarboxylic acid di-Et ester; with (:CCO$_2$Et)$_2$ → 7-(Ph_3Sn)bicyclo[2,2,1]hepta-2,5-diene-2,3-dicarboxylic acid di-Et ester (182).

(30) Rxn. with Br_2 (1:1) in CCl at 0° → $Ph_2(n-C_6H_{13})SnBr$ (460a).
(31) Rxn. with PhLi (1:2) → Ph_4Sn and $p-MeC_6H_4Li$ (584).
(32) Rxn. with I_2 (1:1) in C_6H_6 → $Ph_2(PhCH_2CH_2)SnI$ (682).
(33) Rxn. with HCl → $Ph_2R'SnCl$ (43, 594).

HALOGEN-, CYANO-, AND THIOCYANO-SUBSTITUTED COMPOUNDS

Halogen-substituted tetraorganotin compounds are formed:

1. By reacting organotin halides, containing one to three organic groups, at least one of which is halogen-substituted, with arylmagnesium halides.
2. By reacting organotin halides with halogen-substituted phenylmagnesium halides.
3. By reacting organotin halides with diazoalkanes.
4. By treating vinyltriorganotin compounds with chloroform or with symmetrical or asymmetrical carbon tetrahalides ($CBrCl_3$) in the presence of benzoyl peroxide at elevated temperature under pressure. Polyhalogenated compounds are formed.
5. By reacting organotin hydrides with halogenated olefins.
6. By reacting lithium triorganostannides with dihalobenzenes containing two different halogens.

The halogen-substituted compounds are compiled in Table I.

Cyano-Substituted tetraorganotin compounds are formed:

1. By reacting organotin hydrides with olefinic carbonitriles or with diazoalkyl carbonitriles.
2. By pyrolysis of triorganotin cyanoacetates.

Thiocyano-substituted tetraorganotin compounds are formed:

1. By treating halogen-substituted tetraorganotin compounds with sodium thiocyanate.

The cyano- and thiocyano-substituted compounds are compiled in Table II.

TABLE I

HALOGEN-SUBSTITUTED TETRAORGANOTIN COMPOUNDS

Compound	Prepared from	Yield %	Properties	Ref.
Me_3SnCH_2Cl	$Me_2(ClCH_2)SnCl + MeMgBr$	73	b. 44-48°/15mm; n^{25}_D 1.4860; d^{25}_{25} 1.556 (1,2,3)	486
Me_3SnCH_2Br	$Me_2(BrCH_2)SnBr + MeMgBr$	71	b. 46.2-50°/(1)mm; n^{25}_D 1.5070; d^{25}_{25} 1.722 (2)	486
Me_3SnCH_2I	$Me_3SnCH_2Cl + NaI$	78	b. 53-54.5°/6.5mm; n^{25}_D 1.5510; colorless, turns yellow in light	486
$Me_3SnC_6H_4Br$-p	$Me_3SnBr + p$-BrC_6H_4MgBr		b. 124°/15mm	90
	p-$BrC_6H_4SnBr_3 + MeMgI$	62.5	b. 129-130°/9mm; d_{15} 1.6489 (4)	583
Et_3SnCH_2Cl	$Et_3SnCl + N_2CH_2$		b. 66°/3mm; d^{20}_{20} 1.7917; n^{20}_D 1.5443	574
			Decomp.	574
$Et_3SnCHClMe$	$Et_3SnCl + MeCHN_2$		Decomp. -20°	491
$Et_3SnCH_2CH_2CCl_3$	$Et_3SnCH:CH_2 + CHCl_3$		b. 74°/0.25mm; n^{25}_D 1.5086	491
$Et_3SnCHClCH_2CCl_3$	$Et_3SnCH:CH_2 + CCl_4$		b. 100°/0.3mm; n^{25}_D 1.5230	491
$Et_3SnCHBrCH_2CCl_3$	$Et_3SnCH:CH_2 + CBrCl_3$		b. 115-119°/0.65-0.9mm; n^{25}_D 1.5425	491
$Et_3SnC_6H_4Br$-p	$Et_3SnBr + p$-BrC_6H_4MgBr	48.2	b. 150-151°/6mm; $(4)d_{15}$ 1.4964	583
			b. 165-50/14mm	90
$Et_3SnC_6H_4I$-p	p-$IC_6H_4SnBr_3 + EtMgBr$	16.8	b. 174-175°/10mm; d_{15} 1.5475	583
$Pr_3SnC_6H_4Br$-p	$Pr_3SnCl + p$-BrC_6H_4MgBr	36.3	b. 168°/4mm; d_{15} 1.3722	583
Bu_3SnCH_2Cl	$Bu_2(ClCH_2)SnCl + BuMgBr$	80	b. 108-112°/0.5mm; n^{25}_D 1.4801; d^{25}_{25} 1.135	486
$Ph_3SnC_6H_4F$-p	$Ph_3SnCl + p$-FC_6H_4MgBr	78.6	m. 171° (5)	583
$Ph_3SnC_6H_4Cl$-p	$Ph_3SnCl + p$-ClC_6H_4MgBr	88.2	m. 141° (6)	583
$Ph_3SnC_6H_4Br$-p	$Ph_3SnLi + p$-IC_6H_4Br	19.7		167
	$Ph_3SnCl + p$-IC_6H_4MgI		m. 133-135°	174
	$Ph_3SnCl + p$-BrC_6H_4MgI	68.1	m. 130° (7,8,9,10)	583
	$Ph_3SnCl + p$-BrC_6H_4MgBr	69.2		583
p-$BrC_6H_4SnBr_3 + PhMgBr$		32.1	m. 143° (11)	583
$Ph_3SnC_6H_4I$-p	p-$IC_6H_4SnBr_3 + PhMgBr$			583
$Et_2Sn(C_6H_2I)_2$	$(ClCH_2)_2SnCl_2 + EtMgBr$		b. 96°/2mm; d^{20}_{20} 1.414	575
$Et_2Sn(CHClMe)_2$	$Et_2SnCl_2 + MeCHN_2$		b. 114-115°/5mm; d^{20}_{20} 1.5083	574
$Bu_2Sn(CF_2CF_2H)_2$	$Bu_2SnH_2 + F_2C:CF_2$	28	b. 46-47°/0.2mm; n^{20}_D 1.5083	574, 299a
$(ClCH_2)_2Sn(CHClMe)_2$	$(ClCH_2)_2SnCl_2 + MeCHN_2$		b. 141-142°/5mm; d^{20}_{20} 1.675; n^{20}_D 1.5478	574, 575

(1) Rxn. with NaI in $Me_2CO \rightarrow Me_3SnCH_2I$ (486).
(2) Rxn. with NaSCN $\rightarrow Me_3SnCH_2SCN$ (486).
(3) Rxn. with Mg followed by $BF_3 \rightarrow (Me_3SnCH_2)_3B$ (772).
(4) Rxn. with Mg followed by $CO_2 \rightarrow Me_3SnC_6H_4CO_2H$ (90, 585).
(5) Rxn. with PhLi in $Et_2O \rightarrow Ph_4Sn + p-FC_6H_4Li$ (584).
(6) Rxn. with PhLi in $Et_2O \rightarrow Ph_4Sn + p-ClC_6H_4Li$ (584).
(7) Rxn. with Mg in $Et_2O-C_6H_6$ under reflux $\rightarrow Ph_4Sn$ and small amts. of $(p-Ph_3SnC_6H_4)_2$ (585).
(8) Rxn. with BuLi (1:1) $\rightarrow Ph_4Sn$ and in a ratio 1:2 $\rightarrow p-BrC_6H_4Li$ is formed (162, 584).
(9) Rxn. with PhLi in Et_2O under reflux $\rightarrow Ph_4Sn$ and $p-BrC_6H_4Li$ (584).
(10) Rxn. with $p-BrC_6H_4Li \rightarrow Ph_4Sn$ (584).
(11) Rxn. with PhLi $\rightarrow (Ph_3SnC_6H_4-)_2$, Ph_4Sn, and $p-IC_6H_4Li$ (584).

TABLE II

CYANO- AND THIOCYANO-SUBSTITUTED TETRAORGANOTIN COMPOUNDS

Compound	Prepared from	Yield %	Properties	Ref.
$Pr_3SnCH_2CH_2CN$	$Pr_3SnH + CH_2:CHCN$	70	b. 157-160°/12mm (1,2,3)	271
$Pr_3SnCH_2CH_2CH_2CN$	$Pr_3SnH + CH_2:CHCH_2CN$	53	b. 78-79°/0.001mm (4)	273a
$Pr_3SnCH(CN)CH_2Ph$	$Pr_3SnH + PhCH:CHCN$	83	b. 130-141°/0.0003mm (5)	273a
Bu_3SnCH_2CN	$Bu_3SnOCOCH_2CN$ (pyrolysis)	21	b. 140-144°/1.3mm (6,7)	270
	$Bu_3SnH + N_2CHCN$		b. 145-148°/1.4mm; n_D^{18} 1.4814	321
$Bu_3SnCH_2CH_2CN$	$Bu_3SnH + CH_2:CHCN$	70	b. 132-138°/0.2mm;	271
			b. 126-134°/0.2mm	274
Ph_3SnCH_2CN	$Ph_3SnOCOCH_2CH$ (pyrolysis)	49	m. 106-109° (8,9,10)	270
$Ph_3SnCH_2CH_2CN$	$Ph_3SnH + CH_2:CHCN$	85	m. 93-94 (11,12,13)	271, 274, 276
$Ph_3Sn(CH_2)_3CN$	$Ph_3SnH + CH_2:CHCH_2CN$	73	m. 80-81° (14,15,16)	271, 274, 276
$Ph_3SnCHMeCH_2CN$	$Ph_3SnH + MeCH:CHCN$	89	m. 103-104°	271, 274
$Pr_2Sn(CH_2CH_2CN)_2$	$Pr_2SnH_2 + CH_2:CHCN$	43	b. 113-117°/0.0004mm	273a
Me_3SnCH_2SCN	Mixt. of Me_3SnCH_2Cl and Me_3SnCH_2Br + NaSCN	69	b. 104-105°/4mm; n_D^{25} 1.5247; d_{25} 1.491	486

(1) Rxn. with $LiAlH_4 \to Pr_3Sn(CH_2)_3NH_2$ (681).
(2) Rxn. with MeMgI $\to Pr_3SnCH_2CH_2COMe$ along with Pr_3SnMe (681).
(3) Rxn. with PhMgBr $\to Pr_3SnCH_2CH_2COPh$ (681).
(4) Rxn. with Br_2 (1:1), followed by KOH and AcOH $\to Pr_2Sn(CH_2CH_2CH_2CN)OAc$ (682).
(5) Rxn. with Br_2 (1:1) $\to Pr_2SnBr$ and $PhCHBrCH_2CN$ (682).
(6) Rxn. with $LiAlH_4 \to Bu_3SnH$ (270).
(7) Rxn. with alk. $\to Bu_3SnOH$ (270).
(8) Rxn. with I_2 (1:1) in C_6H_6 at 60° $\to Ph_2Sn(CH_2CN)I$ (270).
(9) Rxn. with $LiAlH_4 \to Ph_3SnH$ (270).
(10) Rxn. with alk. $\to Ph_3SnOH$ (270).
(11) Rxn. with ethanolic NaOH $\to Ph_3SnCH_2CH_2CO_2H$ (273b).
(12) Rxn. with PhMgBr $\to Ph_3SnCH_2CH_2COPh$ (681).
(13) Rxn. with I_2 (1:1) $\to Ph_2Sn(CH_2CH_2CN)I$ (682).
(14) Rxn. with I_2 (1:1) $\to Ph_2Sn(CH_2)_3CO_2H$ (273b).
(15) Rxn. with ethanolic NaOH $\to Ph_3Sn(CH_2)_3CO_2H$ (273b).
(15) Rxn. with I_2 (1:1), followed by KF in EtOH-$H_2O \to Ph_2Sn(CH_2CH_2CH_2CN)F$ (682).
(16) Rxn. with Br_2 (1:1) $\to PhSn(CH_2CH_2CH_2CN)Br_2$ (682).

CARBOXY-, CARBALKOXY-, CARBAMYL-, AND DICARBOXYANHYDRO-SUBSTITUTED COMPOUNDS

Asymmetrical, carboxy-substituted tetraorganotin compounds are prepared:

1. By reacting tetraorganotin compounds containing the haloaryl groups with magnesium then treating the resulting Grignard compounds with carbon dioxide and hydrolyzing the product.
2. By oxidizing the hydroxyhydrocarbyl-substituted organotin compounds with potassium permanganate.
3. By hydrolyzing the carbalkoxyhydrocarbyl-substituted organotin compounds.
4. By the alkaline hydrolysis of cyanohydrocarbyl-substituted organotin compounds.

Carbalkoxyhydrocarbyl-substituted organotin compounds are prepared:

1. By reacting organotin hydrides with esters of unsaturated carboxylic acids or with esters of diazoalkylcarboxylic aicds.
2. By reacting organotin halides in the presence of sodium alkoxides with esters of substituted or unsubstituted malonic acid or with esters of β-ketocarboxylic acids.
3. By treating carboxyhydrocarbyl-substituted organotin compounds with diazometane.
4. By reacting organotin compounds containing the cyclopentadienyl group with maleic acid esters or with esters of acetylenedicarboxylic acid.

Carbamidohydrocarbyl-substituted organotin compounds are prepared by reacting organotin hydrides with amides of unsaturated acids.

Dicarboxyanhydrohydrocarbyl-substituted organotin compounds are prepared by reacting organotin compounds containing the cyclopentadienyl group with maleic anhydride.

CARBOXY-SUBSTITUTED ORGANOTIN COMPOUNDS

Compound	Prepared from	Yield %	Properties	Ref.
$Me_3SnC_6H_4CO_2H$-p	$Me_3SnC_6H_4Br$-p + Mg + CO_2		m. 131-132° (1)	90
$Et_3SnC_6H_4CO_2H$-p	$Et_3SnC_6H_4Br$-p + Mg + CO_2		m. 44.5-46.5 (2,3,4)	90
			m. 86°	585
$Pr_3SnCH_2CH_2CO_2Na$	$Pr_3SnCH_2CH_2CO_2Me$ + NaOH in 75% EtOH		Amorphous, hydroscop. solid, sol. in H_2O and org. solvents (5)	273b
$Bu_3SnCH_2CH_2CO_2H$	$Bu_3SnCH_2CH_2CO_2Me$ + NaOH in 75% EtOH, followed by HCl	68	Colorless oil; at 120°/0.2mm → $Bu_2SnCH_2CH_2COO$	273b
$Ph_3SnCH_2CH_2CO_2Na$	$Ph_3SnCH_2CH_2CN$ + alc. NaOH	63	Amorphous solid (6)	273b
	$Ph_3SnCH_2CH_2CO_2Et$ + NaOH	74		273b
$Ph_3SnCH_2CH_2CO_2Et$			Sternutatory effect	338
$Ph_3Sn(CH_2)_3CO_2H$	$Ph_3Sn(CH_2)_3CN$ + NaOH in aq. alc., followed by 1N HCl	88	Shiny needles, m. 130-131° (7)	273b
$Ph_3Sn(CH_2)_4CO_2H$	$Ph_3SnH + CH_2{:}CHCH_2CH_2CO_2Et$ and hydrolyzing the prod.	41	m. 100-102°; (8)	273b
$Ph_3SnC_6H_4CO_2H$-p	$Ph_3SnC_6H_4CH_2OH + KMnO_4$	44	m. 166-168°	34, 157
$Pr_2Sn(CH_2CH_2CO_2Na)_2$	$Pr_2Sn(CH_2CH_2CO_2Me)_2$ + NaOH in 75% EtOH		Amorphous, hydroscopic solid	273b

(1) Dissocn. const.: 1.05×10^{-6}; strength of the $Sn-C_{ar}$ bond (90, 91, 641).
(2) The compd. is decomp. by H_2O; its alkali salts are also instable (585).
(3) Dissocn. const.: 1.17×10^{-6}; strength of the $Sn-C_{ar}$ bond (90, 91, 641).
(4) Rxn. with Br_2 in Et_2O → Et_3SnBr + p-$BrC_6H_4CO_2H$ (585).
(5) Rxn. with Br_2 in $CHCl_3$ → $Pr_2SnCH_2CH_2COO$ (273b).
(6) Rxn. with 1N HCl → $Ph_2SnCH_2CH_2COO$ (273b).
(7) Rxn. with CH_2N_2 → $Ph_3Sn(CH_2)_3CO_2Me$ (273b).
(8) Rxn. with $MeCHN_2$ → $Ph_3Sn(CH_2)_4CO_2Et$ (273b).

Compound	Prepared from	Yield %	Properties	Ref.
$Pr_3SnCH_2CO_2Et$	$Pr_3SnH + N_2CHCO_2Et$	63	b. 136-140°/11mm; (b) 217-219° (1,2)	321
$Pr_3SnCH_2CH_2CO_2Me$	$Pr_3SnH + CH_2:CHCO_2Me$	40	b. 145-150°/12mm (3)	271
$Pr_3SnCHMeCH_2CO_2Et$	$Pr_3SnH + MeCH:CHCO_2Et$		b. 111-112°/0.6mm (4)	273a
$Pr_3Sn(CH_2)_3CO_2Et$	$Pr_3SnH + CH_2:CHCH_2CO_2Et$	30	b. 117-119°/0.7mm	273a
$Bu_3SnCH_2CO_2Et$	$Bu_3SnH + N_2CHCO_2Et$		b. 159-163°/10mm; decomp. 210°	321
$Bu_3SnCH_2CH_2CO_2Me$	$Bu_3SnH + CH_2:CHCO_2Me$	84	b. 140-141°/0.4mm; (5,6,7,8)	271
			b. 140-142°/0.4mm	274
$Bu_3SnCH(CO_2Bu)_2$	$Bu_3SnCl + CH_2(CO_2Bu)_2$		d^{20}_4 1.0964; n^{20}_D 1.4634 (9)	341, 706, 708
$Bu_3SnCEt(CO_2Et)_2$	$Bu_3SnCl + EtCH(CO_2Et)_2$		d^{20}_4 1.1170; n^{20}_D 1.4576 (9)	341, 706, 708
$Bu_3SnCH(Ac)CO_2-CHMeCH_2CHMe_2$	$Bu_3SnCl + AcCH_2CO_2CHMe-CH_2CHMe_2$		(9)	706, 708
$Ph_3SnCH_2CH_2CO_2Me$	$Ph_3SnH + CH_2:CHCO_2Me$	85	m. 46.5-47° (10,11,12)	271, 274, 276
$Ph_3Sn(CH_2)_3CO_2Me$	$Ph_3Sn(CH_2)_3CO_2H + CH_2N_2$		Colorless oil (13)	273b
$Ph_3SnCH_2CH(Me)-CO_2Me$	$Ph_3SnH + CH_2:CMeCO_2Me$	56	b. 173-176°/0.0002	273a
$Ph_3Sn(CH_2)_4CO_2Me$	$Ph_3Sn(CH_2)_4CO_2H + CH_2N_2$		Shiny platelets, m. 58-59° (13)	273b

(1) Rxn. with NaOH (1:1) in 75% EtOH → $(Pr_3Sn)_2O$ (273b).
(2) Rxn. with Br_2 (1:1) in $CHCl_3$ → $Pr_3SnBr + BrCH_2CO_2Et$ (682).
(3) Rxn. with NaOH in 75% EtOH → $Pr_3SnCH_2CH_2CO_2Na$ (273b).
(4) Rxn. with Br_2 (1:1) in $CHCl_3$ → $Pr_2Sn(CHMeCH_2CO_2Et)Br$ (682).
(5) Rxn. with NaOH in 75% EtOH → $Bu_3SnCH_2CH_2CO_2H$ (273b).
(6) Rxn. with $LiAlH_4$ in Et_2O → $Bu_3Sn(CH_2)_3OH$ (681).
(7) Rxn. with MeMgI (1:3) in Et_2O → $Bu_3SnCH_2CH_2C(OH)Me_2$ (681).
(8) Rxn. with Br_2 (1:1) in $CHCl_3$ → $Bu_2Sn(CH_2CH_2CO_2Me)Br$ (682).
(9) Useful as stabilizer for halogen contg. polymers (341, 706, 708).
(10) Rxn. with NaOH in aq. EtOH → $Ph_3SnCH_2CH_2CO_2Na$ (273b).
(11) Rxn. with I_2 (1:1) → $Ph_2Sn(CH_2CH_2CO_2Me)I$ (682).
(12) Rxn. with Br_2 (1:2) → $PhSn(CH_2CH_2CO_2Me)Br_2$ (632).
(13) Rxn. with $LiAlH_4$ → $Ph_3Sn(CH_2)_xIg$ (681).

CARBALKOXY-SUBSTITUTED ORGANOTIN COMPOUNDS (Cont'd.)

Compound	Prepared from	Yield %	Properties	Ref.
$Ph_3SnCHCH:CHCH-CH(CO_2Et)CHCO_2Et$	$Ph_3SnC_5H_5 + (:CHCO_2Et)_2$	49.9	m. 109.5-111°	182
$Ph_3SnCHCH:CHCH-C(CO_2Et):CCO_2Et$	$Ph_3SnC_5H_5 + (:CCO_2Et)_2$	44.5	m. 107-109°	182
$Pr_2Sn(CH_2CH_2-CO_2Me)_2$	$Pr_2SnH_2 + CH_2:CHCO_2Me$	60	b. 119-121°/0.001mm (14,15,16)	273a
$Bu_2Sn[CH(CO_2Et)_2]_2$	$Bu_2SnCl_2 + CH_2(CO_2Et)_2$		d_{20} 1.1610; n^{20}_D 1.4665 (9)	341, 706, 708
$Bu_2Sn[CH(CO_2Bu)_2]_2$	$Bu_2SnCl_2 + CH_2(CO_2Bu)_2$		d_{20} 1.1055; n^{20}_D 1.4541 (9,19)	341, 349, 706, 708
$Bu_2Sn[CH(CO_2CH_2CH-EtBu)_2]_2$	$Bu_2SnCl_2 + CH_2(CO_2CH_2CH-EtBu)_2$		d_{20} 1.0164; n^{20}_D 1.4557 (9)	706, 341 708
$Bu_2SnC(CH_2Ph)-(CO_2Et)_2$	$Bu_2SnCl_2 + PhCH_2CH(CO_2Et)_2$		(9)	706, 341 708
$Bu_2Sn[C(CO_2Et)_2-CH_2]_2$	$Bu_2SnCl_2 + CH_2CH(CO_2Et)_2]_2$		d_{20} 1.1844; n^{20}_D 1.4710 (9)	341, 706, 708
$Bu_2Sn[CH(CO_2Bu)Ac]_2$	$Bu_2SnCl_2 + AcCH_2CO_2Bu$		Liquid (9)	706, 708
$Bu_2Sn[CH(CO_2Et)Ac]_2$	$Bu_2SnCl_2 + AcCH_2CO_2Et$		(9)	706, 708
$Bu_2Sn[CH(CO_2Et)Bz]_2$	$Bu_2SnCl_2 + BzCH_2CO_2Et$		(9)	706, 708
$Bu_2Sn[C(CO_2Et)AcPr]_2$	$Bu_2SnCl_2 + PrCH(Ac)CO_2Et$		(9)	706, 708
$Bu_2Sn[C(Ac)_2CO_2Et]_2$	$Bu_2SnCl_2 + Ac_2CHCO_2Et$		(9)	706, 708
$Ph_2Sn(CH_2CH_2CO_2-Me)_2$	$Ph_2SnH_2 + CH_2:CHCO_2Me$	34	b. 191-194°/0.003	273a
$PrSn(CH_2CH_2CO_2Me)_3$	$PrSnH_3 + CH_2:CHCO_2Me$	11	b. 136-140°/0.0008mm (17)	273a
$BuSn(CH_2CH_2CO_2Me)_3$	$BuSnH_3 + CH_2:CHCO_2Me$	82	b. 139-141°/0.0004mm (18)	273a

(14) Rxn. with NaOH in 75% EtOH → $Pr_2Sn(CH_2CH_2CO_2Na)_2$ (273b).
(15) Rxn. with Br_2 (1:1) in $CHCl_3$ → $PrSn(CH_2CH_2CO_2Me)_2Br$ (682).
(16) Rxn. with Br_2 (1:2) in $CHCl_3$ → $(MeO_2CCH_2CH_2)_2SnBr_2$ (682).
(17) Rxn. with NaOH in EtOH → $PrSn(CH_2CH_2CO_2Na)_3$ (273b).
(18) Rxn. with Br_2 (1:1) in $CHCl_3$ → $(MeO_2CCH_2CH_2)_3SnBr$ (682).
(19) $Bu_2Sn[CH(CO_2Bu)_2]_2$ prepd. by reacting $Bu_2Sn(OMe)_2$ with $CH_2(CO_2Bu)_2$ at 140°, had d_{20} 1.133 and n^{20}_D 1.4650 (349).

CARBAMYL- AND DICARBOXYANHYDRO-SUBSTITUTED COMPOUNDS

Compound	Prepared from	Yield %	Properties	Ref.
$Pr_3SnCH_2CH_2CONH_2$	$Pr_3SnH + CH_2{:}CHCONH_2$	77	m. 44-47°; b. 155-161°/0.4mm	271
$Ph_3SnCH_2CH_2CONH_2$	$Ph_3SnH + CH_2{:}CHCONH_2$	90	m. 123-124°	271, 276
$Ph_3Sn\underline{CHCHCH{:}CH\underline{CHCHCH\underline{COOCO}}}$	$Ph_3SnC_5H_5 + ({:}CHCO)_2O$	59	m. 144-145°	182

COMPOUNDS CONTAINING THE CARBONYL GROUPS OR THEIR DERIVATIVES

The keto-substituted tetraorganotin compounds are prepared:

1. By reacting organotin hydrides with ketones containing a vinylic group or with diazoketones.
2. By treating the cyanoalkyl-substituted tetraorganotin compounds with Grignard compounds and hydrolyzing the resulting ketimines.
3. By reacting triorgano(alkoxy)- or diorganodi(alkoxy)tin compounds with conjugated diketones, or with enol acetates of cycloketones, or with methallyl acetate.
4. By reacting diorganotin dihalides with ketones or conjugated diketones in the presence of sodium alkoxides.

Acetals of triorganotin-substituted aldehydes are prepared by reacting triorganotin halides with aldehyde acetals containing a vinyl group.

COMPOUNDS CONTAINING CARBONYL GROUPS AND THEIR DERIVATIVES

Compound	Prepared from	Yield %	Properties	Ref.
Et_3SnCH_2COMe	$Et_3SnOMe + AcOCMe:CH_2$	95	b. 100.5-101°/6mm; n_D^{20} 1.4991; d_{20} 1.2875; IR and Raman spectra (1)	727
$Et_3Sn(2$-ketocyclohexyl$)$	$Et_3SnOMe + \overline{CH_2(CH_2)_3-CH:C}OAc$	72	b. 116-117°/4mm; n_D^{20} 1.4865; d_{20} 1.1983; IR and Raman spectra	727
Pr_3SnCH_2COMe	$Pr_3SnOMe + AcOCMe:CH_2$	78	b. 98-100°/1mm; n_D^{20} 1.4865; d_{20} 1.1983; IR and Raman spectra	727
$Pr_3SnCH_2CH_2COMe$	$Pr_3SnCH_2CH_2CN + MeMgI$	49	b. 84-89°/0.2mm (2)	727, 681
Pr_3SnCH_2COPh	$Pr_3SnH + N_2CHCOPh$		b. 155-160° (3)	321
$Pr_3SnCH_2CH_2COPh$	$Pr_3SnCH_2CH_2CN + PhMgI$		b. 158-165°/0.05-0.3mm (4)	681
Bu_3SnCH_2COMe	$Bu_3SnH + N_2CHCOMe$		b. 130-134°/14mm (5)	321
Bu_3SnCH_2COMe	$Bu_3SnOMe + AcOCMe:CH_2$	85	b. 130-132°/2mm; n_D^{20} 1.4842; d_{20} 1.12550; IR and Raman spectra (6)	727
Bu_3SnCH_2COPh	$Bu_3SnH + N_2CHCOPh$		Yellow liquid, b. 200-205°	321
$Bu_3Sn(2$-ketocyclohexyl$)$	$Bu_3SnOMe + \overline{CH_2(CH_2)_3-CH:C}OAc$	70	b. 155-156°/1mm; n_D^{20} 1.4805; d_{20} 1.1290	727
$Bu_3SnCHAc_2$	$Bu_3SnOMe + CH_2Ac_2$		Liquid (7)	341, 349
$Ph_3Sn(CH_2)_4COMe$	$Ph_3SnH + CH_2:CHCH_2CH_2COMe$	39	m. 70-71° (8,9)	273a
$Ph_3SnCH_2CH_2COPh$	$Ph_3SnCH_2CH_2CN + PhMgBr$	33	Yellow crystals, m. 79-80° (10)	681
$Ph_3SnCH_2CH_2CH(OEt)_2$	$Ph_3SnH + CH_2:CHCH(OEt)_2$	90	m. 35.5-37.5°; b. 168-178°/0.002-0.005mm (11)	271, 274, 276
$Bu_2Sn[CH(Ac)SO_2Ph]_2$	$Bu_2SnCl_2 + MeCOCH_2SO_2Ph$		m. 85-87° (7)	341, 706, 708
$Bu_2Sn[C(CH_2CHEtBu)Ac_2]_2$	$Bu_2SnCl_2 + BuEtCHCH_2CHAc_2$		Liquid (7)	341, 706, 708
$Bu_2Sn[C(CH_2Ph)Ac_2]_2$	$Bu_2SnCl_2 + PhCH_2CHAc_2$		Liquid (7)	341, 706, 708

$Bu_2Sn(CAc_2Bz)_2$	$Bu_2Sn(OMe)_2 + Ac_2CHBz$	Liquid, n^{20}_D 1.5859 (7) 349
$Bu_2Sn(CHAc_2)_2$	$Bu_2Sn(OMe)_2 + Ac_2CH_2$	n^{20}_D 1.4998; d_{20} 1.2120 (7) 349

(1) Rxn. with $H_2O \to (Et_3Sn)_2O$ (727).
(2) Derivs.: 2,4-dinitrophenylhydrazone, m. 48-50°; oxime, b. 110°/0.2mm; semicarbazone, m. 78-80° (681).
(3) Rxn. with $LiAlH_4 \to Pr_3SnH$ (321).
(4) Derivs.: 2,4-dinitrophenylhydrazone, orange-red plates, m. 106° (681).
(5) Rxn. with $LiAlH_4 \to Bu_3SnH$ (321).
(6) Semicarbazone, m. 80° (321).
(7) Useful as stabilizer for vinyl halide polymers (349, 706, 708).
(8) 2,4-Dinitrophenylhydrazone, m. 108-110° (273a).
(9) Rxn. with NaOH in 90% EtOH → no. rxn. (273b).
(10) 2,4-Dinitrophenylhydrazone, orange platelets, m. 174-176° (681).
(11) 2,4-Dinitrophenylhydrazone, golden platelets, m. 174-175° (271).

COMPOUNDS CONTAINING THE HYDROXY, ETHER, OR ESTER GROUPS

Hydroxy-substituted organotin compounds are prepared:

1. By reacting organotin halides with hydroxy- and hydroxyalkyl-substituted arylmagnesium halides.
2. By reducing triorganostannyl-substituted carboxylic acid esters with lithium aluminum hydride or with methylmagnesium halides.
3. By hydrolyzing esters of triorganostannyl-substituted alcohols.
4. By reacting hexaorganoditin compounds with bis(hydroxyphenyl)-mercury.
5. By reacting alkali metal triorganostannides with ethylene oxide and hydrolyzing the reaction product; 2-hydroxyethyl derivatives are formed.
6. By reacting triorganotin hydrides with alcohols containing a vinyl group.
7. By hydrolyzing tris(trialkylstannylmethyl)borine in 95% ethanol in the presence of hydrogen peroxide.

Organotin compounds containing the ether groups are prepared:

1. By reacting organotin halides with the Grignard derivatives of ethers.
2. By reacting alkali metal triorganostannides with halogen-substituted ethers.
3. By reacting organotin hydrides with ethers containing a vinyl group.
4. By reacting sodium salts of (hydroxyphenyl)triorganotin compounds with halocarboxylic acid esters.

Organotin compounds containing ester groups are prepared:

1. By reacting organotin hydrides with the unsaturated carboxylic acid esters.

COMPOUNDS CONTAINING THE HYDROXY GROUP

Compound	Prepared from	Yield %	Properties	Ref.
Me_3SnCH_2OH	$(Me_3SnCH_2)_3B + NaOH + H_2O_2$	13	b. 58-59°/6.5mm	772
$Et_3SnC_6H_4OH$-o	Et_3Sn halide + o-HOC_6H_4MgBr	54	b. 155-156°/15mm; d^{25}_4 1.3150; $n^{25}D$ 1.5379	34, 157
	$(Et_3Sn)_2 + (o$-$HOC_6H_4)_2Hg$		b. 197-200°/3mm; d^{25}_4 1.3229; $n^{25}D$ 1.5377 (1)	283, 387
$Bu_3Sn(CH_2)_3OH$	$Bu_3SnCH_2CH_2CO_2Me + LiAlH_4$	56	b. 117-119°/12mm	681
$Bu_3SnCH_2CH_2C(OH)Me_2$	$Bu_3SnCH_2CH_2CO_2Me + MeMgBr$	89	b. 133-138°/0.7mm	681
$Ph_3SnCH_2CH_2OH$	$Ph_3SnLi + \overline{CH_2CH_2O}$	44.8	m. 66-67°	170
	$Ph_3SnNa + \overline{CH_2CH_2O}$	59	m. 66-68°	157
	$Ph_3SnCH_2CH_2OAc + $ alc. KOH	25	m. 68-69°	273b
$Ph_3Sn(CH_2)_3OH$	$Ph_3SnH + CH_2$:CHCH$_2OH$	93	m. 105° (2)	271, 276
$Ph_3Sn(CH_2)_4CH$	$Ph_3(CH_2)_3CO_2Me + LiAlH_4$	34	m. 75-76°	681
$Ph_3SnCH_2CHMeCH_2OH$	$Ph_3SnH + CH_2$:CMeCH$_2OH$	41	m. 100-102°	273a
$Ph_3Sn(CH_2)_5OH$	$Ph_3Sn(CH_2)_4CO_2Me + LiAlH_4$	77	m. 64-66°	681
$Ph_3SnC_6H_4OH$-m	$Ph_3SnCl + m$-HOC_6H_4MgBr	49.5	m. 207-208° (3, 4)	185
$Ph_3SnC_6H_4OH$-o	$Ph_3SnCl + o$-HOC_6H_4MgBr	57	m. 176-179° (5)	157
	$Ph_3SnBr + o$-HOC_6H_4MgBr	10	m. 201-203°	34
$Ph_3SnC_6H_4OH$-p	Ph_3SnCl or $Ph_3SnBr + p$-HOC_6H_4MgBr	10	m. 201-203°	34, 157
	$Ph_3SnCl + o$-$HOCH_2C_6H_4MgBr$	64	m. 158-159° (6)	34, 157
	$Ph_3SnCl + p$-$HOCH_2C_6H_4MgBr$	66	m. 98-100° (7)	34, 157
$Ph_2Sn(C_6H_4OH$-m$)_2$	$Ph_2SnCl_2 + m$-HOC_6H_4MgBr	35	m. 201-203° (at 4°/min.) 189-190° (at 10°/min.)	185
$Ph_2Sn(C_6H_4OH$-o$)_2$	$Ph_2SnCl_2 + o$-HOC_6H_4MgBr	68	m. 136-138°	34
$Ph_2Sn(C_6H_4OH$-p$)_2$	$Ph_2SnCl_2 + p$-HOC_6H_4MgBr	68	m. 136-138°	157
$Ph_2Sn(C_6H_4C(OH)Ph_2)_2$	$(p$-$EtO_2CC_6H_4)_2SnCl_2 + PhMgCl$	72	m. 265-266°	123
$PhSn(C_6H_4OH$-m$)_3$	$PhSnCl_3 + m$-HOC_6H_4MgBr	40.5	m. 203-205°	185

(1) Rxn. with $HgCl_2 \rightarrow o$-HOC_6H_4HgCl (387).
(2) Rxn. with Br_2 in $CCl_4 \rightarrow Ph_2Sn(CH_2CH_2CH_2OH)Br$ (682).
(3) Rxn. with Na followed by $BrCH_2CO_2Et \rightarrow Ph_3SnC_6H_4OCH_2CO_2Et$ (185).
(4) Rxn. with aryl-$N_2BF_4 \rightarrow$ no azo-tin compd. was obtd. (185).
(5) Rxn. with alc. NaOH $\rightarrow$ cleavage of the Sn-C bond (157).
(6) Rxn. with $KMnO_4 \rightarrow Ph_2SnC_6H_4COO$ (?) (157).
(7) Rxn. with $KMnO_4 \rightarrow Ph_3SnC_6H_4CO_2H$ (34, 157).

COMPOUNDS CONTAINING ETHER AND/OR ESTER GROUPS

Compound	Prepared from	Yield %	Properties	Ref.
$Ph_3SnC_6H_4OMe-o$	$Ph_3SnLi + o-MeOC_6H_4I$	24.3	m. 129-130°	167
$Ph_3SnC_6H_4OMe-p$	$Ph_3SnLi + p-MeOC_6H_4Br$	30	m. 151-152°	167
$Ph_3SnC_6H_4CH_2OMe-o$	$Ph_3SnBr + o-MeOCH_2C_6H_4-MgBr$	35	m. 94.5-95.5°	34
	$Ph_3SnCl + o-MeOCH_2C_6H_4-MgBr$	38		158
$Ph_3SnCH_2CH_2OPh$	$Ph_3SnH + CH_2:CHOPh$	89	m. 82.5-83°	271, 276
$Ph_3Sn(CH_2)_3O-CH_2CH_2CN$	$Ph_3SnH + CH_2:CHCH_2O-CH_2CH_2CN$	50	m. 59-61°	273a
$Ph_2Sn(C_6H_4OMe-p)_2$	$Ph_2SnCl_2 + p-MeOC_6H_4MgBr$	88.5	m. 125-126° (1)	57
$(1-C_{10}H_7)_2Sn-(C_6H_4OMe-p)_2$	$(1-C_{10}H_7)_2SnCl_2 + p-MeOC_6H_4MgBr$	68	(2)	57
$Ph_3SnC_6H_4OCH_2-CO_2Et$	$Ph_3SnC_6H_4ONa-m + BrCH_2CO_2Et$	32	m. 97-98°	185
$Ph_3SnCH_2CH_2OAc$	$Ph_3SnH + CH_2:CHOAc$	100	m. 45.5-46.5° (3)	271, 276
$Ph_3Sn(CH_2)_3OAc$	$Ph_3SnH + CH_2:CHCH_2OAc$	42	m. 65-66°	273a, 276

(1) Rxn. with HCl → Ph_2SnCl_2 (57).
(2) Rxn. with HCl → $(1-C_{10}H_7)_2SnCl_2$ (57).
(3) At 180°/0.001mm → Ph_4Sn, $CH_2:CH_2$, and Ph_3SnOAc (271).

AMINO- AND AMIDO-SUBSTITUTED COMPOUNDS

Amino- and amido-substituted tetraorganotin compounds are prepared:

1. By reacting organotin hydrides with amino- or amido-substituted α-olefins.
2. By reacting organotin halides with amino- or amido-arylmagnesium halides or with amino- or amido-aryllithium compounds.
3. By reacting lithium triorganostannides with aminoalkyl halides.
4. By reducing cyano-substituted tetraorganotin compounds with lithium aluminum hydride.
5. By reacting hexaorganoditin compounds with bis(aminoaryl)-mercury compounds.

AMINO- AND AMIDO-SUBSTITUTED COMPOUNDS

Compound	Prepared from	Yield %	Properties	Ref.
p-$Me_2NC_6H_4SnEt_3$	$(Et_3Sn)_2 + (p\text{-}Me_2NC_6H_4)_2Hg$		b. 172–173°/3mm; d^{20}_4 1.2425; n^{22}_D 1.5610 (1,2,3,)	283, 387
$H_2N(CH_2)_3SnPr_3$	$Pr_3SnCH_2CH_2CN + LiAlH_4$	56	m. 143–144°/12mm (4,5)	681
$Et_2N(CH_2)_3SnPh_3$	$Ph_3SnLi + Cl(CH_2)_3NEt_2$		Oily liquid (4,5)	174
$AcNH(CH_2)_3SnPh_3$	$Ph_3SnH + CH_2{:}CHCH_2NHAc$	35	m. 95–98°	273a
p-$AcNHC_6H_4CH_2CH_2SnPh_3$	$Ph_3SnH + CH_2{:}CHC_6H_4NHAc$	98	m. 154–155°	273a
o-$Me_2NC_6H_4SnPh_3$	$Ph_3SnI + o\text{-}Me_2NC_6H_4MgBr$ or $o\text{-}Me_2NC_6H_4Li$	64	m. 110–112°	34, 157
m-$Me_2NC_6H_4SnPh_3$	$Ph_3SnCl + m\text{-}Me_2NC_6H_4Li$	63.4	m. 90–91° (6)	164
p-$H_2NC_6H_4SnPh_3$	$Ph_3SnI + p\text{-}H_2NC_6H_4MgBr$		m. 167–169°	157
p-$Me_2NC_6H_4SnPh_3$	Ph_3Sn halide $+ p\text{-}Me_2NC_6H_4Li$	62	Crystals, m. 132–134° (7,8,9)	34, 157, 174
$(2\text{-}Me_2N\text{-}9\text{-}fluorenyl)SnPh_3$	$Ph_3SnCl + 2\text{-}Me_2N\text{-}9\text{-}fluorenyllithium$	60	m. 150–151°	593, 594
$(p\text{-}Me_2NC_6H_4)_2SnPh_2$	$Ph_2SnCl_2 + p\text{-}Me_2NC_6H_4Li$		(10,11)	174

(1) Rxn. with 5% HCl → Et_3SnCl + $PhNMe_2{\cdot}HCl$ (283).
(2) Rxn. with $HgCl_2$ → Et_3SnCl + $p\text{-}Me_2NC_6H_4HgCl$ (283, 387).
(3) Rxn. with Br_2 in $CHCl_3$ → Et_3SnBr + $p\text{-}Me_2NC_6H_4Br$ (283, 387).
(4) $Ph_3Sn(CH_2)_3NMe_3I$, colorless shiny plates, m. 173–175°, sol. in hot H_2O, EtOH, and $CHCl_3$ (174).
(5) $Ph_3Sn(CH_2)_3NMe_3SO_4Me$, white solid, m. 118–132° (decompn.), sol. in H_2O, EtOH, MeOH, and AcOH (174).
(6) Rxn. with aryl-N_2BF_4 → (6-arylazo-3-dimethylaminophenyl)triphenyltin (164).
(7) $Ph_3SnC_6H_4NMe_3I$, colorless solid, m. 167–169°, turns yellow on exposure to light (174).
(8) $Ph_3SnC_6H_4NMe_3SO_4Me$, white solid, m. 240–243°, slightly sol. in H_2O, sol. in org. solvents (174).
(9) Rxn. with aryl-N_2Cl → (3-aryldiazo-4-Me_2N-phenyl)triphenyltin (34, 157, 164).
(10) $Ph_2Sn(C_6H_4NMe_3I)_2$, tan solid, m. 164–168° (decompn.), sol. in $(CH_2OH)_2$ and MeOH (174).
(11) $Ph_2Sn(C_6H_4NMe_3SO_4Me)_2$, fine powder, decomp. 125° (174).

COMPOUNDS CONTAINING OLEFINIC SUBSTITUENTS

Tetraorganotin compounds containing olefinic groups are prepared:

1. By reacting organotin halides or bis(triorganotin) oxides with alkenyl- or alkenylarylmagnesium halides. Vinylmagnesium halides are prepared and used in tetrahydrofuran.
2. By reacting alkali metal triorganostannides with alkenyl halides.
3. By reacting organotin hydrides with substituted acetylenes.
4. By reacting organotin halides with alkenyllithium compounds.
5. By reacting alkenyltin halides with organomagnesium halides.

TETRAORGANOTIN COMPOUNDS CONTAINING OLEFINIC GROUPS

Compound	Prepared from	Yield %	Properties	Ref.
$Me_3SnCH:CH_2$	$Me_3SnCl + CH_2:CHMgBr$ in C_4H_8O under reflux	68	b. 99-100°; d_{25} 1.265; $n^{25}D$ 1.4544 (1)	487, 740 487
$Me_3SnCH_2CH:CH_2$	$Me_3SnBr + CH_2:CHCH_2MgBr$	47.6	b. 125.5°/642mm; $n^{20}D$ 1.4734 d_{20} 1.2549 b. 128-129°/767mm (2) d_{25} 1.248 Raman spectrum	412 487 121
$Et_3SnCH:CH_2$	$Et_3SnCl + CH_2:CHMgBr$		b. 174-175°; $n^{20}D$ 1.4780; d_{20} 1.2133	413
	$Et_3SnBr + CH_2:CHMgBr$	85	b. 60-62°/13mm (3, 4, 5) 1.198; $n^{25}D$ 1.4738 (6)	487
$Et_3SnCH_2CH:CH_2$	$Et_3SnCl + CH_2:CHCH_2MgCl$		b. 76-77°/10mm	256
$Et_3SnC_6H_4-$ $p-CMe:CH_2$	$Et_3SnCl + p-CH_2:CMeC_6H_4MgBr$	50	b. 129-130°/2mm; $d^{25}_{25,8}$ 2311; $n^{25}D$ 1.5441-1.5448 (7, 8)	34a; 688a
$Pr_3SnCH:CH_2$	$Pr_3SnCl + CH_2:CHMgBr$	87	b. 90°/8.2mm; $n^{25}D$ 1.4776; d_{25} 1.131 (9)	463
	$Pr_3SnH + CH:CCH_2OH$	34	b. 120-122°/0.006mm	273a
$Pr_3SnCH:CHCH_2OH$	$Pr_3SnH + CH:CCBu$	82	b. 75-80°/0.1mm	273a
$Pr_3SnCH:CHBu$	$Pr_3SnH + CH:CPh$	75	b. 119-120°/0.001mm (10)	273a
$Pr_3SnCH:CHPh$	$(Bu_3Sn)_2O + CH_2:CHMgBr$	85	b. 114°/3mm; $n^{25}D$ 1.4761; d_{25} 1.085	459
$Bu_3SnCH:CH_2$	$Bu_3SnCl + CH_2:CHMgBr$	84.5	b. 95°/1.5mm; $n^{25}D$ 1.4751; d_{25} 1.081; (11)	487
	$Bu_3SnCl + CH_2:CHCH_2MgBr$	83	b. 155°/17mm	256
$Ph_3SnCH:CH_2$	$Ph_3SnCl + CH_2:CHMgBr$	93	m. 45.2-45.4° (12, 13)	487
	$Ph_3SnCl + CH_2:CHMgCl$	87.5	m. 39-40°	459
$Ph_3SnCH_2CH:CH_2$	$Ph_3SnCl + CH_2:CHCH_2MgBr$		m. 73.5-74.5°; decomp. 304-310° m. 73-74°	178 298
$Ph_3SnCH:CHCH_2OH$	$Ph_3SnNa + CH_2:CHCH_2Br$	57	m. 73.8-74.2° (14, 15, 16)	273a
	$Ph_3SnH + CH:CCH_2OH$	51	m. 115-125°	273a
$Ph_3SnCH:CHPh$	$Ph_3SnH + CH:CPh$	94	m. 119-120° (17)	273a, 276
$Me_2EtSnCH:CH_2$	$Et_2MeSnBr + CH_2:CHMgBr$	77	b. 56-59°/26mm; d_{25} 1.222; $n^{25}D$ 1.4697	487

Me$_2$Sn(CH:CH$_2$)$_2$	Me$_2$SnCl$_2$ + CH$_2$:CHMgCl or CH$_2$:CHMgBr	79	b. 120-121°/760mm; n^{25}_D 1.4720 b. 118.5-120°/759mm; n^{20}_D 1.4701 (18,19)	459 487 256
Et$_2$Sn(CH$_2$CH:CH$_2$)$_2$	Et$_2$SnCl$_2$ + CH$_2$:CHCH$_2$Mg halides		b. 99-100°/17mm	
Bu$_2$Sn(CH:CH$_2$)$_2$	Bu$_2$SnCl$_2$ + CH$_2$:CHMgCl	89	b. 78-80°/2mm; n^{25}_D 1.4824	459

(1) One mole of Br cleaves the vinyl group (740).
(2) Rxn. with CH$_2$:CMeCO$_2$Me at 120°/6000 atm. in the presence of (Me$_3$CO)$_2$ yields a copolymer (688).
(3) Polymerization at 120°/6000 atm. in the presence of (Me$_3$CO)$_2$ yields a liquid, low mol. wt. prod. (688).
(4) Carboxylic acids, mercaptans, mercuric halides, and trihalides of As, Sb, and Bi cleave the vinyl group (488, 711).
(5) Rxn. with CHCl$_3$, CCl$_4$, CBrCl$_3$, or HSiCl$_3$ in the presence of Bz$_2$O$_2$ yield the following compds.: Et$_3$SnCH$_2$CH$_2$CCl$_3$, Et$_3$SnCHClCH$_2$CCl$_3$, Et$_3$SnCHBrCH$_2$CCl$_3$, and Et$_3$SnCH$_2$CH$_2$SiCl$_3$, respectively (491).
(6) Rxn. with HSCH$_2$CH$_2$OH → Et$_3$SnSCH$_2$CH$_2$OH + CH$_2$:CH$_2$ (368).
(7) Polymerization at 80-130°/6000 atm. in the presence of Me$_2$C(CN)N:2, Bz$_2$O$_2$, or (Me$_3$CO)$_2$ yields 11.5-52% of colorless, powdery polymers, sol. in org. solvents. If the monomer contains some (CH$_2$:CMeC$_6$H$_4$-)$_2$, the polymeric prod. is insol. (688a).
(8) Copolymerization with C$_6$H$_6$ → crumbly rubber (34a).
(9) Rxn. with RCO$_2$H → Pr$_3$SnOCOR (463).
(10) Rxn. with Br$_2$ (1:1) in CHCl$_3$ → Pr$_3$SnBr (682).
(11) Rxns. with halogens → Bu$_3$Sn halides (488).
(12) Rxns. with I$_2$ or Br$_2$ → Ph$_2$(CH$_2$:CH)Sn iodide or bromide (460a, 488).
(13) Rxn. with tetracyclone in PhBr → Ph$_4$Sn, Ph$_3$SnBr, PhC:CPhCHPhCHPhCO, and tetraphenylbenzene (760).
(14) Rxn. with I$_2$ in xylene → Ph$_3$SnI (298).
(15) Rxn. with alc. HCl → SnO$_2$, C$_3$H$_6$, and C$_6$H$_6$ (298).
(16) HCO$_2$H → HCO$_2$Sn(OH)$_3$, C$_3$H$_6$, and C$_6$H$_6$ (298).
(17) Rxn. with I$_2$ → Ph$_3$SnI (682).
(18) Rxns. with carboxylic acids → Me$_2$Sn dicarboxylates (463) and Me$_2$(CH$_2$:CH)Sn carboxylates (488).
(19) Rxns. with halogens or hydrogen halides → cleavage of one vinyl group (488).

TETRAORGANOTIN COMPOUNDS CONTAINING OLEFINIC GROUPS (Cont'd.)

Compound	Prepared from	Yield %	Properties	Ref.
$Bu_2Sn(CH:CH_2)_2$	$Bu_2SnCl_2 + CH_2:CHMgBr$	85	b. 54-55°/0.35mm; d_{25} 1.122 (20,21,22)	487 256
$Bu_2Sn(CH_2CH:CH_2)_2$	$Bu_2SnCl_2 + CH_2:CHCH_2Mg-Br$ (or Cl)		b. 145-146°/17mm (23)	460a
$Ph_2Sn(CH:CH_2)_2$	$Ph_2SnCl_2 + CH_2:CHMgCl$		b. 153-154°/5mm; n^{25}_D 1.5949 (24)	459
	$Ph_2SnCl_2 + CH_2:CHMgBr$		b. 121.5°/8.4mm; d_{25} 1.334	488
$Ph_2Sn(CH_2CH:CH_2)_2$	$Ph_2SnCl_2 + CH_2:CHCH_2MgBr$	70.3	b. 173-174°/5.5mm; d_{20} 1.6025	178
			b. 170-173°/5mm; n^{20} 1.608	298
	$Ph_2SnCl_2 + CH_2:CHCH_2MgCl$	60	b. 120-124°/0.005mm (25-28)	460a
$Ph_2Sn[(CH_2)_3-CH:CH_2]_2$	$Ph_2SnCl_2 + CH_2:CH(CH_2)_3Li$	57.2	b. 173-175°/1.5mm; n^{20}_D 1.5598	178
$BuSn(CH_2CH:CH_2)_3$	$BuSnCl_3 + CH_2:CHMgCl$	83	b. 116-119°/10mm (28); d^{25}_4 1.1335; n^{25}_D 1.5162;	460a
$BuSn(CH:CH_2)_3$	$BuSnCl_3 + CH_2:CHMgBr$	84	b. 77-78°/8.6mm (29); d_{25} 1.174; n^{25}_D 1.4851;	487
$n-C_6H_{13}Sn(CH:CH_2)_3$	$(CH_2:CH)_3SnCl + n-C_6H_{13}MgCl$	80	b. 57-61°/0.03mm (30); d^{25}_4 1.1266; n^{25}_D 1.4851;	460a
$n-C_8H_{17}Sn(CH:CH_2)_3$	$(CH_2:CH)_3SnCl + n-C_8H_{17}MgCl$	86	b. 90-93°/0.2mm; n^{25}_D 1.4819 (30)	460a
$n-C_{10}H_{21}Sn-(CH:CH_2)_3$	$(CH_2:CH)_3SnCl + n-C_{10}H_{21}MgCl$	83	b. 90-94°/0.05mm; d^{25}_4 1.0672 (30)	460a
$Ph_3Sn(CH:CH_2)_3$	$PhSnCl_3 + CH_2:CHMgBr$	38	b. 73-75°/0.45mm (31); d_{25} 1.282; n^{25}_D 1.5478	487

(20) Rxns. with halogens or hydrogen halides → cleavage of one or two vinyl groups (460, 488, 771).
(21) Rxn. with $HgCl_2$ in Et_2O → cleavage of one vinyl group (489, 771).
(22) Rxns. with organic acids, mercaptans or halides of As, Bi, and Sb → cleavage of the vinyl groups (350, 771).
(23) Rxn. with Br_2 → cleavage of the allyl group (460a).
(24) Rxn. with halogens or hydrogen halides → cleavage of one phenyl group and disproportionation to Ph_3Sn halides and $(CH_2:CH)_2Sn$ halides (488, 771).
(25) Above 160° decomp. to Ph_4Sn (298).

(26) Rxn. with alc. HCl → Ph$_2$SnO, C$_3$H$_6$, and C$_6$H$_6$ (298).
(27) Rxn. with HCO$_2$H → HCO$_2$Sn(OH)$_3$, C$_3$H$_6$, and C$_6$H$_6$ (298).
(28) Rxn. with Br$_2$ (1:1) → cleavage of one allyl group (460a).
(29) Rxn. with Hg halides → cleavage of one vinyl group (489).
(30) Rxns. with Br$_2$ (1:1) → cleavage of one vinyl group (460a).
(31) Rxn. with SnCl$_4$ (1:3) → Ph SnCl$_3$ (487).

HETEROCYCLE-SUBSTITUTED COMPOUNDS

Organotin compounds containing at least one heterocyclic group are prepared:

1. By reacting organotin halides with heterocyclyl- or heterocyclyl-alkyllithium compounds.
2. By reacting organotin hydrides with vinyl derivatives of heterocyclic compounds.
3. By reacting organotin halides with the Grignard derivatives of heterocyclic compounds.
4. By reacting heterocyclyltin halides with lithium or Grignard derivatives of aromatic or aliphatic compounds.
5. By reacting dialkoxydialkyltin compounds with butyraldoxime; probably bis(propyloxaziridine) derivative is formed.

HETEROCYCLE-SUBSTITUTED COMPOUNDS

Compound	Prepared from	Yield %	Properties	Ref.
Et_3SnCH_2- $C:CHCH:CHCH:N$	Et_3SnCl + pyridyl-2-CH_2Li	40	b. 120-121°/3-4mm; n^{20}_D 1.531 (1)	596
Et_3SnCH_2- $CHCH(Bu):NCH:CH$	Et_3SnCl + 2-butylpyridyl-4-methyllithium	64	b. 144-145°/3-4mm; n^{20}_D 1.5364	596
$Pr_3SnCH_2CH_2$- $C:CHCH:NCH:CH$	Pr_3SnH + 4-vinylpyridine		b. 121-125°/0.0009mm; (2,3)	
$Ph_3SnC:CHCH:CHCH:N$	Ph_3SnCl + 2-pyridyllithium	12	m. 178-179°	162
$Ph_3SnC:CHCH:CHN:CH$	Ph_3SnCl + 3-pyridyllithium	58	m. 220° (4)	162
$Ph_3SnCH_2CH_2$- $C:CHCH:NCH:CH$	Ph_3SnH + 4-vinylpyridine		m. 112-113° (5)	273a, 276
$Ph_3SnCH_2CH_2$- $NC_8H_4-C_6H_4$	Ph_3SnH + N-vinylcarbazol	93	m. 207°	273a
$Ph_3SnC:CHCH:CHO$	Ph_3SnCl + 2-furyllithium of 2-furyl-MgI	65	m. 158-159 (6)	162
$Ph_3SnCH_2CH_2$- $NCH_2CH_2CH_2CO$	Ph_3SnH + 1-vinylpyrrolidin-2-one	78	m. 74-76°	271
$Ph_3SnC:CHCH:CHS$			Useful as corrosion inhibiter for dielectrics	122
$Bu_2Sn(NCHOPr)_2$ (?)	$Bu_2Sn(OMe)_2$ + PrCH:NOH		Liquid, n^{20}_D 1.4998; d_{20} 1.2120 (7)	349
$(p-MeOC_6H_4)_2Sn$- $(C:CHCH:CHS)_2$	$(2-C_4H_3S)_2SnCl_2$ + $p-MeOC_6H_4MgBr$	18	m. 89-93° (7)	57
$(1-C_{10}H_7)_2Sn$- $(C:CHCH:CHS)_2$	$(1-C_{10}H_7)_2SnCl$ + $2-C_4H_3SMgI$	80	m. 145-146° (7)	57
$Ph_2Sn(C:CHCH:CHS)_2$	$(2-C_4H_3S)_2SnCl_2$ + PhLi	30	m. 202-210° (8)	57

(1) Rxn. with air → Et_3SnOH (596);
(2) Picrate, yellow needles, m. 147-148° (273a).
(3) Rxn. with Br_2 (1:1) → cleavage of one propyl group (682).
(4) Methiodide, m. 183-184° (162).
(5) Picrate, m. 149-150° (373a).
(6) Attempted rxn. with (:CHCO)$_2$O (182).
(7) Rxn. with HCl → cleavage of both thienyl groups (57).
(8) Rxn. with HCl → cleavage of the phenyl and thienyl groups (57).

COMPOUNDS CONTAINING ACETYLENIC GROUPS

Tetraorganotin compounds containing a triple bond are prepared by reacting organotin halides with alkynyl- or aralkynyl- lithium, -silver, or -magnesium halide compounds.

ACETYLENIC ORGANOTIN COMPOUNDS

Compound	Prepared from	Properties	Ref.
$Et_3SnC\!:\!CPh$	$Et_3SnCl + PhC\!:\!CMgBr$	b. 162°/15mm (1)	214
$Bu_3SnC\!:\!CPh$	$Bu_3SnCl + PhC\!:\!CLi$	b. 174°/4mm; sensitive to moisture (1)	666
$Ph_3SnC\!:\!CH$	$Ph_3SnBr + HC\!:\!CMgBr$	Crystals, m. 34° (1)	215
$Ph_3SnC\!:\!CMe$	$Ph_3SnBr + MeC\!:\!CMgBr$	m. 43° (1)	215
$Ph_3Sn\text{-}C\!:\!CCH(OEt)_2$	$Ph_3SnBr + (EtO)_2CHC\!:\!CAg$	Grayish ppt., m. 58-60°; IR spectrum (2-5)	251
$Ph_3SnC\!:\!CPh$	$Ph_3SnCl + PhC\!:\!CMgBr$	m. 62° (1)	214
$(p\text{-}MeC_6H_4)_3SnC\!:\!CPh$	$(p\text{-}MeC_6H_4)_3SnCl + PhC\!:\!CMgBr$	m. 106° (1)	214
$(p\text{-}ClC_6H_4)_3SnC\!:\!CPh$	$(p\text{-}ClC_6H_4)_3SnBr + PhC\!:\!CMgBr$	m. 132° (1)	214

(1) Rxn. with alkali yields $R_3SnOH + CH\!:\!CR$ (214, 215, 666).
(2) Rxn. with H_2 over Pd-C → reduction of the acetylenic bond (251).
(3) Rxn. with 16% aq. HCl → Ph_3SnCl (251).
(4) Rxn. with Br_2 in CCl_4 → reduction of the acetylenic bond (251).
(5) Rxn. with $KMnO_4$ → reduction and formation of Ph_3SnOH (251).

HETERO-POLYFUNCTIONALLY SUBSTITUTED COMPOUNDS

m-(CARBETHOXYMETHOXY)PHENYLTRIPHENYLTIN $m\text{-}(EtO_2CCH_2O)\text{-}C_6H_4SnPh_3$
Prepn.: By reacting $Ph_3SnC_6N_4ONa$ with $BrCH_2CO_2Et$; 32% yield (185).
Props.: m. 97-98° (185).

[3-(2'-CYANOETHOXY)PROPYL]TRIPHENYLTIN $Ph_3SnCH_2CH_2CH_2OCH_2CH_2CN$
Prepn.: By heating Ph_3SnH with $CH_2\!:\!CHCH_2OCH_2CH_2CN$ at 160°; 50% yield (273a).
Props.: m. 59-61° (273a).

(3,3-DIETHOXYPROPYNYL)TRIPHENYLTIN $(EtO)_2CHC\!:\!CSnPh_3$
Prepn.: By reacting Ph_3SnBr with $AgC\!:\!CCH(OEt)_2$ in Me_2CO (251).
Props.: Grayish ppt., m. 58-60°; IR spectrum (251).
Rxns. with: Br_2 in CCl_4 → redn. (251).
$KMnO_4$ → redn. and Ph_3SnOH (251).
H_2/Pd-C → redn. of the acetylenic bond (251).
16% aq. HCl → Ph_3SnCl (251).

(3-CHLORO-2-HYDROXYPROPYL)TRIPHENYLTIN $Ph_3SnCH_2CH(OH)CH_2Cl$
Prepn.: By reacting Ph_3SnLi with $CH_2ClCHCH_2O$; 27% yield (170).
Props.: m. 97-99° (170).

DIBUTYLSTANNYLENEBIS(CYANOACETIC ACID) DIMETHYL ESTER
$$Bu_2Sn(CHCNCO_2Me)_2$$
Prepn.: By reacting Bu_2SnCl_2 with $NCCH_2CO_2Me$ in MePh in the presence of NaOMe (341).
By heating $Bu_2Sn(OMe)_2$ with $NCCH_2CO_2Me$ (1:2) under N (349).
Props.: Yellow, viscous liquid (341, 349).
Use: Stabilizer for halogen-contg. resins (341).
Stabilizer for vinyl chloride polymer (349).

KETOCARBOXY-SUBSTITUTED ORGANOTIN COMPOUNDS

The following compounds were prepared by reacting organotin halides with ketocarboxylic esters in the presence of sodium or sodium alkoxides.

Compound	Prepared from	Properties	Ref.
$Bu_2Sn(CHAcCO_2Et)_2$	Bu_2SnCl_2 + $AcCH_2CO_2Et$	Liquid (1)	341
$Bu_2Sn(CHAcCO_2Bu)_2$	Bu_2SnCl_2 + $AcCH_2CO_2Bu$	Liquid, decomp. at temp. (1)	341
$Bu_2Sn(CHBzCO_2Et)_2$	Bu_2SnCl_2 + $BzCH_2CO_2Et$	(1)	341
$Bu_2Sn(CAc_2CO_2Et)_2$	Bu_2SnCl_2 + Ac_2CHCO_2Et	(1)	341
$Bu_2Sn(CBuAcCO_2Et)_2$	Bu_2SnCl_2 + $BuCHAcCO_2Et$	(1)	341
$Bu_3SnCHAcCO_2CHMeCH_2CHMe_2$	Bu_3SnCl + $AcCH_2CO_2CHMeCH_2CHMe_2$	Liquid (1)	341

(1) Useful as stabilizer for halogen-contg. resins (341).

3-ARYLAZO DERIVATIVES OF 4-DIMETHYLAMINOPHENYLTRIPHENYLTIN

The following compounds were prepared by coupling $4\text{-Me}_2\text{NC}_6\text{H}_4\text{SnPh}_3$ with substituted aryldiazonium chlorides (RN_2Cl):

Compound	RN_2Cl	Yield %	Properties	Ref.
$3\text{-}(p\text{-}O_2NC_6H_4N_2)\text{-}4\text{-Me}_2NC_6H_3SnPh_3$	$p\text{-}O_2NC_6H_4N_2Cl$	24	m. 187-189°	34, 157, 164
$3\text{-}(p\text{-}BrC_6H_4N_2)\text{-}4\text{-Me}_2NC_6H_3SnPh_3$	$p\text{-}BrC_6H_4N_2Cl$	16.8	m. 170-172°	164
$3\text{-}(p\text{-}ClC_6H_4N_2)\text{-}4\text{-Me}_2NC_6H_3SnPh_3$	$p\text{-}ClC_6H_4N_2Cl$	4.5	m. 162-165°	164

6-ARYLAZO DERIVATIVES OF 3-DIMETHYLAMINOPHENYLTRIPHENYLTIN

The following compounds were prepared by coupling $3\text{-Me}_2\text{NC}_6\text{H}_4\text{SnPh}_3$ with substituted aryldiazonium fluoroborates (RN_2BF_4):

Compound	RN_2BF_4	Yield %	Properties	Ref.
$6\text{-}(p\text{-}O_2NC_6H_4N_2)\text{-}3\text{-Me}_2NC_6H_3SnPh_3$	$p\text{-}O_2NC_6H_4N_2BF_4$	66.6	m. 205°	164
$6\text{-}(p\text{-}BrC_6H_4N_2)\text{-}3\text{-Me}_2NC_6H_3SnPh_3$	$p\text{-}BrC_6H_4N_2BF_4$	48.6	m. 199-200°	164
$6\text{-}(p\text{-}HO_2CC_6H_4N_2)\text{-}3\text{-Me}_2NC_6H_3SnPh_3$	$p\text{-}HO_2CC_6H_4N_2BF_4$	47.8	Decomp. 358°	164

For other hetero-polyfunctionally substituted organotin compounds see pp.: 102, 107, 108, 114, and 118.

2. ALKALI METAL ORGANOSTANNIDES

The compounds are prepared:

1. By reacting tin dihalides with organo-alkali metal compounds in a molar ratio of 1:3 or by combining diorganotin compounds with organo-alkali metal compounds in a molar ratio of 1:1.
2. By treating tetraorganotin compounds with sodium.
3. By reacting triorganotin hydrides with organolithium compounds.
4. By treating triorganotin halides or hydroxides with alkali metals in liquid ammonia.
5. By reacting diorganotin dihalides with sodium in a molar ratio of 1:4.

LITHIUM TRIBUTYLSTANNIDE $LiSnBu_3$

Prepn.: By reacting $SnCl_2$ with BuLi (1:3) in Et_2O (172).
Rxn. with: RI → Bu_3SnR (172).

LITHIUM TRIPHENYLSTANNIDE $LiSnPh_3$

Prepn.: By reacting powdered anhyd. $SnCl_2$ with PhLi (1:3) in Et_2O at $-10°$ (163, 184).
By reacting Ph_2Sn with PhLi (1:1) in $C_6H_6-Et_2O$; the prod. was isolated as monoetherate (571).
By reacting Ph_3SnH with MeLi in Et_2O (571).
Props.: $Ph_3SnLi \cdot Et_2O$, pale yellow powder (571); The compd. dissoc.: $Ph_3SnLi \rightleftharpoons Ph_2Sn + PhLi$, therefore it is considered to be a complex of bivalent Sn (28, 591).
Rxns. with: Br_2 → Ph_2SnBr_2 (28).
I_2 in Et_2O → Ph_4Sn and $(Ph_3Sn)_2$ (170).
I_2 in xylene under reflux → Ph_4Sn and Ph_3SnI (170).
Br_2 and fluorene → Ph_2Sn combines with Br_2 to form Ph_2SnBr_2 and PhLi metalates fluorene (591).
RX (X= halogen) → Ph_3SnR and Ph_4Sn (163, 167, 174, 591).
Ph_3SnBr (1:1) in Et_2O → $(Ph_3Sn)_2$ and some Ph_4Sn (571).
Ph halide → Ph_4Sn and Li halide (591).
Ph_3SiCl → $Ph_3SnSiPh_3$ (184).
OCH_2CHCH_2Cl → $Ph_3SnCH_2CH(OH)CH_2Cl$ and Ph_4Sn (170).
CH_2CH_2O → Ph_4Sn (5.2%) and $Ph_3SnCH_2CH_2OH$ (44.8%) (170).
CO_2 → Ph_4Sn and $(Ph_3Sn)_2$ (170).
Ph_2CO → Ph_4Sn and $(Ph_3Sn)_2$ (170).
Et_2CO_3 and with $ClCO_2Et$ → $(Ph_3Sn)_2$, Ph_4Sn, and CO (184).
PhCH:CHBz → 7.8% Ph_4Sn and unreacted starting matl. (170).
RLi → Ph_3SnR (167).
NH_4Br in liq. NH_3 → Ph_3SnH (571).
H_2O in Et_2O → $(Ph_3Sn)_2$ and Ph_2SnO (571).

LITHIUM TRI-o-TOLYLSTANNIDE $LiSn(C_6H_4-o-Me)_3$

Prepn.: By reacting $SnCl_2$ with $o-MeC_6H_4Li$ (1:3) in Et_2O at $-10°$ (170).
Rxn. with: $PhCH_2Cl$ → $[(o-MeC_6H_4)_3Sn]_2$, $(o-MeC_6H_4)_4Sn$, and $(o-MeC_6H_4)_3SnCH_2Ph$ (170).

LITHIUM TRI-p-TOLYLSTANNIDE $LiSn(C_6H_4-p-Me)_3$

Prepn.: By reacting $SnCl_2$ with $p-MeC_6H_4Li$ (1:3) at Et_2O at $-10°$ (170).
Rxn. with: $p-MeC_6H_4I$ in Et_2O → $(p-MeC_6H_4)_4Sn$ (170)

SODIUM TRIETHYLSTANNIDE $NaSnEt_3$

Prepn.: By reacting Et_3Sn halides or Et_3SnOH with 1 at. equiv. Na in liq. NH_3 and treating the Et_3Sn radical with Na in small portions (195).
By treating Et_4Sn with Na (195).

SODIUM TRIPHENYLSTANNIDE   NaSnPh$_3$
Prepn.: By adding Na to Ph$_3$SnCl (3:1) in liq. NH$_3$ (273a).
Rxns. with: CH$_2$:CHCH$_2$Br → Ph$_3$SnCH$_2$CH:CH$_2$ (273a).

SODIUM DIPHENYL(TRIPHENYLMETHYL)STANNIDE   NaSn(CPh$_3$)Ph$_2$
Prepn.: By adding Ph$_3$CNa to Ph$_2$Sn (1:1) in Et$_2$O; isolated as monoetherate (571).
Props.: Monoetherate, deep red crystals (571).

SODIUM DIETHYLSTANNIDE(2$^-$)   Na$_2$SnEt$_2$
Prepn.: By reacting Et$_2$SnBr$_2$ with Na (195).

SODIUM TETRAETHYLDISTANNIDE   (Et$_2$SnNa)$_2$
Prepn.: By reacting Et$_2$SnBr$_2$ with Et$_2$Sn and Na in liq. NH$_3$ (195).

3. ORGANOTIN HYDRIDES

Organotin hydrides R$_{3-n}$SnH$_{1+n}$ are formed by reducing organotin halides with lithium aluminum hydride, or with aluminum amalgam and water in ether, or with sodium, lithium, or ammonium bromide in liquid ammonia.

METHYLTIN HYDRIDE   MeSnH$_3$
Prepn.: By reducing MeSnCl$_3$ with LiAlH$_4$ (1:3) in Et$_2$O (126).
Props.: b. 0°(calcd.) (126). Microwave spectrum; dipole moment; mol. structure; and intramolecular distances (326).

ETHYLTIN HYDRIDE   EtSnH$_3$
Prepn.: By reducing EtSnCl$_3$ with LiAlH$_4$ in Et$_2$O at r.t. (113).
By-prod. in the prepn. of Et$_2$SnH$_2$ from Et$_2$SnCl$_2$ and LiAlH$_4$ (113).

VINYLTIN HYDRIDE   CH$_2$:CHSnH$_3$
Prepn.: By treating CH$_2$:CHSnCl$_3$ with LiAlH$_4$ (626a).
Props.: Very instable compd., IR spectrum (626a).
Rxn. with: CF$_3$CO$_2$H → cleavage of CH$_2$:CH$_2$ (626a).

PROPYLTIN HYDRIDE   PrSnH$_3$
Rxn. with: CH$_2$:CHR (1:3) → PrSn(CH$_2$CH$_2$R)$_3$ (273a).

BUTYLTIN HYDRIDE $BuSnH_3$
Prepn.: By reducing $BuSnCl_3$ with $LiAlH_4$ in Et_2O; 37% yield (272).
Props.: Colorless liquid, b. 99-101°, instable on storage (272).
Rxn. with: $CH_2:CHR$ (1:3) → $BuSn(CH_2CH_2R)$ (273a).

DIMETHYLTIN HYDRIDE Me_2SnH_2
Prepn.: By reducing Me_2SnCl_2 with $LiAlH_4$ (1:2) in Et_2O; 72% yield (126, 126a).
Props.: b. 33-40°; $n^{20}D$ 1.4480; d_{20} 1.4766; MR_D 36.17 (126).

DIETHYLTIN HYDRIDE Et_2SnH_2
Prepn.: By reducing Et_2SnCl_2 with $LiAlH_4$ in Et_2O at r.t. (113).
Rxn. with: CH_2N_2 in the presence of Cu → Et_2SnMe_2 (321).

DIPROPYLTIN HYDRIDE Pr_2SnH_2
Prepn.: By reducing Pr_2SnCl_2 with $LiAlH_4$ in Et_2O; 72% yield (272).
Props.: b. 39.0-40.5°/12mm (272); IR spectrum (365).
Rxns. with: $PhC:CH$ or p-divinylbenzene → Sn-contg. polymers (276).
$RCH:CH_2$ (1:3) → $Pr_2Sn(CH_2CH_2R)_2$ (273a).

DIBUTYLTIN HYDRIDE Bu_2SnH_2
Prepn.: By reducing Bu_2SnCl_2 with $LiAlH_4$ in Et_2O; 66% yield (272, 299a).
Props.: b. 75-76° (272); IR spectrum (365).
Rxn. with: $F_2C:CF_2$ at 90° → $Bu_2Sn(CF_2CF_2H)_2$ (299a).

DIPHENYLTIN HYDRIDE Ph_2SnH_2
Rxns. with: $CH_2:CHCO_2Me$ (1:2) → $Ph_2Sn(CH_2CH_2CO_2Me)_2$ (273a).
$CH_2:CHAc$ → $Ph_2SnCH_2:CHCHOHMe$ (692).
α,β-Unsatd. ketones and aldehydes → Ph_2Sn and α,β-unsatd. carbinols (692).

TRIMETHYLTIN HYDRIDE Me_3SnH
Prepn.: By reducing Me_3Sn halide with $LiAlH_4$ (1:1) in Et_2O (126).
Props.: b. 59°(calcd); thermodynamic data (126).
Biol. props.: Inhibition of the oxidative phosphorylation induced by various enzymes (3).

TRIETHYLTIN HYDRIDE Et_3SnH

Prepn.: By reducing Et_3SnCl or Et_3SnI with $LiAlH_4$ in Et_2O; 66% yield (25, 113, 272).
By reducing Et_3SnCl with Al-Hg and H_2O in Et_2O at 10°; 30% yield (275).
By-prod. in the redn. of Et_2SnCl_2 with $LiAlH_4$ (113).
Props.: b. 51.5-52°/20mm; n^{20}_D 1.4700; d_{20} 1.258 (113); b. 149°; d_{22} 1.259; n^{20}_D 1.4725 (25); b. 79-81°/92mm (272); Dipole moment (352).
Rxns. with: Inorg. halides → Et_3Sn halides (25).
Inorg. oxides → $(Et_3Sn)_2O$ (25).
Biol. props.: Inhibition of the oxidative, enzyme-induced phosphorylation (3) Toxicity (639).

TRIPROPYLTIN HYDRIDE Pr_3SnH

Prepn.: By reducing Pr_3SnCl with $LiAlH_4$ in Et_2O; 75% yield (272).
By reducing Pr_3SnCl with Al-Hg in Et_2O at 10°; 65% yield (275).
Props.: b. 79-81°/92mm (272); IR spectrum (365).
Rxns. with: $CH_2{:}CHR$ → $Pr_3SnCH_2CH_2R$ (271, 273a).
$CH_2{:}CHCO_2H$ → $Pr_3SnOCOCH{:}CH_2$ (271).
CH_2N_2 in the presence of a trace of Cu powder → Pr_3GeMe (321).
N_2CHCO_2Et in C_6H_6 → $Pr_3SnCH_2CO_2Et$ (321).
N_2CHBz in C_6H_6 → Pr_3SnCH_2Bz (321).
$RC{:}CH$ (1:1) → $Pr_3SnCH{:}CHR$ (273a).
$RC{:}CH$ (2:1) → $Pr_3SnCH_2CHRSnPr_3$ (273a).
Biol. props.: Inhibition of oxidative phosphorylation induced by various enzymes (3) Toxicity (639).

TRIISOPROPYLTIN HYDRIDE $i\text{-}Pr_3SnH$

Biol. props.: Inhibition of the oxidative, enzyme-induced phosphorylation (3).

TRIBUTYLTIN HYDRIDE Bu_3SnH

Prepn.: By reducing Bu_3SnCl with $LiAlH_4$ in Et_2O; 74% yield (272).
By reducing Bu_3SnCl with Al-Hg + H_2O in Et_2O at 10°; 60% yield (275).
Props.: b. 76-81°/0.7-0.9mm (272).
Rxns. with: $CH_2{:}CHR$ → $Bu_3SnCH_2CH_2R$ (271)
CH_2N_2 in the presence of Cu → Bu_3SnMe (321).
N_2CHCO_2Et → $Bu_3SnCH_2CO_2Et$ (321).
N_2CHAc → Bu_3SnCH_2Ac (321).
N_2CHBz → Bu_3SnCH_2Bz (321).
N_2CHCN → Bu_3SnCH_2CN and SnH_4 (321).
Biol. props.: Inhibition of oxidative phosphorylation induced by various enzymes (3).

TRIHEXYLTIN HYDRIDE $(n-C_6H_{13})_3SnH$
<u>Biol. props.</u>: Inhibition of the oxidative, enzyme-induced phosphorylation (3).

TRIOCTYLTIN HYDRIDE $(n-C_8H_{17})_3SnH$
<u>Biol. props.</u>: Inhibition of the oxidative, enzyme-induced phosphorylation (3).

TRIPHENYLTIN HYDRIDE Ph_3SnH
<u>Prepn.</u>: By reacting Ph_3SnCl with Na in liq. NH_3 and treating the resulting Na deriv. with NH_4Br; 42% yield (154).
By reducing Ph_3SnCl with Al-Hg and H_2O in Et_2O at 10°; 65% yield (275).
By reducing Ph_3SnCl or Ph_3SnBr with $LiAlH_4$ in Et_2O; up to 70% yield (178, 272, 571).
By reacting Ph_3SnBr with Li in NH_3; 35% yield (571).
By reacting Ph_3SnLi with NH_4Br in liq. NH_3 (571).
<u>Props.</u>: b. 164-165°/0.3mm (178); b. 168-172°/0.5mm; m. 26-28° (272); b. 155-157°/0.1mm (571); On standing disproportionates to Sn and Ph_4Sn (178); Dipole moment (352).
<u>Rxns. with</u>: HCl in $Et_2O \rightarrow Ph_3SnCl$ (571).
MeLi $\rightarrow Ph_3SnMe$, Ph_4Sn and LiH (173).
MeLi in $Et_2O \rightarrow CH_4$ and Ph_3SnLi (571).
Bz_2O_2 in C_6H_{14} in light $\rightarrow$ org. compd. m. 184-185° and Ph_3SnOBz (178).
Ph_3SiH and Bz_2O_2 in $C_6H_{14} \rightarrow Ph_4Sn$, Sn, Ph_3SnH and Ph_3SiH (178).
$Ca\overline{H}Fe(CO)_4 \rightarrow (Ph_3Sn)_2Fe(CO)_4$ (225).
$C_6H_{13}CH:CH_2$ in the presence of Bz_2O_2 or in UV light $\rightarrow Ph_4Sn$ but no $Ph_3SnC_8H_{17}$ (138).
$RC:CH$ (1:1) $\rightarrow Ph_3SnCH:CHR$ (273a, 276).
$RC:CH$ (2:1) $\rightarrow Ph_3SnCH_2CHRSnPh_3$ (273a, 276).
$HC:CH$ (2:1) $\rightarrow (Ph_3SnCH_2-)_2$ (273a).
$CH_2:CHR$ at 80°$\rightarrow Ph_3SnCH_2CH_2R$ (271, 273a, 273b, 274).
$CH_2:CHCO_2H \rightarrow Ph_2SnCH_2CH_2COO$ and C_6H_6 (271).
$CH_2:CHCH_2Br$ at 80° $\rightarrow Ph_3SnBr$ and C_3H_6 (271, 274).
$CH_2:CHCH_2CN$ (2:1) $\rightarrow Ph_3Sn(CH_2)_3CN$ (271).
$CH_2:CRCH_2X$ (X= Br or Cl) $\rightarrow Ph_3SnX$ and $CH_2:CRMe$ (732).
$CH_2:CHCH_2SH \rightarrow (Ph_3Sn)_2S$, $CH_2:CHMe$ and H_2S (732).
$CH_2:CHCOR$ (2:1) $\rightarrow (Ph_3Sn)_2$ and $CH_2:CHCHOHR$ (732).
$CH_2:CHCH_2CH_2COMe \rightarrow Ph_3Sn(CH_2)_4COMe$ (732).
Diolefins $\rightarrow$ the corresponding satd. bis(triphenylstannyl) compds. (273a).
$RBr \rightarrow Ph_3SnBr$ and RH (732, 760).
RNH_2 (R= $n-C_4H_9$ or $n-C_6H_{13}$) $\rightarrow (Ph_3Sn)_2$, RH, and NH_3 (732).
$PhNO_2$ (Ph_3SnH in excess) $\rightarrow PhNH_2$ (732).

4. ORGANOTIN HALIDES

TRIORGANOTIN HALIDES

HOMOGENEOUS, NON-FUNCTIONALLY SUBSTITUTED TRIORGANOTIN HALIDES

Homogeneous trihydrocarbyltin chlorides, bromides, and iodides are prepared:

1. By reacting homogeneous tetrahydrocarbyltin compounds with tin tetrahalides, other metal halides, or hydrogen halides in a molar ratio of 3:1, or with dihydrocarbyltin dihalides (1:1).
2. By reacting tin tetrahalides with hydrocarbylmercury halides.
3. By treating bis(trihydrocarbyltin) oxides or sulfides with hydrogen or metal halides.
4. By reacting trihydrocarbyltin hydroxides with hydrogen halides.
5. By cleaving hexahydrocarbylditin compounds with halogens or metal halides
6. By reacting trihydrocarbyltin hydrides with alkyl or inorganic halides.
7. By treating trihydrocarbyltin carboxylates or other oxygenated salts with silver halides.
8. By reacting tin or tin alloys with hydrocarbylmercury halides, hydrocarbyltin trihalides, or dihydrocarbyltin dihalides.
9. By reacting tin tetrahalides with hydrocarbyl-alkali metal compounds.
10. By reacting tin, tin oxide, or tin alloys with hydrocarbyl halides in the presence or absence of metallic catalysts.

Homogeneous trihydrocarbyltin fluorides are formed:

1. By treating bis(trihydrocarbyltin) oxides or sulfides with hydrofluoric acid or silver fluoride.
2. By treating trihydrocarbyltin chlorides, bromides, iodides, cyanides, cyanates, and isothiocyanates with silver or potassium fluorides.

TRIMETHYLTIN FLUORIDE Me_3SnF
__Props.:__ Soly. in H_2O and C_6H_6 (616); Magnetic susceptibility (260).

TRIMETHYLTIN CHLORIDE Me_3SnCl

Prepn.: By treating Me_4Sn with dry HCl in $CHCl_3$ under reflux; 70% yield (361).
By reacting Me_4Sn with $HgCl_2$ in EtOH; 70% yield (361).
By passing MeCl through a mixt. of SnO and Cu (10:1) at 300° (571).
By treating $Me_3SnOP(O)(OMe)Me$ with 5% HCl at r.t. or with Cl_2 or AgCl in $CHCl_3$ (30).
By-prod. in the rxn. of MeCl with Sn at elevated temp. in the presence or absence of Cu; Cu_3Sn may be used as the source of Sn (418).

Props.: m. 42°; b. 45-47°/10mm (30); m. 37-38° (361); m. 37.5-39.5° (517); Molecular structure and bond length (514); Raman and IR spectra (800); Conductance in non-aqueous solvents (788, 789).

Rxns. with: Alkali metals in $NH_3 \rightarrow (Me_3Sn)_2$ (63).
$LiAlH_4 \rightarrow Me_3SnH$ (127).
Me_3SnOH (1:1) $\rightarrow Me_3SnOH \cdot Me_3SnCl \cdot H_2O$ (195, 198).
Me_3SnOH (1:2) $\rightarrow (Me_3SnOH)_2 \cdot Me_3SnCl$ (198).
RMg halides $\rightarrow Me_3SnR$ (269, 487, 460a).
RSH in the presence of Na_2CO_3 or NaOH $\rightarrow Me_3SnSR$ (429).
ROONa $\rightarrow Me_3SnOOR$ (442).
$CH_2:CMeCO_2K \rightarrow Me_3SnOCOCMe:CH_2$ (304).
KF $\rightarrow Me_3SnF$ (633).
$NH_3 \rightarrow Me_3SnCl \cdot 2NH_3$, insol. ppt. (633).

Uses: Insecticide and fungicide (691).

TRIMETHYLTIN BROMIDE Me_3SnBr

Prepn.: By bromination of Me_4Sn in $CHCl_3$ at -40° (487).
By passing MeBr through a mixt. of SnO and Cu (10:1) at 300° (517).

Props.: m. 26-27°; b. 163-165° (517); Molecular structure and bond length (514); Magnetic susceptibility (260).

Rxns. with: RMgBr $\rightarrow Me_3SnR$ (90, 493).
$(Me_3Sn)_2O \rightarrow (Me_3Sn)_3OBr$ (199, 200).
$Ph_3PCH_2 \rightarrow Ph_3PCH_2SnMe_3 \; Me_3SnBr_2$ (663a).

TRIMETHYLTIN IODIDE Me_3SnI

Prepn.: By treating Me_4Sn with I_2 (1:1) (357).
By passing MeI through a mixt. of SnO and Cu at 350° (517).

Props.: b. 69°/15mm (357); m. -5 to -3°; b. 160-170° (517).
Raman and IR spectra (331); Molecular structure and bond length (514).

Rxns. with: $(MeO)_3P \rightarrow Me_3SnPO(OMe)_2$ (?) (30).
$(EtO)_2PONa \rightarrow (Me_3SnOH)_2 \cdot Me_3SnI$ (30).
$PhPO(OMe)_2 \rightarrow Me_3SnP(:O)Ph(OMe)$ (?) (30).
$Ph_2POR \rightarrow Me_3SnP(:O)Ph_2$ (?) (32).
$(Me_3Sn)_2O \rightarrow (Me_3Sn)_3OI$ (199, 200).
$Na_2S \rightarrow (Me_3Sn)_2S$ (204).
RMg halides $\rightarrow Me_3SnR$ (358, 359).
Polarographic redn. (765).

TRIETHYLTIN FLUORIDE Et_3SnF

Prepn.: By treating $(Et_3Sn)_2O$ with HF, AgF, Ph_2SiF_2, or Pr_3GeF (18, 19).
By treating Et_3SnX (X= Cl, Br, I, CN, NCS, NCO) with AgF (18).
By reacting Et_3SnSMe or $(Et_3Sn)_2S$ with AgF (18).
Props.: m. 302° (19); Soly. data (616).
Rxns. with: AcOH → Et_3SnOAc (?) (18).
Electrochem. redn. (103).

TRIETHYLTIN CHLORIDE Et_3SnCl

Prepn.: From $(Et_3Sn)_2O$ by reacting with: a) hydrochloric acid (18); b) $EtOBCl_2$ under reflux; 37% yield (19); c) $MeSiCl_3$ under reflux; 77% yield (19); d) $GeCl_4$, $SnCl_4$, n-$BuCl_3$, $SnCl_2$, Et_2SnCl_2, PCl_3, $AsCl_3$, or $SbCl_3$ under reflux; 74-95% yield (19).
From $(Et_3Sn)_2S$ by refluxing with: a) $GeCl_4$, $SnCl_4$, $SnCl_2$, PCl_3, or $AsCl_3$, 74-92% yield (19) and b) AgCl in CCl_4 or without solvent (18).
From Et_3SnSMe by refluxing with AgCl in CCl_4 or without solvent (18).
By reacting Et_3SnH with inorganic chlorides ($KAuCl_4$, $CdCl_2$, $HgCl_2$, $TiCl_4$, $PdCl_2$, K_2PtCl_6, $GeCl_4$, $SnCl_4$, $SnCl_2$, $AsCl_3$, S_2Cl_2, etc.) (25).
By reacting Et_3SnX (X= I, Br, CN, NCS) with AgCl under reflux (18).
From Et_4Sn by reacting with: a) $SnCl_4$ (3:1) under reflux (113); b) $HgCl_2$ in EtOH or $CHCl_3$ under reflux; the prod. contains Et_3SnCl (35% yield) and Et_2SnCl_2 (10% yield) (361); c) $FeCl_3$ in $CHCl_3$ under reflux, 68% yield (362); and d) $BiCl_3$ in $CHCl_3$ under reflux, 30% yield (363).
By treating $Et_3SnOP(O)(OEt)Et$ with AgCl at r.t. (30).
By heating $(Et_3Sn)_2$ with $HgCl_2$ at 150°; 74% yield (387).
By-prod. in the rxn. of $(Et_3Sn)_2$ with PhHgCl at 150-160° (284).
By treating Et_3SnOH with HCl soln. (743).
Props.: m. 15°, b. 210°, $n^{20}D$ 1.508 (19); b. 98-100°/12mm (30); m. 15°, b. 206°, d_{20} 1.429; $n^{20}D$ 1.5055 (113); m. 12-14°; b. 89-91°/12mm (268); b. 200-205°/735mm (363); Thermodynamic data (113); UV spectrum (439); Dipole moment (520, 521).
Rxns. with: AgX (X= F, AcO, NCO) → Et_3SnX (18).
Ag_2O → $(Et_3Sn)_2O$ (18).
$(:CMgBr)_2$ → $Et_3SnC:CSnEt_3$ (50).
Electrochemical redn. (103).
$LiAlH_4$ → Et_3SnH (113, 272).
RMgBr → Et_3SnR (269, 360, 413, 688a).
Et_3SnOH (1:2) → $(Et_3SnOH)_2 \cdot Et_3SnCl \cdot H_2O$ (195).
Aq. KOH → Et_3SnOH (268, 387).
$MeSO_2NHNa$ → Et_3SnNHO_2SMe (268).
Al-Hg + H_2O in Et_2O → Et_3SnH (275).
Na → $(Et_3Sn)_2$ (283, 387).
ROONa → Et_3SnOOR (442).
RLi → Et_3SnR (596).

TRIETHYLTIN CHLORIDE Et_3SnCl (Cont'd.)

Rxns. with (Cont'd.): KF → Et_3SnF (633).
Anhyd. NH_3 → $Et_3SnCl \cdot 2NH_3$, insol. ppt. (633).
NaOR → Et_3SnOR (707).
NaOH in EtOH followed by heating at 200-220° → $(Et_3Sn)_2O$ (731)
EtOH, solvolysis, conductivity data (743).
H_2O-dioxane, hydrolysis, conductivity data (743).
Polarographic redn. (765).
p-$(CH_2:CMe)C_6H_4Li$ → $Et_3SnC_6H_4CMe:CH_2$ (34a).
Biol. props.: Fungicidal effect (267).
Uses: Manuf. of antifouling marine paints (549).
Insecticide and fungicide (691).

TRIETHYLTIN BROMIDE Et_3SnBr

Prepn.: By reacting $(Et_3Sn)_2O$ with hydrobromic acid (18).
By reacting $(Et_3Sn)_2S$ or Et_3SnSMe with AgBr in xylene or without solvent (18).
By reacting Et_3SnI with AgBr under reflux (18).
By refluxing $(Et_3Sn)_2O$ or $(Et_3Sn)_2S$ with $SiBr_4$ or $SnBr_4$; 84-91% yield (19).
By reacting Et_3SnH with inorg. bromides (AgBr, $CuBr_2$) (25).
By reacting Et_4Sn with Br; 88% yield (157, 487).
By reacting Mg_2Sn with a very large excess EtBr in cyclohexane at 160°/3 hrs. under pressure; 67% Et_3SnBr and 19% Et_2SnBr_2; When petr. ether or isoöctane is used, Et_3SnBr is predominantly formed (266).
By heating Et_4Sn with PBr_3 (1:1) at 150° (764).
Props.: b. 221°; $n^{20}D$ 1.531 (19); b. 105-107°/15mm; $n^{20}D$ 1.5240 (157); b. 95-98°/13mm (764); Magnetic susceptibility (260).
Rxns. with: NaOH, followed by CO_2 → $(Et_3SnO)_2CO$ (585).
Ag salts → the corresponding Et_3Sn salts (18).
Ag_2O → $(Et_3Sn)_2O$ (18).
RMgBr → Et_3SnR (34, 90, 269, 583, 740).
$(:CMgBr)_2$ → $Et_3SnC:CSnEt_3$ (50).
NaC:CH → $Et_3SnC:CSnEt_3$ (50).
Electrochem. redn. (103).
Na in C_6H_6 under reflux in the presence of air → Et_2SnO (195).
Na in NH_3 → $(Et_3Sn)_2$ (195).
Et_3SnOH (1:2) → $(Et_3SnOH)_2Et_3SnBr \cdot H_2O$ (195).
Aq. KCN in Et_2O → Et_3SnCN (268).

TRIETHYLTIN IODIDE Et_3SnI

Prepn.: By reacting $(Et_3Sn)_2O$ with HI (18).
By reacting $(Et_3Sn)_2S$ or Et_3SnSMe with AgI in xylene or without solvent (18).
By refluxing $(Et_3Sn)_2O$ or $(Et_3Sn)_2S$ with $C_{12}H_{25}SiI_3$; up to 88% yield (19).
By reacting Et_4Sn with I_2 (195, 357).
By refluxing Et_3SnMe with I_2 in Et_2O; 91.8% yield (359).
By reacting Et_4Sn with white P and I_2 in CS_2 (764).
Props.: b. 234° (19, 195); b. 117-118°/15mm (357); b. 116-120°/13-14mm (764); b. 226-228°/748mm; $n^{18}D$ 1.5633 (359); $n^{20}D$ 1.566 (19).
Rxns. with: Ag salts → the corresponding Et_3Sn salts (18).
Ag carboxylates → the corresponding Et_3Sn esters (26).
Ag_2O → $(Et_3Sn)_2O$ (18).
$(EtO)_3P$ → $Et_3SnPO(OEt)_2$ (?) (30).
$NaOP(OEt)_2$ → Et_3SnOEt (?) (30).
$PhP(OEt)_2$ → $Et_3SnP(O)Ph(OEt)$ (?) (32).
Ph_2POEt → $Et_3SnP(O)Ph_2$ (?) (32).
$EtP(O)(OEt)_2$ → $Et_3SnOP(O)(OEt)Et$ (354).
RMgBr → Et_3SnR (358, 359, 360).
$LiAlH_4$ → Et_3SnH (25).
NaOH → Et_3SnOH (195).
Na_2S → $(Et_3Sn)_2S$ (204).
Et_3SnOH → $(Et_3SnOH)_2 \cdot Et_3SnI \cdot H_2O$ (195).
Electrochem. redn. (103).
Polarographic redn. (765).

TRIPROPYLTIN FLUORIDE Pr_3SnF

Props.: Soly. data (616).

TRIPROPYLTIN CHLORIDE Pr_3SnCl

Prepn.: By reacting Pr_4Sn with $SnCl_4$ (3:1) at 200°; 88% yield (463).
By reacting PrCl with NaSnZn at 140-145° under N in an autoclave; the prod. consists of a mixt. of Pr_4Sn and Pr_3SnCl (590).
Props.: IR spectrum (365).
Rxns. with: $LiAlH_4$ → Pr_3SnH (272).
Al-Hg + H_2O in Et_2O → Pr_3SnH (275).
$CH_2:CHMgBr$ in C_4H_8O → $Pr_3SnCH:CH_2$ (463).
RMgBr → Pr_3SnR (583).
Electrochem. redn. (103).
Use: Insecticide and fungicide (691).

TRIPROPYLTIN BROMIDE Pr_3SnBr

Prepn.: By bromination of Pr_4Sn (211).
By reacting $Pr_3SnCH_2CO_2Et$ or $Pr_3SnCH:CHPh$ with Br_2 (1:1) in $CHCl_3$ at $0°$ (682).
By-prod. in the rxn. of $PrBr$ with Mg_2Sn in xylene under reflux (741).
Props.: b. 252-255° (207); b. 126-127°/12mm (682); b. 251-256° (741).
Rxns. with: $MeCHN_2$ → instable liquid, b. 130-135°/3mm (160).
$NaOH$ in $EtOH$ → Pr_3SnOH (211).

TRIPROPYLTIN IODIDE Pr_3SnI

Prepn.: By reacting Pr_4Sn with I_2 (211, 357).
Props.: b. 259-264° (slight decompn.); b. 147-8°/20mm (211); b. 140-141°/4mm (357).
Rxns. with: Na_2S → $(Pr_3Sn)_2S$ (204).
RMg halide → Pr_3SnR (359).

TRIISOPROPYLTIN CHLORIDE $i-Pr_3SnCl$

Prepn.: By treating $i-Pr_3SnOH$ with concd. HCl (743).
By-prod. in the rxn. of $SnCl_4$ with $i-PrLi$ (1:3) (743)
Rxns. with: $EtOH$ or $PrOH$, solvolysis (743).
H_2O-dioxan, hydrolysis, conductivity data (743).

TRIISOPROPYLTIN BROMIDE $i-Pr_3SnBr$

Prepn.: By heating $i-Pr_4Sn$ with $SnBr_4$ (3:1) at 125-130° for 45 min. and subsequently at 200° for 2 hrs.; the prod. contains some dibromide (268); at 125-165° about 90% yield of the title compd. was obtd. (743).
Props.: b. 112-113.5°/12mm, a prod. contaminated with $i-Pr_2SnBr_2$ (268).
Rxn. with: Aq. KOH in Et_2O → $(i-Pr_3Sn)_2O$ (268).

TRIBUTYLTIN CHLORIDE Bu_3SnCl

Prepn.: By reacting $SnCl_4$ with excess Bu_4Sn at 230° and cooling to 30° (256, 268, 270).
By reacting $SnCl_4$ with Bu_4Sn (1:1) at 0-20°; Bu_3SnCl and $BuSnCl_3$ are formed in about equal amts. (609).
By reacting a mixt. of Bu_2SnCl_2, Bu_3SnCl and Bu_4Sn with $SnCl_4$ at 0°; $BuSnCl_3$ is also formed (192).
By treating Bu_4Sn with $AlCl_3$ in $CHCl_3$; the prod. contains 22% Bu_3SnCl and 28.5% Bu_2SnCl_2 (362).
By reacting $BuMgCl$ with $SnCl_4$ (1.15:0.5) in C_6H_6 and refluxing the rxn. prod. contg. some Bu_2SnCl_2 with required amt. of Bu_4Sn to convert the Bu_2SnCl_2 to Bu_3SnCl (117).

TRIBUTYLTIN CHLORIDE Bu_3SnCl (Cont'd.)

Prepn. (Cont'd.): By heating BuCl with Na-Sn at 140-155°, or with Na-Sn-Zn at 160-170°; or with Na-K-Sn at 160-165° under N in an autoclave; the prod. contains Bu_3SnCl and Bu_4Sn (590).
By treating $(Bu_3Sn)_2$ with Cl_2 at r.t.; 84% yield (265).
By reacting $Bu_3SnCH:CH_2$ with Cl_2 (1:1) at -78° to r.t.; 86.4% yield (488).
By reacting a mixt. of Mg (treated with EtBr and BuCl in Et_2O) and $BuSnCl_3$ in MePh with BuCl; the prod. consists mostly of Bu_3SnCl along with small mats. of Bu_2SnCl_2 and Bu_4Sn (430).
By-prod. in the rxn. of BuCl with $SnCl_4$ and Mg (4:1:4) in MePh-Et_2O (427).

Props.: Colorless liquid, b. 120°/3-4mm (117); b. 172°/25mm (256); b. 140-152°/10mm (270); b. 145-147°/5mm; d^{20}_4 1.2105; $n^{22}D$ 1.4908 (362); b. 98-100°/0.45-0.35mm; $n^{25}D$ 1.4903 (488); IR spectrum (365).

Rxns. with: $Na_2P_3O_{10}$ and Bu_2SnCl_2 → $(Bu_3Sn)_3BuSnP_3O_{10}$ (94).
$(C_8H_{17}O)_2POONa$ → $Bu_3SnOPO(OC_8H_{17})_2$ (95).
RCO_2M' (M= Na or K) → Bu_3SnO_2CR (116, 270, 309).
BuONa → Bu_3SnOBu (97).
RSH in the presence of Na_2CO_3 or NaOH → Bu_3SnSR (429, 559).
ROONa → Bu_3SOOR (442).
RMg halide → Bu_3SnR (256, 372a).
$HS(CH_2)_xCONXY$ → $Bu_3SnS(CH_2)_xCONXY$ (310).
$RSO_2NR'Na$ → $Bu_3SnNR'O_2SR$ (342).
$(EtO)_2CHC:CMgBr$ → Bu_3SnBr (251).
$SnCl_4$ (2:1) → $3Bu_2SnCl_2$ (265).
Aq. NaOH at 90-95° → $(Bu_3Sn)_2O$ (268).
$LiAlH_4$ → Bu_3SnH (272).
Al-Hg + H_2O in Et_2O → Bu_3SnH (275).
Electrochem. redn. (103).
Compds. contg. active CH_2 group → Bu_3SnCH- (341).
PhC:CLi in $Et_2O-C_6H_6$ — $Bu_3SnC:CPh$ (666).
NaOR → Bu_3SnOR (707).
20% NaOH in the presence of EtOH, followed by heating at 200-220° → $(Bu_3Sn)_2O$ (731).

TRIBUTYLTIN BROMIDE Bu_3SnBr

Prepn.: By reacting $Bu_3SnCH:CH_2$ with Br_2 (1:1-1:2) at -78° to r.t.; 76% yield (488).

Props.: b. 120-122°/1.8-1.6mm; d_{25} 1.338; $n^{25}D$ 1.5022 (488).

TRIBUTYLTIN IODIDE Bu_3SnI

Prepn.: By refluxing $Bu_3Sn(sec.-Bu)$ with I_2 in xylene (360).
Props.: b. 168°/8mm; $n^{29}D$ 1.5345 (360).
Rxns. with: $PhP(OEt)_2$ → $Bu_3SnP(Ph)O(OEt)$ (?) (32).
RMg halide → Bu_3SnR (359, 360).

TRIISOBUTYLTIN FLUORIDE   i-Bu$_3$SnF
Props.: Soly. in H$_2$O and C$_6$H$_6$ (616).

TRI-tert-BUTYLTIN CHLORIDE   t-Bu$_3$SnCl
Prepn.: By reacting SnCl$_4$ with t-BuLi (1.05:3); t-Bu$_2$SnCl$_2$ is also formed (743).
Rxns. with: EtOH or PrOH, solvolysis (743).
H$_2$O-dioxan, hydrolysis, conductivity data (743).

TRIPENTYLTIN CHLORIDE   (C$_5$H$_{11}$)$_3$SnCl
Prepn.: By heating C$_5$H$_{11}$Cl with NaSnZn (2% Zn) at 162° under N in an autoclave; the prod. contains a mixt. of (C$_5$H$_{11}$)$_3$SnCl and (C$_5$H$_{11}$)$_4$Sn (590).

TRIISOPENTYLTIN FLUORIDE   (i-C$_5$H$_{11}$)$_3$SnF
Props.: Soly. in H$_2$O and C$_6$H$_6$ (616).

TRIISOPENTYLTIN IODIDE   (i-C$_5$H$_{11}$)$_3$SnI
Prepn.: By reacting (i-C$_5$H$_{11}$)$_4$Sn with I$_2$ (357).
By refluxing MeSn(i-C$_5$H$_{11}$)$_3$ with I$_2$ in xylene; 87% yield (359).
Props.: b. 178-179°/11mm; n^{18}D 1.5209 (359).
Rxn. with: RMg halides → (i-C$_5$H$_{11}$)$_3$SnR (359, 360).

TRICYCLOHEXYLTIN CHLORIDE   (C$_6$H$_{11}$)$_3$SnCl
Rxn. with: NaHAl(OEt)$_3$ → (C$_6$H$_{11}$)$_3$SnHAl(OEt)$_3$ (475).

TRIHEXYLTIN FLUORIDE   (C$_6$H$_{13}$)$_3$SnF
Props.: Soly. in H$_2$O and C$_6$H$_6$ (616).

TRIHEPTYLTIN IODIDE   (n-C$_7$H$_{15}$)$_3$SnI
Prepn.: By refluxing (n-C$_7$H$_{15}$)$_4$Sn with I$_2$ in MePh or xylene; 60% yield (360).
Props.: b. 235-240°/12mm; n^{30}D 1.4732 (360).

TRIOCTYLTIN CHLORIDE   (n-C$_8$H$_{17}$)$_3$SnCl
Rxn. with: MeONa → (C$_8$H$_{17}$)$_3$SnOMe (97; 125).

TRIOCTYLTIN IODIDE   (n-C$_8$H$_{17}$)$_3$SnI
Prepn.: By refluxing (n-C$_8$H$_{17}$)$_4$Sn with I$_2$ in MePh or xylene; 60% yield (360).
Props.: b. 215-220°/5mm; d^{20}$_4$ 1.3205; n^{20}D 1.5181 (360).

TRIDODECYLTIN CHLORIDE $(n-C_{12}H_{25})_3SnCl$
Prepn.: By treating an ether soln. of $(n-C_{12}H_{25})_4Sn$ with dry HCl (369).
Props.: m. 33° (369).

TRIS(TETRADECYL)TIN CHLORIDE $(n-C_{14}H_{25})_3SnCl$
Prepn.: By treating $(n-C_{14}H_{25})_4Sn$ with dry HCl in Et_2O; 85% yield (369).
Props.: m. 46-47° (369).

TRIS(OCTADECYL)TIN CHLORIDE $(n-C_{18}H_{37})_3SnCl$
Prepn.: By treating $(n-C_{18}H_{37})_4Sn$ with dry HCl in Et_2O; 62% yield (369).
Props.: m. 61-62° (369).

TRIPHENYLTIN FLUORIDE Ph_3SnF
Props.: Conductance in $HCONMe_2$ (788).

TRIPHENYLTIN CHLORIDE Ph_3SnCl
Prepn.: By treating Ph_4Sn with HCl in $CHCl_3$ (36).
By reacting Ph_4Sn with $SnCl_4$ (3:1) at 180-215°; 72% yield (154, 270). A similar rxn. at 220° with a molar ratio 2.76-1.28 gave a yield of 60-70% (164, 487).
By reacting Sn with PhHgCl in alc.; 47% yield (382).
By reacting $PhSnCl_3$ with Sn-Na; 75% yield (382).
By reacting Ph_2SnCl_2 with Sn or Sn-Na; 76.6-83.3% yield (382).
By treating $Ph_2Sn(CH:CH_2)_2$ with HCl (1:1) in boiling $CHCl_3$ (redistribution) (488).
By treating Ph_3SnH with HCl in Et_2O (571).
By reacting Ph_3SnH with $CH_2:CRCH_2Cl$ (732).
By reacting $SnCl_2$ with Ph_2Hg (1.6:0.84) in ligroin at 60-65°; 75% yield (729).
Props.: m. 106° (36, 571); m. 103-104° (154); m. 105.5-107° (270); m. 104-106° (164, 382); m. 102-104° (729); IR spectrum (251); UV spectrum (663); Dipole moment (520, 521); Conductance in non-aqueous solvents (788, 789); Analytical method (100, 512).
Rxns. with: RLi → Ph_3SnR (34, 154, 164, 593, 594, 595).
NaC:CH → $Ph_3SnC:CSnPh_3$ (50).
Na in liq. NH_3 → Ph_3SnNa (157, 273a).
Na in liq. NH_3, followed by NH_4Br → Ph_3SnH (154).
Et_2Ba → Ph_3SnEt (153).
Ph_3SnLi → $(Ph_3Sn)_2$ (163).
Ph_3GeK → $Ph_3SnGePh_3$ (183).
Alc. KOH followed by H_2O → Ph_3SnOH (36, 382).
Aq. NaOH in Et_2O → Ph_3SnOH (282a).
NaCN in liq. NH_3 → Ph_3SnCN (50).
$LiAlH_4$ → Ph_3SnH (178, 272).
$NaHAl(OEt)_3$ → $Ph_3SnHAl(OEt)_3$ (475).

TRIPHENYLTIN CHLORIDE Ph_3SnCl (Cont'd.)
Rxns. with (Cont'd.): Al-Hg + H_2O in Et_2O → Ph_3SnH (275).
Sn-Na → Ph_4Sn (382).
RSH in the presence of Na_2CO_3 or NaOH → Ph_3SnSR (429).
$C_{12}H_{25}SC(S)SK$ → $Ph_3SnSC(S)SC_{12}H_{25}$ (704).
RMg halide → Ph_3SnR (138, 157, 158, 162, 174, 178, 182, 185, 214, 319, 459, 460a, 487, 583).
$p-BrC_6H_4MgBr$ → $Ph_3SnC_6H_4Br$ along with $(p-Ph_3SnC_6H_4-)_2$ and some Ph_4Sn (583).
$(EtO)_2CHC\mathord:CMgBr$ → Ph_3SnBr (251).
$(\mathord:CMgBr)_2$ → $Ph_3SnC\mathord:CSnPh_3$ (50).
$NCCH_2CO_2K$ at 70-80° → $Ph_3SnOCOCH_2CN$ (270).
$EtO_2CCH_2CO_2K$ → $CH_2(CO_2SnPh_3)_2$ (270).
Slow neutrons → radioactive Ph_3SnCl (concentration method) (722).
EtOH or PrOH → solvolysis (743).
H_2O–dioxan → hydrolysis; conductivity data (743).
Polarographic redn. (765).
N_2O_3 (1:6) → $3PhN_2NO_3$ (350a).
$TlCl_3$ → Ph_2TlCl (136a).

TRI-p-TOLYLTIN CHLORIDE $(p-MeC_6H_4)_3SnCl$
Prepn.: By treating $(p-MeC_6H_4)_4Sn$ with $SnCl_4$ (3:1) (63).
By reacting $p-MeC_6H_4Cl$ with Sn; 44.2% yield (382).
Props.: m. 97.5-98° (382).
Rxns. with: Alkali → $(p-MeC_6H_4)_3SnOH$ (63, 282a, 382).
K (1:1) in $(CH_2OMe)_2$ → $(Me_3Sn)_2$ (63).
$(\mathord:CMgBr)_2$ → $[(p-MeC_6H_4)_3SnC]_2$ (214).
RMgBr → $(p-MeC_6H_4)_3SnR$ (214).

TRIPHENYLTIN BROMIDE Ph_3SnBr
Prepn.: By reacting $C_5H_5SnPh_3$ with Br in CCl_4; 52% yield (182).
By reacting $Ph_2Sn(CH\mathord:CH_2)_2$ with HBr at -40° and distg. the resulting $Ph(CH_2\mathord:CH)_2SnBr$ (redistribution) (488).
By reacting Ph_3SnH with $CH_2\mathord:CHCH_2Br$, n-BuBr, or PhBr (732).
By heating Ph_4Sn with $SnBr_4$ (3:1) at 205-215° (740).
By-prod. in the rxn. of $Ph_3SnCH\mathord:CH_2$ with tetracyclone in PhBr (760).
By refluxing Ph_3SnH with PhBr (760).
By reacting SnBr with Ph_2Hg (1.06:0.56) in ligroin at 60-70°; 23% yield (729).
Props.: m. 120-121° (182, 740); IR spectrum (251); m. 121.5-122.5° (760); m. 124-125° (729).
Rxns. with: Br_2 → Ph_2SnBr_2 (28).
RMgBr → Ph_3SnR (34, 215).
RLi → Ph_3SnR (34).
$(EtO)_3P$ or $NaOP(OEt)_2$ → Ph_4Sn (30).
$AgC\mathord:CCH(OEt)_2$ → $Ph_3SnC\mathord:CCH(OEt)_2$ (251).
Li in liq. NH_3 → $(Ph_3Sn)_2$ and Ph_3SnH (571).
$LiAlH_4$ in Et_2O → Ph_3SnH (571).

TRIPHENYLTIN IODIDE Ph_3SnI

Prepn.: By reacting Ph_4Sn with I_2 in $CHCl_3$; 70% yield (157).
By reacting Ph_3SnLi with I_2 in xylene under reflux; 56.7% yield (170).
By reacting $CH_2{:}CHCH_2SnPh_3$ with I_2 in hot xylene (298).
By refluxing $Ph_2(CH_2{:}CH)_2Sn$ with I_2 (1:1) in Et_2O; the prod. contains a mixt. of Ph_3SnI and $(CH_2{:}CH)_3SnI$ formed by redistribution of $Ph(CH_2{:}CH)_2SnI$ (488).
By treating $Ph_3SnCH{:}CHPh$ with I_2 (1:1) in CCl_4; 62% yield (682).
Props.: m. 119-121° (157); m. 117-119° (170); m. 115-117° (298); Analytical method (100).
Rxns. with: RMg halides → Ph_3SnR (34, 157).
RLi → Ph_3SnR (34, 157).
BzONa → Ph_3SnOBz (178).
Slow neutrons → radioactive Ph_3SnI (concentration method) (722).
Biol. props.: Sternutatory effect (338).

TRI-o-TOLYLTIN CHLORIDE $(o{-}MeC_6H_4)_3SnCl$

Rxns. with: Na in C_6H_6-xylene → $(o{-}MeC_6H_4)_3Sn_2$ (170).
$PhCH_2MgCl$ → $(o{-}MeC_6H_4)_3SnCH_2Ph$ (170).
$({:}CMgBr)_2$ → $[(o{-}MeC_6H_4)_3SnC{:}]_2$ (214).

TRI-o-TOLYLTIN BROMIDE $(o{-}MeC_6H_4)_3SnBr$

Prepn.: By reacting $(o{-}MeC_6H_4)_4Sn$ with I_2 in $CHCl_3$ and treating the resulting $(o{-}MeC_6H_4)_3SnI$ with HBr (380).
Props.: m. 94-96° (380).
Rxn. with: Na in EtOH → $[(o{-}MeC_6H_4)_3Sn]_2$ (380).

TRI-p-TOLYLTIN BROMIDE $(p{-}MeC_6H_4)_3SnBr$

Prepn.: By treating $(p{-}MeC_6H_4)_3SnOH$ with HBr (63).
Rxns. with: Me_3SnNa → $(p{-}MeC_6H_4)_3SnSnMe_3$ (63).
Na-K alloy → $[(p{-}MeC_6H_4)_3Sn]_2$ (63).

TRIBENZYLTIN FLUORIDE $(PhCH_2)_3SnF$

Prepn.: By treating $(PhCH_2)_3Sn_2O$ with HCl in Et_2O and pptg. the $(PhCH_2)_3Sn^+$ with F^- (38).
Props.: m. 242° (38).

TRIBENZYLTIN CHLORIDE $(PhCH_2)_3SnCl$

<u>Prepn.</u>: By reacting $PhCH_2HgCl$ with Sn in xylene under reflux; 71.6 yield (382).
By boiling $PhCH_2Cl$ with Sn in H_2O; 84.5% yield (506).
<u>Props.</u>: m. 127-130° (382).
<u>Rxns. with</u>: $(:CMgBr)_2 \rightarrow [(PhCH_2)_3SnC]_2$ (214).
Aq. $NH_3 \rightarrow (PhCH_2)_3SnOH$ (382).
RCOONa in $C_6H_6 \rightarrow (PhCH_2)_3SnOOCR$ (507).
HSRSH (2:1) $\rightarrow (PhCH_2)_3SnSRSSn(CH_2Ph)_3$ (702).
Polarographic redn. (765).
<u>Use</u>: Stabilizer for vinyl halide polymers (777).

TRI(p-METHYLBENZYL)TIN CHLORIDE $(p-MeC_6H_4CH_2)_3SnCl$

<u>Prepn.</u>: By reacting $p-MeC_6H_4CH_2Cl$ with Sn in H_2O under reflux (506).
<u>Props.</u>: m. 190° (506).

TRI(2-BIPHENYLYL)TIN BROMIDE $(2-PhC_6H_4)_3SnBr$

<u>Prepn.</u>: By reacting $2-PhC_6H_4MgBr$ with $SnBr_4$ (4:1) (39).
<u>Props.</u>: m. 167.5-168° (39).
<u>Rxns. with</u>: Na in xylene $\rightarrow [(2-PhC_6H_4)_3Sn]_2$ (41).

HOMOGENEOUS, FUNCTIONALLY SUBSTITUTED TRIORGANOTIN HALIDES

Tris(haloalkyl)tin halides are formed in the reactions of tin tetrahalides with diazoalkanes in a molar ratio of 1:3.

Tris(haloalkenyl)tin halides are obtained by reacting tin or tin alloys with haloalkenylmercury halides or with bis(haloalkenyl)mercury compounds.

Tris(haloaryl- or haloaralkyl)tin halides were prepared from hexakis(haloaryl- or haloaralkyl)tin compounds by treating them with halogens and by reacting tris(haloaralkyl)tin oxides with hydrogen halides. The fluorides are prepared from other halides by treating them with metal fluorides.

Trialkenyltin halides are formed from tetraalkenyltin compounds by treating them with halogens or by the redistribution reaction of dialkenylhydrocarbyltin halides at elevated temperatures. The fluorides were obtained metathetically from other halides and metal fluorides.

Tris(carboalkoxyhydrocarbyl)tin halides were prepared by reacting tin metal with carboalkoxyhydrocarbylmercury halides or by cleaving tris(carboalkoxyhydrocarbyl)alkyltin compounds with halogens.

TRIS(CHLOROMETHYL)TIN CHLORIDE $(ClCH_2)_3SnCl$
Prepn.: By-prod. in the rxn. of $SnCl_4$ with CH_2N_2 (3:2) in
C_6H_6 at 3-5° (375).
Props.: b. 138-140°/5mm; $d^{20}20$ 2.03; $n^{20}D$ 1.593 (574, 575).
Rxn. with: Alkali in aq. soln → ppt. sol. in excess of alk. (575).

TRIS(BROMOMETHYL)TIN BROMIDE $(BrCH_2)_3SnBr$
Prepn.: By-prod. in the rxn. of $SnBr_4$ with CH_2N_2 (1:2) in C_6H_6
at 1-2° (575).
Props.: b. (crude) 165°/5mm (575).

TRIS(1-CHLOROETHYL)TIN CHLORIDE $(MeCHCl)_3SnCl$
Prepn.: By-prod. in the rxn. of $SnCl_4$ with $MeCHN_2$ (2:1) in
C_6H_6 at 2-4° (575).
Props.: b. 130°/3mm; $d^{20}20$ 1.684; $n^{20}D$ 1.5450 (574, 575).
Rxn. with: $MeCHN_2$ (1:3) in C_6H_6 at 2-15° → $(MeCHCl)_4Sn$ (575).

TRI(trans-CHLOROVINYL)TIN CHLORIDE $(ClCH:CH)_3SnCl$
Prepn.: By heating trans-ClCH:CHHgCl with Na-Sn (15% Na) in
C_6H_6 at 45-48°; 40% yield (391).
By heating trans-$(ClCH:CH)_2Hg$ with Sn in EtOH contg. HCl at
50°; 29% yield (391).
Props.: m. 120-121°; the trans-isomer is converted to cis-
isomer by UV irradiation (391) or by heating with Bz_2O_2 in
xylene at 95-100° (725).
Rxn. with: $HgCl_2$ in EtOH → C_2H_2 and ClCH:CHHgCl (391).

TRI(cis-CHLOROVINYL)TIN CHLORIDE $(ClCH:CH)_3SnCl$
Prepn.: By heating cis-ClCH:CHHgCl with Na-Sn (15% Na) in
C_6H_6 at 45-48°; 58% yield (391).
By UV irradiation of the trans-isomer (391).
By heating cis-$(ClCH:CH)_2Hg$ with Sn in EtOH contg. HCl to
50° (391).
By heating trans-$(ClCH:CH)_3SnCl$ with Bz_2O_2 in xylene at
95-100° (725).
Props.: b. 119.5°/1mm; $n^{20}D$ 1.5821; $d^{20}4$ 1.8058 (391).
Rxn. with: $HgCl_2$ in EtOH → C_2H_2 and ClCH:CHHgCl (391).

TRI-p-CHLOROPHENYLTIN BROMIDE $(p-ClC_6H_4)_3SnBr$
Rxns. with: $(:CMgBr)_2$ → $[(p-ClC_6H_4)_3SnC:]_2$ (214).
PhC:CMgBr → $(p-ClC_6H_4)_3SnC:CPh$ (214).
Aq. KOH, followed by AcOH → $(p-ClC_6H_4)_3SnOAc$ (268).

TRI-p-CHLOROPHENYLTIN IODIDE $(p\text{-}ClC_6H_4)_3SnI$
Prepn.: By reacting $(p\text{-}ClC_6H_4)_3Sn_2$ with I_2 in C_6H_6 (382).
Props.: m. 94-95° (382).

TRI-o-CHLOROBENZYLTIN FLUORIDE $(o\text{-}ClC_6H_4CH_2)_3SnF$
Prepn.: By treating $[(o\text{-}ClC_6H_4CH_2)_3Sn]_2O$ with HCl in Et_2O and pptg. the $(o\text{-}ClC_6H_4CH_2)_3Sn^+$ with F^- (38).
Props.: m. 186-187° (38).

TRIVINYLTIN FLUORIDE $(CH_2{:}CH)_3SnF$
Prepn.: By reacting $(CH_2{:}CH)_3SnCl$ or $(CH_2{:}CH)_2SnBr$ with aq.-alc. KF (487, 488).
Props.: m. >300° (488).

TRIVINYLTIN CHLORIDE $(CH_2{:}CH)_3SnCl$
Prepn.: By adding $SnCl_4$ to $(CH_2{:}CH)_4Sn$ (1:3) and completing the rxn. on a steam bath; 43.6% yield along with some $(CH_2{:}CH)_2SnCl_2$ (487); 96% yield resulted when the rxn. was carried out at 30° (460).
Props.: b. 90-96°/26mm; n^{25}_D 1.5337 (487); b. 59-60°/6mm; n^{25}_D 1.5235; d_{20} 1.5139 (460).
Rxns. with: Aq.-alc. KF → $(CH_2{:}CH)_3SnF$ (487).
Aq. NaOH → $(CH_2{:}CH)_3SnOH$ (460).
RMg halide → $(CH_2{:}CH)_3SnR$ (460a).

TRIVINYLTIN BROMIDE $(CH_2{:}CH)_3SnBr$
Prepn.: By reacting $Ph_2Sn(CH{:}CH_2)_2$ with HBr (1:1) at -40°; the prod. contains $(CH_2{:}CH)_3SnBr$ and Ph_3SnBr formed by redistribution of $Ph(CH_2{:}CH)_2SnBr$ during distn. (488).
Props.: b. 53-54°/0.3-0.4mm (488).
Rxns. with: K_2F_2 → $(CH_2{:}CH)_3SnF$ (488).

TRIVINYLTIN IODIDE $(CH_2{:}CH)_3SnI$
Prepn.: By refluxing $Ph_2Sn(CH{:}CH_2)_2$ with I_2 (1:1) in Et_2O; the prod. contains a mixt. of Ph_3SnI and $(CH_2{:}CH)_3SnI$ formed by redistribution of $Ph(CH_2{:}CH)_2SnI$ (488).
By refluxing $(CH_2{:}CH)_4Sn$ with I_2 (1:1) in Et_2O (488).
Props.: b. 60-62°/1.8mm; n^{25}_D 1.5835 (488).

TRIS(CARBOMETHOXYETHYL)TIN BROMIDE $(MeO_2CCH_2CH_2)_3SnBr$
Prepn.: By reacting $BuSn(CH_2CH_2CO_2Me)_3$ with Br_2 (1:1) in $CHCl_3$ at 0°; 84% yield (682).
Props.: b. 154-158°/0.005mm (682).

TRI(p-CARBETHOXYPHENYL)TIN CHLORIDE $(p\text{-}EtO_2CC_6H_4)_3SnCl$
__Prepn.:__ By reacting $p\text{-}EtO_2CC_6H_4HgCl$ with Sn (382).
__Props.:__ Oil (382).
__Rxn. with:__ H_2S in alc. → $[(p\text{-}EtO_2CC_6H_4)_3Sn]_2S$ (382).

For silicon-containing trialkyltin halides see p. 232.

HETEROGENEOUS, NON-FUNCTIONALLY SUBSTITUTED TRIORGANOTIN HALIDES

Trihydrocarbyltin halides of the general formula $R_2R'SnX$, where R is an alkyl, cycloalkyl, aryl, or aralkyl group, R' is one of the preceding groups but different from R, and X is a halogen atom, are prepared by treating heterogeneous tetraorganotin compounds, $R_2SnR'_2$ or R_3SnR', with halogens in benzene, ether, chloroform, or carbon tetrachloride; or with hydrogen halides in ether; or with metal halides in alcohol.

HETEROGENEOUS, NON-FUNCTIONALLY SUBSTITUTED TRIORGANOTIN HALIDES

$R_2R'SnX$	Prepared from	Yield %	Properties	Ref.
$Me_2EtSnCl$	$Me_3SnEt + HCl$	80	b. 166-168°; d^{20}_4 1.6024; n^{20}_D 1.5082	361
Me_2EtSnI	$Me_3SnEt + HgCl_2$	81		361
$Me_2(i-Pr)SnI$	$Me_3SnEt + I_2$		b. 78°/13mm; n^{18}_D 1.5705	359
Me_2BuSnI	$i-Pr_2SnMe_2 + I_2$	50.6	b. 77-78°/9.2mm; n^{25}_D 1.5553 (1)	492
	$Me_3SnBu + I_2$	67.5	b. 100-103°/12mm; Rxn. (1)	269
		95	b. 118-120°/25mm; d^{20}_4 1.7817; n^{20}_D 1.5478	
$Me_2(1-Bu)SnI$	$Me_2SnBu_2 + I_2$	27.6	b. 88-89°/5.4mm; n^{25}_D 1.5467	358
	$1-BuSnMe_3 + I_2$	91	b. 95°/15mm; d^{20}_4 1.7803; n^{20}_D 1.5475	492
$Me_2(s-Bu)SnI$	$s-Bu_2SnMe_2 + I_2$	25	b. 84°/5.5mm; n^{25}_D 1.5510	358
$Me_2(cyclo-C_5H_9)SnI$	$(C_5H_9)_2SnMe_2 + I_2$	33.8	b. 77°; n^{25}_D 1.5758	492
$Me_2(n-C_5H_{11})SnI$	$n-C_5H_{11}SnMe_3 + I_2$	93	b. 132-133°/23mm; d^{18}_4 1.7192; n^{18}_D 1.5440	492
$Me_2(1-C_5H_{11})SnI$	$1-C_5H_{11}SnMe_3 + I_2$	86	b. 115°/15mm; d^{21}_4 1.7027; n^{21}_D 1.5410	358
$Me_2(cyclo-C_6H_{11})SnI$	$(C_6H_{11})_2SnMe_2 + I_2$	44.2	b. 86°/0.65	358
$Me_2(n-C_8H_{17})SnBr$	$C_8H_{17}SnMe_3 + Br_2$	86	b. 77-81°/0.01mm; Rxn. (1)	492
$Me_2(n-C_8H_{17})SnI$	$C_8H_{17}SnMe_3 + I_2$	36	b. 102-118°/12mm; Rxn. (1)	460a
$Me_2(n-C_{10}H_{21})SnBr$	$C_{10}H_{21}SnMe_3 + Br_2$	80	b. 110-114°/0.2mm; Rxn. (1)	269
$Me_2(n-C_{12}H_{25})SnBr$	$C_{12}H_{25}SnMe_3 + Br_2$	80	b. 124-128°/0.05mm; Rxn. (1)	460a
$Me_2(n-C_{12}H_{25})SnI$	$C_{12}H_{25}SnMe_3 + I_2$	56.5	b. 156-158°/1mm; Rxn. (1)	460a
$Et_2MeSnBr$	$Et_2SnMe_2 + Br_2$		Rxn. with: RMg halide → Et_2MeSnR	269
Et_2PrSnI	$Et_3SnPr + I_2$		b. 132-134°/16mm; d^{20}_4 1.7730; n^{20}_D 1.5582 (1)	487
Et_2BuSnI	$Et_3SnBu + I_2$	46	b. 124-128°/12mm; Rxn. (1)	359
		86	b. 134-135°/13mm; d^{20}_4 1.6485; n^{20}_D 1.5460	268, 269
$Et_2(n-C_6H_{13})SnI$	$C_6H_{13}SnEt_3 + I_2$ in C_5H_5N	66	b. 98-100°/0.25mm; Rxn. (1)	358
$Et_2(n-C_8H_{17})SnI$	$C_8H_{17}SnEt_3 + I_2$	69	b. 132-133°/0.9mm; Rxn. (1)	269
$Et_2(n-C_{12}H_{25})SnI$	$C_{12}H_{25}SnEt_3 + I_2$	56	b. 182-184°/1.5mm; Rxn. (1)	269

(1) Rxns.: with 20% aq. KOH, followed by AcOH → $R_2R'SnOAc$.

HETEROGENEOUS, NON-FUNCTIONALLY SUBSTITUTED TRIORGANOTIN HALIDES (Cont'd.)

$R_2R'SnX$	Prepared from	Yield %	Properties	Ref.
Pr_2BuSnI	$Pr_3SnBu + I_2$		b. 159-160°/24mm; d^{20}_4 1.5624; n^{20}_D 1.5320	359
$i-Pr_2MeSnI$	$i-Pr_2SnMe_2 + I_2$	23.4	b. 96°/7.6mm; n^{25}_D 1.5518	492
Bu_2MeSnI	$Bu_2SnMe_2 + I_2$	65.5	b. 82°/0.35mm; n^{25}_D 1.5375	492
$Bu_2(cyclo-C_5H_9)SnI$	$(C_5H_9)_2SnBu_2 + I_2$	39.3	b. 125°/0.4mm; n^{25}_D 1.5497	492
$Bu_2(cyclo-C_6H_{11})SnI$	$(C_6H_{11})_2SnBu_2 + I_2$	54.2	b. 136°/0.6mm; n^{25}_D 1.5494	492
$Bu_2(1-C_5H_{11})SnI$	$1-C_5H_{11}SnBu_3 + I_2$		b. 195°/12mm; d^{20}_4 1.4310; n^{20}_D 1.5254	359
$Bu_2(n-C_6H_{13})SnI$	$C_6H_{13}SnBu_3 + I_2$		b. 180°/8mm; d^{18}_4 1.403; n^{18}_D 1.5246	360
$Bu_2PhSnBr$	$Bu_2SnPh_2 + Br_2$	76	b. 120-124°/0.001mm	460a
$s-Bu_2MeSnI$	$s-Bu_2SnMe_2 + I_2$	56	b. 71°/0.25mm; n^{25}_D 1.5498	492
$(cyclo-C_5H_9)_2MeSnI$	$(C_5H_9)_2SnMe_2 + I_2$	45.6	b. 115°/0.7mm; n^{25}_D 1.5836	492
$(n-C_5H_{11})_2EtSnI$	$C_5H_{11})_2SnEt_2 + I_2$	68.5	b. 160-164°/16mm; Rxn.(1)	269
$(cyclo-C_6H_{11})_2MeSnI$	$(C_6H_{11})_2SnMe_2 + I_2$	45.1	b. 136-140°/0.65mm	492
$(cyclo-C_6H_{11})_2BuSnI$	$(C_6H_{11})_2SnBu_2 + I_2$	36.8	b. 160°/0.7mm; n^{25}_D 1.5630	492
$Ph_2BuSnBr$	$Ph_3SnBu + Br_2$	63	b. 155-159°/0.2mm	460a
$Ph_2(n-C_6H_{13})SnBr$	$Ph_3SnC_6H_{13} + Br_2$	91	Instable at elevated temp.	460a
$Ph_2(PhCH_2CH_2)SnI$	$Ph_2CH_2CH_2SnPh_3 + I_2$	55	m. 59-60°	682
$Ph_2(Ph_3C)SnCl$	$Ph_3CSnPh_3 + HCl$		m. 210°	43
$Ph_2(9-fluorenyl)-SnCl$	$9-fluorenyl\ SnPh_3 + HCl$		m. 140-141°	594
$(9-Fluorenyl)_2-PhSnCl$	$(9-fluorenyl)_2SnPh_2 + HCl$		Fine needles, m. 143°	594

(1) Rxns: with 20% aq. KOH, followed by AcOH → $R_2R'SnOAc$.

HETEROGENEOUS, FUNCTIONALLY-SUBSTITUTED TRIORGANOTIN HALIDES

Heterogeneous halogen-substituted triorganotin halides are obtained by reacting homogeneous or heterogeneous diorganotin dihalides or monoorganotin trihalides with calculated amounts of diazoalkanes (CH_2N_2 and $MeCHN_2$). Compounds containing halogen-substituted aryl groups are prepared by reacting diorganotin dihalides with halogen-substituted arylmagnesium halides in a molar ratio of 1:1. Triorganotin fluorides of this class are obtained by treating the corresponding iodides with potassium fluoride.

Cyanoalkyldiaryltin halides are prepared by treating the corresponding cyanoalkyltriaryltin compounds with halogens in solvents. The fluorides are obtained from other halides by treating them with potassium fluoride.

Alkenyldialkyltin halides are formed in the reactions of dialkenyldialkyltin compounds with hydrogen halides or halogens.

Alkenyldiaryltin halides are prepared from alkenyltriaryltin compounds by treating them with halogens in solvents. The fluorides are obtained by treating the corresponding iodides with potassium fluoride solutions.

Dialkenylalkyl- and dialkenylaryltin halides are prepared by a controlled cleavage of dialkenyldialkyl-, dialkenyldiaryl-, and trialkenylalkyltin compounds with halogens, hydrogen halides, or mercury halides and by treating alkyltin trihalides with alkenylmagnesium halides in a molar ratio of 1:2. Vinyl derivatives are prepared from vinylmagnesium halides in tetrahydrofuran.

Bis(carboalkoxyalkyl)alkyltin halides are prepared by a controlled cleavage of bis(carboalkoxyalkyl)dialkyltin compounds with halogens.

Dialkyl- and diaryl(carboalkoxyalkyl)tin halides are prepared from trialkyl- and triaryl(carboalkoxyalkyl)tin compounds by a controlled cleavage with halogens.

o-Carboxyphenyldiphenyltin chloride resulted from hydrolysis of $Ph_2SnC_6H_4\text{-}o\text{-}\overline{COO}$ with concentrated hydrochloric acid, and the carbomethoxy derivative was obtained by treating the carboxy compounds with diazomethane.

Hydroxyalkyldiaryltin halides are prepared by treating hydroxyalkyltriaryltin compounds with halogens in molar ratio of 1:1.

HALO-SUBSTITUTED TRIORGANOTIN HALIDES

Compounds	Prepared from	Yield %	Properties	Ref.
$Me_2(CH_2I)SnF$	$Me_2(CH_2I)SnI + KF$		Decomp. 290° (1)	486
$Me_2(CH_2Cl)SnCl$	$Me_2SnCl_2 + CH_2N_2$	80	b. 76.7°/11mm; n^{25}_D 1.5263; Rxn. (2)	486
$Me_2(CH_2Br)SnBr$	$Me_2SnBr_2 + CH_2N_2$	73	b. 75-79°/3.5-4mm; n^{25}_D 1.5683; Rxn. (3)	486
$Me_2(CH_2I)SnI$	$Me_2SnI_2 + CH_2N_2$	78	b. 111-111.2°/5mm; n^{25}_D 1.6690; Rxn.	486
$Bu(CH_2Cl)_2SnCl$	$BuSnCl_3 + CH_2N_2$ (1:2)	57.9	b. 82-87°/0.18-0.2mm; n^{25}_D 1.5394; d_{25} 1.649	486
$Bu_2(CH_2Cl)SnCl$	$Bu_2SnCl_2 + CH_2N_2$ (1:1)	74.5	b. 106-110°/0.3mm; n^{25}_D 1.5095; d_{25} 1.378; Rxn. (4)	486
$ClCH_2(MeCHCl)_2$-SnCl	$ClCH_2SnCl_3 + MeCHN_2$ (1:2)		b. 128°/3mm; d^{20}_{20} 1.765; n^{20}_D 1.555	574, 575
$Ph_2(CH_2Br)SnF$	$Ph_2(CH_2Br)SnBr + KF$		m. 260°	486
$Ph_2(CH_2Br)SnBr$	$Ph_2SnBr_2 + CH_2N_2$ (1:1)			486
$(p-ClC_6H_4)Et_2-SnBr$	$Et_2SnBr_2 + p-ClC_6H_4$-MgBr	9	b. 122°/0.2mm; Rxn. (5)	268
$(p-BrC_6H_4)Et_2-SnBr$	$Et_2SnBr_2 + p-BrC_6H_4$-MgBr	12	b. 150-160°/1mm; Rxn. (6)	268

Rxns.:
(1) $Me_2(CH_2Cl)SnCl + MeMgBr \rightarrow Me_3SnCH_2Cl$ (486).
(2) $Me_2(CH_2Br)SnBr + MeMgBr \rightarrow Me_3SnCH_2Br$ (486).
(3) $Me_2(CH_2I)SnI + KF \rightarrow Me_2(CH_2I)SnF$ (486).
(4) $Bu_2(CH_2Cl)SnCl + BuMgBr \rightarrow Bu_3SnCH_2Cl$ (486).
(5) $(p-ClC_6H_4)Et_2SnBr + 15\% KOH \rightarrow (p-ClC_6H_4)Et_2SnOH$ (268).
(6) $(p-BrC_6H_4)Et_2SnBr + 15\% KOH$, followed by $AcOH \rightarrow (p-BrC_6H_4)Et_2SnOAc$ (268).

CYANOALKYLDIARYLTIN HALIDES

Compound	Prepared from	Yield %	Properties	Ref.
$Ph_2Sn(CH_2CN)I$	$Ph_3SnCH_2CN + I_2$	53	m. 99-100°	269
$Ph_2Sn(CH_2CH_2CN)I$	$Ph_3SnCH_2CH_2CN + I_2$	62	m. 121-122.5°	682
$Ph_2Sn[(CH_2)_3CN]I$	$Ph_3Sn(CH_2)_3CN + I_2$			682
$Ph_2Sn[(CH_2)_3CN]F$	$Ph_2Sn[(CH_2)_3CN]I + KF$	44	Solid, shrinks and darkens at 280°	682

ALKENYLDIALKYL- AND ALKENYLDIARYLTIN HALIDES

Compound	Prepared from	Yield %	Properties	Ref.
$Me_2(CH_2:CH)SnCl$	$Me_2Sn(CH:CH_2)_2$ + HCl at 50-60°		b. 73-75°/27mm; d_{20} 1.575; n^{25}_D 1.5105	488, 771 488
$Me_2(CH_2:CH)SnBr$	$Me_2Sn(CH:CH_2)_2$ + HBr at -40°	77	b. 59-61°/9.5mm; d_{20} 1.838; n^{25}_D 1.5350	488, 771 488
$Me_2(CH_2:CH)SnI$	$Me_2Sn(CH:CH_2)_2$ + I_2 in Et_2O	67	b. 57.5-59°/5.2-5.3mm; d_{28} 2.033; n^{20}_D 1.5762	488, 771 488
$Bu_2(CH_2:CH)SnCl$	$Bu_2Sn(CH_2:CH_2)_2$ + HCl	79.3	b. 82-83°/0.6mm; d_{20} 1.273; n^{20}_D 1.4973	488, 771 488
$Bu_2(CH_2:CH)SnBr$	$Bu_2Sn(CH:CH_2)_2$ + Br_2 $Bu_2Sn(CH:CH_2)_2$ + HBr	78	b. 72-73°/0.03mm b. 96°/0.65-0.45mm; d_{20} 1.416; n^{25}_D 1.5102; Rxn. (1)	460 488, 771
$Bu_2(CH_2:CH)SnI$	$Bu_2Sn(CH:CH_2)_2$ + I_2		b. 108.5-109.8°/1.75mm d_{25} 1.556; n^{25}_D 1.5384	488, 771
$Ph_2(CH_2:CH)SnF$	$Ph_3SnCH:CH_2$ + I_2, followed by KF		m. 300°	488
$Ph_2(CH_2:CH)SnBr$	$Ph_3SnCH:CH_2$ + Br_2	95	Instable at elevated temp.	460a
$Bu_2(CH_2:CHCH_2)-SnBr$	$Bu_2Sn(CH_2CH:CH_2)_2$ + Br_2	87	b. 85-89°/0.03mm	460a
$Ph_2(CH_2:CHCH_2)-SnBr$	$Ph_2Sn(CH_2CH:CH_2)_2$ + Br_2	93	Instable at elevated temp.	460a

Rxn.: (1) $Bu_2(CH_2:CH)SnBr$ + 20% KOH in $Et_2O \rightarrow [Bu_2(CH_2:CH)Sn]_2O$ (460a).

DIALKENYLALKYL- AND DIALKENYLARYLTIN HALIDES

Compound	Prepared from	Yield %	Properties	Ref.
$(CH_2:CH)_2BuSnCl$	$BuSnCl_3 + CH_2:CHMgCl$		b. 112-114°/4mm; n_D^{20} 1.4987; d_{20} 1.266	459
$(CH_2:CH)_2BuSnBr$	$Bu_2Sn(CH:CH_2)_2 + HgBr_2$		b. 72°/0.75mm; d_{25} 1.529; n_D^{25} 1.5221	489
$(CH_2:CH)_2(n-C_6H_{13})SnBr$	$(CH_2:CH)_3SnC_6H_{13} + Br_2$	90	b. 70-74°/0.05mm	460a
$(CH_2:CH)_2(n-C_8H_{17})SnBr$	$(CH_2:CH)_3SnC_8H_{17} + Br_2$	80	b. 82-86°/0.01mm	460a
$(CH_2:CH)_2(n-C_{10}H_{21})SnBr$	$(CH_2:CH)_3SnC_{10}H_{21} + Br_2$	75	b. 96-100°/0.01mm	460a
$(CH_2:CH)_2PhSnF$	$(CH_2:CH)_2PhSnBr + KF$		m. >300°	488
$(CH_2:CH)_2PhSnBr$	$(CH_2:CH)_2SnPh_2 + HBr$			488
$(CH_2:CHCH_2)_2BuSnBr$	$(CH_2:CHCH_2)_3SnBu + Br_2$	99	Instable at elevated temp.	460a

ALKYL- AND ARYL(CARBOALKOXY ALKYL)TIN HALIDES

Compound	Prepared from	Yield %	Properties	Ref.
$Pr(MeO_2CCH_2CH_2)_2SnBr$	$Pr_2Sn(CH_2CH_2CO_2Me)_2 + Br_2$	82	b. 120-122°/0.002mm	682
$Pr_2(EtO_2CCH_2CHMe)SnBr$	$Pr_3SnCHMeCH_2CO_2Et + Br_2$	62	b. 119-120°/0.02mm	682
$Bu_2(MeO_2CCH_2CH_2)SnBr$	$Bu_3SnCH_2CH_2CO_2Me + Br_2$	90	b. 125-128°/0.002mm	682
$Ph_2(MeO_2CCH_2CH_2)SnI$	$Ph_3SnCH_2CH_2CO_2Me + I_2$	82	Colorless needles, m. 76°	682

(o-CARBOXYPHENYL)DIPHENYLTIN CHLORIDE (?) $Ph_2Sn(C_6H_4\text{-}o\text{-}CO_2H)Cl$

Prepn.: By treating $Ph_2SnC_6H_4\text{-}o\text{-}COO$ with concd. HCl (157).
Props.: Sinters at 200°, but doesn't melt at 340° (157).
Rxns. with: $CH_2N_2 \rightarrow Ph_2Sn(C_6H_4\text{-}o\text{-}CO_2Me)Cl$ (157).

(o-CARBOMETHOXYPHENYL)DIPHENYLTIN CHLORIDE $Ph_2Sn(C_6H_4\text{-}o\text{-}CO_2Me)Cl$

Prepn.: By treating $Ph_2Sn(C_6H_4\text{-}o\text{-}CO_2H)Cl$ with CH_2N_2 (157).
Props.: m. 168-169° (157).

(3-HYDROXYPROPYL)DIPHENYLTIN BROMIDE $Ph_2Sn(CH_2CH_2CH_2OH)Br$

Prepn.: By reacting $Ph_3Sn(CH_2)_3OH$ with Br_2 (1:1) in CCl_4; 36% yield (682).
Props.: m. 127-128.5° (682).

SILICON-CONTAINING TRIORGANOTIN HALIDES

Alkylbis(silylmethyl)- and dialkyl(silylmethyl)tin halides are formed beside each other in the reactions of dialkylbis(silylmethyl)tin compounds with halogens or hydrogen halides.

Compound	Prepared from	Yield %	Properties	Ref.
$(Me_3SiCH_2)Me_2SnBr$	$(Me_3SiCH_2)_2SnMe_2$ + Br_2 (1:1)	31.4	b. 50°/0.35mm; $n^{25}D$ 1.5101	493
	$(Me_3SiCH_2)_2SnMe_2$ + HBr (1:1)	52		493
$(Me_3SiCH_2)Bu_2SnBr$	$(Me_3SiCH_2)_2SnBu_2$ + Br_2 (1:1)		b. 91°/0.2mm; $n^{25}D$ 1.5037	493
	$(Me_3SiCH_2)_2SnBu_2$ + HBr (1:1)	97		493
$(Me_3SiCH_2)_2MeSnBr$	$(Me_3SiCH_2)_2SnMe_2$ + Br_2 (1:1)	56	b. 73°/0.18mm; $n^{25}D$ 1.5018	493
	$(Me_3SiCH_2)_2SnMe_2$ + HBr (1:1)	43		493
$(Me_3SiCH_2)_2MeSnI$	$(Me_3SiCH_2)_2SnMe_2$ + I_2 (1:1)	81.4	b. 85-86°/0.35mm; $n^{25}D$ 1.5268	493
$(Me_3SiCH_2)_2BuSnI$	$(Me_3SiCH_2)_2SnBu_2$ + I_2 (1:1)	93.5	b. 105°/0.4mm; $n^{25}D$ 1.5257	493

DIORGANOTIN DIHALIDES

DIALKYL- AND DIARYLTIN DIHALIDES

Dialkyl-, dialkenyl-, diaryl-, and diaralkyltin dichlorides, dibromides, and diiodides are formed:

1. By reacting tin tetrahalides with the corresponding organomagnesium halides.
2. By treating triorganotin halides or tetraorganotin compounds with tin tetrahalides or mercury dihalides.
3. By cleaving tetraorganotin compounds with hydrogen halides or with halogens.

Dialkyl- and diaryltin dichlorides, dibromides, and diiodides are formed:

1. By reacting tin dihalides with dialkyl- or diarylmercury or with dialkyl- or diaryllead dihalides.
2. By reacting dialkyl- or diaryltin oxides with hydrogen halides.

Dialkyltin dihalides are formed:

1. By treating tin powder, tin oxide, or tin dihalides with alkyl halides at elevated temperature in the presence or absence of metallic catalysts or under -radiation.

Dialkenyl- and diaryltin dihalides are formed:

1. By treating tin dihalides with alkenyl or aryl derivatives of other metals.

Bis(haloalkyl)tin dihalides are formed:

1. By reacting tin tetrahalides with diazoalkanes.

Diaryltin dihalides are formed:

1. By decomposing aryldiazonium salt complexes with tin tetrahalides, $(RN_2X)_2 \cdot SnX_4$, in the presence of tin powder.

Diorganotin difluorides are formed:

1. By precipitation of other diorganotin dihalides with KF.
2. By treating dihydrocarbyldi(hydrocarbyloxy)tin with AcF.

DIMETHYLTIN DICHLORIDE Me_2SnCl_2

Prepn.: By bubbling MeCl through molten Sn contg. catalytic amts. of Zn or Cu at 300-450°; 70% yield along with some $MeSnCl_3$ and Me_3SnCl (456, 516, 518); Cu_3Sn may be used as the source of Sn (518). The presence of Na increases the yield of Me_3SnCl (516).
By bubbling MeCl through $SnCl_2$ at 365° (517).
By passing MeCl over solid Sn contg. 40% Cu, Ag, or Au at 250-450° (560).
By bubbling MeCl through SnO_2 in the presence of CuO at 300° (487, 492, 517).
By refluxing Me_2PbCl_2 with $SnCl_2$ in EtOH; 47% yield (289, 290).
By treating Me_4Sn with excess $HgCl_2$ in EtOH; 86% yield (361).
By hydrolyzing $Me_3SnPO(OMe)_2$ with concd. HCl at 100° (30).

Props.: m. 107° (30); m. 106° (456); m. 107.5-108° (289); m. 107-108°, b. 185-190° (361); Molecular structure and bond length (514); Raman and IR spectra (800); Polarographic studies (550); Conductance in non-aqueous solvents (788, 789); Pyrolysis at 554° and 688° → $SnCl_2$ and hydrocarbons formed from the Me radical (423) and kinetics of the rxn. of the Me radical with MePh (424).

Rxns. with: $LiAlH_4$ (1:2) → Me_2SnH_2 (126, 126a).
Cationic exchange resin → retention of Me_2Sn^{++} which can be eluted with various acids; thus Me_2Sn tungstate, molybdate, sulfide, oxalate, succinate, naphthionate, salicylate, phthalate, benzoate, ferricyanide, ferrocyanide, iodate, arsenate, vanadate, cyanate, and antimonate were prepd. (189).
Me_2SnO → $Me_2SnCl_2 \cdot 2Me_2SnO$ (195).
$KCo(CO)_4$ contg. KCN → $Me_2SnO(H_2O)_x \cdot Me_2Sn[Co(CO)_4]_2$ (?), yellow substance stable in air (227).
H_2O, electrolytic dissocn. (454).
NaOH → Me_2SnO (454).
Na or K salts of chromic, arsenic, oxalic, salicylic, phthalic, succinic, molybolic, tungstic, ferrocyanic, ferricyanic, iodic, and benzoic acid → the corresponding Me_2Sn salts and complexes: $Me_2SnO \cdot Me_2SnCrO_4$; $Me_2SnHAsO_4$; $Me_2SnC_2O_4 \cdot H_2O$; $Me_2Sn(C_7H_5O_3)_2$; $Me_2SnC_8H_4O_4$; $Me_2SnO \cdot Me_2SnC_4H_4O_4$; Me_2SnMoO_4; Me_2SnWO_4; $(Me_2Sn)_2Fe(CN)_6 \cdot Me_2SnO$; $(Me_2Sn)_3[Fe(CN)_6]_2$; $Me_2Sn(IO_3)_2$; and $Me_2SnO \cdot Me_2Sn(C_7H_5O_2)_2$ (455).
RMgCl (1:2) → Me_2SnR_2 (459, 487, 492, 564, 372a).
RMgCl (1:1) → Me_2SnR_2 (460a).
CH_2N_2 at -5° in Et_2O → $Me_2(CH_2Cl)SnCl$ (486).
$PhPO(OH)_2$ → $Me_2SnO_2P(O)Ph$ (487).
$PhP(OH)_2$ → Me_2SnO_2PPh (487).
Na_2HPO_4 → $Me_2SnO_2P(O)H$ (487).
$NaH_2PO_2 \cdot H_2O$ → $Me_2Sn[OP(O)H_2]_2 \cdot 2H_2O$ (487).
Me_3SiOK → $Me_2Sn(OSiMe_3)_2$ (542).
$Ph_3P{:}CH_2$ followed by $HgBr_2$ → $[(Ph_3P \cdot CH_2)_2SnMe_2][HgBr_2Cl]_2$ (663a).

Addn. Compds.: $Me_2SnCl_2 \cdot 2C_5H_5N$, m. 163° (284); Conductance in non-aqueous solvents (789).
$Me_2SnCl_2 \cdot 2Me_2SnO$, prepd. from Me_2SnCl_2 and excess Me_2SnO in hot H_2O (195).

DIMETHYLTIN DIBROMIDE Me_2SnBr_2
Prepn.: By passing MeBr through Sn at 395° (517).
Props.: m. 75-77° (517); Molecular structure and bond length (514).
Rxns. with: NH_4OH — Me_2SnO (195).
$Me_2SnO \rightarrow (Me_2SnO)_3 \cdot Me_2SnBr_2$ (195).
CH_2N_2 in Et_2O at -5° $\rightarrow Me_2(CH_2Br)SnBr$ (486).
$Ph_3P:CH_2 \rightarrow [(Ph_3PCH_2)_2SnMe_2][Me_2SnBr_4]$ (663a).
Addn. Compds.: $Me_2SnBr_2 \cdot 2C_5H_5N$, m. 172° (284).

DIMETHYLTIN DIIODIDE Me_2SnI_2
Prepn.: By heating powdered Sn with MeI (1:2 by wt.) at 140-145° (262).
By stepwise cleavage of the Me groups from Me_4Sn by I_2 (357).
Props.: Monoclinic crystals, m. 30° (262); m. 44° (357); Raman and IR spectra (331); Molecular structure and bond length (514).
Rxns. with: $(RO)_3P \rightarrow Me_2Sn[PO(OR)_2]_2$? (31).
$(MeO)_2PONa \rightarrow Me_2SnO$ (31).
$PhP(OMe)_2 \rightarrow Me_2Sn[P(O)Ph(OMe)]_2$? (32).
$Ph_2POEt \rightarrow [Me_2SnP(O)Ph_2]_2$? (32).
Me_2SnO in H_2O or EtOH $\rightarrow HO(SnMe_2O)_3H \cdot Me_2SnI_2$ and $EtO(SnMe_2O)_3H \cdot Me_2SnI_2$, respectively (195).
KHS $\rightarrow Me_2SnS$ (205).
Et_2SnO in $H_2O \rightarrow HO(SnEt_2O)_3H \cdot Me_2SnI_2$ (208).
Alc. in the presence of aromatic or heterocyclic bases $\rightarrow$
    $Me_2SnO \cdot Me_2SnIOH$ (263).
CH_2N_2 in Et_2 at -5° $\rightarrow Me_2(CH_2I)SnI$ (486).
Polarographic redn. (765).
Addn. compds.: $Me_2SnI_2 \cdot 2C_5H_5N$, white crystals, m. 151-152°, sol. in H_2O (262).
$Me_2SnI_2 \cdot 2PhNH_2$, white prisms, m. 109-110° (262).
$Me_2SnI_2 \cdot 2\ o\text{-}MeC_6H_4NH_2$, white needles, m. 69-70° (262).
$Me_2SnI_2 \cdot 2\ MeC_5H_4N$, deliquescent crystals (262).
$Me_2SnI_2 \cdot 2\ PhNEt_2$, white prisms, m. 88-89° (262).
$Me_2SnI_2 \cdot 2\ MeC_9H_6N$, white needles turning yellow in air, m. 110-111° (262).
$Me_2SnI_2 \cdot Et_2SnO$, m. 103-106° (211); m. 100-106° (209).

ETHYLMETHYLTIN DICHLORIDE $EtMeSnCl_2$
Prepn.: By reacting Et_2SnMe_2 with gaseous HCl; 92% yield (321).
Props.: m. 51.5° (321).

DIETHYLTIN DIFLUORIDE Et_2SnF_2
Prepn.: By treating $Et_2Sn(OEt)_2$ with AcF (1:2) in Et_2O; 85% yield (577).
Props.: Crystals (577).

DIETHYLTIN DICHLORIDE Et_2SnCl_2

Prepn.: By reacting Et_4Sn with $SnCl_4$ (1:1) under reflux or at 220-230°; 52% yield (113, 249).
By reacting Et_4Sn with $AlCl_3$ in $CHCl_3$; 71% yield (362).
By heating Et_2SnMe_2 with $HgCl_2$ in EtOH; 85% yield (361).
By heating Et_2Sn with $HgCl_2$ on steam bath; 80% yield (283, 387).
By refluxing Et_2PbCl_2 with $SnCl_2$ in EtOH; 51% yield (289, 290).
By treating Et_2SnO with HCl in Ac_2O (197).
By reacting Et_4Pb with $SnCl_4$ (2.5:1) (74a).
By treating a fine spray of Sn at 350° with preheated EtCl (517).
By passing EtCl over solid Sn contg. 40% Cu or Au or Ag at 250-450° (560).
By reacting $SnCl_2$ with Et_2Hg in ligroin at 85-90°; 52% yield (729).
By reacting $Et_2SnPO(OEt)_2$ with 20% HCl at 160-170° (30).
By-prod. in the rxn. of Et_4Sn with $HgCl_2$ in EtOH or $CHCl_3$ under reflux (361).
By-prod. in the rxn. of Et_4Sn with $BiCl_3$ in $CHCl_3$ (363).
Props.: m. 84.5-85° (30); m. 84° (113, 283, 289, 290, 363, 389); m. 83-84° (517); m. 84-85° (361); m. 75° (197); m. 89° (249); b. 227° (113); Thermodynamic data (113); UV spectrum (439); Raman spectrum (769); Dipole moment (520, 521); Polarographic studies (550, 765).
Rxns. with: RMgBr (1:2) → Et_2SnR_2 (113, 256, 269, 361).
$LiAlH_4$ → Et_3SnH, Et_2SnH_2, and $EtSnH_3$ (113).
Et_2SnO → $(Et_2SnO)_3 \cdot Et_2SnCl_2$ (195).
Et_2SnO in ROH → $RO(Et_2SnO)_3R \cdot Et_2SnCl_2$ (340).
Et_3SnOH → $(Et_3SnOH)_2 \cdot Et_2SnCl_2$ (195).
Et_3N in EtOH → $(Et_2SnCl)_2$ (249).
$BuCHEtCH_2OH$ in the presence of C_5H_5N → $BuCHEtCH_2O(Et_2SnO)_{11.7}$-$CH_2CHEtBu$ (343).
RCO_2Na and NaOR, (1:1:1) → $Et_2Sn(OR_1)OCOR$ (347).
NaOR → $Et_2Sn(OR)_2$ (348).
$NaOP(OEt)_2$ → Et_2SnO (31).
R_2SiCl_2 in aq. NH_3-MePh → $[(-OR_2Si-)_m-OSnEt_2]_n$ (27a).
Dithizone → colored complex (colorimetry) (1).
Polarographic redn. (765).
Electrolysis in MeOH → decompn. (438).
Electrolysis in aq. KCl → Et_2Sn radical (?) (441).
Biol. props.: Inhibition of the oxidative, enzyme-induced phosphorylation (3); Inhibition of the α-ketoacid oxidases (4); Toxicity and symptoms (532).
Addn. Compds.: $Et_2SnCl_2 \cdot 2\ PhN_2Cl$, decomp. 80-81° (437).
$Et_2SnCl_2 \cdot 2\ p\text{-}ClC_6H_4N_2Cl$, decomp. 98-99° (437).
$Et_2SnCl_2 \cdot 2\ p\text{-}BrC_6H_4N_2Cl$, decomp. 110° (437).
$Et_2SnCl_2 \cdot 2\ p\text{-}EtOC_6H_4N_2Cl$, decomp. 80° (437).
$Et_2SnCl_2 \cdot 2\ p\text{-}MeC_6H_4N_2Cl$, decomp. 88-89° (437).

DIETHYLTIN DIBROMIDE Et_2SnBr_2

Prepn.: By reacting Et_4Sn with Br_2 (195).
By γ-irradiation of a mixt. of Sn powder and EtBr (613).
Props.: m. 63°; b. 230-235° (195); Magnetic susceptibility (260); Polarographic studies (550, 765).
Rxns. with: $NH_4OH \to Et_2SnO$ (195).
Et_2Sn and Na in $NH_3 \to (Et_2SnNa)_2$ (195).
Na $\to Et_2SnNa_2$ (195).
$Et_2SnO \to (Et_2SnO)_3 \cdot Et_2SnBr_2$ (195).
$Et_3SnOH \to (Et_3SnOH)_2 Et_2SnBr_2$ (195).
RMgBr (1:1) $\to Et_2SnRBr$ (268).
Polarographic redn. (765).
Addn. Compds.: $Et_2SnBr_2 \cdot 2\ C_5H_5N$, m. 140° (284).
$Et_2SnBr_2 \cdot Pr_2SnO$, m. 103-105° (211); m. 104-105 (209).
$Et_2SnBr_2 \cdot Et_2SnO$, insol. in H_2O dimeric in C_6H_6, trimeric in $C_{10}H_8$ (195).

DIETHYLTIN DIIODIDE Et_2SnI_2

Prepn.: By heating powdered Sn with EtI at 165-170° (262).
By heating Et_4Sn with calcd. amt. I_2 at 140-150°; 70% yield (354).
By stepwise cleavage of the Et groups from Et_4Sn with I_2 (357).
Props.: m. 42-44.5° (262); m. 44°, decomp. in sunlight (354). m. 30-31° (357); Polarographic studies (550).
Rxns. with: $(RO)_3P \to Et_2Sn[PO(OR)_2]_2$ (?) (31) or $Et_2SnOP(:O)-(OEt)Et_2$ (33, 354).
$(EtO)_2PONa \to Et_2SnO + Et_2Sn[PO(OEt)_2]_2$ (?) (31).
$PhP(OR)_2 \to Et_2SnPhPO(OR)_2$ (?) (32).
$Ph_2POEt \to Et_2Sn[P(:O)Ph_2]_2$ (?) (32) or $Et_2Sn[OP(:O)Ph_2]_2$ (33).
$EtPO(OEt)_2 \to Et_2SnOP(:O)(OEt)Et_2$ (33, 354).
$PhEtP(:O)OEt \to Et_2SnOP(:O)EtPh_2$ (33).
$Et_2SnO \to (Et_2SnO)_3 \cdot Et_2SnI_2$ (195).
$Et_3SnOH \to (Et_3SnOH)_2 \cdot Et_2SnI_2$ (195).
Me_2SnO in EtOH $\to Et(Me_2SnO)_3OEt \cdot Et_2SnI_2$ (208).
Alc. in the presence of org. bases $\to Et_2SnO \cdot Et_2SnIOH$ (263).
Polarographic redn. (765).
Biol. props.: Toxicity and symptoms (685).
Addn. Compds.: $Et_2SnI_2 \cdot 2\ C_5H_5N$, light yellow, prismatic crystals, m. 115-116° (262).
$Et_2SnI_2 \cdot Me_2SnO$, m. 102-103° (209).

DIPROPYLTIN DICHLORIDE Pr_2SnCl_2

Prepn.: By heating Pr_4Sn with $SnCl_4$ (1:1) at 220-230° (249).
By passing PrCl over solid Sn contg. >40% Cu or Cu, Ag, or Au as catalyst at 250-450° (preferably 300°) (560).
Props.: IR spectrum (365); Polarographic studies (550, 765).
Rxns. with: Org. bases and EtOH in $Et_2O \to (Pr_2SnCl)_2$ (249).
$LiAlH_4 \to Pr_2SnH_2$ (272).
Polarographic redn. (765).

DIPROPYLTIN DIBROMIDE Pr_2SnBr_2

Prepn.: By reacting Pr_4Sn with Br_2 (211).
By γ-irradiation of a mixt. of Sn powder and PrBr (613).
Props.: b. 263-265°; m. 49° (207).
Rxns. with: Pr_2SnO or $Pr_2Sn(OH)_2$ (1:1) → $Pr_2SnO \cdot Pr_2SnBr_2$ (209).
NaOH → Pr_2SnO (211).
Na_2S → Pr_2SnS (211).
Addn. compds.: $Pr_2SnBr_2 \cdot 2\ C_5H_5N$, m. 128° (284).
$Pr_2SnBr_2 \cdot Et_2SnO$, m. 85-87° (209, 211).
$Pr_2SnBr_2 \cdot Me_2SnO$, m. 106-107° (209, 211).

DIPROPYLTIN DIIODIDE Pr_2SnI_2

Prepn.: By reacting Pr_4Sn with I_2 (211).
By heating powdered Sn with PrI at 135-140° (262).
Props.: b. 270-273°/76mm; b. 166-167°/1mm (262).
Rxns. with: $(RO)_3P$ → $Pr_2Sn[PO(OR)_2]_2$ (31).
Pr_2SnO in H_2O → $H(Pr_2SnO)_3OH \cdot Pr_2SnI_2$ (208).
Pr_2SnO in EtOH → $Et(Pr_2SnO)_3OEt \cdot Pr_2SnI_2$ (208).
Addn. Compds.: $Pr_2SnI_2 \cdot 2\ C_5H_5N$, prismatic cyrstals, m. 64-65° (262).
$Pr_2SnI_2 \cdot 2\ PhNH_2$, octahedral crystals, sol. in CCl_4 (262).
$Pr_2SnI_2 \cdot 2\ PhNEt_2$, large bright crystals, m. 63-64° (262).
$Pr_2SnI_2 \cdot 2\ MeC_9H_6N$, prismatic crystals, m. 71-72° (262).

DIISOPROPYLTIN DICHLORIDE $i-Pr_2SnCl_2$

Prepn.: By-prod. in the rxn. of $SnCl_4$ with i-PrLi (1:3) (743).

DIISOPROPYLTIN DIIODIDE $i-Pr_2SnI_2$

Rxn. with: Alc. in the presence of org. bases →
$i-Pr_2SnO \cdot i-Pr_2SnIOH$ (263).

BUTYLPROPYLTIN DICHLORIDE $BuPrSnCl_2$

Prepn.: By treating Pr_3SnBu with $BiCl_3$ at 120°; 37% yield (363).
Props.: m. 67-68° (363).

DIBUTYLTIN BROMIDE CHLORIDE $Bu_2SnBrCl$

Prepn.: By reacting $(Bu_2SnCl)_2$ with Br_2 (1:1) in CCl_4;
72% yield (248, 249).
Props.: Long prisms, m. 34-34.5°; b. 104-106°/0.55mm; $n^{27}D$ 1.5252
(248); White needles, m. 32.5-35.5°; b. 98-104°/1mm (249).
Rxns. with: Et_3N (1:1) and EtOH in Et_2O → $(Bu_2SnCl)_2$ (249).
Cl_2 in CCl_4 → Bu_2SnCl_2 (248).

DIBUTYLTIN DICHLORIDE Bu_2SnCl_2

Prepn.: By reacting $SnCl_4$ with BuMgI (0.8:1.15) in C_6H_6 and refluxing the rxn. prod. contg. Bu_3SnCl with a calcd. amt. $SnCl_4$ (117).

By heating Bu_4Sn with $SnCl_4$ (1:1) at 135-140°; $BuSnCl_3$ and Bu_3SnCl, which are also formed, are returned to the rxn. zone to increase the yield of the dichloride (240). In a continuous process the rxn. zone is maintained at 200° (242); A batch process at 200-215°, 50% yield (248) and at 240°, 95% yield (246, 265, 417).

By treating Bu_4Sn with $AlCl_3$ in $CHCl_3$; 28.5% yield (362).

By reacting Bu_4Sn with $BiCl_3$ at 120° (363).

By reacting a mixt. of $SnCl_2$ and Bu_2SnCl_2 with a Bu-Sn compd. having a melting point below that of Bu_2SnCl_2 (for example, compd. obtd. from BuCl, $SnCl_2$, and Na) (408).

By passing BuCl over solid Sn contg. 40% Cu or Cu, Ag or Au as catalyst at 250-450° (560).

By heating a mixt. of Bu_4Sn and Bu_3SnCl (obtd. from Na-Sn, Na-Sn-Zn, or Na-K-Sn and BuCl) with calcd. amt. $SnCl_4$ at 160-207°; 94% yield (590).

By refluxing $(Bu_2SnO)_x$ with concd. HCl (590).

By adding granular, extruded, or liq. Na to a mixt. of BuCl and $SnCl_2$ (3-4:1) preheated to 95-100° (738).

Props.: Crystals, m. 42° (117); m. 39.5-40° (590); m. 40.5 (256); m. 41-42° (362); m. 44-45 (363); m. 39° (417); b. 140-143°/10mm (117, 590); b. 135°/10mm (242); b. 89-91°/0.2mm (248); b. 153-156°/5mm (362); IR spectrum (365); Polarography (550).

Rxns. with: KOAc in EtOH → $Bu_2Sn(OAc)_2$ (116).

$NaCo(CO)_4$ → an oily prod. which reacts with Ph_3P in Et_2O to form $Bu_2Sn[Co(CO)_3PPh_3]_2$ (227).

$Fe(CO)_5$ + KOH → $Bu_2SnFe(CO)_4$ (227).

Na (1:1) in EtOH → $(Bu_2SnCl)_2$ (248).

Org. bases + EtOH → $(Bu_2SnCl)_2$ (249).

RMg halide → Bu_2SnR_2 (256, 459, 487, 492, 493, 564, 460a, 372a).

BuCl and Na (1:2:4) → Bu_4Sn and some $(Bu_3Sn)_2$ (265).

$LiAlH_4$ → Bu_2SnH_2 (272, 299a).

RSH → $Bu_2Sn(SR)_2$ (311, 312, 429).

Bu_2SnO in ROH → $RO(Bu_2SnO)_xR \cdot Bu_2SnCl_2$ (340).

$CHR(CO_2R)_2$ in MePh in the presence of NaOMe → $Bu_2Sn[CR(CO_2R)_2]_2$ (341).

Compd. contg. a reactive -CH_2- group → $Bu_2Sn(CH\lt)_2$ (341).

$RSO_2NR'Na$ → $Bu_2Sn(NR'O_2SR)_2$ (342).

NaOR in MePh → $RO(Bu_2SnO)_xR$ (343).

Na carboxylates + NaOH (1:1:1) → $Bu_2Sn(OR)$·carboxylate (345, 347).

NaOR → $Bu_2Sn(OR)_2$ (348).

Monoglycerides and NaOH → $Bu_2SnOCH_2CHOCH_2O_2CR$ (745).

RSH in the presence of Na_2CO_3 → $Bu_2Sn(SR)_2$ (429).

CH_2N_2 (1:1) in Et_2O → $Bu_2(CH_2Cl)SnCl$ (486).

NaSCN in Me_2CO → $Bu_2Sn(SCN)_2$ (486).

Bu_2SnO at 110° → $Bu_2SnCl_2 \cdot Bu_2SnO$ (610).

$KPO_2(SR)_2$ → $Bu_2Sn[OP(O)(SR)_2]_2$ (703).

$NaCR(CO_2R)_2$ → $Bu_2Sn[CR(CO_2R)_2]_2$ (706).

MeC(ONa):CHAc → $Bu_2Sn(OCMe:CHAc)_2$ (706).

HOROH in the presence of Na_2CO_3 → Bu_2SnORO and $HOR(OSnBu_2OR)_xOH$ (746).

Polarographic redn. (765).

DIBUTYLTIN DICHLORIDE Bu_2SnCl_2 (Cont'd.)
Biol. props.: Anthelmintic efficacy (649, 683, 684).
Addn. compds.: $Bu_2SnCl_2 \cdot Bu_2SnO$, m. 109-110°, prepd. by heating equimolar quantities of Bu_2SnCl_2 and Bu_2SnO at 110° (610).
Use: Catalyst for the prepn. of polyesters (67).
Stabilizer for vulcanized rubber (791).

DIBUTYLTIN DIBROMIDE Bu_2SnBr_2
Prepn.: By exposing a mixt. of Sn powder and BuBr to γ-radiation (613).
By-prod. in the rxn. of $Bu_2Sn(CH:CH_2)_2$ with Br_2 (1:1) (460).
Props.: $n^{25}D$ 1.4970; d_{25} 1.3913 (460).
Rxn. with: $NaSC(S)OR \rightarrow Bu_2SnSC(S)OR_2$ (597).

DIISOBUTYLTIN DIIODIDE $i-Bu_2SnI_2$
Prepn.: By heating powdered Sn with i-BuI at 170° (262).
Props.: b. 290-295° (262).
Rxn. with: Alc. in the presence of arom. or heterocyclic amines $\rightarrow i-Bu_2SnO \cdot i-Bu_2SnIOH$ (263).
Addn. compd.: $i-Bu_2SnI_2 \cdot 2\ PhNEt_2$ (262).

DIISOAMYLTIN DIIODIDE $(i-C_5H_{11})_2SnI_2$
Prepn.: By heating powdered Sn with $i-C_5H_{11}I$ (262).
Props.: Oily liquid, b. 202-205°/8mm (262).
Rxn. with: Alc. in the presence of an arom. or heterocyclic amines $\rightarrow i-Am_2SnO \cdot i-Am_2SnIOH$ (263).

DI-n-HEXYLTIN DICHLORIDE $(n-C_6H_{13})_2SnCl_2$
Prepn.: By heating $(n-C_6H_{13})_4Sn$ with $SnCl_4$ (1:1) for 1½ hrs. at 100° and for 3 hrs. at 220-225°; 87.5% yield (273).
Props.: m. 44-46°, b. 134-148°/1.5mm (273).
Rxn. with: $NaOH \rightarrow (C_6H_{13})_2SnO$ (273).

DI-n-OCTYLTIN DICHLORIDE $(C_8H_{17})_2SnCl_2$
Prepn.: By heating $(C_8H_{17})_4Sn$ with $SnCl_4$ (1:1) at 110° for 1 hr. and at 230-240° for 2 hrs.; 83.5% yield (273).
Props.: m. 45.5-46.5°; b. 164-165°/0.16mm (273).
Rxn. with: KOH in $EtOH-H_2O \rightarrow (C_8H_{17})_2SnO$ (273).

BIS(2-ETHYL-n-HEXYL)TIN DICHLORIDE $(BuCHEtCH_2)_2SnCl_2$
Prepn.: By heating $(BuCHEtCH_2)_4Sn$ with $SnCl_4$ (1:1) at 225-230° for 3 hrs.; 65% yield (273).
Props.: Yellow oil, b. 154-162°/0.3mm (273).
Rxn. with: KOH in $EtOH-H_2O \rightarrow (BuCHEtCH_2)_2SnO$ (273).

BIS(3,5,5-TRIMETHYL-n-HEXYL)TIN DICHLORIDE $(Me_3CCH_2CHMeCH_2CH_2)_2SnCl_2$

<u>Prepn.</u>: By heating $(Me_3CCH_2CHMeCH_2CH_2)_4Sn$ with $SnCl_4$ at 225-230° for 3 hrs; 75% yield (273).
<u>Props.</u>: b. 154-157°/0.1mm (273).
<u>Rxn. with</u>: KOH in H_2O-Me_2CO → $(Me_3CCH_2CHMeCH_2CH_2)_2SnO$ (273).

DIDODECYLTIN DICHLORIDE $(C_{12}H_{25})_2SnCl_2$

<u>Prepn.</u>: By reacting $SnCl_4$ with $C_{12}H_{25}MgBr$ (522).
<u>Props.</u>: m. 19.4°; b. 218° (decompn.) (522).
<u>Rxns. with</u>: $HSCH_2CH_2CH_2CO_2Bu$ → $(C_{12}H_{25})_2Sn(SCH_2CH_2CH_2CO_2Bu)_2$ (308).
NaOH in Et_2O → $(C_{12}H_{25})_2SnO$ (522).
$KPO_2(SCHMe_2)_2$ → $(C_{12}H_{25})_2Sn[OP(O)(SCHMe_2)_2]_2$ (703).
$NaSC(S)NR_2$ → $(C_{12}H_{25})_2Sn[SC(S)NR_2]_2$ (705).

DICYCLOHEXYLTIN DICHLORIDE $(C_6H_{11})_2SnCl_2$

<u>Prepn.</u>: By heating $(C_6H_{11})_2SnPh_2$ with HCl; 85.5% yield (57).
<u>Props.</u>: m. 87-89° (57).

DIPHENYLTIN DICHLORIDE Ph_2SnCl_2

<u>Prepn.</u>: By heating $(p\text{-}MeOC_6H_4)_2SnPh_2$ or $(1\text{-}C_{10}H_7)_2SnPh_2$ with HCl; 90% yield (57).
By reacting Ph_2Hg with $SnCl_2$ in EtOH and Me_2CO under reflux or without any solvent at 250-260°; 71% yield (123)
By heating Ph_4Sn with $SnCl_4$ (1:1) at 180°; 83% yield (183); at 220° (594).
By heating Ph_4Sn with $SnCl_4$ (1:1) at 190°; $PhSnCl_3$ and Ph_3SnCl, which are also formed, are returned to the rxn. zone to be converted to the title compd. (240, 492). In a similar process the rxn. is conducted at 240° (417).
By refluxing Ph_2PbCl_2 with $SnCl_2$ in EtOH; 13.6% yield (289, 290).
By decompg. $(PhN_2Cl)_2 \cdot SnCl_4$ with Cu, Zn, or Sn in boiling EtOAc; 10-25% yield (386).
By UV irradiation of $Ph_4Sn + SnCl_4$ (1:1); theoret. yield (435).
By heating Ph_2TlCl with $SnCl_2$ (1:2) at 200°; 69% yield (730).
<u>Props.</u>: m. 42° (123, 289, 290, 435, 594); m. 42-44° (185); m. 38° (417); m. 40-41° (730); b. 180-185°/5mm (594); UV spectrum (663).
<u>Rxns. with</u>: RMgX → Ph_2SnR_2 (57, 157, 178, 182, 185, 434, 459, 460a, 492, 493).
H_2S → Ph_2SnS (123).
RLi → Ph_2SnR_2 (174, 178, 594).
H_2NNH_2 and EtOH → $(Ph_2SnCl)_2$ (249).
NaOMe at 80° in MePh → $MeO(SnPh_2O)_{1.52}Me$ (343).
Na oleate in BuOH → $Ph_2Sn(OBu)$ oleate (345).
NaOR in ROH at 0° → $Ph_2Sn(OR)_2$ (348).
Sn or Sn-Na → Ph_3SnCl (382).
Electrolysis in MeOH → $(Ph_3SnCl)_x$ (438).
$NaHAl(OEt)_3$ → $Ph_2SnHAl(OEt)_{3,2}$ (475).

DIPHENYLTIN DICHLORIDE Ph_2SnCl_2 (Cont'd.)
Rxns. with (Cont'd.): N_2O_3 (1:4) → $2PhN_2NO_3$ (350a).
$TlCl_3 \cdot 4H_2O$ in EtOH — Ph_2TlCl (136a).
Biol. props.: Efficacy in the treatment of pathological conditions in animals (683, 684).
Addn. compd.: $Ph_2SnCl_2 \cdot 4C_5H_5N$, unstable complex (284).
Use: Stabilizer for vulcanized rubber (791).

DIPHENYLTIN DIBROMIDE Ph_2SnBr_2
Prepn.: By reacting Ph_3SnLi with Br_2 at -25-25° (28).
By reacting Ph_3SnBr with Br_2 (the yield depends on the rxn. temp.) (28).
By heating Ph_4Sn with $SnBr_4$ (1:1) at 220° (740).
Props.: m. 37° (740).
Rxns. with: RMgX → Ph_2SnR_2 (34, 319).
CH_2N_2 (1:1) → $Ph_2(CH_2Br)SnBr$ (486).
Addn. compd.: $Ph_2SnBr_2 \cdot 4C_5H_5N$, instable complex (284).

DI-o-TOLYLTIN DICHLORIDE $(o\text{-}MeC_6H_4)_2SnCl_2$
Prepn.: By heating $(o\text{-}MeC_6H_4)_4Sn$ with $SnCl_4$ (1:1) on a steam bath and then at 200-205° (280).
By decompg. $(o\text{-}MeC_6H_4N_2Cl)_2 \cdot SnCl_4$ in the presence of Sn in boiling EtOAc; 20% yield (281, 386).
By reacting $SnCl_2$ with $(o\text{-}MeC_6H_4)_2Hg$ (2.5:0.78) in ligroin at 70°; 66% yield (729).
Props.: m. 49-50° (280); m. 49.5-50° (729).
Rxns. with: NH_4OH in EtOH → $(o\text{-}MeC_6H_4)_2SnO$ (386).

DI-p-TOLYLTIN DICHLORIDE $(p\text{-}MeC_6H_4)_2SnCl_2$
Prepn.: By heating $(p\text{-}MeC_6H_4)_4Sn$ with $SnCl_4$ (1:1) on a water bath and then at 200-205° (280).
By reacting $SnCl_2$ with $(p\text{-}MeC_6H_4)_2Hg$ (2.5:0.78) in ligroin at 90-95°; 72% yield (729).
By heating $(p\text{-}MeC_6H_4)_2TlCl$ with $SnCl_2$ over free flame; 60% yield (730).
Props.: m. 49-50° (280); m. 49.5° (729); m. 49-49.5° (730).
Rxns. with: $HgCl_2$ (1:4) → $p\text{-}MeC_6H_4HgCl$, and with (1:1) $HgCl_2$ + KOH → $(p\text{-}MeC_6H_4)_2Hg$ (280).
$SnCl_4$ (1:1) → $p\text{-}MeC_6H_4SnCl_3$ (280).
$TlCl_3 \cdot 4H_2O$ and 20% NaOH → $(p\text{-}MeC_6H_4)_2TlCl$ (289, 290).

DI-p-TOLYLTIN DIBROMIDE $(p\text{-}MeC_6H_4)_2SnBr_2$
Prepn.: By pouring $(p\text{-}MeC_6H_4)_2SnO$ dissolved in AcOH into dil. aq. HBr (425).
Props.: m. 74° (425).

DIBENZYLTIN DICHLORIDE $(PhCH_2)_2SnCl_2$

Prepn.: By irradiating $(PhCH_2)_2SnPh_2$ in $CHCl_3$ or CCl_4; 30% yield (434).

By reacting $(PhCH_2)_2SnPh_2$ with alc. HCl; 80% yield (434).

By heating $(PhCH_2)_2SnPh_2$ with $(CH_2CO)_2NH$ or $(CH_2CO)_2NBr$ and treating the rxn. prod. with alc. HCl (434).

Props.: m. 163° (434).

Rxn. with: RCO_2Na in C_6H_6 → $(PhCH_2)_2Sn(O_2CR)_2$ (507).

Biol. props.: Efficacy in treatment of pathological conditions in animals (684).

DI-1-NAPHTHYLTIN DICHLORIDE $(1-C_{10}H_7)_2SnCl_2$

Prepn.: By heating $(p-MeOC_6H_4)_2Sn(1-C_{10}H_7)_2$ with HCl; 89% yield (57).

By reacting $1-C_{10}H_7MgBr$ with $SnCl_4$ in $Et_2O-C_6H_6$ and treating the resulting pale yellow powder with additional $SnCl_4$ in a sealed tube in xylene at 150° (414).

By reacting $SnCl_2$ with $(1-C_{10}H_7)_2Hg$ in ligroin at 90-95°; 53% yield.

Props.: m. 136-137° (57, 729); m. 137.5° (414).

Rxns. with: RMgBr → $(1-C_{10}H_7)_2SnR_2$ (57).

$1-C_{10}H_7MgBr$ → pale yellow powder (414).

$HgCl_2$ in alc. → $1-C_{10}H_7HgCl$ (414).

Aq. NH_4OH → $(1-C_{10}H_7)_2SnO$ (414).

5% KOH satd. with H_2S in EtOH → $(1-C_{10}H_7)_2SnS$ (414).

NaI in Me_2CO → $(1-C_{10}H_7)_2SnI_2$ (414).

$SnCl_4$ (1:1) → $1-C_{10}H_7SnCl_3$ (414).

Br_2 in CHCl → $(1-C_{10}H_7)_2SnBr_2$ (414).

DI-1-NAPHTHYLTIN DIBROMIDE $(1-C_{10}H_7)_2SnBr_2$

Prepn.: By reacting $1-C_{10}H_7MgBr$ with $SnCl_4$ in $Et_2O-C_6H_6$ and treating the resulting pale-yellow powder with Br_2 in $CHCl_3$ (414).

By treating $(1-C_{10}H_7)_2SnO$ with HBr (414).

By heating $(1-C_{10}H_7)_2TlBr$ with $SnBr_2$ at 200°; 50% theory (730).

Props.: m. 142° (414); m. 140.5-142° (730).

DI-1-NAPHTHYLTIN DIIODIDE $(1-C_{10}H_7)_2SnI_2$

Prepn.: By treating $(1-C_{10}H_7)_2SnCl_2$ with NaI in Me_2CO (414).

Props.: m. 160° (414).

DI-2-BIPHENYLYLTIN DICHLORIDE $(2-PhC_6H_4)_2SnCl_2$

Prepn.: By treating $(2-PhC_6H_4)_4Sn$ with HCl (39).

DI-4-BIPHENYLYLTIN DICHLORIDE $(p\text{-}PhC_6H_4)_2SnCl_2$
Prepn.: By heating $(PhC_6H_4)_4Sn$ with $SnCl_4$ (1:1) in a sealed tube
   at 145-185°; 80% yield (539).
Props.: m. 140°; soly. data (539).
Rxns. with: 5% KOH in EtOH satd. with $H_2S \rightarrow (PhC_6H_4)_2SnS$ (539).
NH_4OH in EtOH $\rightarrow (PhC_6H_4)_2SnO$ (539).

DI-4-BIPHENYLYLTIN DIBROMIDE $(p\text{-}PhC_6H_4)_2SnBr_2$
Prepn.: By heating $(PhC_6H_4)_4Sn$ with $SnBr_4$ (1:1) in a sealed tube
   at 90-210°; 70.6% yield (539).
Props.: m. 144-145°; soly. data (539).

DI(9-PHENANTHRYL)TIN DICHLORIDE $(C_{14}H_9)_2SnCl_2$
Prepn.: By treating $(C_{14}H_9)_4Sn$ with $HClCHCl_3$ under reflux (40).
Props.: White powder (40).

FUNCTIONALLY SUBSTITUTED DIORGANOTIN HALIDES

DIVINYLTIN DICHLORIDE $(CH_2{:}CH)_2SnCl_2$
Prepn.: By adding $SnCl_4$ to $(CH_2{:}CH)_4Sn$ (1:1) and completing the
   rxn. on a steam bath; 70% yield (487).
By reacting $(CH_2{:}CH)_4Sn$ with $SnCl_4$ (1:1) at 30°; 98% yield (460).
By heating $(CH_2{:}CH)_2TlCl$ with $SnCl_2$ in Me_2CO (726).
Props.: Needles, m. 74.5-75.5° (487); m. 13.2°, b. 54-56°/3mm;
$n^{25}D$ 1.5490; d_{20} 1.7645 (460); b. 53-55°/2mm; $n^{20}D$ 1.5500;
d_{20} 1.7621 (726).
Rxns. with: $NaSCN \rightarrow (CH_2{:}CH)_2Sn(SCN)_2$ (487).
Aq. $NH_3 \rightarrow (CH_2{:}CH)_2SnO$ (487).
$o\text{-}HO_2CC_6H_4CO_2K \rightarrow (CH_2{:}CH)_2SnO_2CC_6H_4\text{-}o\text{-}COO$ (487).
$PhPO(OH)_2 \rightarrow (CH_2{:}CH)_2SnO_2PPhO$ (487).
Aq. $NaOH \rightarrow (CH_2{:}CH)_2SnO$ (460).
$RMgBr \rightarrow (CH_2{:}CH)_2SnR_2$ (493).

DIVINYLTIN DIBROMIDE $(CH_2{:}CH)_2SnBr_2$
Prepn.: By heating $(CH_2{:}CH)_2TlBr$ with $SnBr_2$ in Me_2CO (726).
Props.: b. 47°/2mm; $n^{20}D$ 1.5920; d_{20} 2.1924 (726).

DIALLYLTIN DIBROMIDE $(CH_2{:}CHCH_2)_2SnBr_2$
Prepn.: By reacting tetraallyltin with $SnBr_4$ in a sealed tube
   at 50°/20mm (553).
Props.: b. 77-79°/2mm (553).
Addn. compd.: $(CH_2{:}CHCH_2)_2SnCl_2 \cdot 2C_5H_5N$, b. 99° (553).

DI-(cis-PROPENYL)TIN DICHLORIDE   (MeCH:CH)$_2$SnCl$_2$
Prepn.: By reacting (cis-MeCH:CH)$_2$TlCl with SnCl$_2$ (395).
Rxns. with: HgBr$_2$ → (cis-MeCH:CH)$_2$Hg and cis-MeCH:CHHgBr (395).
TlCl$_3$ → (cis-MeCH:CH)$_2$TlCl (395).

DI-(trans-PROPENYL)TIN DICHLORIDE   (MeCH:CH)$_2$SnCl$_2$
Prepn.: By reacting (trans-MeCH:CH)$_2$TlCl with SnCl$_2$ (395).
Rxns. with: HgBr$_2$ → (trans-MeCH:CH)$_2$Hg (395).
TlCl$_3$ → (trans-MeCH:CH)$_2$TlCl (395).

DI-(cis-PROPENYL)TIN DIBROMIDE   (MeCH:CH)$_2$SnBr$_2$
Prepn.: By reacting (cis-MeCH:CH)$_2$Hg with SnBr$_2$ (395).
By heating (cis-MeCH:CH)$_2$TlBr with SnBr at 200-220°; 70% yield (730).
Props.: b. 125°/10mm; n^{20}D 1.5845 (730).
Rxns. with: HgBr$_2$ → (cis-MeCH:CH)$_2$Hg and cis-MeCH:CHHgBr (395).
TlCl$_3$ → (cis-MeCH:CH)$_2$TlCl (395).

DI-(trans-PROPENYL)TIN DIBROMIDE   (MeCH:CH)$_2$SnBr$_2$
Prepn.: By reacting (trans-MeCH:CH)$_2$Hg with SnBr$_2$ (395).
By heating (trans-MeCH:CH)$_2$TlBr with SnBr$_2$ at 200-220°; 62% yield (730).
Props.: b. 123-124°/8mm; n^{20}D 1.5792 (730).
Rxns. with: HgBr$_2$ → (trans-MeCH:CH)$_2$Hg (395).
TlCl$_3$ → (trans-MeCH:CH)$_2$TlCl (395).

DIPROPENYLTIN DIBROMIDE   (MeCH:CH)$_2$SnBr$_2$
Prepn.: By reacting (MeCH:CH)$_2$TlBr with SnBr$_2$ in C$_6$H$_6$ at 50° (728, 730).
Props.: b. 120°/10mm; n^{20}D 1.5840 (728, 730); b. 121-122°/10mm; n^{20}D 1.5823 (730).

DIISOPROPENYLTIN DIBROMIDE   [CH$_2$:C(Me)]$_2$SnBr$_2$
Prepn.: By reacting SnBr$_2$ with [CH$_2$:C(Me)]$_2$TlBr at 200°; 75% yield (728, 730).
By reacting SnBr$_2$ with [CH$_2$:C(Me)]$_2$TlBr in Me$_2$CO 5 hrs. at r.t. and 3 hrs. at 50°, extracting the residue obtd. from the Me$_2$CO-sol. fraction with petroleum ether, treating the prod. with 20% KOH, and reacting the resulting solid with HBr (728, 729, 730).
By reacting SnBr$_2$ with [CH$_2$:C(Me)]$_2$Hg in petr. ether at r.t. or at 65°, concentrating the petr. ether soln., treating the residue with 20% KOH, and reacting the resulting ppt. with HBr; the alk. soln. retained tetraisopropenyltin. If the rxn. is carried out in Me$_2$CO or without solvent, tetraisopropenyltin is the main prod. (728, 729).
Props.: b. 100-101°/10mm; n^{20}D 1.5665; d$^{20}_4$ 1.9363; MR 60.77 (728, 730); b. 102.5°/9mm; n^{20}D 1.5667; d$^{20}_4$ 1.9360; MR 60.84 (729).

BUTYLVINYLTIN DICHLORIDE $CH_2{:}CHSnBuCl_2$
Prepn.: By reacting a soln. of $BuSnCl_3$ in heptane with a soln. of $CH_2{:}CHMgCl$ in C_4H_8O (1:1) at 25-40°; 53.5% yield (459).
Props.: b. 82-84°/3mm; $n^{25}D$ 1.4970; d_{25} 1.370 (459).

DI(cis-2-CHLOROVINYL)TIN DICHLORIDE $(ClCH{:}CH)_2SnCl_2$
Prepn.: By reacting (cis-$ClCH{:}CH)_2Hg$ with $SnCl_2$ in EtOH contg. HCl at 45-50°; 50% yield (388).
By reacting $(ClCH{:}CH)_2PbCl_2$ with $SnCl_2$ in EtOH at 50-55°; 67% yield (390).
By-prod. in the rxn. of (cis-$ClCH{:}CH)_2Hg$ with Sn in EtOH contg. HCl at 50° (391).
Props.: b. 100-102°/3mm, $n^{20}D$ 1.5675; d^{20}_4 1.7494 (388); m. 77-79° (390).
Rxns. with: 50% KOH → C_2H_2 (388).
$HgCl_2$ → cis-$ClCH{:}CHHgCl$ (388).

DI-(trans-2-CHLOROVINYL)TIN DICHLORIDE $(ClCH{:}CH)_2SnCl_2$
Prepn.: By reacting (trans-$ClCH{:}CH)_2Hg$ with $SnCl_2$ in slightly acidified (HCl) EtOH at 45-50°; 58% yield (388).
By heating (trans-$ClCH{:}CH)_2Hg$ with Sn in EtOH contg. HCl at 50° (391).
By reacting (trans-$ClCH{:}CH)_2TlCl$ with $SnCl_2$ in MeOH under reflux; 0.36% yield (136a).
Props.: m. 77.5-78.5° (391); m. 76.5-78° (136a).
Rxns. with: $TlCl_3 \cdot 4H_2O$ and Na_2CO_3 → $(ClCH{:}CH)_2TlCl$ (290).
15% KOH → $SnCl_4$ and C_2H_2 (388, 391).
$HgCl_2$ → $ClCH{:}CHHgCl$ (388).
Addn. Compds.: $(ClCH{:}CH)_2SnCl_2 \cdot 2C_5H_5N$, slightly sol. in org. solvents (391).

BIS(CHLOROMETHYL)TIN DICHLORIDE $(ClCH_2)_2SnCl_2$
Prepn.: By-prod. in the rxn. of $SnCl_4$ with CH_2N_2 (3:2) in C_6H_6 at 3-5° (575).
Props.: m. 89.5-90° (574, 575).
Rxn. with: $MeCHN_2$ (1:4) in C_6H_6 → $(ClCH_2)_2Sn(CHClMe)_2$ (575).

BIS(BROMOMETHYL)TIN DIBROMIDE $(BrCH_2)_2SnBr_2$
Prepn.: By reacting $SnBr_4$ with CH_2N_2 (1:2) in C_6H_6 at 1-2° (575).
Props.: m. 87° (574, 575).
Rxn. with: CH_2N_2 (1:3) in C_6H_6 → $(BrCH_2)_4Sn$ (575).

BIS(1-CHLOROETHYL)TIN DICHLORIDE $(MeCHCl)_2SnCl_2$
Prepn.: By-prod. in the rxn. of $SnCl_4$ with $MeCHN_2$ (1:2) in C_6H_6 at 2-4° (575).
Props.: m. 12°; b. 112°/4mm; $d^{20}{}_{20}$ 1.829; $n^{20}{}_D$ 1.5535 (574, 575).
Rxn. with: NH_4OH in EtOH → $(MeCHCl)_2SnO$ (575).

BIS(2-CHLOROETHYL)TIN DIIODIDE (?) $(ClCH_2CH_2)_2SnI_2$
Prepn.: By refluxing Sn powder with $ClCH_2CH_2I$ (160).
Props.: Instable, red liquid, b. 65-70°/1mm (160).

BIS(1-CHLOROBUTYL)TIN DICHLORIDE $(PrCHCl)_2SnCl_2$
Prepn.: By reacting $SnCl_4$ with $PrCHN_2$ (1:2) in C_6H_6 at 1° (575).
Props.: m. 53°; b. 134°/5mm (574, 575).

BIS(p-CHLOROPHENYL)TIN DICHLORIDE $(p\text{-}ClC_6H_4)_2SnCl_2$
Prepn.: By reacting $(p\text{-}ClC_6H_4N_2Cl)_2 \cdot SnCl_4$ with Sn powder in EtOAc (281, 386).
By pouring $(p\text{-}ClC_6H_4)_2SnO$ dissolved in AcOH into dil. aq. HCl (425).
Props.: m. 88° (425).
Rxns. with: $TlCl_3 \cdot 4H_2O$ in the presence of 20% NaOH → $(p\text{-}ClC_6H_4)_2TlCl$ (289, 290).
NH_4OH in EtOH → $(p\text{-}ClC_6H_4)_2SnO$ (386).

BIS(p-IODOPHENYL)TIN DICHLORIDE $(p\text{-}IC_6H_4)_2SnCl_2$
Prepn.: By adding a soln. of $(p\text{-}IC_6H_4)_2SnO$ in AcOH to dil. aq. HCl (425).
Props.: m. 147° (425).

BIS(p-BROMOPHENYL)TIN DICHLORIDE $(p\text{-}BrC_6H_4)_2SnCl_2$
Prepn.: By reacting $(p\text{-}BrC_6H_4N_2Cl)_2 \cdot SnCl_4$ with Sn powder in EtOAc (281, 386).
By pouring $(p\text{-}BrC_6H_4)_2SnO$ dissolved in AcOH into dil. aq. HCl (425).

BIS(o-METHOXYPHENYL)TIN DICHLORIDE $(o\text{-}MeOC_6H_4)_2SnCl_2$
Prepn.: By reacting $(o\text{-}MeOC_6H_4N_2Cl)_2 \cdot SnCl_4$ with Sn powder in EtOAc, EtOH, or MeOH (281, 386).
Props.: m. 113° (281, 386).

BIS(p-METHOXYPHENYL)TIN DICHLORIDE (p-MeOC$_6$H$_4$)$_2$SnCl$_2$
Prepn.: By heating (p-MeOC$_6$H$_4$)$_2$Sn(α-C$_4$H$_3$S)$_2$ with HCl (57).
By heating (p-MeOC$_6$H$_4$)$_4$Sn with SnCl$_4$ in a sealed tube at 80-185°; 87% yield (539).
Props.: m. 76° (539).
Rxns. with: 5% KOH and H$_2$S in EtOH → (p-MeOC$_6$H$_4$)$_2$SnS (539).
NH$_4$OH in EtOH → (p-MeOC$_6$H$_4$)$_2$SnO (539).

BIS(p-METHOXYPHENYL)TIN DIBROMIDE (p-MeOC$_6$H$_4$)$_2$SnBr$_2$
Prepn.: By heating (p-MeOC$_6$H$_4$)$_4$Sn with SnBr$_4$ in a sealed tube at 80-185°; 74% yield (539).
Props.: m. 102° (539).

DIPHENETYLTIN DICHLORIDE (p-EtOC$_6$H$_4$)$_2$SnCl$_2$
Prepn.: By heating (p-EtOC$_6$H$_4$)$_4$Sn with SnCl$_4$ in a sealed tube at 80-185°; 66% yield (539).
Props.: m. 46°; soly. data (539).
Rxns. with: 5% KOH in EtOH satd. with H$_2$S → (p-EtOC$_6$H$_4$)$_2$SnS (539).
NH$_4$OH in EtOH → (p-EtOC$_6$H$_4$)$_2$SnO (539).

BIS(CARBETHOXYMETHYL)TIN DIBROMIDE (EtO$_2$CCH$_2$)$_2$SnBr$_2$
Prepn.: By reacting Sn powder with BrCH$_2$CO$_2$Et; 15.5% yield (160, 306).
Props.: m. 139° (160, 306).

BIS(1-CARBETHOXYETHYL)TIN DIBROMIDE (EtO$_2$CCHMe)$_2$SnBr$_2$
Prepn.: By heating Sn powder with MeCHBrCO$_2$Et at 150° (160).
Props.: b. 82-85°/1.5mm (160).

BIS(2-CARBOMETHOXYETHYL)TIN DIBROMIDE (MeO$_2$CCH$_2$CH$_2$)$_2$SnBr$_2$
Prepn.: By reacting Pr$_2$Sn(CH$_2$CH$_2$CO$_2$Me)$_2$ with Br$_2$ (1:2) in CHCl$_3$; 46% yield (682).
Props.: White needles, m. 138-140° (682).

BIS-(p-CARBETHOXYPHENYL)TIN DICHLORIDE (EtO$_2$CC$_6$H$_4$)$_2$SnCl$_2$
Prepn.: By reacting (p-EtO$_2$CC$_6$H$_4$)$_2$Hg with SnCl$_2$ in EtOH and Me$_2$CO under reflux or without solvent at 250-260°; 86% yield (123).
Props.: m. 102-103° (123).
Rxns. with: HgCl$_2$ → p-EtO$_2$CC$_6$H$_4$HgCl (123).
PhMgCl (1:1) → Ph$_2$Sn(C$_6$H$_2$Ph$_2$·OH)$_2$ (123).
H$_2$S in 5% alc. KOH → (p-EtO$_2$CC$_6$H$_4$)$_2$SnS (123).
8-Hydroxyquinoline → (p-EtO$_2$CC$_6$H$_4$)$_2$Sn(OC$_9$H$_6$N)$_2$ (123).
TlCl$_3$·4H$_2$O and Na$_2$CO$_3$ → (p-EtO$_2$CC$_6$H$_4$)$_2$TlCl (289, 290).

BIS(p-CARBETHOXYPHENYL)TIN DIBROMIDE $(EtO_2CC_6H_4)_2SnBr_2$
Prepn.: By heating $SnBr_2$ with $(EtO_2CC_6H_4)_2Hg$ at 225-235°; 57% yield (123).
Props.: m. 69-69.5° (123).
Rxns. with: $NH_4OH \rightarrow (EtO_2CC_6H_4)_2SnO$ (123).
8-Hydroxyquinoline $\rightarrow (EtO_2CC_6H_4)_2Sn(OC_9H_6N)_2$ (123).

BIS(p-CARBETHOXYPHENYL)TIN DIIODIDE $(EtO_2CC_6H_4)_2SnI_2$
Prepn.: By reacting SnI_2 with $(EtO_2CC_6H_4)_2Hg$ at 250-260° or in EtOH under reflux (123).
Props.: Oily product (123).
Rxn. with: 8-Hydroxyquinoline $\rightarrow (EtO_2CC_6H_4)_2Sn(OC_9H_6N)_2$ (123).

(2-CARBOMETHOXYETHYL)PHENYLTIN DIBROMIDE $Ph(MeO_2CCH_2CH_2)SnBr_2$
Prepn.: By treating $Ph_3SnCH_2CH_2CO_2Me$ with Br_2 (1:2) in C_6H_6; 86% yield (682).
Props.: m. 111-114° (682).

(3-CYANOPROPYL)PHENYLTIN DIBROMIDE $Ph(NCCH_2CH_2CH_2)SnBr_2$
Prepn.: By treating $Ph_3Sn(CH_2)_3CN$ with Br_2 (1:2) in $CHCl_3$; 71% yield (682).
Props.: m. 101-102° (682).

SILYL- AND HETEROCYCLE-SUBSTITUTED DIORGANOTIN DIHALIDES

BIS[(TRIMETHYLSILYL)METHYL]TIN DIBROMIDE $(Me_3SiCH_2)_2SnBr_2$
Prepn.: By treating $(Me_3SiCH_2)_2SnPh_2$ with HBr (1:2) in CH_2Cl_2 at -78°; 98% yield (493).
Props.: m. 38.6-39.8° (493).
Rxns. with: Aq. KOH (1:2) $\rightarrow (Me_3SiCH_2)_2SnO$ (493).

BIS[(TRIMETHYLSILYL)METHYL]TIN DIIODIDE $(Me_3SiCH_2)_2SnI_2$
Prepn.: By heating $(Me_3SiCH_2)_2SnPh_2$ with I_2 (1:2) in C_6H_6 under reflux; 83% yield (493).
Props.: m. 34.6-35.4° (493).

DI-α-THIENYLTIN DICHLORIDE $(\alpha-C_4H_3S)_2SnCl_2$
Prepn.: By reacting $(\alpha-C_4H_3S)_4Sn$ with $SnCl_4$ (57).
By heating $(\alpha-C_4H_3S)_2Sn(\alpha-C_{10}H_7)_2$ with HCl; 90.2% yield (57).
Rxns. with: RLi $\rightarrow (\alpha-C_4H_3S)_2SnR_2$ (57).
RMgBr $\rightarrow (\alpha-C_4H_3S)_2SnR_2$ (57).

ORGANOTIN TRIHALIDES

ALKYL- AND ARYLTIN TRIHALIDES

Organotin trihalides are prepared by the following methods:

1. By treating dihydrocarbyl- or trihydrocarbyltin halides or tetrahydrocarbyltin compounds with calculated amounts of tin tetrahalides.
2. By treating organostannonic acids with hydrogen halides.
3. By reacting potassium trihalostannites with alkyl or aryl halides.
4. By treating tin tetrahalides with diazoalkanes (haloalkyl derivatives are formed).
5. By decomposing tin tetrahalide-aryldiazonium salt complexes in the presence of Sn powder.
6. By reacting alkyl halides with tin dihalides at elevated temperature.

METHYLTIN TRICHLORIDE $MeSnCl_3$

Prepn.: By reacting $MeSnPh_3$ with $SnCl_4$ (1:3) at 185-190°; the prod. contains $MeSnCl_3$ and $PhSnCl_3$ (409).
By passing MeCl through SnI_2 at 365° (517).
Props.: m. 45-46° (409); Raman and IR spectra (800); Molecular structure and bond length (514).
Rxns. with: $RSH \rightarrow MeSn(SR)_3$ (429).
$PhLi$ in $Et_2O \rightarrow MeSnPh_3$ (409).
Addn. Compds.: $MeSnCl_3 \cdot 2\ C_5H_5N$, non-fusible substance (284).
The following addn. compds. were prepd. by treating ArN_2Cl or $ArN_2Cl \cdot FeCl_3$ with $MeSnO_2H$ (2:1) in concd. HCl and MeOH satd. with HCl, respectively: $MeSnCl_3 \cdot 2\ p\text{-}MeC_6H_4N_2Cl$, decomp. 112° (437).
$MeSnCl_3 \cdot 2\ p\text{-}BrC_6H_4N_2Cl$, decomp. 147° (437).
$MeSnCl_3 \cdot 2\ p\text{-}EtOC_6H_4N_2Cl$, decomp. 108° (437).
$MeSnCl_3 \cdot 2\ o\text{-}O_2NC_6H_4N_2Cl$, decomp. 106° (437).

METHYLTIN TRIBROMIDE $MeSnBr_3$

Props.: Molecular structure and bond length (514).
Addn. compds.: $MeSnBr_3 \cdot 2\ C_5H_5N$, m. 203° (284).

METHYLTIN TRIIODIDE $MeSnI_3$

Prepn.: By treating Sn with MeI at 385° (517).
By heating $KSnCl_3$ with MeI in a sealed tube at 90° (544).
Props.: m. 84-85.5° (517); m. 85° (544); Molecular structure and bond length (514); Polarographic studies (550).

METHYLTIN TRIIODIDE   MeSnI$_3$ (Cont'd.)

Rxns. with: (EtO)$_3$P (1:2) → MeISn[P(OEt)$_3$]$_2$ (31).
(MeO)$_3$P → MeSn[PO(OMe)$_2$]$_3$? (31).
H$_2$O → MeSnO$_2$H (544).
Addn. compd.: MeSnI[P(OEt)$_3$]$_2$, golden-yellow solid, m. 161-162°, prepd. by reacting MeSnI$_3$ with (EtO)$_3$P (1:2) at 85-90° (31).

ETHYLTIN TRICHLORIDE   EtSnCl$_3$

Prepn.: By refluxing Et$_4$Sn with SnCl$_4$ (1:3) (113, 256).
Props.: b. 196-198° (256); m. -10°; d$_{20}$ 1.965; n^{20}D 1.5408; b. 192° (calcd.); thermodynamic data (113).
Rxn. with: LiAlH$_4$ → EtSnH$_3$ (113).

ETHYLTIN TRIIODIDE   EtSnI$_3$

Prepn.: By heating KSnCl$_3$ with excess EtI in a sealed tube at 110°; 37% yield (544).
Props.: b. 181-184°/19mm (544); Polarographic studies (550).
Rxn. with: H$_2$O → EtSnO$_2$H (544).

PROPYLTIN TRIIODIDE   PrSnI$_3$

Prepn.: By heating KSnCl$_3$ with excess PrI in a sealed tube at 130°; 25% yield (543).
Props.: b. 200°/16mm (decompn.) (544).
Rxn. with: H$_2$O → PrSnO$_2$H (544).

ISOPROPYLTIN TRICHLORIDE   i-PrSnCl$_3$

Prepn.: By heating KSnCl$_3$ with excess i-PrI in a sealed tube at 130°; 40% yield (544).
Props.: b. 75°/16mm (544).
Rxn. with: H$_2$O → i-PrSnO$_2$H (544).

BUTYLTIN TRICHLORIDE   BuSnCl$_3$

Prepn.: By reacting SnCl$_4$ with BuMgCl (1:1) in Et$_2$O-C$_6$H$_6$; 13% yield (272).
By reacting a mixt. of Bu$_2$SnCl$_2$, Bu$_3$SnCl and Bu$_4$Sn with SnCl$_4$ at 0° (192).
By reacting Bu$_4$Sn with SnCl$_4$ (1:1) at 0-20°; the prod. contains about equiv. amts. of BuSnCl$_3$ and Bu$_3$SnCl (609).
Props.: Colorless liquid, b. 102-103°/12mm (272).
Rxns. with: LiAlH$_4$ → BuSnH$_3$ (272).
RSH in the presence of Na$_2$CO$_3$ or NaOH → BuSn(SR)$_3$ (429).
RMgCl (1:1) → BuRSnCl$_2$ (459).
RMg halide (1:4) → BuSnR$_3$ (487, 460a).

BUTYLTIN TRIIODIDE   BuSnI$_3$
Prepn.: By heating KSnCl$_3$ with BuI at 90°; 25% yield (160, 306).
Props.: Instable, red liquid, b. 154°/5mm (160, 306).

PHENYLTIN TRICHLORIDE   PhSnCl$_3$
Prepn.: By heating Ph$_4$Sn with SnCl$_4$ (1:3) at 150°; 71-78% yield (185, 594).
By reacting PhSn(CH:CH$_2$)$_3$ with SnCl$_4$ (487).
By heating MeSnPh$_3$ with SnCl$_4$ (1:3) at 185-190°; the prod. contains PhSnCl$_3$ and MeSnCl$_3$ (409).
Props.: b. 96-97°/1.4mm (185); b. 128°/15mm (594); b. 142-143°/25mm (409); UV spectrum (663).
Rxns. with: RMg halides → PhSnR$_3$ (182, 185, 319, 487).
LiAlH$_4$ → BuSnH$_3$ (272).
Na-Sn → Ph$_3$SnCl (382).
NaHAl(OEt)$_3$ → PhSn[HAl(OEt)$_3$]$_3$ (475).
RLi → PhSnR$_3$ (594).
N$_2$O$_3$ (1:2) → PhN$_2$NO$_3$ (350a).
ClCH:CHI·Cl$_2$ → SnCl$_4$ and ClCH:CHPhICl (59a).
ICH:CHI·Cl$_2$ → SnCl$_4$Ph$_2$ICl, and C$_2$H$_2$ (59a).
Ph$_2$ICl in dil. HCl → SnCl$_4$ and Ph$_2$ICl (665a).
PhICl$_2$ in dil. HCl → SnCl$_4$ and Ph$_2$ICl (655a).
Addn. compds.: Ph$_3$CSnCl$_4$·Ph, brown yellow complex (284).
Ph$_3$SnCl$_3$·2 C$_5$H$_5$N (284).

PHENYLTIN TRIIODIDE   PhSnI$_3$
Prepn.: By heating KSnCl$_3$ with excess PhI in a sealed tube at 210° (544).
Props.: Decomp. 220° (544).
Rxn. with: H$_2$O → PhSnO$_2$H (544).

o-TOLYLTIN TRICHLORIDE   o-MeC$_6$H$_4$SnCl$_3$
Prepn.: By heating (o-MeC$_6$H$_4$)$_4$Sn with SnCl$_4$ (1:3) at 210-215° in a sealed tube (280).
Props.: b. 157-158°/20mm; d^{20}_4 1.7719 (280).
Rxns. with: H$_2$S in H$_2$O → (o-MeC$_6$H$_4$SnS)$_2$S (280).
HgCl$_2$ → o-MeC$_6$H$_4$HgCl (280).
20% KOH → o-MeC$_6$H$_4$SnO$_2$H (280).
HgCl$_2$ (1:1) in 5N KOH → (o-MeC$_6$H$_4$)$_2$Hg (280).
Addn. compds.: o-MeC$_6$H$_4$SnCl$_4$·Ph$_3$C, brown-yellow (284).
p-MeC$_6$H$_4$SnCl$_3$·2 C$_5$H$_5$N (284).

p-TOLYLTIN TRICHLORIDE   p-MeC$_6$H$_4$SnCl$_3$
Prepn.: By heating (p-MeC$_6$H$_4$)$_4$Sn with SnCl$_4$ (1:3) in a sealed tube at 210-215°; 40% yield (280).
By heating (p-MeC$_6$H$_4$)$_2$SnCl$_2$ with SnCl$_4$ (1:1) at 100-215°; 45% yield (280).
Props.: b. 152-153°/15mm; d^{20}$_4$ 1.7512 (280).
Rxn. with: 20% KOH → p-MeC$_6$H$_4$SnO$_2$H (280).
Addn. compds.: p-MeC$_6$H$_4$SnCl$_4$·CPh$_3$, brown-yellow complex (284).
p-MeC$_6$H$_4$SnCl$_3$·2 C$_5$H$_5$N (280).

1-NAPTHTHYLTIN TRICHLORIDE   1-C$_{10}$H$_7$SnCl$_3$
Prepn.: By heating (1-C$_{10}$H$_7$)$_2$SnCl$_2$ with SnCl$_4$ in a sealed tube at 150°; 91% yield (414).
Props.: m. 77-78° (414).
Rxns. with: Alkali → 1-C$_{10}$H$_7$SnO$_2$H (414).
HgCl$_2$ → 1-C$_{10}$H$_7$HgCl (414).
RMgBr → 1-C$_{10}$H$_7$SnR$_3$ (414).

FUNCTIONALLY SUBSTITUTED ALKYL- AND ARYLTIN TRIHALIDES

CHLOROMETHYLTIN TRICHLORIDE   ClCH$_2$SnCl$_3$
Prepn.: By reacting SnCl$_4$ with CH$_2$N$_2$ (3:2) in C$_6$H$_6$ at 3-5°; (ClCH$_2$)$_2$SnCl$_2$ and (ClCH$_2$)$_3$SnCl are also formed (575).
Props.: b. 72.5-73°/5mm; d^{20}$_{20}$ 2.21; n^{20}$_D$ 1.5689 (574, 575).
Rxns. with: H$_2$O → hydrolysis (575).
CH$_2$N$_2$ (1:6) in C$_6$H$_6$ at 3-5° → (ClCH$_2$)$_4$Sn (575).
MeCHN$_2$ (1:2) in C$_6$H$_6$ at 2-3° → (MeCHCl)$_2$ClCH$_2$SnCl (575).

BROMOMETHYLTIN TRIBROMIDE   BrCH$_2$SnBr$_3$
Prepn.: By reacting SnBr$_4$ with CH$_2$N$_2$ (1:2) in C$_6$H$_6$ at 1-2°; the prod. contains also (BrCH$_2$)$_2$SnBr$_2$ and (BrCH$_2$)$_3$SnBr (575).
Props.: b. 109°/5mm; d^{20}$_{20}$ 3.251 (574, 575).

1-CHLOROETHYLTIN TRICHLORIDE   (MeCHCl)SnCl$_3$
Prepn.: By reacting SnCl$_4$ with MeCHN$_2$ (2:1) in C$_6$H$_6$ at 2-4° (575).
Props.: b. (crude) 69-71°/4mm (575).

p-BROMOPHENYLTIN TRIBROMIDE   p-BrC$_6$H$_4$SnBr
Prepn.: By heating (p-BrC$_6$H$_4$)$_2$SnBr$_2$ with SnBr$_4$ (1:1); 90.5% yield (583).
Props.: m. 49-50° (583).
Rxn. with: RMgI (1:3) in Et$_2$O → p-BrC$_6$H$_4$SnR$_3$ (583).

p-IODOPHENYLTIN TRIBROMIDE $p\text{-}IC_6H_4SnBr_3$
<u>Rxn. with</u>: $RMgBr$ (1:3) → $p\text{-}IC_6H_4SnR_3$ (583).

2-CHLOROVINYLTIN TRICHLORIDE $ClCH:CHSnCl_3$
<u>Prepn.</u>: By-prod. in the rxn. of trans-$(ClCH:CH)_2Hg$ with Sn in EtOH (391).
<u>Props.</u>: b. 63-65°/4mm; $n^{29}D$ 1.5602; $d^{29}4$ 2.0362 (391).

VINYLTIN TRICHLORIDE $CH_2:CHSnCl_3$
<u>Prepn.</u>: By adding $(CH_2:CH)_4Sn$ to $SnCl_4$ (1:3) and completing the rxn. on a steam bath; 77% yield (487); 86% yield obtd. by carrying the rxn. at 30-70° (460).
By reacting $SnCl_4$ with $(CH_2:CH)_2Hg$ (626a).
<u>Props.</u>: b. 48-55°/5.2mm (487); b. 64-65°/15mm; $n^{25}D$ 1.5361; d_{25} 1.9981 (460).
<u>Rxn. with</u>: $LiAlH_4$ → $CH_2:CHSnH_3$ (626a).

VINYLTIN TRIBROMIDE TRIMER $(CH_2:CHSnBr_3)_3$
<u>Prepn.</u>: By treating $CH_2:CHSnO_2H$ with hot, concd. HBr (523).
<u>Props.</u>: White crystals, m. 119° (523).

ALLYLTIN TRIBROMIDE COMPLEX $CH_2:CHCH_2SnBr_3 \cdot 2\ HBr$
<u>Prepn.</u>: By reacting allylstannoic acid with hydrobromic acid (256).
<u>Props.</u>: Pale-yellow cubes sol. in H_2O, insol. in org. solvents; decomp. on heating (256).

ALLYLTIN TRIBROMIDE TRIMER $(CH_2:CHCH_2SnBr_3)_3$
<u>Prepn.</u>: By reacting $CH_2:CHCH_2SnO_2H$ with hot, concd. HBr (523).
<u>Props.</u>: White crystals, m. 109 (523).

(1,2-DIPHENYLVINYL)TIN TRICHLORIDE $PhCH:CPhSnCl_3$
<u>Prepn.</u>: By reacting cis-$(PhCH:CPh)_2Hg$ with $SnCl_2$ in Me_2CO (394).
By treating $PhCH:CPhSnO_2H$ with $SOCl_2$ (394).
<u>Props.</u>: m. 108-110° (394).
<u>Rxns. with</u>: $HgCl_2$ in EtOH → cis-$PhCH:CPhHgCl$ (394).
Alc. HgO → cis-$(PhCH:CPh)_2Hg$ (394).
Dioxane·Br_2 complex → $PhCH:CPhBr$ (394).

o-CARBOMETHOXYPHENYLTIN TRICHLORIDE $o\text{-}MeO_2CC_6H_4SnCl_3$
<u>Prepn.</u>: By reacting $(o\text{-}MeO_2CC_6H_4N_2Cl)_2 \cdot SnCl_4$ with Sn powder in EtOAc (281, 386).
<u>Props.</u>: m. 164° (281).

5. ORGANOTIN-OXYGEN COMPOUNDS

ORGANOSTANNONIC ACIDS AND THEIR DERIVATIVES

Organostannonic acids are prepared:

1. By hydrolyzing monoorgano-substituted tin trihalides.
2. By reacting alkali metal stannites with hydrocarbyl halides or sulfates.
3. By reacting aryldiazonium salts with $SnCl_2 \cdot 2\ H_2O$ and tin powder and treating the reaction product with alkali.

METHANESTANNONIC ACID   MeSnOOH
Prepn.: By reacting 5.5 moles of NaOH with 1 mole of $SnCl_2$ in H_2O and 3 moles MeCl at 20-30° under pressure; 82.7% yield (118).
By reacting Me_2SO_4 with K_2SnO_2 (523).
By hydrolyzing $MeSnI_3$ (544, 546).
Rxns. with: Mercaptans → $MeSn(SR)_3$ (52, 429, 680).
ArN_2Cl in concd. HCl → $MeSnCl_3 \cdot 2\ ArN_2Cl$ (437).
2-Mercaptothiazole (TSH) → $MeSn(ST)_3$, useful as stabilizer for vinyl chloride resins (530).
H_2O → $(MeSn)_2Se_3$ (546).

ETHANESTANNONIC ACID   EtSnOOH
Prepn.: By hydrolysis of $EtSnI_3$ (544, 546).
By reacting ethyl halides with K_2SnO_2 in aq. EtOH (523).
Rxns. with: Mercaptans → $EtSn(SR)_3$ (52, 429, 680).
H_2Se → $(EtSn)_2Se_3$ (546).

ETHYLENESTANNONIC ACID $CH_2:CHSnO_2H$
Prepn.: By bubbling $CH_2:CHBr$ into a soln of Na_2SnO_2 in aq. EtOH at 0° (523).
By reacting polyvinyl chloride of a low mol. wt. with R_2SnO_2 (523).
Rxns. with: AcOH under reflux → $(CH_2:CHSnO_2Ac)_3$, trimeric anhydride of $CH_2:CHSnO_2H$ with AcOH, which decomp. at 240° (523).
HBr (hot concd. soln.) → $(CH_2:CHSnBr_3)_3$ (523).

2-PROPENESTANNONIC ACID $CH_2:CHCH_2SnO_2H$
Prepn.: By reacting allyl halides with R_2SnO_2 in aq.-EtOH (523).
Props.: Amorphous, infusible solid (523).
Rxn. with: HBr → $(CH_2:CHCH_2SnBr_3)_3$ (523).

BUTANESTANNONIC ACID   BuSnOOH
Rxns. with: Mercaptans (RSH) → BuSn(SR)$_3$ (52, 308, 559).
2-Mercaptothiazoles (TSH) → BuSn(ST)$_3$, useful as stabilizers
   for vinyl chloride resins (530, 781).

DODECANESTANNONIC ACID $C_{12}H_{23}SnO_2H$
Prepn.: By reacting $C_{12}H_{23}Cl$ with K_2SnO_2 in aq. EtOH (523).
Rxns. with: Glycerol at 190-200° in the presence of $ZnCl_2$ →
   Dodecanestannonic triglyceride (523).
Glycerol monoacetate at 140-150° in the presence of $ZnCl_2$ →
   dodecanestannonic triglyceride and AcOH (523).
Glycerol diacetate at 130-140° in the presence of $ZnCl_2$ →
   $C_3H_5OSn(O)C_{12}H_{52}OAc$ (523).

BENZENESTANNONIC ACID   PhSnOOH
Rxns. with: Mercaptans (RSH) → PhSn(SR)$_3$ (52).
2-Mercaptothiazoles (TSH) → PhSn(ST)$_3$, useful as stabilizers
   for vinyl chloride resins (530, 781).

o-TOLUENESTANNONIC ACID   o-MeC$_6$H$_4$SnO$_2$H
Prepn.: By hydrolyzing o-MeC$_6$H$_4$SnCl$_3$ with 20% KOH and
   neutralizing the soln. with AcOH and then with CO_2 (280).
By-prod. in the rxn. of o-MeC$_6$H$_4$N$_2$BF$_4$ with Sn-SnCl$_2$·2 H$_2$O
   in Me$_2$CO, followed by a treatment with NH$_4$OH (392).
Props.: Infusible at 295° (280).
Rxn. with: K$_3$Fe(CN)$_6$ → decomp. to K$_2$SnO$_3$ (280).

p-TOLUENESTANNONIC ACID   p-MeC$_6$H$_4$SnO$_2$H
Prepn.: By hydrolyzing p-MeC$_6$H$_4$SnCl$_3$ with 20% KOH and
   neutralizing the soln. with AcOH and then with CO_2 (280).
By-prod. in the rxn. of p-MeC$_6$H$_4$N$_2$BF$_4$ with Sn-SnCl$_2$·2 H$_2$O in
   Me$_2$CO, followed by a treatment with NH$_4$OH (392).
Props.: Infusible at 295° (280).

p-CHLOROBENZENESTANNONIC ACID   p-ClC$_6$H$_4$SnO$_2$H
Prepn.: By-prod. in the rxn. of p-ClC$_6$H$_4$N$_2$BF$_4$ with Sn-SnCl$_2$·2 H$_2$O
   in Me$_2$CO, followed by a treatment with NH$_4$OH; 2.4% yield (392).

p-BROMOBENZENESTANNONIC ACID   p-BrC$_6$H$_4$SnO$_2$H
Prepn.: By-prod. in the rxn. of p-BrC$_6$H$_4$N$_2$BF$_4$ with Sn-SnCl$_2$· H$_2$O
   in Me$_2$CO at 14° and treating the prod. with NH$_4$OH; 10.8 yield
   (392).

o-CARBOMETHOXYBENZENESTANNONIC ACID $MeO_2CC_6H_4SnO_2H$
Prepn.: By-prod. in the rxn. of $MeO_2CC_6H_4N_2BF_4$ with
 Sn + $SnCl_2 \cdot 2 H_2O$ in Me_2CO at 11° and treating the rxn. prod.
 with NH_4OH (392).

1,2-DIPHENYLETHYLENESTANNONIC ACID $PhCH:CPhSnO_2H$
Prepn.: By-prod. in the rxn. of cis-$(PhCH:CPh)_2Hg$ with $SnCl_2$
 in Me_2CO followed by hydrolysis (394).
Props.: m. 157-160° (394).
Rxns. with: $SOCl_2$ → $PhCH:CPhSCl_3$ (394).
Dioxane·Br_2 complex → PhCH:CHBr (394).

1-NAPHTHALENESTANNONIC ACID $1-C_{10}H_7SnO_2H$
Prepn.: By hydrolyzing $1-C_{10}H_7SnCl_3$ (414).

ETHANEDISTANNONIC ACID $(-CH_2SnO_2H)_2$
Prepn.: By passing $BrCH_2CH_2Br$ into a soln. of K_2SnO_2 in
 aq.-EtOH (523).
Props.: Amorphous, infusible solid (523).
Rxns. with: $SOCl_2$ and C_5H_5N → $(-CH_2SnOCl)_2$ (523).
$SOCl_2$, $H_2NCH_2CH_2NH_2$, and C_5H_5N → $(H_2NCH_2CH_2NHOSnCH_2-)_2$ (?) (523).

ETHANEDISTANNONYL CHLORIDE $(-CH_2SnOCl)_2$
Prepn.: By refluxing $(-CH_2SnO_2H)_2$ with $SOCl_2$ and C_5H_5N (523).
Props.: Brown-yellow crystals decompg. without fusion (523).

1,2-ETHANEDISTANNONYL BIS(ETHYLENDIAMIDE) (?)
$(-CH_2SnONHCH_2CH_2NH_2)_2$
Prepn.: By reacting $(-CH_2SnOCl)_2$ with $H_2NCH_2CH_2NH_2$ in the
 presence of C_5H_5N (523).

1,3-PROPANEDISTANNONIC ACID
Prepn.: By reacting aq. K stannite soln. with $CH_2(CH_2Br)_2$ in
 EtOH at 15-25°, passing CO_2, distg. the alc., isolating the
 ppt., and adding it to 1000 part H_2O at 90° (228).
Props.: Amorphous solid which doesn't melt at 320°, sol. in
 dil. HCl and NaOH, insol. in H_2O and org. solvents (228).
Rxn with: Alkali → polymeric stannones, $[-(CH_2)_3SnO-]_n$ (228).
Use: Intermediate in the preparation of polymers (228).

1,4-BUTANEDISTANNONIC ACID $(-CH_2CH_2SnO_2H)_2$
Prepn.: By reacting $(-CH_2CH_2Br)_2$ with K_2SnO_2 in aq. EtOH (523).
Props.: Amorphous, infusible solid (523).

ORGANOTIN TRICARBOXYLATE

ISOPROPYLTIN TRIS(BENZYLMERCAPTOISOBUTYRATE)
$$Me_2CHSn(O_2CCHMeCH_2SCH_2Ph)_3$$

Prepn.: By refluxing Me_2CHSnO_2H with benzylmercaptoisobutyric acid in C_6H_6 in the presence of $p\text{-}MeC_6H_4SO_3H$ and removing the H_2O formed by azeotropic distn. (309).
Props.: Oily prod. which solidifies in the ice box and melts at r.t. (309).
Use: Stabilizer for the Cl-contg. resins (309).

DIORGANOTIN OXIDES

Dialkyltin oxides are prepared:

1. By hydrolyzing dialkyltin halides.
2. By reacting tin alloys with alkyl halides and aerating the reaction product.
3. By pyrolysis of trialkyltin hydroxides.

Diaryltin oxides are prepared:

1. By reacting aryl- or substituted aryldiazonium fluoroborate with tin dichloride dihydrate and powdered tin in acetone and hydrolyzing the product with aqueous ammonia.
2. By reacting tin dichloride with bisaryl chloroiodides in acetone and hydrolyzing the reaction product.
3. By hydrolyzing diaryltin dihalides.

The molecular structure of diorganotin oxides has not yet been elucidated; however, there are strong indications that the compounds exist as low molecular weight polymers. Most of the compounds are amorphous powders without any definite melting point.

Dialkyltin oxides form crystalline complexes with dialkyltin dihalides or diacetates.

DIMETHYLTIN OXIDE Me_2SnO

Prepn.: By treating Me_2SnBr_2 with aqueous NH_3 (195).
Rxns. with: Aldehydes → crystalline adducts (93).
Me_2Sn dihalides → $(Me_3SnO)_3 \cdot Me_2Sn$ dihalides (195).
RSH → $Me_2Sn(SR)_2$ (559, 607, 686, 781).
RCO_2R' (2:1) → $[-OC(OR')(R)-O-SnMe_2-O-SnMe_2-]_x$ (242a).
$HS(CH_2)_nO_3B$ (3:2) → $[Me_2Sn]_3[S(CH_2)_nO_3B]_2$ (748).

DIETHYLTIN OXIDE Et_2SnO

Prepn.: By treating Et_2SnBr_2 with aqueous NH_3 (195).
By prolonged boiling of Et_3SnBr with Na in C_6H_6 in the presence of air (195).
By pyrolyzing Et_3SOH at 200-220° in a sealed tube (197).
By heating NaSnZn alloy with 2.5 l. EtCl at 89°, removing the excess of EtCl, adding $n-C_6H_{14}$, and aerating the rxn. mixture; 45% yield (590).

Props.: Decomp. at elevated temp. into $(Et_3Sn)_2$ and $(Et_3Sn)_2O$ (203).

Rxns. with: HCl in Ac_2O → Et_2SnCl_2 (197).
Et_2Sn dihalides → $(Et_2SnO)_3 \cdot Et_2Sn$ dihalides (195).
Me_2SnI_2 in H_2O → $HO(Et_2SnO)_3H \cdot Me_2SnI_2$ (208).
Et_2SnCl_2 in ROH → $RO(Et_2SnO)_3R \cdot Et_2SnCl_2$ (340).

DIBUTYLTIN OXIDE $(Bu_2SnO)_x$

Prepn.: By heating NaSn with BuCl at 162°, filtering the rxn. mixt., washing the residue with C_6H_6, and passing air into the filtered soln.; after pptn. of $(Bu_2SnO)_x$ the C_6H_6 mother liquor contains Bu_3SnCl (590).

Rxns. with: ROH → $Bu_2Sn(OR)_2$ (66, 599, 745).
Aldehydes and ketones in MePh under reflux → crystalline adducts useful as stabilizers for chlorinated org. mols. and vinyl halide polymers (92, 93).
Bu_2SnCl_2 in ROH → $RO(Bu_2SnO)_xR \cdot Bu_2SnCl_2$ (340).
Bu_2SnCl_2 at 110° → $Bu_2SnCl_2 \cdot Bu_2SnO$ (616).
RCO_2H → $Bu_2Sn(O_2CR)_2$ (27, 309).
RCO_2H and $(:CHCO)_2O$ (2:2:1) → $(:CHCO_2SnBu_2O_2CR)_2$ (661).
Esters of O-contg. acids having at least one O in an acid radical attached to a replaceable acid H → compds. useful as stabilizers for vinyl and vinylidene polymers (632).
$HS(CH_2)_xCO_2R$ → $Bu_2Sn[S(CH_2)_xCO_2R]_2$ (307, 557, 598, 750).
$HSCH_2CONXY$ — $Bu_2Sn(SCH_2CONXY)_2$ (310).
RSH (1:2) → $Bu_2Sn(SR)_2$ (429, 431, 607, 680, 781).
$HSCH_2CH_2OH$ (1:1) → $Bu_2SnSCH_2CH_2O$ (431).
RCO_2R' (2:1) → $[-OC(OR')(R)OSnBu_2OSnBu_2-]_x$ (242a).
$[HS(CH_2)_xCO_2(CH_2)_y]_2$ → $Bu_2Sn[S(CH_2)_xCO_2(CH_2)_y]_2$ (702, 750).
$Bu_2Sn(O_2CC_{11}H_{23})_2$ and $(:CHCO)_2O$ (1:1:1) → $(:CHCO_2SnBu_2O_2CC_{11}H_{23})_2$ (620).
RSO_3H → $Bu_2Sn(O_3SR)_2$ (487).
Concd. HCl → Bu_2SnCl_2 (590).
$[HS(CH_2)_xO_3B]$ → $(Bu_2Sn)_3[S(CH_2)_xO_3B]_2$ (748).
$HS(CH_2)_xOB(OR)_2$ (1:2) → $Bu_2Sn[S(CH_2)_xOB(OR)_2]_2$ (749).
Polycarboxylic acids or anhydrides and polyhydric alcs. at 160-170° → oily prods. useful as stabilizers for polyvinyl halide resins (801).
Esters of organic hydroxy acids, $HOR'(CO_2)_mR_m$ → $Bu_2Sn[OR'(CO_2)_mR_m]_2$ (804).

DIBUTYLTIN OXIDE $(Bu_2SnO)_x$ (Cont'd.)
Biol. props.: Anthelmintic efficacy (649).
Efficacy in treatment of pathological conditions in animals (684).
Use: Catalyst for the prepn. of polyesters (67).
Catalyst for ester interchange between α,ω-glycols and alkyl
    diurethanes of aromatic m- or p-diamines (629).
Antioxidant for vulcanized rubber (790a).
Anticrazing agent for poly(dichlorostyrene) (641a).

DIDODECYLTIN OXIDE $[Me(CH_2)_{11}]_2SnO$
Prepn.: By reacting $(C_{12}H_{25})_2SnCl_2$ with 1N NaOH in Et_2O (522).
Props.: Yellowish white, amorphous wax, decomp. ~110°; sol. in
CS_2; the oxide is monomeric (522).
Rxns. with: BuCHEtCHO → crystalline adduct (93).
Concd. HCl → $(C_{12}H_{25})_2SnCl_2$ (522).
RSH (1:2) → $(C_{12}H_{25})_2Sn(SR)_2$ (559, 680).
$(HSRO)_3B$ (3:2) → $[(C_{12}H_{25})_2Sn]_3(SRO)_3H_2$ (748).
Use: Antioxidant for vulcanized rubber (790a).

OTHER DIALKYL- AND DIALKENYLTIN OXIDES

Compound	Prepared from	Solvent	Yield %	Properties	Ref.
$(CH_2{:}CH)_2SnO$	$(CH_2{:}CH)_2SnCl_2$ + NH_4OH or NaOH	H_2O	80	White solid, m. > 345°	460, 487
$(MeCHCl)_2SnO$	$(MeCHCl)_2SnCl_2$ + NH_4OH	EtOH		m. 180°	575
Pr_2SnO	Pr_2SnBr_2 + NaOH	EtOH		(1, 2)	211
$(n\text{-}C_6H_{13})_2SnO$	$(n\text{-}C_6H_{13})_2SnCl_2$ + NaOH	H_2O	100	(3)	273
$(n\text{-}C_8H_{17})_2SnO$	$(n\text{-}C_8H_{17})_2SnCl_2$ + KOH	EtOH-H_2O	96	White powder (3)	273
$(BuEtCHCH_2)_2SnO$	$(BuEtCHCH_2)_2SnCl_2$ + KOH	EtOH-H_2O	94	(3)	273
$(Me_3CCH_2CHMeCH_2CH_2)_2SnO$	$(Me_3CCH_2CHMeCH_2CH_2)_2SnCl_2$ + KOH	Me_2CO-H_2O	96.5	(3)	273

(1) Rxn. with Pr_2SnI_2 in H_2O → $HO(Pr_2SnO)_3H \cdot Pr_2SnI_2$ (208).
(2) Rxn. with Pr_2SnI_2 in EtOH → $EtO(Pr_2SnO)_3Et \cdot Pr_2SnI_2$ (208).
(3) Rxns. with carboxylic acids → dialkyltin dicarboxylates (273).

DIPHENYLTIN OXIDE Ph_2SnO

Prepn.: By reacting PhN_2BF_4 with $SnCl_2 \cdot 2\ H_2O$ mixed with Sn in Me_2CO at 18° and treating the rxn. prod. with 25% NH_4OH; 41% yield (392).

By adding $SnCl_2$ and powd. Sn to Ph_2ICl in Me_2CO, heating the mixt. under reflux, and pouring the prod. into 10% NaOH; 76% yield (425).

Rxns. with: $P_2O_5 \rightarrow (Ph_2Sn)_2P_2O_7$ (94).

$SOCl_2 \rightarrow Ph_2SnCl_2$ (392).

RCO_2R' (2:1) $\rightarrow [-OC(OR')(R)-O-SnPh_2-O-SnPh_2-]_x$ (242a).

K metal in $C_4H_8O \rightarrow$ red-colored addn. compd. (669).

RSH (1:2) $\rightarrow Ph_2Sn(SR)_2$ (607, 680).

$[HS(CH_2)_xO]_3B$ (3:2) $\rightarrow (Ph_2Sn)_3[S(CH_2)_xO]_3B_2$ (748).

Aldehydes $\rightarrow$ crystalline adducts, useful as stabilizers for halogenated polymers.

Biol. props.: Anthelminthic efficacy (684).

Use: Stabilizer for vinyl resins (578).

Antioxidant for vulcanized rubber (790a).

OTHER DIARYL- AND DI(ARYLALKENYL)TIN OXIDES

Compound	Prepared from	Yield %	Properties	Ref.
$(o\text{-MeC}_6H_4)_2\text{SnO}$	$o\text{-MeC}_6H_4N_2BF_4 + \text{Sn} + \text{SnCl}_2 \cdot 2 H_2O$ (1)	27.3	(2)	392
$(p\text{-MeC}_6H_4)_2\text{SnO}$	$p\text{-MeC}_6H_4N_2BF_4 + \text{Sn-SnCl}_2 \cdot 2 H_2O$ (1)	35.8		392
	$(p\text{-MeC}_6H_4)_2\text{ICl} + \text{Sn} + \text{SnCl}_2$ (1)	67	m. 74° (3)	425
$(p\text{-ClC}_6H_4)_2\text{SnO}$	$(p\text{-ClC}_6H_4)_2\text{SnCl}_2 + \text{NH}_4\text{OH}$		m. 74° (3)	386
	$p\text{-ClC}_6H_4N_2BF_4 + \text{Sn} + \text{SnCl}_2 \cdot 2 H_2O$ (1)	13.1		392
	$(p\text{-ClC}_6H_4)_2\text{ICl} + \text{Sn} + \text{SnCl}_2$ (1)	82	m. 88° (2,4,5)	425
$(p\text{-BrC}_6H_4)_2\text{SnO}$	$p\text{-BrC}_6H_4N_2BF_4 + \text{Sn} + \text{SnCl}_2 \cdot 2 H_2O$ (1)	13.4		392
	$(p\text{-BrC}_6H_4)_2\text{ICl} + \text{Sn} + \text{SnCl}_2$ (1)	47	m. 104° (2,5)	425
$(p\text{-IC}_6H_4)_2\text{SnO}$	$(p\text{-IC}_6H_4)_2\text{ICl} + \text{Sn} + \text{SnCl}_2$ (1)		(5)	425
$(p\text{-O}_2NC_6H_4)_2\text{SnO}$	$p\text{-O}_2NC_6H_4N_2BF_4 + \text{Sn} + \text{SnCl}_2 \cdot 2 H_2O$ (1)	2.5	(2)	392
$(o\text{-MeO}_2CC_6H_4)_2\text{SnO}$	$o\text{-MeO}_2CC_6H_4N_2BF_4 + \text{Sn} + \text{SnCl}_2 \cdot 2 H_2O$ (1)	5		392
$(p\text{-EtO}_2CC_6H_4)_2\text{SnO}$	$(p\text{-EtO}_2CC_6H_4)_2\text{SnBr}_2 + \text{NH}_4\text{OH}$	100	m. 300°	123
$(p\text{-MeOC}_6H_4)_2\text{SnO}$	$(p\text{-MeOC}_6H_4)_2\text{SnCl}_2 + \text{NH}_4\text{OH}$	100	White, amorph. powder, m. 250°	539
$(p\text{-EtOC}_6H_4)_2\text{SnO}$	$(p\text{-EtOC}_6H_4)_2\text{SnCl}_2 + \text{NH}_4\text{OH}$	100	White, amorph. powder, m. 250° (3,4)	539
$(1\text{-}C_{10}H_7)_2\text{SnO}$	$(1\text{-}C_{10}H_7)_2\text{SnCl}_2 + \text{NH}_4\text{OH}$		Amorph. powder (2)	414
$(2\text{-}C_{10}H_7)_2\text{SnO}$	$2\text{-}C_{10}H_7N_2BF_4 + \text{Sn} + \text{SnCl}_2 \cdot 2 H_2O$	11		392
$(p\text{-PhC}_6H_4)_2\text{SnO}$	$(p\text{-PhC}_6H_4)_2\text{SnCl}_2 + \text{NH}_4\text{OH}$	100	m. 250°; insol. in common solvents	539
$(\text{PhCH:CPh})_2\text{SnO}$	$\text{SnCl}_2 + \text{cis-}(\text{PhCH:CPh})_2\text{Hg}$ (1)		m. 300° (6)	394

(1) The rxn. is carried out in Me_2CO and the rxn. prod. is hydrolyzed with NH_4OH (392) or with NaOH (425).
(2) Rxns. with $SOCl_2 \to R_2SnCl_2$ (392).
(3) Rxns. with HBr $\to R_2SnBr_2$ (425).
(4) Rxns. with $HgCl_2 \to R_2Hg$ (386).
(5) Rxns. with HCl $\to R_2SnCl_2$ (425).
(6) Rxn. with dioxane·Br_2 complex $\to$ PhCH:CPhBr (394).

COMPLEX COMPOUNDS OF DIALKYLTIN OXIDES WITH DIALKYLTIN DIHALIDES OR DIACETATES

The complexes with dialkyltin dihalides are formed by reacting dialkyltin oxides or dihydroxides with dialkyltin dihalides in hot alcohol, water, or benzene.

The complexes with dialkyltin diacetates are formed by dissolving dialkyltin oxides in diluted acetic acid and neutralizing the solutions with sodium hydroxide.

Complex	Prepared from	Solvent	Properties	Ref.
$(Me_2SnO)_2 \cdot Me_2SnCl_2$	$Me_2SnO + Me_2SnCl_2$	H_2O		195
$Me_2SnO \cdot Me_2SnI_2$	Me_2SnO or $Me_2Sn(OH)_2 + Me_2SnI_2$	EtOH	m. 68–70°	211
$Me_2SnO \cdot Et_2SnI_2$	$Me_2SnO + Et_2SnI_2$	EtOH	m. 102–103°	209
$Me_2SnO \cdot Pr_2SnBr_2$	$Me_2SnO + Pr_2SnBr_2$	EtOH	m. 68–70°	209
$Et_2SnO \cdot Me_2SnI_2$	Et_2SnO or $Et_2Sn(OH)_2 + Me_2SnI_2$	EtOH	m. 100–106°	209
$Et_2SnO \cdot Et_2SnBr_2$	$Et_2SnO + Et_2SnBr_2$	C_6H_6	m. 103–106°; Insol. in H_2O; dimeric in C_6H_6; trimeric in $C_{10}H_8$	211; 195
$Et_2SnO \cdot Pr_2SnBr_2$	Et_2SnO or $Et_2Sn(OH)_2 + Pr_2SnBr_2$	EtOH	m. 85–87°	209, 211
$Pr_2SnO \cdot Et_2SnBr_2$	$Pr_2SnO + Et_2SnBr_2$	EtOH	m. 104–105°; m. 103–105°	209, 211
$Pr_2SnO \cdot Pr_2SnBr_2$	$Pr_2SnO + Pr_2SnBr_2$	EtOH	Colorless plates, m. 106–107°	209, 211
$Me_2SnO \cdot Me_2Sn(OAc)_2$	$Me_2SnO + AcOH$, followed by NaOH		m. 231°. Trimeric structure	195, 203, 206
$Et_2SnO \cdot Et_2Sn(OAc)_2$	$Et_2SnO + AcOH$, followed by NaOH		Decomp. 165°; mol. wt. detn. in $C_{10}H_8$ and PhOH show different values; Mol. structure	195, 203

COMPLEX COMPOUNDS OF DIALKYLTIN OXIDES WITH DIALKYLTIN HYDROXIDE IODIDES

The following complex compounds were prepared by reacting the mixtures of 2 vols. dialkyltin diiodides and 3 vols. of an arom. or heterocyclic amine with small amounts of EtOH (363):

$Me_2SnO \cdot Me_2Sn(OH)I$, white crystals, sol. in H_2O, insol. in alc.
$Et_2SnO \cdot Et_2Sn(OH)I$, prismatic crystals, m. 140-141°.
$i\text{-}Pr_2SnO \cdot i\text{-}Pr_2Sn(OH)I$, crystalline powder, m. 187°.
$i\text{-}Bu_2SnO \cdot i\text{-}Bu_2Sn(OH)I$, cubic crystals, m. 215°.
$(i\text{-}C_5H_{11})_2SnO \cdot (i\text{-}C_5H_{11})_2Sn(OH)I$, crystals decompg. in air.

DIORGANOTIN BIS(ORGANOOXIDES)

Dialkyl- and diaryl- bis(organoöxy)tin compounds are formed:

1. By reacting dialkyl- and diaryltin dihalides, R_2SnX_2, with sodium organooxides, NaOR, or with alcohols or enols, ROH, in the presence of basic compounds.
2. By the metathetic reaction of dialkyl- and diarylbis-(organooxy)tin compounds with alcohols or enols at elevated temperatures.
3. By treating dialkyl- and diaryltin oxides with alcohols or enols at elevated temperatures and, if desired, with azeotropic removal of water.

Numerous compounds computed in the following table marked with asterisks are useful as stabilizers for vinyl halide polymers.

DIORGANOTIN BIS(ORGANOOXIDES)

$R_2Sn(OR')_2$	Prepared from	Properties	Ref.
$Me_2Sn(OSiMe_3)_2$	$Me_2SnCl_2 + Me_3SiOK$	Small glittering needles	542
$Et_2Sn(OMe)_2$	$Et_2SnCl_2 + NaOMe$	b$_{20}$ 124-126°/3mm; d$_{20}$ 1.4804; n^{20}D 1.4206	348
$Et_2Sn(OEt)_2$		Rxn. with AcF (1:1) → $Et_2(EtO)$ SnF; with AcF (1:2) → Et_2SnF_2	577
$Et_2Sn(OCHMeCO_2Bu)_2$*	$Et_2Sn(OMe)_2 + MeCHOHCO_2Bu$	d$_{20}$ 1.2041; n^{20}D 1.4652	349
$Et_2Sn(OCH_2Ph)_2$	$Et_2Sn(OMe)_2 + PhCH_2OH$		349
$Bu_2Sn(OMe)_2$*	$Bu_2SnCl_2 + NaOMe$	b$_{20}$ 136-139°/1.2mm; d$_{20}$ 1.2862; n^{20}D 1.4831	348
$Bu_2Sn(OCH_2CH:CH_2)_2$*	$Bu_2SnCl_2 + NaOCH_2CH:CH_2$		348
$Bu_2Sn(OBu)_2$*	$Bu_2SnO + BuOH$	m. 110-115°	66, 539, 834
	$Bu_2SnCl_2 + NaOBu$	(1)	343, 348
$Bu_2Sn(O-n-C_8H_{17})_2$*			834
$Bu_2Sn(O-n-C_{12}H_{25})_2$*	$Bu_2Sn(OMe)_2 + n-C_{12}H_{25}OH$	d$_{20}$ 1.025; n^{20}D 1.4730	349
$Bu_2Sn[(OCH_2CH_2)_3OBu]_2$*	$Bu_2Sn(OMe)_2 + BuO(CH_2CH_2)_3H$		349
$Bu_2Sn(OCH_2CHCH_2OCHMeO)_2$*	$Bu_2Sn(OMe)_2 + OCHMeOCH_2CHCH_2OH$		349
$Bu_2Sn(OCMe:CHCOMe)_2$	$Bu_2SnCl_2 + MeC(ONa):CHAc$	b. 130-132°/1.6mm	341, 706, 708
$[Bu_2Sn(OCH_2)_2]_2C$*	$Bu_2SnO + C(CH_2OH)_4$	m. 125-128°	66, 599
$Bu_2Sn(O-n-C_{10}H_{21})_2$	$Bu_2SnO + n-C_{10}H_{21}OH$		599
$Bu_2Sn(O-furfuryl)_2$	$Bu_2SnO + OCH:CHCH:CCH_2OH$		599
$Bu_2Sn(OC_6H_4CO_2Me-o)_2$*	$Bu_2Sn(OMe)_2 + o-HOC_6H_4CO_2Me$		349
	$Bu_2SnCl_2 + o-HOC_6H_4CO_2Me$	Soft solid	808
$Bu_2Sn(OPh)_2$	$Bu_2SnO + PhOH$		599
$Bu_2Sn(OCH_2Ph)_2$*			826
$Bu_2Sn[OCH(n-C_6H_{13})CH:CH-(CH_2)_2CO_2CH_2-]_2$*	$Bu_2SnCl_2 + [n-C_6H_{13}CHOH-CH:CH(CH_2)_2CO_2CH_2-]_2$		808

(1) Useful as catalyst in the preparation of polyesters (67).

DIORGANOTIN BIS(ORGANOOXIDES) (Cont'd.)

$R_2Sn(OR')_2$	Prepared from	Properties	Ref.
$Bu_2Sn[OCH(n-C_6H_{13})CH:CH-(CH_2)_2CO_2Me]_2$*	$Bu_2SnCl_2 + n-C_6H_{13}CHOHCH:CH(CH_2)_2CO_2Me$		808
$Bu_2Sn[OCH(CO_2Bu)CHOHCO_2Bu]_2$	$Bu_2SnCl_2 + $ [$CHOHCO_2Bu]_2$		808
$Bu_2Sn[OC(CH_2CO_2Bu)_2CO_2Bu]_2$	$Bu_2SnCl_2 + (BuO_2CCH_2)_2COHCO_2Bu$		808
$Bu_2Sn(OCHMeCO_2Et)_2$*	$Bu_2SnCl_2 + MeCHOHCO_2Et$		808
$Bu_2Sn(OCMe_2CO_2Et)_2$*	$Bu_2SnCl_2 + Me_2COHCO_2Et$		804, 808
$Bu_2Sn[OC(CH_2CO_2Et)_2CO_2Et]_2$	$Bu_2SnCl_2 + (EtO_2CCH_2)_2COHCO_2Et$		804, 808
$Bu_2SnOCH(n-C_6H_{13})CH:CH-(CH_2)_7CO_2CH_2CH_2O$	$Bu_2SnCl_2 + [n-C_6H_{13}CHOH-CH:CH(CH_2)_7CO_2CH_2CH_2]O$	(2)	790, 804
$Bu_2SnOCH(n-C_6H_{13})CH:CH-(CH_2)_7CO_2Me_2$*	$Bu_2SnCl_2 + n-C_6H_{13}CHOH-CH:CH(CH_2)_7CO_2Me$		804
$Bu_2SnOC(CH_2CO_2Bu)_2CO_2Bu_2$*	$Bu_2SnCl_2 + (BuO_2CCH_2)_2COHCO_2Bu$	(2)	790
$Bu_2SnOCH_2CH(O_2CC_{11}H_{23})-CH_2O_2CC_{11}H_{23}_2$			348
$Ph_2Sn(OMe)_2$		White solid, decomp. 270°	679
$Ph_2Sn(OEt)_2$		(3)	123
$(EtOCOC_6H_4)_2Sn(OC_9H_6N)_2$	$(EtOCOC_6H_4)_2Sn$ dihalide + 8-hydroxyquinoline	m. 216-217°	66, 599
$Bu_2SnOCH_2CH_2O$*	$Bu_2SnO + (-CH_2OH)_2$	m. 195-200° (2)	746
Bu_2SnOCH_2CHMeO*	$Bu_2SnO + (-CH_2OH)_2$		746
$Bu_2SnO(CH_2)_2CHMeO$*	$Bu_2SnCl_2 + MeCHOHCH_2OH$	m. 182°	746
$Bu_2Sn(OCH_2CH_2)_2O$*	$Bu_2SnCl_2 + MeCHOHCH_2CH_2OH$	m. 130°	66, 599
$Bu_2SnOCH_2CHOHCH_2OCO-(CH_2)_7CH:CH(CH_2)_7Me$*	$Bu_2SnO + O(CH_2CH_2OH)_2$	m. 120-130°	745
	$Bu_2SnO +$ glyceryl monooleate		745
	$Bu_2SnCl_2 +$ glyceryl monooleate		
$Ph_2Sn(OCH_2C:CHCH:CHO)_2$*	$Ph_2Sn(OMe)_2 +$ furyl alc.		349, 826
$[Bu_2Sn(OMe)_2]_x$*	$Bu_2SnCl_2 + MeOH(NH_3)$	Polymer	343

(2) Antioxidant for vulcanized rubber (790).
(3) Corrosion inhibitor for transformers (679).

DIALKYL- AND DIARYLALKOXYTIN CARBOXYLATES

Dialkyl- and diarylalkoxytin carboxylates, $R_2Sn(OR")OCOR'$, are formed by reacting dialkyl- and diaryltin dihalides, R_2SnX_2, with a mixture of sodium carboxylate, $NaOCOR'$, and sodium alkoxide, $NaOR"$, in an alcohol $R"OH$ or by reacting dialkyldialkoxytin compounds with a carboxylic acid in molar ratio of 1:1.

The compounds are useful as stabilizers for halogen-containing polymers. They react with moist air at 95-100° to form polymeric compounds having the general formula $R'CO_2(R_2SnO)_xR"$.

DIALKYL- AND DIARYLALKOXYTIN CARBOXYLATES

$R_2Sn(OR")OCOR'$	Prepared from	Properties	Ref.
$Et_2Sn(OCH_2CH_2OEt)OCO-CH:CHMe$	Et_2SnCl_2	Liquid	345
$Et_2Sn(OEt)OCO(CH_2)_8CO_2Et$	Et_2SnCl_2		345, 347
$Bu_2Sn(OMe)OAc$	Bu_2SnCl_2	Liquid	345, 347
$Bu_2Sn(OMe)OCOCH_2CH_2CO_2Me$	Bu_2SnCl_2	d_{20} 1.350; n^{20}_D 1.4980	345, 347
$Bu_2Sn(OMe)OCO-n-C_{11}H_{23}$	Bu_2SnCl_2	d_{20} 1.126; n^{20}_D 1.4860	345, 347
$Bu_2Sn(OMe)OCOCH:CHCO_2Me$	Bu_2SnCl_2; $Bu_2Sn(OMe)_2$ + $(:CHCO)_2O$	d_{20} 1.374; n^{20}_D 1.5108	345, 347; 349
$Bu_2Sn(OMe)OCOCH:CHCO_2Bu$	Bu_2SnCl_2	d_{20} 1.325; n^{20}_D 1.5068	345, 347
$Bu_2Sn(OMe)OCOCH:CHCO_2CH_2-CH:CH_2$	Bu_2SnCl_2	d_{20} 1.291; n^{20}_D 1.5011	345, 347
$Bu_2Sn(OMe)OCOCH:CHCO_2CH_2-CHCH_2CH_2CH_2O$	Bu_2SnCl_2	d_{20} 1.273; n^{20}_D 1.5036	345, 347
$Bu_2Sn(OMe)$oleate	Bu_2SnCl_2	d_{25} 1.129; n^{25}_D 1.4882	347
$Bu_2Sn(OMe)OCOC_6H_4CO_2Me-o$	Bu_2SnCl_2	d_{20} 1.3700; n^{20}_D 1.5448	345, 347
$Bu_2Sn(OMe)$sorbate	$Bu_2Sn(OMe)_2$ + sorbic acid	Solid	349
$Ph_2Sn(OBu)$oleate	Ph_2SnCl_2	Waxy solid	345
$Ph_2Sn(OMe)OCOCH_2C(:CH_2)-CO_2Bu$	Ph_2SnCl_2		345

DIORGANOTIN DICARBOXYLATES, DISULFONATES AND ARSONATES

Diorganotin dicarboxylates are prepared:

1. By reacting dialkyl-, diaryl-, and diaralkyltin dihalides with alkali metal carboxylates.
2. By metathetic reaction of dialkyl-, diaryl-, and diaralkyltin dicarboxylates with other carboxylic acids.
3. By reacting dialkyl-, diaryl-, and diaralkyltin oxides with carboxylic acid anhydrides or with carboxylic acids either in a solvent, with azeotropic removal of water, or without solvents at elevated temperatures in vacuo.
4. By reacting homogeneously and heterogeneously substituted tetrahydrocarbyltin compounds with carboxylic acids.

Diorganotin disulfonates are prepared by reacting diorganotin oxides with sulfonic acids.

Diorganotin arsonates are prepared by reacting diorganotin dicarboxylates with arsonic acids.

$R_2Sn(OCOR')_2$	Prepared from	Properties	Ref.
$Me_2Sn(OOC)_2 \cdot H_2O$	$Me_2SnCl_2 + (COOK)_2$	White solid, darkens 310°	455
$Me_2Sn(OCOCH_2Br)_2$	$Me_2Sn(CH:CH_2)_2 + BrCH_2CO_2H$	m. 159-160°	463
$Me_2Sn(OCOCH_2Cl)_2$	$Me_2Sn(CH:CH_2)_2 + ClCH_2CO_2H$	m. 162-165°	463
$Me_2Sn(OCOCHCl_2)_2$	$Me_2Sn(CH:CH_2)_2 + Cl_2CHCO_2H$	m. 226-228°	463
$Me_2Sn(OCOCCl_3)_2$	$Me_2Sn(CH:CH_2)_2 + Cl_3CCO_2H$	m. 190-191°	463
$Me_2Sn(OCOCF_3)_2$	$Me_2Sn(CH:CH_2)_2 + F_3CCO_2H$	m. 240-243°	463
$Me_2Sn(OCOC_6H_4OH\text{-}o)_2$	$Me_2SnCl_2 + Na$ salicylate	m. 205-206°	455
$Me_2SnOCOC_6H_4COO$	$Me_2SnCl_2 + Na_2$ pthalate	White crystals, darkening >340°	455
$Me_2Sn(OCOCH_2\text{-})_2 \cdot Me_2SnO$	$Me_2SnCl_2 + (CH_2CO_2Na)_2$	White powder, decomp. >314°	455
$Me_2Sn(OBz)_2 \cdot Me_2SnO$	$Me_2SnCl_2 + BzONa$	White crystals, m. 236-237°	455
$(CH_2:CH)_2Sn OCOC_6H_4\text{-}o\text{-}COO$	$(CH_2:CH)_2SnCl_2 +$ $o\text{-}HOOCC_6H_4CO_2K$	Fine crystals, m. 283° (decompn.)	455
$i\text{-}Pr_2Sn(OCOCHCl_2)_2$	$i\text{-}Pr_4Sn + CHCl_2CO_2H$	m. 69-71°	487
$i\text{-}Pr_2Sn(OCOCH_2Cl)_2$	$i\text{-}Pr_4Sn + ClCH_2CO_2H$	m. 54-55°	468
$Bu_2Sn(OAc)_2 \cdot H_2O$	$Bu_2Sn(OAc)_2 + (COOH)_2$		468
$Bu_2Sn(OAc)_2$	$Bu_2SnCl_2 + KOAc$	Insol. solid, decomp. 195°	27
$Bu_2Sn(OCO\text{-}n\text{-}C_{11}H_{23})_2$		b. 142-145°/10mm (1-3) (4-6)	116
$Bu_2Sn(OCOCHMeCH_2S\text{-}n\text{-}C_{12}H_{25})_2$*	$Bu_2SnO + n\text{-}C_{12}H_{25}SCH_2\text{-}CHMeCO_2H$		336 309

(1) Uses: Antioxidant for vulcanized rubber (807); Catalyst in the preparation of polyesters (67, 614a); Catalyst for the ester interchange between α-ω-glycols and alkyldiarethans of aromatic amines (629); Stabilizer for halogen-contg. polymers (116, 850); Stabilizer for polymeric organosilicon compds. (759).

(2) Biol. props.: Anthelmintic efficacy (684).

(3) Rxns. with: Dicarboxylic acids → $Bu_2SnOCO(CH_2)_x 2$ (27).
$HSCH_2CO_2H → Bu_2SnOCOCH_2S$ (193).
$H_2O → AcO(Bu_2SnO)_xAc$ (344).
$RAsO_3H_2 → Bu_2Sn[OAs(O)(OH)R]_2$ (796).

(4) Uses: Antioxidant for vulcanized rubber (790a, 807); Anticrazing agent for poly-(dichlorostyrene) (641a); Catalyst for the prepn. of polyesters (67); Stabilizer for polyvinyl chloride (135, 336, 779); Stabilizer for chlorinated fatty acid esters (856). Reduces characteristic carbonyl and polyene inflections of the light transmittance curves (135).

(5) Biol. props.: Anthelmintic props. (277, 612, 648, 683, 684).

(6) Rxn. with: $Bu_2SnO + (:CHCO)_2O → Bu_2Sn[OCOC_{11}H_{23})OCOCH]_2$ (620).

* Useful as stabilizer for halogen-contg. polymers.

DIORGANOTIN DICARBOXYLATES, DISULFONATES, AND ARSONATES (Cont'd.)

$R_2Sn(OCOR')_2$	Prepared from	Properties	Ref.
$Bu_2Sn(OCOC_6H_4OH-o)_2$*		Crystals, m. 187-187.5°	761
[$Bu_2SnOCO(CH_2)_8COO-]_4$	$Bu_2SnO + (-CH_2CO_2)_2O$	(7,8)	27
$Bu_2Sn(OCOCH:)_2$ (maleate)			850
$Bu_2Sn(OCOCH:)_2$ (fumarate)*			856
$Bu_2Sn(OCOCH:CHCO_2Bu)_2$*			
$Bu_2Sn[OCOCH:CHCO_2(CH_2)_5CHMe_2]_2$		Antioxidant for vulcanized rubber	807
$\{Bu_2Sn[OCO(CH_2)_8]_2\}_x$	Bu_2SnO + sebacic acid	White solid, m. 78-82°; mol. wt. 3000	27
$Bu_2SnOCOC_6H_4-p-COO$ (?)	$Bu_2Sn(OAc)_2 + p-C_6H_4(CO_2H)_2$	Insol. powder, m. 245°	27
[$Bu_2SnOCOCH_2CHMe(CH_2)_2COO-]_2$	$Bu_2Sn(OAc)_2$ + 3-Me-adipic acid	Crystals, m. 143-144-5°; mol. wt. 783-800	27
[$Bu_2Sn(OCO-n-C_{11}H_{23})OCOCH:]_2$*	$Bu_2SnO + Bu_2Sn(OCOC_{11}H_{23})_2$ + (:CHCO)$_2$O		620, 661
[$Bu_2Sn(OAc)OCOCH:]_2$*	$Bu_2SnO + AcOH + (CHCO)_2O$		661
[$Bu_2Sn(OCOC_{21}H_{43})OCOCH:]_2$*	Bu_2SnO + behenic acid + (:CHCO)$_2$O		661
$Bu_2Sn(O_3SC_6H_4Me-p)_2$	$Bu_2SnO + p-MeC_6H_4SO_3H$	m. 320°	487
$(n-C_6H_{13})_2Sn(OOCH)_2$	$(n-C_6H_{13})_2SnO + (:CHCO)_2O$	Yellow syrup	336
$(n-C_6H_{13})_2Sn(OCOCH:)_2$*	$(n-C_6H_{13})_2SnO + n-C_{11}H_{23}-CO_2H$	m. ~13°	273, 336
$(n-C_6H_{13})_2Sn(OCO-n-C_{11}H_{23})_2$			273, 336
$(n-C_8H_{17})_2Sn(OOCH)_2$	$(n-C_8H_{17})_2SnO + HCOOH$	m. 85-86°	273
$(n-C_8H_{17})_2Sn(OAc)_2$	$(n-C_8H_{17})_2SnO + AcOH$	Liquid	273
$(n-C_8H_{17})_2Sn(OCOPr)_2$*	$(n-C_8H_{17})_2SnO + PrCO_2H$	Liquid	273, 336
$(n-C_8H_{17})_2Sn(OCOCHEtBu)_2$*	$(n-C_8H_{17})_2SnO + BuEtCHCO_2H$	Waxy solid	273, 336
$(n-C_8H_{17})_2Sn(OCO-n-C_{11}H_{23})_2$*	$(n-C_8H_{17})_2SnO$ + lauric acid	Waxy solid	273, 336
$(n-C_8H_{17})_2Sn(OCO-n-C_{17}H_{35})_2$*	$(n-C_8H_{17})_2SnO$ + stearic acid	Waxy solid	273, 336
$(n-C_8H_{17})_2Sn(OCOCH:CH-CO_2Bu)_2$*	$(n-C_8H_{17})_2SnO$ + monobutyl maleate	Liquid	273, 336
$(n-C_8H_{17})_2Sn(OCOCH:)_2$*	$(n-C_8H_{17})_2SnO + (:CHCO_2H)_2$	Crystalline compd.	273, 336

Compound		Properties	Ref.
(BuEtCH$_2$)$_2$Sn(OCO-n-C$_{11}$H$_{23}$)$_2$*	(BuEtCH$_2$)$_2$SnO + lauric acid	Colorless oil	273, 336
(Me$_3$CCH$_2$CHMeCH$_2$CH$_2$)$_2$Sn(OCO-n-C$_{11}$H$_{23}$)$_2$*	(Me$_3$CCH$_2$CHMeCH$_2$CH$_2$)$_2$SnO + lauric acid	Colorless oil	273, 336
Ph$_2$Sn(OCO-n-C$_{17}$H$_{35}$)$_2$*		Anthelmintic efficacy	903
(PhCH$_2$)$_2$Sn(OAc)$_2$		Rxn. with RAsO$_3$H$_2$ → Ph$_2$Sn(OAs(O)(OH)R)$_2$	684
(PhCH$_2$)$_2$Sn(OCOCH:)$_2$			796
{(PhCH$_2$)$_2$Sn(OAc)$_2$}		Anthelmintic efficacy	684
(PhCH$_2$)$_2$Sn(OCO-n-C$_{11}$H$_{23}$)$_2$**	(PhCH$_2$)$_2$SnCl$_2$ + n-C$_{11}$H$_{23}$CO$_2$Na	Anthelmintic efficacy m. 88-92°	684, 507
(PhCH$_2$)$_2$Sn(OCO-n-C$_{17}$H$_{35}$)$_2$**	(PhCH$_2$)$_2$SnCl$_2$ + n-C$_{17}$H$_{35}$CO$_2$Na	m. 99-100°	507
(PhCH$_2$)$_2$Sn(ricinoleate)$_2$**	(PhCH$_2$)$_2$SnCl$_2$ + Na ricinoleate		507
(PhCH$_2$)$_2$Sn(OCOCH:CHCO$_2$Me)$_2$**	(PhCH$_2$)$_2$SnCl$_2$ + MeOCOCH:CHCO$_2$Na	m. 114-117°	507
Bu$_2$SnOCOCH$_2$C(:CH$_2$)COO (?)	Bu$_2$Sn(OAc)$_2$ + itaconic acid	Waxy or plastic, satd. prod.	27
[Bu$_2$Sn(OCOCH$_2$CH$_2$-)$_2$]$_x$	Bu$_2$Sn(OAc)$_2$ + adipic acid	m. 136-137° mol. wt. 1205	27
Bu$_2$Sn[OAs(O)(OH)C$_6$H$_4$NO$_2$-p]$_2$	Bu$_2$Sn(OAc)$_2$ + p-O$_2$NC$_6$H$_4$AsO(OH)$_2$	Decomp. 165-172° (9)	796
	Bu$_2$Sn(OAc)$_2$ + 3-O$_2$N-4-H$_2$N-C$_6$H$_3$AsO(OH)$_2$	Decomp. 189-193° (9)	796
	Bu$_2$Sn(OAc)$_2$ + 3-O$_2$N-4-HO-C$_6$H$_3$AsO(OH)$_2$	Decomp. 164-170° (9)	796
Bu$_2$Sn[O$_2$As(O)C$_6$H$_4$NO$_2$-p]$_2$	Bu$_2$Sn(OAc)$_2$ + p-O$_2$NC$_6$H$_4$AsO(OH)$_2$	Decomp. 249° (9)	796
Bu$_2$Sn[O$_2$As(O)C$_6$H$_3$-3-NO$_2$-4-OH]$_2$	Bu$_2$Sn(OAc)$_2$ + 3-O$_2$N-4-HO-C$_6$H$_3$AsO(OH)$_2$	Decomp. 191° (9)	796
Bu$_2$Sn[O$_2$As(O)C$_6$H$_3$-3-NO$_2$-4-NH$_2$]	Bu$_2$Sn(OAc)$_2$ + 3-O$_2$N-4-H$_2$N-C$_6$H$_3$AsO(OH)$_2$	Decomp. 262° (9)	796
Bu$_2$Sn[O$_2$As(O)C$_6$H$_4$NH$_2$-p]	Bu$_2$Sn(OAc)$_2$ + p-H$_2$NC$_6$H$_4$AsO(OH)$_2$	Decomp. 175-185° (9)	796
Ph$_2$Sn[OAs(O)(OH)Bu]$_2$	Bu$_2$Sn(OAc)$_2$ + BuAs(O)(OH)$_2$	Decomp. 296° {1}	796
Ph$_2$Sn[O$_2$As(O)Bu]	Ph$_2$Sn(OAc)$_2$ + BuAs(O)(OH)$_2$	Decomp. 297° (9)	796

* Useful as stabilizer for halogen-contg. polymers.
** Useful as plasticizer for vinyl halide polymers (507).
(7) Antioxidant for vulcanized rubber (807); Anticrazing agent for poly(dichlorostyrene) (641a); Stabilizer for polyvinyl halides polymers and copolymers (336, 779, 798).
{8} Biol. props: Anthelmintic props. (649, 683, 684).
(9) Useful in treatment of protozoal infections (296).

POLYMERIC DIORGANOTIN DIALKOXIDES, DICARBOXYLATES, AND THEIR COMPLEXES

The following compounds having the general formulae: $R'OR_2SnO_xR'$, $R'CO_2R_2SnO_xOCR'$, and $R'OR_2SnO_xR' \cdot R''SnX_2$ (X= Cl, Br, or I), respectively, were prepared by:

1. Reacting R_2SnX_2 with NaOR' in a solvent (Method I).
2. Reacting excess R_2SnO with $R''SnX_2$ in R'OH (Method II).
3. Exposing R_3SnX to air in sunlight (Method III).
4. Treating R_2Sn with $CHCl_3$ or EtBr in the presence of H_2O (Method IV).
5. Bubbling moist air through $R_2Sn(O_2CR')_2$ at 140° (Method V).

POLYMERIC DIORGANOTIN DIALKOXIDES, DICARBOXYLATES, AND THEIR COMPLEXES

Compound	Starting matls.	Solvent	Method	Properties	Ref.
$MeO(Bu_2SnO)_2Me$				m. 92-94° (1)	699
$MeO(Bu_2SnO)_{\bar{x}}Me$ ($\bar{x}=1.4-9.15$)	$Bu_2SnCl_2 + NaOMe$	MePh	I		343
$CH_2:CHCH_2O(Bu_2SnO)_{2.29}CH_2CH:CH_2$	$Bu_2SnCl_2 + NaOCH_2CH:CH_2$	MePh	I	Viscous oil (1)	343
$C_6H_{11}O(Bu_2SnO)_{8.74}C_6H_{11}$	$Bu_2SnCl_2 + NaOC_6H_{11}$	MePh	I	(1)	343
$BuOCH_2CH_2O(Bu_2SnO)_{11.29}CH_2CH_2OBu$	$Bu_2SnCl_2 + NaOCH_2CH_2OBu$	xylene	I	(1)	343
$PhCH_2O(Bu_2SnO)_{6.83}CH_2Ph$	$Bu_2SnCl_2 + NaOCH_2Ph$	MePh	I	(1)	343
$BuO(Bu_2SnO)_{0.66}Bu$	$Bu_2SnCl_2 + NaOBu$		I	(1)	343
$BuCHEtCH_2O(Bu_2SnO)_{2.42}CH_2CHEtBu$	$Bu_2SnCl_2 + NaOCH_2CHEtBu$	MePh	I	(1)	343, 699
$BuCHEtCH_2OEt_2SnO_{11.7}CH_2CHEtBu$	$Et_2SnCl_2 + BuCHEtCH_2OH$ in the presence of C_5H_5N	MePh	I	(1)	343
$PrO(1-C_5H_{11})_2SnO_{7.5}Pr$	$(1-C_5H_{11})_2SnCl_2 + NaOPr$	MePh	I	(1)	343, 699
$MePh_2SnO_7.52Me$	$Ph_2SnCl_2 + NaOMe$	MePh	I	(1)	343, 699
$HOMe_2SnO_3H \cdot Me_2SnI_2$	$Me_2SnI_2 + Me_2SnO$	cold H_2O	II		195
$EtO(Me_2SnO)_3Et \cdot Me_2SnBr_2$	$Me_2SnBr_2 + Me_2SnO$	hot EtOH	II	m. 210-215° (dec.) mol. structure (2)	195, 203
$EtO(Me_2SnO)_3Et \cdot Me_2SnI_2$	$Me_2SnI_2 + Me_2SnO$	hot EtOH	II	m. 215-218° (dec.) mol. structure (3)	195, 203
$EtO(Me_2SnO)_3Et \cdot Et_2SnI_2$	$Et_2SnI_2 + Me_2SnO$	hot EtOH	II	m. 127-162° (dec.)	208
$HOEt_3SnO_3H \cdot Me_2SnI_2$	$Me_2SnI_2 + Et_2SnO$	H_2O	II	m. 180-198° (dec.)	208
$HC(Et_2SnO)_3H \cdot Et_2SnCl_2$	$Et_2SnO + Et_2SnCl_2$	moist C_6H_6	II	m. 215-217° (dec.)	195
	Et_3SnCl		III		195
	$Et_2Sn + CHCl_3 + H_2O$		IV	m. 216° (1)	210, 340

(1) Useful as stabilizer for vinyl halide polymers.
(2) Rxn. with $H_2O - HO(Me_2SnO)_3H \cdot Me_2SnBr_2$ (196).
(3) Rxn. with $H_2O - HO(Me_2SnO)_3H \cdot Me_2SnI_2$ (196).
(4) Rxn. with hot EtOH — $EtO(Et_2SnO)_3Et \cdot Et_2SnCl_2$ (195).

POLYMERIC DIORGANOTIN DIALKOXIDES, DICARBOXYLATES, AND THEIR COMPLEXES (Cont'd.)

Compound	Starting matls.	Solvent	Method	Properties	Ref.
$HO(Et_2SnO)_3H \cdot Et_2SnBr_2$	$Et_2SnO + Et_2SnBr_2$	moist C_6H_6	II	m. 206-218° (decompn.)	195
	Et_3SnBr		III	(5)	195
	$Et_2Sn + EtBr + H_2O$			m. 210°	210
$HO(Et_2SnO)_3H \cdot Et_2SnI_2$	$Et_2SnO + Et_2SnI_2$	moist C_6H_6	II	m. 205-213° (decompn.)	195
	Et_3SnI		III		195
$EtO(Et_2SnO)_3Et \cdot Et_2SnCl_2$			(6)	crystals (1)	340
$HO(Pr_2SnO)_3H \cdot Pr_2SnI_2$	$Pr_2SnO + Pr_2SnI_2$	H_2O	II	m. 188-205° (decompn.)	208, 211
$EtO(Pr_2SnO)_3Et \cdot Pr_2SnI_2$	$Pr_2SnO + Pr_2SnI_2$	EtOH	II	m. 122-140°	208
	$Pr_2SnO + Pr_2SnI_2$	EtOH	II	m. 104-122° (decompn.)	211
$PrO(Pr_2SnO)_3Pr \cdot Pr_2SnI_2$	$Pr_2SnO + Pr_2SnI_2$	PrOH	II	m. 113-150°	211
	$Pr_2SnO + Pr_2SnI_2$	PrOH	II	m. 151-178° (decompn.)	208
$MeO(Bu_2SnO)_3Me \cdot Bu_2SnCl_2$	$Bu_2SnO + Bu_2SnCl_2$	MeOH	II	crystals (1)	340
$BuO(Bu_2SnO)_5Bu \cdot Bu_2SnCl_2$	$Bu_2SnO + Bu_2SnCl_2$	MePh	II	crystals (1,7)	340
$EtO(Me_2SnO)_3Et \cdot Et_2SnI_2$	$Me_2SnO + Et_2SnI_2$	EtOH	II	m. 127-162° (decompn.)	211
$EtO(Et_2SnO)_3Et \cdot Me_2SnI_2$	$Et_2SnO + Me_2SnI_2$	EtOH	II	m. 150-178° (decompn.)	211
$HO(Et_2SnO)_3H \cdot Me_2SnI_2$	$Et_2SnO + Me_2SnI_2$	H_2O	II	m. 180-198° (decompn.)	211
$EtCO_2(Bu_2SnO)_4OCEt$	$Bu_2Sn(OCOEt)_2$		V	x̄ = 3 and 4 (1)	344
$BuCHEtCO_2(Bu_2SnO)_4OCCHEtBu$	$Bu_2Sn(O_2CCHEtBu)_2$		V	(1)	344, 699
$HO(CH_2CH_2OSnBu_2O)_2CH_2CHOH$	$Bu_2SnCl_2 + (CH_2ONa)_2$	MePh	I	m. 215-218° (1)	746

(1) Useful as stabilizer for vinyl halide polymers.
(5) Rxn. with hot EtOH – $EtO(Et_2SnO)_3Et \cdot Et_2SnBr_2$ (195).
(6) Prepared by recrystallizing $HO(Et_2SnO)_3H \cdot Et_2SnCl_2$ from EtOH (340).
(7) The crystalline residue obtd. after removing MePh was treated with BuOH (340).

TRIS(DIBENZYLSTANNOXANE)DIOL MONOACETATE $HO(PhCH_2)_2SnO_3Ac$
Prepn.: By refluxing $(PhCH_2)_2SnO$ with $PhCH_2OAc$ (344).
Props.: m. 155-170° (344).
Use: Stabilizer for polyvinyl chloride resins (344).

POLY(DIBUTYLSTANNOXANE)DIOL DIACETATE $AcO(Bu_2SnO)_xAc$
Prepn.: By distilling $Bu_2Sn(OAc)_2$ in vacuo at 134-135°/5mm $AcO(Bu_2SnO)_2Ac$ is formed; distn. with steam yields $AcO(BuSnO)_3Ac$; and blowing moist air through $Bu_2Sn(OAc)_2$ at 140° yields $AcO(Bu_2SnO)_7Ac$ (344).
Use: Stabilizer for polyvinyl chloride resins (344).

OTHER DIORGANOTIN SALTS

Diorganotin salts of inorganic oxy-acids are prepared by reacting diorganotin halides with sodium or potassium salts of inorganic oxy-acids.

OTHER DIORGANOTIN SALTS

Compounds	Prepared from	Properties	Ref.
Me_2SnSO_3		Antioxidant for vulcanized rubber	790a
$Me_2SnCrO_4 \cdot Me_2SnO$	$Me_2SnCl_2 + K_2CrO_4$	Yellow powder, darkens at 265°	455
$Me_2SnO_2As(O)OH$	$Me_2SnCl_2 + K_2HAsO_4$	White powder	455
Me_2SnMoO_4	$Me_2SnCl_2 + K_2MoO_4$	White solid, decomp. 343°	455
Me_2SnWO_4	$Me_2SnCl_2 + K_2WO_4$	White solid	455
$Me_2Sn(IO_3)_2$	$Me_2SnCl_2 + KIO_3$	White crystals, discolor on standing	455
$Me_2Sn(VO_3)_2$ (?)	$Me_2SnCl_2 + NH_4VO_3$	Yellow ppt.	455

TRIORGANOTIN HYDROXIDES

The hydroxides are prepared by the hydrolysis of triorganotin halides or oxides.

TRIMETHYLTIN HYDROXIDE Me$_3$SnOH

Prepn.: By reacting Na$_2$Sn with MeBr in NH$_3$ and treating the rxn. prod. with alc., H$_2$O, and Et$_2$O (210).
By reacting Me$_4$Sn with SnBr$_4$ (3:1) and distg. with steam the resulting Me$_3$SnBr in the presence of KOH (268).
By hydrolyzing Me$_3$Sn phosphonates (30, 32).
Props.: m. 118° (30); m. 118-118.8° (268).
Rxns. with: Na in C$_6$H$_6$ → (Me$_3$Sn)$_2$O (199).
Me$_3$Sn halides (1:1) and (2:1) → Me$_3$SnOH·Me$_3$Sn halide·H$_2$O and (Me$_3$SnOH)$_2$·Me$_3$Sn halide, respectively (195, 198, 210).
AcOH → Me$_3$SnOAc (268).
RSH → Me$_3$SnSR (419, 680).
Addn. compds.: Me$_3$SnOH·Me$_3$SnCl·H$_2$O, prepd. by refluxing Me$_3$SnOH with Me$_3$SnCl (1:1) in moist C$_6$H$_6$ (195, 198) or by recrystallizing (Me$_3$SnOH)$_2$·Me$_3$SnCl from H$_2$O (198); m. ~98° (decompn.) (195); m. 81-95° (decompn.) (198); on drying over CaCl$_2$ loses H$_2$O (198); Reacts with Me$_3$SnOH (1:1) to form (Me$_3$SnOH)$_2$·Me$_3$SnCl (198) and with alc. Ag$_2$O forms AgCl and Me$_3$SnOH (198).
(Me$_3$SnOH)$_2$·Me$_3$SnCl, prepd. by reacting Me$_3$SnOH with Me$_3$SnCl (2:1) in hot C$_6$H$_6$ or by treating Me$_3$SnOH·Me$_3$SnCl·H$_2$O with Me$_3$SnOH (1:1) in C$_6$H$_6$; m. 85-91° (decompn.) (198).
(Me$_3$SnOH)$_2$·Me$_3$SnI, prepd. by heating (EtO)$_2$PONa with Me$_3$SnI under reflux (30) or by dissolving (Me$_3$Sn)$_3$OI in moist C$_6$H$_6$ (200, 203); Decomp. at 145-155° (30); Reacts with Na in NH$_3$ in the presence of H$_2$O to form Me$_3$SnNa (200).

TRIETHYLTIN HYDROXIDE Et$_3$SnOH

Prepn.: By refluxing (Et$_3$Sn)$_2$O with i-Pr$_3$CeOH (19).
By hydrolyzing Et$_3$SnI or Et$_3$SnCl with alkali; 96% yield (195, 268, 387).
By heating (Et$_3$Sn)$_2$ with HgCl$_2$ (1:1) at 130° and treating the prod. with KOH; 70% yield (283).
Props.: m. 49-50° (268); m. 43-44° (283, 387); m. 43-47°; b. 272°; on distg. in vacuo loses H$_2$O and forms (Et$_3$Sn)$_2$O (195, 203); At 200-220° in a sealed tube → Et$_2$SnO + C$_2$H$_6$ (197, 203); Dissocn. const. (532).
Rxns. with: H$_2$O → (Et$_3$Sn)$_2$O·2 H$_2$O (195).
Na in NH$_3$ → (Et$_3$Sn)$_2$ (195).
Et$_3$Sn halides → (Et$_3$SnOH)$_2$·Et$_3$Sn halide·H$_2$O (195).
AcOH at 100-150° → Et$_3$SnOAc (268, 387).
ROH → Et$_3$SnOR (268).
RSO$_2$NH$_2$ → Et$_3$SnNHSO$_2$R (268).
HCl → Et$_3$SnCl (743).

TRIETHYLTIN HYDROXIDE Et_3SnOH (Cont'd.)

Biol. props.: Metabolism (107); Fungicidal activity (267, 643); Toxicity and symptoms (532, 625).
Addn. compds.: $(Et_3SnOH)_2 \cdot Et_3SnCl \cdot H_2O$, prepd. by reacting Et_3SnOH with Et_3SnCl (2:1) in moist C_6H_6; m. 77° (195).
$(Et_3SnOH)_2 \cdot Et_3SnBr \cdot H_2O$, prepd. in the same manner as the preceding complex (195).
$(Et_3SnOH)_2 \cdot Et_3SnI \cdot H_2O$, m. 108-110°, prepd. in the same manner as the preceding two complexes (195).

TRIVINYLTIN HYDROXIDE $(CH_2:CH)_3SnOH$

Prepn.: By reacting $(CH_2:CH)_3SnCl$ with aq. NaOH (460).
Props.: m. 67.5-69.0° (460).

TRIPROPYLTIN HYDROXIDE Pr_3SnOH

Prepn.: By heating Pr_3SnBr with aq. NaOH or KOH in EtOH (211, 268).
Props.: m. 34-35° (268); On distg. in vacuo loses H_2O and forms $(Pr_3Sn)_2O$ (211).
Rxns. with: AcOH → Pr_3SnOAc (268).
$CH_2:CHCO_2H$ → $Pr_3SnOCOCH:CH_2$ (271).

TRIBUTYLTIN HYDROXIDE Bu_3SnOH

Rxns. with: $POCl_3$ → $(Bu_3Sn)_3PO_4$ (94).
RSH → Bu_3SnSR (429, 680).

TRIPHENYLTIN HYDROXIDE Ph_3SnOH

Prepn.: By hydrolyzing Ph_3SnCl with alc. KOH or with NaOH in Et_2O; 75-95% yield (36, 282a, 382).
By oxidizing $(Ph_3Sn)_2$ with $KMnO_4$ in Me_2CO and diluting the prod. with H_2O; theoret. yield (36).
By hydrolyzing $Ph_3SnCH_2CH_2OAc$ with alc. KOH; 60% yield along with some $Ph_3SnCH_2CH_2OH$ (273b)
Props.: m. 118° (282a, 382); m. 119-121° (273b).
Rxns. with: AcOH → Ph_3SnOAc (268).
RSH → Ph_3SnSR (429).
Mild dehydration (recrystallization from AcCN) → $(Ph_3Sn)_2O$ (474a).
Use: Stabilizer for vinyl halide polymers (902).

TRI-p-TOLYLTIN HYDROXIDE $(p-MeC_6H_4)_3SnOH$

Prepn.: By hydrolyzing $(p-MeC_6H_4)_3SnCl$; 80% yield (63, 282a, 382).
By-prod. in the rxn. of $p-MeC_6H_4N_2BF_4$ with $Sn-SnCl_2 \cdot 2 H_2O$ in Me_2CO followed by a treatment with NH_4OH (392).
Props.: m. 108° (382); m. 108-109° (282a).
Rxn. with: HBr → $(p-MeC_6H_4)_3SnBr$ (64).

TRIBENZYLTIN HYDROXIDE $(PhCH_2)_3SnOH$

Prepn.: By hydrolyzing $[(PhCH_2)_3Sn]_2O$ in EtOH under reflux; theoret. yield (38).
By hydrolyzing $(PhCH_2)_3SnCl$ (382).
Props.: Long needles sintering at $102°$ (38); m. $122-124°$ (382).

The following three compounds were obtained as by-products in the reactions of RN_2BF_4 with $Sn-SnCl_2 \cdot 2\ H_2O$ in Me_2CO followed by a treatment with NH_4OH:

$(p-ClC_6H_4)_3SnOH$, 5.5% yield (392).
$(p-BrC_6H_4)_3SnOH$, 6% yield (392).
$(o-MeO_2CC_6H_4)_3SnOH$ (392).

BIS(TRIORGANOTIN) OXIDES

The oxides are formed by a direct or indirect dehydration of triorganotin hydroxides and by the oxidation of triorganotin hydrides with metal oxides.

BIS(TRIMETHYLTIN) OXIDE $(Me_3Sn)_2O$

Prepn.: By treating Me_3SnOH with Na in C_6H_6 (199).
Props.: Colorless liquid, b. $84°/22mm$; b. $86°/24mm$; sol. in all org. solvents; absorbs moisture from air and forms Me_3SnOH (199).
Rxn. with: Me_3Sn halides $\rightarrow (Me_3Sn)_3O$-halides (199, 200).

BIS(TRIETHYLTIN) OXIDE $(Et_3Sn)_2O$

Prepn.: By hydrolyzing triethyltin halides with aq. alkali and dehydrating the prod. at elevated temp. (18, 465, 731).
By distilling Et_3SnOH under reduced pressure (195).
By reacting Et_3SnSMe or $(Et_3Sn)_2S$ with Ag_2O (18).
By reacting Et_3SnH with metal oxides (HgO, ZnO, Fe_2O_3, PbO, As_4O_6, V_2O_5, $KMnO_4$) (25).
Props.: b. $272°$; b. $100°/1mm$; d_{20} 1.319; $n^{20}D$ 1.529 (18); b. $154°/10mm$; b. $158°/15mm$ (195); b. $110-118°/1mm$; b. $270°$ (465); b. $146-147°/12mm$; $n^{20}D$ 1.4975 (731); Forms dihydrate, m. $123-145°$ (decompn.) (195); At $270° \rightarrow Et_2SnO$, Et_2Sn and unidentified prods. (197).
Rxns. with: $RCO_2H \rightarrow Et_3SnOCOR$ (18, 26, 365, 508).
$ROH \rightarrow Et_3SnOR$ (195, 465, 508).
$RSH \rightarrow Et_3SnSR$ (18, 365, 488).
$H_2S \rightarrow (Et_3Sn)_2S$ (18, 365).
Aqueous H halides $\rightarrow Et_3Sn$ halides (18, 19).
$HCN \rightarrow Et_3SnCN$ (18).
$HNCS \rightarrow Et_3SnNCS$ (18).

BIS(TRIETHYLTIN) OXIDE $(Et_3Sn)_2O$ (Cont'd.)

<u>Rxns. with</u> (Cont'd.): Esters or halogenoids of an element other than Sn or of Sn in an oxidation state of 2,3, or 4 → the corresponding Et_3Sn esters, halides, and halogenoids, respectively (19).

$EtOBCl_2$ → Et_3SnCl (19).
$PO(NCO)_3$ → $(Et_3Sn)_3PO_4$ (?) (19).
$PO(NCS)_3$ → Et_3SnNCS (19).

<u>Use</u>: Stabilizer for vinyl halide polymers (731).

BIS(TRIPROPYLTIN) OXIDE $(Pr_3Sn)_2O$

<u>Prepn.</u>: By distg. Pr_3SnOH in vacuo (211).
By reacting $SnCl_2$ with PrMgCl (2:4.5) in C_6H_6, decompg. the mixt. with 10% HCl, and treating it with 3N NaOH (466).
By hydrolyzing $Pr_3SnCH_2CO_2Et$ with NaOH in 75% EtOH at r.t.; 92% yield (273b).

<u>Props.</u>: b. 195-198°/21mm (207); Absorbs moisture from air and forms Pr_3SnOH (211); b. 142-144°/1mm (466).

<u>Rxns. with</u>: RCO_2H (1:2) → Pr_3SnO_2CR (466, 273b).
ROH (1:2) → Pr_3SnOR (466).
RSH (1:2) → Pr_3SnSR (466).

BIS(TRIISOPROXYLTIN) OXIDE $(i-Pr_3Sn)_2O$

<u>Prepn.</u>: By shaking $i-Pr_3SnBr$ with aq. KOH in Et_2O and drying the prod. in vacuo over H_2SO_4; 75% yield (268).

<u>Rxn. with</u>: AcOH (1:2) → $i-Pr_3SnOAc \cdot H_2O$ (268).

BIS(TRIBUTYLTIN) OXIDE $(Bu_3Sn)_2O$

<u>Prepn.</u>: By reacting Bu_3SnCl with aq. NaOH at 90-95°; 88% yield (268).
By shaking Bu_3SnCl with NaOMe in MePh and heating the resulting prod. with small amt. of H_2O (631).
By reacting Bu_3SnCl with 20% NaOH in the presence of EtOH, distilling off the EtOH, and heating the residue at 200-220° (731).

<u>Props.</u>: Waxy substance (631); b. 200°/4mm (731).

<u>Rxns. with</u>: AcOH → Bu_3SnOAc (268).
RMgCl → Bu_3SnR (459).

<u>Use</u>: Stabilizer for vinyl halide polymers (731).

BIS(TRIISOBUTYLTIN) OXIDE $(i-Bu_3Sn)_2O$

<u>Prepn.</u>: By-prod. in the rxn. of $SnCl_4$ with $i-Bu_3Al$ in $n-C_7H_{16}$ at 40-48°, followed by a treatment with NaOH (581).

<u>Props.</u>: b. 197-198°/12mm; n^{21}_D 1.4850; d_{20} 1.1547 (581).

BIS(TRIHEXYLTIN) OXIDE   [(C$_6$H$_{13}$)$_3$Sn]$_2$O

Prepn.: By shaking (C$_6$H$_{13}$)$_3$SnBr with aq. NaOH in Et$_2$O; 51% yield (268).
Rxn. with: AcOH → (C$_6$H$_{13}$)$_3$SnOAc (268).

BIS(TRIOCTYLTIN) OXIDE   [(C$_8$H$_{17}$)$_3$Sn]$_2$O

Prepn.: By brominating (C$_8$H$_{17}$)$_4$Sn at -40°, shaking the resulting (C$_8$H$_{17}$)$_3$SnBr with aq. 33% NaOH in Et$_2$O, and drying the product after removal of the solvent at 100°/12mm; 83.5% yield (268).
Rxn. with: AcOH (1:2) → (C$_8$H$_{17}$)$_3$SnOAc (268).

BIS(TRIPHENYLTIN) OXIDE   (Ph$_3$Sn)$_2$O

Prepn.: By refluxing C$_5$H$_5$SnPh$_3$ with 95% EtOH, evapg. the mixt. to dryness, and distg. the residue with C$_6$H$_6$; 92% yield (182).
By treating C$_5$H$_5$SnPh$_3$ with 95% EtOH at 20° and pouring the mixt. into H$_2$O; 81% yield (182).
By recrystallizing Ph$_3$SnOH from hot anhydr. AcCN (474a).
Props.: m. 123-124° (182); Colorless, double-refractive plates (474a).
Rxns. with: RAsO$_3$H$_2$ → Ph$_3$SnOAs(O)(OH)R (796).
HCl → Ph$_3$SnCl (474a).
H$_2$O → Ph$_3$SnOH (474a).
Use: Stabilizer for vinyl halide polymers (731).

BIS(DIBUTYLVINYLTIN) OXIDE   [Bu$_2$(CH$_2$:CH)Sn]$_2$O

Prepn.: By treating Bu$_2$(CH$_2$:CH)SnBr with 20% aq. KOH in Et$_2$O; 95% yield (460a).
Rxn. with: AcOH on steam bath → Bu$_2$(CH$_2$:CH)SnOAc (460a).

OTHER TRIORGANOTIN OXIDES

The following compounds were prepared by reacting SnBr$_4$ with appropriate Grignard compounds in ether and hydrolyzing the reaction product:

(PhCH$_2$)$_3$Sn$_2$O, colorless rhombohedra, m. 120° (38).
(o-BrC$_6$H$_4$CH$_2$)$_3$Sn$_2$O, colorless rhombohedra, m. 158-159° (38).
(o-ClC$_6$H$_4$CH$_2$)$_3$Sn$_2$O, colorless rhombohedra, m. 133.5° (38).
(o-FC$_6$H$_4$CH$_2$)$_3$Sn$_2$O, colorless rhombohedra, m. 113° (38).

The above compounds are converted to the corresponding triorganotin chlorides by treating them with HCl in ether. With aqueous ethanol under reflux they form triorganotin hydroxides (38).

TRIORGANOTIN ALKOXIDES AND PHENOXIDES

The compounds are formed by reacting trialkyltin halides with sodium alkoxides or aryloxides, by treating trialkyltin oxides with alcohols or phenols, by etherifying trialkyltin hydroxides with alcohols or phenols, by cleaving tetraalkyltin compounds with phenols, and by transetherification reactions.

ETHOXYTRIETHYLTIN $\quad$ Et$_3$SnOEt (?)
Prepn.: By treating (EtO)$_2$PONa with Et$_3$SnI in EtOH (30).
By reacting Et$_3$SnCl with NaOEt (707).
By reacting (Et$_3$Sn)$_2$O with abs. EtOH (195).
Props.: Colorless liquid, b. 82-84°/11mm; n^{15}D 1.4842; $d^0{}_0$ 1.3042 (30); b. 190-192° (195, 707).
Use: Stabilizer for vinyl halide resins (707).

TRIETHYLPHENOXYTIN $\quad$ Et$_3$SnOPh
Prepn.: By mixing Et$_3$SnOH with PhOH in Et$_2$O; 77% yield (268).
By refluxing Et$_3$SnSCH$_2$CHMe$_2$ with PhOH (466).
By reacting (Et$_3$Sn)$_2$O with PhOH; theoret. yield (465).
By refluxing Et$_4$Sn with PhOH (1:1); 8% yield (467).
Props.: b. 147.5-148°/12mm (268); b. 112-115°/1mm; b. 259° (466); b. 115-126°/1mm; b. 280°/760mm; d_{25} 1.315; n^{20}D 1.5422 (465). b. 115°/1mm; n^{20}D 1.5415 (467).

TRIETHYL(p-TOLYLOXY)TIN $\quad$ Et$_3$SnOC$_6$H$_4$-p-Me
Prepn.: By reacting (Et$_3$Sn)$_2$O with p-MeC$_6$H$_4$OH (1:1); theoret. yield (465).
By refluxing Et$_4$Sn with p-MeC$_6$H$_4$OH (1:1); 8% yield (467).
Props.: b. 124-127°/1mm; b. 303°/760mm (decompn.); d_{25} 1.283; n^{20}D 1.5365 (465); b. 124-126°/1mm; n^{20}D 1.5360 (467).

TRIETHYL(p-NITROPHENOXY)TIN $\quad$ Et$_3$SnOC$_6$H$_4$NO$_2$
Prepn.: By dissolving p-O$_2$NC$_6$H$_4$OH and Et$_3$SnOH in EtOH (268).
Props.: Intensively yellow oil (268).

PHENOXYTRIPROPYLTIN $\quad$ Pr$_3$SnOPh
Prepn.: By refluxing (Pr$_3$Sn)$_2$O with PhOH; 100% yield (406).
Props.: b. 145-147°/1mm; d_{20} 1.2167; n^{20}D 1.5284 (466).

BUTOXYTRIBUTYLTIN Bu_3SnOBu

<u>Prepn.</u>: By reacting Bu_3SnCl or Bu_3SnBr with BuONa in anhyd. BuOH under reflux (97).

By reacting $SnBr_4$ with BuMgBr (1:2.88) in EtOH and treating the prod. with NaOBu under reflux (125).

<u>Props.</u>: b. 124-128°/3mm; n^{20}_D 1.4685-88 (97); b. 114-127°/2.5mm; n^{20}_D 1.4726 (125).

<u>Use</u>: Stabilizer for vinyl polymers (97).

The following tributyltin alkoxides and homologs, useful as stabilizers for vinyl halide resins were prepared by reacting tributyltin chloride with hydrocarbyloxy-sodium derivatives:

Bu_3SnOMe, liquid (707).
$Bu_3SnOCMe_3$, b. 66-70°/1.2mm (707).
Bu_3SnO-cyclo-C_6H_{11} (707).
Bu_3SnOCH_2Ph, b. 90-95°/1mm (707).
$Bu_3SnOCMe:CHAc$, oil (706, 708).

METHOXYTRI-n-OCTYLTIN $(n-C_8H_{17})_3SnOMe$

<u>Prepn.</u>: By reacting $(n-C_8H_{17})_3SnCl$ with NaOMe in MeOH (97, 125).
<u>Props.</u>: n^{20}_D 1.4781 (125).
<u>Use</u>: Stabilizer for vinyl halide polymers (97).

TRIBUTYL(DODECYLOXY)TIN $Bu_3SnOC_{12}H_{25}$

<u>Prepn.</u>: By heating Bu_3SnOMe with $C_{12}H_{25}OH$ (1:1) at 90° while passing dry N into the mixt. (349).
<u>Props.</u>: Liquid (349).

4-METHYL-4-(TRIBUTYLSTANNYLOXY)-2-PENTANONE $Bu_3SnOCMe_2CH_2Ac$

<u>Prepn.</u>: By heating Bu_3SnOMe with diacetone alc. under N (349).
<u>Use</u>: Stabilizer for vinyl chloride polymers (349).

DIBUTYL-3-INDENYLMETHOXYTIN $Bu_2Sn(OMe)(C_9H_7)$

<u>Prepn.</u>: By heating $Bu_2Sn(OMe)_2$ with indene (1:1) under N (349).
<u>Use</u>: Stabilizer for vinyl chloride polymers (349).

(DIBUTYLMETHOXYSTANNYL)PHENYLACETIC ACID 2-ETHYLHEXYL ESTER
$Bu_2(MeO)SnCHPhCO_2CH_2CHEtBu$

<u>Prepn.</u>: By heating $Bu_2Sn(OMe)_2$ with $PhCH_2CO_2CH_2CHEtBu$ (1:1) under N (349).
<u>Use</u>: Stabilizer for vinyl chloride polymers (349).

1,4-BIS(TRIETHYLSTANNYLOXY)BUTANE, 3,3,10,10-TETRAETHYL-
4,9,3,10-DIOXADISTANNADODECANE   (Et$_3$SnOCH$_2$CH$_2$)$_2$

Prepn.: By fractionating in vacuo a mixt. of (Et$_3$Sn)$_2$O and
1,4-butanediol; 39.4% yield (508).
Props.: b. 184-185°/4.5mm; n$^{20}_D$ 1.4975; d$_{20}$ 1.2010 (508).

TRIETHYL[3-(ETHYLDIMETHYLSILYL)PROPOXY]TIN, 3,3,-DIETHYL-8,8-
DIMETHYL-4-OXA-8-SILA-3-STANNADECANE Me$_3$SnO(CH$_2$)$_3$SiMe$_2$Et

Prepn.: By fractionating in vacuo a mixt. of (Et$_3$Sn)$_2$O and
Me$_2$EtSi(CH$_2$)$_3$OH; 73.3% yield (508).
Props.: b. 131-133°/5mm; n$^{20}_D$ 1.4769; d$_{20}$ 1.1208 (508).

TRIALKYLTIN PEROXY DERIVATIVES

The compounds are formed by reacting trialkyltin halides, R$_3$SnX, with sodium hydrocarbylperoxides, NaOOR', or by reacting trialkyltin methoxides, R$_3$SnOMe, with hydrocarbyl hydroperoxides, R'OOH.

R$_3$SnOOR'	Yield %	Properties	Ref.
Me$_3$SnOOCMe$_3$	>90	b. 56°/12mm (1,2)	442
Et$_3$SnOOCMe$_3$	>90	b. 56-57°/12mm (1,2)	442
Et$_3$SnOOC$_6$H$_4$CHMe$_2$	>90	b. 105-110°/1-2mm (decompn.) (1,2)	442
Bu$_3$SnOOCMe$_3$	>90	b. 105-110°/1mm (decompn.) (1,2)	442

(1) Reactions with H$_2$O, alkali, or acids yield R$_3$SnOH and R'OOH
(2) Reactions with aqueous Na$_2$SO$_3$ yield R$_3$SnOH and R'OH.

TRIORGANOTIN CARBOXYLATES

The compounds are prepared:

1. By reacting trialkyl-, triaryl-, and triaralkyltin halides with alkali metal or silver carboxylates.
2. By reacting trialkyl-, triaryl-, and triaralkyltin hydroxides or hydrides with carboxylic acids.
3. By reacting bis(triorganotin) oxides or sulfides with carboxylic acids, carboxylic acid esters, or metal carboxylates.
4. By treating homogeneously or heterogeneously substituted tetraorganotin compounds with carboxylic acids.

The homogeneously substituted trialkyl-, triaryl-, and triaralkyltin carboxylates are compiled in Table I and heterogeneously substituted organotin carboxylates in Table II.

TABLE I

$R_3SnOCOR'$	Prepared from	Yield %	Properties	Ref.
Me_3SnOAc	$Me_3SnOH + AcOH$		m. 196.5-197.5°	268
$Me_3SnOCOCMe:CH_2$	$Me_3SnCl + CH_2:CMeCO_2H$		m. 100°; copolymerizes with other olefins in the presence of the peroxide catalysts	304
$(Et_3SnO)_2CO$	$Et_3SnOH + CO_2$		Decomp. 138-139°	585
$Et_3SnOOCH$		98	Insecticidal props.	625
Et_3SnOAc	$(Et_3Sn)_2O + AcOH$		m. 119°; b. 224°	18, 465
	$(Et_3Sn)_2O + EtOAc$ or Me_3SiOAc		b. 214°	19
	$Et_3SnOH + AcOH$	95	m. 134-135 (268); m. 131° (387)	268, 387
	$(Et_3Sn)_2S + AgOAc$		m. 121°, b. 217°	18
	$Et_3SnSMe + AgOAc$ or Me_3SiOAc			18, 19
	$Et_3SnSCHMe_2 + AcOH$		b. 223-226°	466
	Et_3Sn halides $+ AgOAc$		Rxn. with $AgF \rightarrow Et_3SnF$	18
	$Et_3SnCH:CH_2 + AcOH$	100	m. 132-134°	488
			Fungicidal props.	267
			Toxicity & symptoms	625
$Et_3SnOCOCH_2OH$	$Et_3SnI + HOCH_2CO_2Ag$	25	m. 106.5-107°	26
$Et_3SnOCOCH_2F$	$(Et_3Sn)_2O + FCH_2CO_2H$	30	m. 155-156	26
$Et_3SnOCOCH_2Cl$	$Et_4Sn + ClCH_2CO_2H$	55	m. 111-112°; b. 118-119°/1mm	467
$Et_3SnOCOCH_2Br$	$(Et_3Sn)_2O + BrCH_2CO_2H$	50	m. 99.5°	26
$Et_3SnOCOCH_2I$	$(Et_3Sn)_2O + ICH_2CO_2H$	60	m. 94.5°	26
$Et_3SnOCOCHCl_2$	$Et_4Sn + Cl_2CHCO_2H$	60	m. 132-133°; b. 121-123°	467
$Et_3SnOCOCF_3$	$(Et_3Sn)_2O + Me_2Si-(OCOCF_3)_2$ ⇌ $MeSi-(OCOCF_3)_3$	63	m. 107-108°; b. 140°/100 mm	19
				19
	$(Et_3Sn)_2O + F_3CCO_2H$	59	m. 107°; b. 218°	465
	$Et_4Sn + F_3CCO_2H$		m. 122-123°; b. 218°	467
$Et_3SnOCOEt$	$(Et_3Sn)_2O + EtCO_2H$		m. 103°; b. 238°	465
$Et_3SnOCOCH_2CH_2Cl$	$(Et_3Sn)_2O + ClCH_2CH_2CO_2H$	70	m. 87.5-88°	26
$Et_3SnOCOCHClMe$	$(Et_3Sn)_2O + MeCHClCO_2H$	70	m. 90.5-91°	26

Compound	Reactants	Yield %	Properties	Ref.
Et$_3$SnOCOCH$_2$CH$_2$Br	(Et$_3$Sn)$_2$O + BrCH$_2$CH$_2$CO$_2$H	60	m. 84.5°	26
Et$_3$SnOCOCHBrMe	(Et$_3$Sn)$_2$O + MeCHBrCO$_2$H	40	m. 88-89°	26
Et$_3$SnOCOCHBrCH$_2$Br	(Et$_3$Sn)$_2$O + BrCH$_2$CHBrCO$_2$H	70	m. 99.5-100°	26
Et$_3$SnOCOC$_2$F$_5$	(Et$_3$Sn)$_2$O + C$_2$F$_5$CO$_2$H	40	m. 94.5°	26
Et$_3$SnOCOC$_3$F$_7$	(Et$_3$Sn)$_2$O + C$_3$F$_7$CO$_2$H	40	m. 75.5°	26
Et$_3$SnOCOCH:CH$_2$	(Et$_3$Sn)$_2$O + CH$_2$:CHCO$_2$H	70	116-117°	26
			Toxicity and symptoms	625
Et$_3$SnOCOCMe:CH$_2$	(Et$_3$Sn)$_2$O + CH$_2$:CMe-CO$_2$Me		m. 75.5°	26
	(Et$_3$Sn)$_2$O + CH$_2$:CMe-CO$_2$H		m. 75°	50
Et$_3$SnOCOPr	(Et$_3$Sn)$_2$O + PrCO$_2$H	36	m. 92°; b. 96-98°/1mm; b. 286°	465
Et$_3$SnOCOPh	(Et$_3$Sn)$_2$O + PhCO$_2$Et		m. 71°; b. 124°/1mm	19
	(Et$_3$Sn)$_2$O + BzOH	18	m. 71°; b. 132-134°/1mm	465
	Et$_4$Sn + BzOH		m. 80°	467
			Toxicity and symptoms	625
			Insecticidal props.	625
Et$_3$SnOCOCH-(C$_6$H$_4$Cl-p)$_2$				
Et$_3$SnOCO-CHCH(CH:CMe$_2$)CMe$_2$			Insecticidal props	625
Et$_3$SnOCOCH:CHPh	(Et$_3$Sn)$_2$O + PhCH:CH-CO$_2$H	55	m. 107-108°	26
Pr$_3$SnOAc	Pr$_3$SnOH + AcOH	16.2	m. 99-100°	268
	Pr$_3$SnCH:CH$_2$ + AcOH	100	m. 100°	463
	(Pr$_3$Sn)$_2$O + AcOH		m. 82°; b. 81-83°/1mm	273b, 466
			Toxicity and symptoms	532
Pr$_3$SnOCOCH$_2$CN	Pr$_3$SnCH:CH$_2$ + CH$_2$CNCO$_2$H	72	m. 91-92°	463
Pr$_3$SnOCOCH$_2$Br	Pr$_3$SnCH:CH$_2$ + CH$_2$BrCO$_2$H	60.5	m. 77-78°	463
Pr$_3$SnOCOCH$_2$Cl	Pr$_3$SnCH:CH$_2$ + CH$_2$ClCO$_2$H	58.1	m. 78-80°	463
	Pr$_4$Sn + CH$_2$ClCO$_2$H	35	m. 78-79°; b. 135-140°/1mm	468
Pr$_3$SnOCOCHCl$_2$	Pr$_3$SnCH:CH$_2$ + CHCl$_2$CO$_2$H	63.3	m. 84-85°	463
	Pr$_4$Sn + CHCl$_2$CO$_2$H	39	m. 83-84°	468
Pr$_3$SnOCOCF$_3$	Pr$_3$SnCH:CH$_2$ + CF$_3$CO$_2$H	64.4	m. 94-95°	463
	(Pr$_3$Sn)$_2$O + CF$_3$CO$_2$H	100	m. 80°; b. 88-90°/1mm	466
			Rxn. with PhSH → Pr$_2$Sn(SPh)$_2$	466

TABLE I (Cont'd.)

$R_3SnOCOR'$	Prepared from	Yield %	Properties	Ref.
$Pr_3SnOCOCCl_3$	$Pr_3SnCH:CH_2 + CCl_3CO_2H$	53.8	m. 96-97°	463
$Pr_3SnOCOEt$	$(Pr_3Sn)_2O + EtCO_2H$	100	m. 70°; b. 88-89°/1mm	466
$Pr_3SnOCOPr$	$(Pr_3Sn)_2O + PrCO_2H$	100	m. 66°; b. 102-104°	466
$Pr_3SnOCOCH:CH_2$	Pr_3SnH or $Pr_3SnOH + CH_2:CHCO_2H$	83	Shiny needles, m. 94-96°	271
$Pr_3SnOCOCHBrMe$	$Pr_3SnCH:CH_2 + MeCHBrCO_2H$	61.4	m. 88-89°	463
Pr_3SnOBz	$(Pr_3Sn)_2O + BzOH$	100	m. 45-46°; b. 158-160°/1mm	466
$i-Pr_3SnOAc \cdot H_2O$	$Pr_3SnOAc + BzOH$ $(i-Pr_3Sn)_2O + AcOH$		Crystals, anhydrous prod. is liquid Toxicity and symptoms	466 268
Bu_3SnOAc	$Bu_3SnCl + KOAc$ $(Bu_3Sn)_2O + AcOH$	98	Asbestos-like, fibrous substance m. 84.5-85° Toxicity and symptoms Fungicidal props. Stabilizer for halogen-contg. polymers	532 116 268 532 267
$Bu_3SnOCOCH_2CN$	$Bu_3SnCl + KOCOCH_2CN$	100	at 140-150°/50mm → Bu_3SnCH_2CN	116
$Bu_3SnOCOCH_2CH_2-SCHEt(n-C_5H_{11})$	$Bu_3SnCl + NaOCOCH_2CH_2-SCHEt(n-C_5H_{11})$		m. r.t.; stabilizer for Cl-congt. resins	270
$(n-C_6H_{13})_3SnOAc$	$(n-C_6H_{13})_3Sn_2O + AcOH$	63	m. 68-69°	309
$(n-C_8H_{17})_3SnOAc$	$(n-C_8H_{17})_3Sn_2O + AcOH$	14	m. 47-48°	268
Ph_3SnOAc	$Ph_3SnOH + AcOH$	62	m. 121-122°	268
	$Ph_3SnCH_2CH_2OAc$ by pyrolysis		m. 122-123°	271
			Fungicidal prop. Toxicity and symptoms	267 532
$Ph_3SnOCOCH_2CN$	$Ph_3SnCl + KOCOCH_2CN$		m. 140-142°; at 140° in vacuo → $Ph_3SnCH_2CN + CO_2$	270
$Ph_3SnOCOCH_2COEt$			Stermutatory effect	338
Ph_3SnOBz	$Ph_3SnH + Bz_2O_2$ $Ph_3SnI + NaOBz$	57	m. 82-84	178 178
$(Ph_3SnOCO)_2CH_2$	$Ph_3SnCl + EtO_2CCH_2CO_2K$		m. 136-137°; on heating → Ph_4Sn	270

(p-ClC$_6$H$_4$)$_3$SnOAc	(p-ClC$_6$H$_4$)$_3$SnOH + AcOH	68	m. 148.5-149.5° Fungicidal props.	268 267
(PhCH$_2$)$_3$SnOAc	(PhCH$_2$)$_3$SnCl + NaOAc		m. 117-118°*	507
(PhCH$_2$)$_3$SnOCO-n-C$_{11}$H$_{23}$	(PhCH$_2$)$_3$SnCl + n-C$_{11}$H$_{23}$CO$_2$Na		m. 90-91°*	507
(PhCH$_2$)$_3$SnOCO-n-C$_{17}$H$_{35}$	(PhCH$_2$)$_3$SnCl + n-C$_{17}$H$_{35}$CO$_2$Na		m. 99-100°*	507
(PhCH$_2$)$_3$Sn ricinoleate	(PhCH$_2$)$_3$SnCl + Na ricinoleate		Oil*	507
(PhCH$_2$)$_3$SnOBz	(PhCH$_2$)$_3$SnCl + NaOBz		m. 98-100°*	507

* Useful as plasticizer for vinyl halide resins.

TABLE II

$R_2R'SnOCOR''$	Prepared from	Yield %	Properties	Ref.
$Me_2(CH_2:CH)Sn-OCOCH_2Cl$	$Me_2Sn(CH:CH_2)_2$ + CH_2ClCO_2H	80	m. 104-107°	463
$Me_2(CH_2:CH)Sn-OCOCHBrMe$	$Me_2Sn(CH:CH_2)_2$ + $MeCHBrCO_2H$	100	m. 79-81°	488
$Me_2BuSnOAc$	$(Me_2BuSn)_2O$ + AcOH	73.5	m. 106-108°	269
$Me_2(n-C_8H_{17})SnOAc$	$[Me_2(n-C_8H_{17})Sn]_2O$ + AcOH	68	Fungicidal props. m. 95-97°	269 269
$Me_2(n-C_{10}H_{21})SnOAc$	$[Me_2(n-C_{10}H_{21})Sn]_2O$ + AcOH	98 98	Fungicidal props. Amorphous, m. 96-97.8° Amorphous, m. 96-98°	269 460a 460a
$Me_2(n-C_{12}H_{25})SnOAc$	$[Me_2(n-C_{12}H_{25})Sn]_2O$ + AcOH	83 96	m. 99-102° Amorphous, m. 97-98.6°	269 460a
$Et_2BuSnOAc$	$(Et_2BuSn)_2O$ + AcOH	80	m. 71-73°; fungicidal props.	269
$Et_2Br(CH_2)_5SnOAc$	$[Et_2Br(CH_2)_5Sn]_2O$ + AcOH		m. 68-69°	268
$Et_2(n-C_6H_{13})SnOAc$	$[Et_2(n-C_6H_{13})Sn]_2O$ + AcOH	95	m. 64-65°; fungicidal props.	269
$Et_2(n-C_8H_{17})SnOAc$	$[Et_2(n-C_8H_{17})Sn]_2O$ + AcOH	66	m. 67-69°; fungicidal props.	269
$Et_2(n-C_{12}H_{25})SnOAc$	$[Et_2(n-C_{12}H_{25})Sn]_2O$ + AcOH	60	m. 68-72°; fungicidal props.	269
$Et_2PhSnOAc$	$(Et_2PhSn)_2O$ + AcOH	43	m. 93.5-94.5°	268
$Et_2(p-ClC_6H_4)SnOAc$	$Et_2(p-ClC_6H_4)SnOH$	21	m. 100-101°	268
$Et_2(p-BrC_6H_4)SnOAc$	$Et_2(p-BrC_6H_4)SnOH$	65	m. 108-109; Fungicidal props.	268 267
$Pr_2[2-(4-pyridyl)-ethyl]SnOAc$	$Pr_2(C_5H_4NCH_2CH_2)SnOH$ + AcOH	24	b. 96-97°	682
$Bu_2(CH_2:CH)SnOAc$ $(n-C_5H_{11})_2EtSnOAc$	$[Bu_2(CH_2:CH)Sn]_2O$ + AcOH $[(n-C_5H_{11})_2EtSn]_2O$ + AcOH	89 86.5	Colorless liquid m. 50-53.5°	460a 269
$Pr_2(CH_2CNCH_2CH_2)-SnOAc$	$Pr_2(CH_2CNCH_2CH_2)SnOH$ + AcOH	75	Fungicidal props. b. 71-72°	269 682
$Pr_2(PhCH_2CH_2)SnOAc$	$Pr_2(PhCH_2CH_2)SnOH$ + AcOH	94	b. 79-80°	682
$Ph_2(CH_2:CHCH_2)-SnOAc$	$[Ph_2(CH_2:CHCH_2)Sn]_2O$ + AcOH	92	Decomp. 260°	460a

ORGANOTIN INNER SALTS

(2-CARBOXYETHYL)DIPROPYLTIN INNER SALT $Pr_2\overline{SnCH_2CH_2CO0}$
Prepn.: From $Pr_3SnCH_2CH_2CO_2Na$ and Br_2 (1:1) in $CHCl_3$; 66% yield (682).
Props.: m. 158-160° (682).

DIBUTYL(2-CARBOXYETHYL)TIN INNER SALT $Bu_2\overline{SnCH_2CH_2CO0}$
Prepn.: By heating $Bu_3SnCH_2CH_2CO_2H$ at 120°/0.2mm; 56% yield (273b).
By treating $Bu_2Sn(CH_2CH_2CO_2Me)Br$ with KOH (1:2) in 75% EtOH at r.t. and acidifying the rxn. mixt. with 1N HCl, 43% yield (273b).
Props.: Shiny needles, m. 110-113° (273b).

(2-CARBOXYETHYL)DIPHENYLTIN INNER SALT $Ph_2\overline{SnCH_2CH_2CO0}$
Prepn.: From Ph_3SnH and excess acrylic acid on a steam bath; 93% yield (271, 274).
By neutralizing $Ph_3SnCH_2CH_2CO_2Na$ with 1N HCl (273b).
Props.: White powder, doesn't melt at 320° (271, 273b, 274).

(o-CARBOXYPHENYL)DIPHENYLTIN INNER SALT $Ph_2\overline{SnC_6H_4CO0}$
Prepn.: By oxidation of $Ph_3SnC_6H_4CH_2OH$ with $KMnO_4$ in Et_2O; 54% yield (157).
Rxn. with: HCl → $Ph_2Sn(C_6H_4CO_2H)Cl$ (?) (157).

2,2-DIBUTYL-1,3,2-OXATHIASTANNOLAN-5-ONE $Bu_2\overline{SnSCH_2CO0}$
Prepn.: By reacting $Bu_2Sn(OAc)_2$ with $HSCH_2CO_2H$ in AcOH (193).
Props.: m. 185-187° (193).
Use: As bactericide and fungicide (193).

5-AMINO-2,2-DIBUTYL-1,3,2-OXATHIASTANNINAN-6-one
$Bu_2\overline{SnSCH_2CH(NH_2)CO0}$
Prepn.: By reacting Bu_2SnO in aq. HNO_3 with cysteine·HCl (193).
Props.: m. 205-207° (193).
Use: As bactericide and fungicide (193).

ORGANOTIN PHOSPHORUS SALTS

ORGANOTIN PHOSPHATES, PYROPHOSPHATES, THIOPHOSPHATES, AND PHOSPHITES

Organotin salts of various phosphorus acids are prepared:

1. By reacting organotin halides with alkali salts of phosphorus acids or esters.
2. By reacting organotin oxides with phosphorus pentoxide or with phosphoryl isocyanate.
3. By reacting organotin hydroxides with phosphorus oxychloride.

ORGANOTIN PHOSPHATES, PYROPHOSPHATES, THIOPHOSPHATES, AND PHOSPHITES

Compound	Prepared from	Yield %	Properties	Ref.
Me_2SnO_2POOH	$Me_2SnCl_2 + Na_2HPO_4 \cdot 12 H_2O$	82	m. 345° (1)	487
$Bu_2Sn[OP(O)(SC_6H_{10}Me)_2]_2$	$Bu_2SnCl_2 + KOP(O)(S\text{-cyclo-}C_6H_{10}Me)_2$		(1)	703
$Bu_2SnOP(O)(SC_6H_4\text{-}C_8H_{17})_2]_2$	$Bu_2SnCl_2 + KOP(O)(SC_6H_4\text{-}n\text{-}C_8H_{17})_2$		(1)	703
$(n\text{-}C_{12}H_{25})_2SnOP(O)\text{-}(SCHMe_2)_2$	$(n\text{-}C_{12}H_{25})_2SnCl_2 + KOP(O)(SCHMe_2)_2$		(1)	703
$(Ph_2SnO_2P)_2O$	$Ph_2SnO + P_2O_5$		Viscous liquid (2)	94
$Ph_2SnOP(O)OC_6H_{13}]_2O$	$Ph_2SnCl_2 + [NaOP(O)(O\text{-}n\text{-}C_6H_{13})]_2O$		Yellow oil (2)	95, 96
$(Et_3Sn)_3PO_4$ (?)	$(Et_3Sn)_2O + PO(NCO)_3$		Non-volatile solid (2)	19
$(Bu_3Sn)_3PO_4$	$Bu_3SnOH + POCl_3$		Yellow oil (2)	94
$Bu_3SnOP(O)(O\text{-}n\text{-}C_8H_{17})_2$	$Bu_3SnCl + NaOP(O)(O\text{-}n\text{-}C_8H_{17})_2$		Viscous liquid (2)	95, 96
$[Bu_3SnOP(O)(OBu)O\text{-}]_2\text{-}P(O)OSnBu_3$	$Bu_3SnCl + Bu_2SnCl_2$ (3:1) $+ Na_2P_3O_{10}$		Yellow oil (2)	94
$[Bu_3SnOP(O)(O\text{-}n\text{-}C_8H_{17})O\text{-}]_2\text{-}P(O)OSnBu_3$	$Bu_3SnCl + Na_3P_3O_5 + (O\text{-}n\text{-}C_8H_{17})_2$			94, 96
$Me_2SnOP(O)H_2]_2 \cdot H_2O$	$Me_2SnCl_2 + NaH_2PO_2 \cdot H_2O$	89	Crystals, m. 213° (decompn.) (3)	487
$Bu_2SnOP(O)HO$				790a

(1) Useful as stabilizer for lubricating oils (703).
(2) Useful as stabilizer for organic, halogen-contg. compounds (94, 95, 96).
(3) Useful as antioxidant for vulcanized rubber (790a).

ORGANOTIN PHOSPHONATES

Organotin phosphonates (listed in the following table) having the general formula $R_{4-n}SnX_n$, in which R is an organic radical, X an $-OP(O)R'R''$ group, and $n = 1, 2,$ or 3, were prepared by reacting organotin halides, $R_{4-n}SnX_n$, with organo-phosphites, phosphinates, and phosphonates. The compounds are formed by the Arbuzov rearrangement; thus, the molecular structure with an Sn-P bond, originally assigned to them, was revised and a structure with an O-bridge between Sn and P was proposed (33, 354). Their reactions with halogens and hydrogen halides yield the corresponding organotin halides and with aqueous alkali hydroxides the corresponding organotin hydroxides and oxides, respectively.

ORGANOTIN PHOSPHONATES

Compound	Incorrect formula	Prepared from	Properties	Rxns	Ref.
MeSn[OP(O)(OMe)Me]$_3$	MeSnP(O)(OMe)$_3$	MeSnI$_3$ + (MeO)$_3$P	White solid	1	31
Me$_2$Sn[OP(O)(OMe)Me]$_2$	Me$_2$SnP(O)(OMe)$_2$	Me$_2$SnI$_2$ + (MeO)$_3$P	m. 245-247°	1,2,3	31
Me$_2$Sn[OP(O)(OEt)Et]$_2$	Me$_2$SnP(O)(OEt)$_2$	Me$_2$SnI$_2$ + (EtO)$_3$P	Glassy solid	1,2,3	31
Et$_2$Sn[OP(O)(OMe)Me]$_2$	Et$_2$SnP(O)(OMe)$_2$	Et$_2$SnI$_2$ + (MeO)$_3$P	m. 263.5-265°	1,2,3	31
Me$_2$Sn[OP(O)(H)Ph]$_2$		Me$_2$Sn(CH:CH$_2$)$_2$ + PhP(O)(H)OH	m. 260° decompn.		463
Et$_2$Sn[OP(O)(OEt)Et]$_2$	Et$_2$SnP(O)(OEt)$_2$	Et$_2$SnI$_2$ + (EtO)$_3$P or NaOP(OEt)$_2$	Monomer, m. 173-174° Dimer, m. 249-251° m. 261-263° m. 234.5°	1,2,3 5	31 31 33 354
Et$_2$Sn[OP(O)(OPr)Pr]$_2$	Et$_2$SnP(O)(OPr)$_2$	Et$_2$SnCl$_2$ + EtP(O)(OEt)$_2$ Et$_2$SnI$_2$ + (PrO)$_3$P	m. 262-264°	1,2,3,4	31
Pr$_2$Sn[OP(O)(OPr)Pr]$_2$	Pr$_2$SnP(O)(OPr)$_2$	Pr$_2$SnI$_2$ + (PrO)$_3$P	m. 251-253°	1,2,3	31
Me$_2$Sn[OP(OMe)Ph]$_2$	Me$_2$SnP(O)(OMe)Ph$_2$	Me$_2$SnI$_2$ + PhP(OMe)$_2$	decomp. 291-293°**	1,2,3	32
Et$_2$Sn[OP(OEt)Ph]$_2$	Et$_2$SnP(O)(OEt)Ph$_2$	Et$_2$SnI$_2$ + PhP(OEt)$_2$	decomp. 296-300°**	1,2,3	32, 33
Et$_2$Sn[OP(OPr)Ph]$_2$	Et$_2$SnP(O)(OPr)Ph$_2$	Et$_2$SnI$_2$ + PhEtP(O)OEt Et$_2$SnI$_2$ + PhP(OPr)$_2$	m. 292-294° decomp. 245-248°**	8 1,2,3	33 32
Me$_3$Sn[OP(OMe)MePh]	Me$_3$SnP(O)(OMe)Ph	Me$_3$SnI + PhP(OMe)$_2$	m. 132-134°**	1,3	32
Et$_3$Sn[OP(O)(OEt)Et]	Et$_3$SnP(O)(OEt)$_2$	Et$_3$SnI + EtP(O)(OEt)$_2$ Et$_3$SnI + (EtO)$_3$P	b. 210-220°/2.5mm n^{14}D 1.4858 n^{14}D 1.4902	1,6 5	30 30, 354
Et$_3$Sn[OP(O)EtPh]	Et$_3$SnP(O)(OEt)Ph	Et$_3$SnI + PhP(OEt)$_2$	m. 294-297°		32

$Bu_3SnOP(O)-$ $EtPh$	$Bu_3SnI + PhP(OEt)_2$	m. 302-305°**		32
$Me_2Sn[OP(O)-$ $Ph_2]_2$	$Me_2SnI_2 + Ph_2P(OEt)$	decomp. 372-375°**	1,2,3	32
$Et_2Sn[OP(O)-$ $Ph_2]_2$	$Et_2SnI_2 + Ph_2P(OEt)$	decomp. 351-353°**	1,2, 3,4	32
$Me_3SnOP(O)Ph_2$	$Me_3SnI + Ph_2P(OEt)$	m. 346-349° decomp. 365-368°	4	32 33 32
$Et_3SnOP(O)Ph_2$	$Et_3SnI + Ph_2P(OEt)$	decomp. 346-348°**		32
$Me_3SnOP(O)-$ $(OMe)Me$	$Me_3SnI + (MeO)_3P$	m. 96°, sol. in Et_2O, $CHCl_3$, Me_2CO, $EtOH$, and H_2O	1,2 3,4 6,7	30 487
$Me_2SnOP(O)-$ $(Ph)O$	$Me_2SnCl_2 + PhP(O)(OH)_2$	m. 350°		
$(CH_2:CH)_2-$ $SnOP(O)(Ph)O$	$(CH_2:CH)_2SnCl_2 + PhP(O)(OH)_2$	m. 355°		487

* Stable in hot water.
** Insoluble in organic solvents; stable in hot water.
(1) Reaction with hydrogen halides → $R_{4-n}SnX_n$.
(2) Reactions with halogens in $CHCl_3$ → $R_{4-n}SnX_n$.
(3) Reactions with aq. alkali → R_3SnOH and R_2SnO, respectively.
(4) Reaction with AcCl → $R_{4-n}SnCl_n$.
(5) Reaction with HCl at 100° or in a sealed tube → $EtP(O)(OH)_2$.
(6) Reaction with AgCl → $R_{4-n}SnCl_n$.
(7) Concd. HCl at 100° → Me_2SnCl_2.
(8) Concd. HCl → $EtPhPO_2H$.

OTHER TRIORGANOTIN SALTS

    TRIMETHYLTIN SULFATE $(Me_3Sn)_2SO_4$
Biol. props.: Toxicity and symptoms (532).

    TRIETHYLTIN SULFATE $(Et_3Sn)_2SO_4$
Prepn.: By mixing $Et_3SnSCH_2CHMe_2$ or $Et_3SnSC_6H_4$-p-Me with 100% H_2SO_4 (466).
Rxn. with: Dithizone → colored complex (1).
Biol. props.: Inhibition of the phosphorylation process (4).
Shift in the phosphate distribution in tissue (531).
Toxicity and symptoms (532).

    TRIPHENYLTIN p-NITROPHENYLARSONATE $Ph_3SnOAs(O)(OH)C_6H_4NO_2$-p
Prepn.: By reacting $(Ph_3Sn)_2O$ with $p-O_2NC_6H_4AsO(OH)_2$ at elevated temp. (796).
Props.: Decomp. 272° (796).
Use: Treatment of protozoal infections (796).

6. ORGANOTIN-SULFUR AND -SELENIUM COMPOUNDS

ORGANOTIN TRISULFIDES AND TRISELENIDES

    BIS(o-TOLYLTIN) TRISULFIDE $(o-MeC_6H_4SnS)_2S$
Prepn.: By reacting $o-MeC_6H_4SnCl_3$ with H_2S in H_2O; theoret. yield (280).
Props.: Yellow, amorphous, infusible powder (280).

    BIS(METHYLTIN) TRISELENIDE $(MeSnSe)_2Se$
Prepn.: By passing H_2Se through a soln. of $MeSnO_2H$ (546).

    BIS(ETHYLTIN) TRISELENIDE $(EtSnSe)_2Se$
Prepn.: By passing H_2Se through a soln. of $EtSnO_2H$ (546).

ORGANOTIN TRIMERCAPTIDES

Alkyl-, aryl-, and aralkyl-tris(organothio)tin compounds, $RSn(SR')_3$, are formed:

1. By reacting alkyl-, aryl-, and aralkylstannonic acids with thiols in the presence of a solvent with azeotropic removal of water or in the absence of solvents at 125-150°.
2. By reacting alkyl-, aryl-, and aralkyltin trihalides with thiols in the presence of Na_2CO_3 or NaOH

Numerous compounds in the following table marked with asterisk are useful as stabilizers for vinyl halide polymers.

ORGANOTIN TRIMERCAPTIDES

$RSn(SR')_3$	Prepared from	Properties	Ref.
$MeSn(SCH_2CH_2OH)_3$	$MeSnO_2H + HSCH_2CH_2OH$	n^{20}_D 1.6168*	52
$MeSn(SBu)_3$	$MeSnO_2H + BuSH$	n^{20}_D 1.5541*	52, 782
$MeSn(SCHMePr)_3$	$MeSnO_2H + PrCHMeSH$	n^{20}_D 1.5452*	52, 782
$MeSn(S-n-C_{12}H_{25})_3$	$MeSnO_2H + n-C_{12}H_{25}SH$	*	52, 782, 429, 680
$MeSn(S-tert-C_{12}H_{25})_3$	$MeSnCl_3 + n-C_{12}H_{25}SH$	*	429, 680
$MeSn(S-tert-C_{12}H_{25})_3$	$MeSnO_2H + tert-C_{12}H_{25}SH$	*	52, 782
$MeSn(S-n-C_{18}H_{37})_3$	$MeSnO_2H + n-C_{18}H_{37}SH$	m 50-60°*	52, 782
$MeSn(S-\alpha-pinenyl)_3$	$MeSnO_2H + \alpha$-pinenemercaptan	n^{20}_D 1.5449*	52, 782
$MeSn(S-\alpha-C_{10}H_7)_3$	$MeSnO_2H + \alpha-C_{10}H_7SH$	n^{20}_D 1.71*	52, 782
$MeSn(SCH_2C_6H_4OH-p)_3$	$MeSnO_2H + p-HOC_6H_4CH_2SH$	m 143-145°*	52, 782
$MeSn(SCH_2O_2CPh)_3$	$MeSnO_2H + PhCO_2CH_2SH$	n^{20}_D 1.6523*	52
$MeSn(SCH_2C_6H_4Cl-p)_3$	$MeSnO_2H + p-ClC_6H_4CH_2SH$	n^{20}_D 1.6523*	52, 782
$MeSn(SC_6H_4OC_6H_4S)_3SnMe$	$MeSnO_2H + (HSC_6H_4)_2O$	*	782
$MeSn(SC_6H_4CO_2Me-p)_3$	$MeSnO_2H + p-MeOCOC_6H_4SH$	n^{20}_D 1.6579*	782
$EtSn(SCH_2CO_2(CH_2)_5CHMe_2)_3$		Antioxidant for vulcanized rubber	805
$EtSn(S-n-C_{12}H_{25})_3$	$EtSnO_2H \cdot H_2O + n-C_{12}H_{25}SH$	*	429, 680
	$EtSnCl_3 + n-C_{12}H_{25}SH$	*	680
$(1-PrSn)_2p-(SCH_2CH_2OCO)_2C_6H_4)_3$	$1-PrSnO_2H + p-C_6H_4(CO_2CH_2CH_2SH)_2$	*	598
$1-PrSn(SCH_2CH_2CH_2CONHBu)_3$	$1-PrSnO_2H + HS(CH_2)_3CONHBu$	*	310
$BuSn(SBu)_3$	$BuSnO_2H + BuSH$	n^{20}_D 1.5420*	52, 782
$BuSn(S-n-C_{12}H_{25})_3$	$BuSnO_2H + n-C_{12}H_{25}SH$	*	429
$BuSn(S-tert-C_{12}H_{25})_3$	$BuSnO_2H + tert-C_{12}H_{25}SH$	*	52, 782
$BuSn(SCH_2C_6H_4OH-p)_3$	$BuSnO_2H + p-HOC_6H_4CH_2SH$	n^{20}_D 1.6540*	52, 782
$BuSn(SC_6H_4O_2Me-p)_3$	$BuSnO_2H + p-MeO_2CC_6H_4SH$	*	782
$BuSn(SCH_2OCOPh)_3$	$BuSnO_2H + PhCO_2CH_2SH$	*	52
$BuSn(SCH_2CO_2-n-C_{10}H_{21})_3$	$BuSnO_2H + n-C_{10}H_{21}O_2CCH_2SH$	*	559
$BuSn(SCH_2CH_2CO_2-n-C_6H_{13})_3$	$BuSnO_2H + HSCH_2CH_2CO_2C_6H_{13}$		308

218

	Antioxidant for vulcanized rubber	
BuSn(SCH$_2$CH$_2$CO$_2$CH$_2$Ph)$_3$		805
BuSn(SC:NC$_6$H$_4$S-o)$_3$	BuSnO$_2$H + o-SC$_6$H$_4$N:CSH	781
PhSn(SBu)$_3$	BuSnO$_2$H + BuSH	52, 782
PhSn(SCH$_2$C$_6$H$_4$OH-p)$_3$	PhSnO$_2$H + p-HOC$_6$H$_4$CH$_2$SH	52, 782
PhSn(SC:NC$_6$H$_4$S-o)$_3$	PhSnO$_2$H + o-SC$_6$H$_4$N:CSH m. 158°	781
PhSn(SC:CHCH:CHS)$_3$	PhSnO$_2$H + SCH:CHCH:CSH Yellow liquid	781
PhCH$_2$Sn[SC(S)OBu]$_3$	PhCH$_2$SnBr$_3$ + BuOC(S)SNa *	597

* Useful as stabilizer for vinyl halide polymers.

For numerous other methyl-, ethyl-, butyl-, phenyl-, furyl-, and thienyltin trimercaptides, useful as stabilizers for vinyl halide polymers, see references 52, 608, and 782.

DIORGANOTIN SULFIDES

The compounds are prepared by reacting diorganotin dihalides with alkali metal sulfides or bisulfides or with hydrogen sulfide in the presence of alkali.

R_2SnS	Prepared from	Solvent	Properties	Ref.
Me_2SnS	Me_2SnI_2 + KHS	EtOH contg. H_2S	m. 149; trimeric form postulated (1)	205
Et_2SnS	Et_2SnBr_2 + Na_2S	EtOH	m. 24°; b. 219-221°; trimeric form postulated	205
Pr_2SnS	Pr_2SnBr_2 + Na_2S	EtOH	b. 254°/16mm (slight decompn.)	207
Bu_2SnS			Anthelmintic props. (1,2)	649, 684
$(n-C_{12}H_{25})_2SnS$			(1)	802
Ph_2SnS	Ph_2SnCl_2 + H_2S	5% alc. KOH	m. 183-184° (2,3)	123
$(p-EtO_2CC_6H_4)_2SnS$	$(p-EtO_2CC_6H_4)_2SnCl_2$ + H_2S	5% alc. KOH	m. 141-142.5	123
$(p-MeO_2CC_6H_4)_2SnS$	$(p-MeO_2CC_6H_4)_2SnCl_2$ + H_2S	5% alc. KOH	m. 95°; soly. data	539
$(p-EtOC_6H_4)_2SnS$	$(p-EtOC_6H_4)_2SnCl_2$ + H_2S	5% alc. KOH	m. 127°; soly. data	539
$(p-PhC_6H_4)_2SnS$	$(p-PhC_6H_4)_2SnCl_2$ + H_2S	5% alc. KOH	White flakes, m. 134.5-135°	539
$(1-C_{10}H_7)_2SnS$	$(1-C_{10}H_7)_2SnCl_2$ + H_2S	5% alc. KOH	Crystalline powder, m. 215°	414

(1) Useful as stabilizer for vinyl resins (802).
(2) Useful as antioxidant for vulcanized rubber (806).
(3) Useful as anthelmintic for animals (684).

BIS(TRIMETHYLSILYL)METHYLTIN SULFIDE POLYMER $[(Me_3SiCH_2)_2SnS]_x$
Prepn.: By adding $Na_2S \cdot H_2O$ to alc. soln. of $(Me_3SiCH_2)_2SnI_2$ (1:1) and satg. the rxn. mixt. with H_2S (493).
Props.: Two fractions were isolated: White fluffy solid, less sol. in alc., m. 150-165°, probably high polymer; large crystals more sol. in alc., m. 74.4-75.5°, probably a mixt. of trimer and tetramer (493).

DIORGANOTIN DIMERCAPTIDES

Dialkyl- and diaryl-bis(organothio)tin compounds, $R_2Sn(SR')_2$, are formed:

1. By reacting dialkyl- and diaryltin oxides with thiols in a molar ratio of 1:2 in the presence of a solvent, with azeotropic removal of water (Method I), or without solvent at 120-150° (Method Ia).
2. By reacting dialkyl- and diaryltin dihalides with thiols in the presence or absence of Na_2CO_3 or NaOH (Method II).
3. By heating dialkyldialkoxytin compounds with thiols (Method III).
4. By heating trialkyltin carboxylates with thiols (Method IV).

The compounds are compiled in Tables I and II. Table II comprises dialkyl- and diaryltindithio derivatives of mercaptoborates. Numerous compounds in the tables marked with one asterisk are useful as stabilizers for vinyl halide polymers, while those marked with two asterisks are useful as antioxidants for vulcanized rubber.

TABLE I

$R_2Sn(SR')_2$	Prepn. Method	Properties	Ref.
$Me_2Sn(SBu)_2$	Ia	n^{20}_D 1.5371*	607
$Me_2Sn(S-n-C_{12}H_{25})_2$	I	*	429, 680
$Me_2Sn(SCH_2CH_2CO_2C_{18}H_{37})_2$		**	805
$Me_2Sn(SR')_2$ (R'= lauryl and myristyl radicals)	Ia	n^{20}_D 1.4978*	607
$Me_2Sn(SC_6H_4Me-p)_2$	Ia	m. 60-61° *	607
$Me_2Sn(SC_6H_4CO_2Me-p)_2$	Ia	n^{20}_D 1.6468	607
$Me_2Sn(S-2-benzothiazolyl)_2$	I	m. 106°	781
$Me_2Sn(S-tetrahydro-2-thiazolyl)_2$	I	m. 122° *	781
$Me_2Sn(SCH_2CO_2C_6H_4Me-p)_2$		**	805
$Me_2Sn(SCH_2CO_2CC_{11}H_{23})_2$	I	*	750
$Me_2Sn(SCHPhCH_2O_2CPr)_2$	I	*	750
$Me_2Sn[SC_6H_4CH_2O_2C(CH_2)_4]_2$	I	*	750
$Pr_2Sn(SPh)_2$	IV	b. 226-230°/1mm; n^{20}_D 1.6298	466
$Bu_2Sn(SEt)_2$	I	Yellow liquid*	429, 680
$Bu_2Sn(SBu)_2$	I	*	429, 680
$Bu_2Sn(SCH_2CHMe_2)_2$	I	*	429, 680
$Bu_2Sn(SCHMeEt)_2$	I	*	429, 680
$Bu_2Sn(SC_5H_{11})_2$	I, Ia	* n^{20}_D 1.5200	429, 607, 680
$Bu_2Sn(SC_7H_{15})_2$	I	*	429, 680
$Bu_2Sn(SC_8H_{17})_2$	I	*	429, 680
$Bu_2Sn(SC_9H_{19})_2$	I	*	429, 680
$Bu_2Sn(SC_{10}H_{21})_2$	I	*	429, 680
$Bu_2Sn(SC_{12}H_{25})_2$	I, Ia	*, ** (745a)	429, 607, 680
$Bu_2Sn(S-tert-C_{12}H_{25})_2$	II	Colorless liquid, d_{20} 1.045; n^{20}_D 1.5177*	311, 312, 429
$Bu_2Sn(SC_{13}H_{27})_2$	III		349
$Bu_2Sn(SC_{14}H_{29})_2$	Ia	n^{20}_D 1.5168*	607
$Bu_2Sn(SC_{16}H_{33})_2$	Ia	n^{20}_D 1.4982**	607
	I	*	429, 680
	I, Ia	n^{20}_D 1.4975	429, 607, 680

Compound		Properties	References
$Bu_2Sn(S\text{-}tert\text{-}C_{16}H_{33})_2$	Ia	n^{20}_D 1.5064*	607
$Bu_2Sn(SC_{18}H_{37})_2$	Ia	m. 26°*	607
$Bu_2Sn(SPh)_2$	I	*	429, 680
$Bu_2Sn(SC_6H_4Me\text{-}p)_2$	Ia	n^{20}_D 1.6105	607
	I	*	429, 680
$Bu_2Sn(SCH_2Ph)_2$	Ia	d^{20}_D 1.6029*	607
$Bu_2Sn(SC_6H_4NH_2\text{-}o)_2$	Ia	d^{20}_D 1.5989*	607
	I	*	429, 680
$Bu_2Sn(S\text{-}1\text{-}C_{10}H_7)_2$	Ia	n^{20}_D 1.6630	607
$Bu_2Sn(SC_6H_4\text{-})_2$	Ia	Waxy solid*	607
$Bu_2Sn(S\text{-}2\text{-}benzothiazolyl)_2$	I	m. 185°*	429, 680, 781
$Bu_2Sn(S\text{-}2\text{-}thiazolenyl)_2$	I	White solid*	781
$Bu_2Sn[S\text{-}(4\text{-}Ph\text{-}thiazolyl\text{-}2)]_2$	I	Brown liquid*	781
$Bu_2Sn(SCH_2CH_2OH)_2$	I	Liquid*	431
$Bu_2Sn(SCH_2CH_2O_2CC_8H_{17})_2$	Ia	*	598
$Bu_2Sn(CH_2CH_2O_2C_6H_4\text{-}2\text{-}C_8H_{17})_2$	I	n^{20}_D 1.5429	607
$Bu_2Sn(SCH_2CH_2O_2C(CH_2)_3\text{-})_2$	I	*	750
$Bu_2Sn(SCH_2CH_2O_2CC_{11}H_{23})_2$	I	*	747, 750
$Bu_2Sn(SCH_2CH_2OBz)_2$	I	*	750
$Bu_2Sn(SCH_2CH_2\text{-}abietate)_2$	I	*	750
$Bu_2Sn(SCH_2CO_2C_6H_{11})_2$	Ia	*	307, 667
$Bu_2Sn(SCH_2CO_2C_8H_{17})_2$	I	**	805
$Bu_2Sn[SCH_2CO_2(CH_2)CHMeCH_2CMe_3]_2$	I	*, **	559, 805
$Bu_2Sn(SCH_2CO_2\text{-}cyclohexyl)_2$	I	*	559
$Bu_2Sn(SCH_2CO_2\text{-}trimethylnonyl)_2$	I	*	559
$Bu_2Sn(SCH_2CO_2C_{18}H_{37})_2$	I	*	559, 805
$Bu_2Sn(SCH_2CO_2\text{-}tetrahydroabietyl)_2$	Ia	*	307, 559
$Bu_2Sn(SCH_2CO_2CH_2OPh)_2$	I	*	559
$Bu_2Sn(SCH_2CO_2CH_2CHBuCH_2Bu)_2$	I	*	805
$Bu_2SnS(CH_2)_2CO_2Bu)_2$	I	*	559
$Bu_2Sn\{(CH_2)_2CO_2CH_2CHBuCH_2Bu\}_2$	I	*	559
$Bu_2Sn(SCH_2CO_2H)$	I	*	559
$Bu_2Sn(SC_6H_4CO_2Me\text{-}o)_2$	I	*, **	559, 805
$Bu_2SnSCH(CO_2CH_2CHEtBu)CH_2CO_2CH_2CHEtBu)_2$	I	**	805
$Bu_2SnSCH(CO_2H)CH_2CO_2H)_2$	I	*	559

---

* Useful as stabilizer for vinyl halide polymers.
** Useful as antioxidant for vulcanized rubber.

TABLE I (Cont'd.)

$R_2Sn(SR')_2$	Prepn. Method	Properties	Ref.
$Bu_2Sn[SCHMeCO_2CH_2CHMeCH_2CMe_3]_2$	I	*	559
$Bu_2Sn[SCHEtCO_2CH_2CH_2CHMeCH_2CMe_3]_2$	I	*	559
$Bu_2Sn[SCHMeCH_2CO_2CH_2CH_2CHMeCH_2CMe_3]_2$	I	*	559
$Bu_2Sn[SCHPrCO_2CH_2CH_2CHMeCH_2CMe_3]_2$	I	*	559
$Bu_2Sn[SCHBuCO_2CH_2CH_2CHMeCH_2CMe_3]_2$	I	*	559
$Bu_2SnS[CH(C_6H_{13})CO_2Bu]_2$	I	*	559
$Bu_2Sn[SCEt(C_6H_{13})CO_2Et]_2$	I	*	559
$Bu_2Sn[SCH(C_8H_{17})CO_2Bu]_2$	I	*	559
$Bu_2Sn[SCH(C_7H_{15})CO_2Bu]_2$	I	*	559
$Bu_2Sn[SCH(C_{10}H_{21})CO_2Bu]_2$	I	*	559
$Bu_2Sn[SCH(C_{14}H_{29})CO_2Bu]_2$	I	*	559
$Bu_2Sn[SCH(C_{16}H_{33})CO_2Bu]_2$	I	*	559
$Bu_2Sn[SCHPhCO_2Et]_2$	I	*	559
$Bu_2Sn[SCPh_2CO_2Et]_2$	I	*	559
$Bu_2SnSCEt_2CO_2CH_2CH_2CHMeCH_2CMe_3]_2$	I	*	559
$Bu_2SnS(CH_2)_4CO_2CH_2CH_2CH_2CHMeCH_2CMe_3]_2$	I	*	559
$Bu_2SnS(CH_2)_5CO_2CH_2CH_2CHMeCH_2CMe_3]_2$	I	*	559
$Bu_2SnS(CH_2)_6CO_2CH_2CH_2CHMeCH_2CMe_3]_2$	I	*	559
$Bu_2SnS(CH_2)_7CO_2Bu]_2$	I	*	559
$Bu_2Sn\{SCH_2CO_2CH_2CH_2OCC_{11}H_{23}\}_2$	I	*	559
$Bu_2Sn(SCH_2CO_2CH_2CH_2-ricinoleate)_2$	I	*	559
$Bu_2Sn(SCH_2CONCH_2CH_2OCH_2CH_2)_2$	I	*	310
$Bu_2Sn(SCH_2CONHC_5H_{11})_2$	I	*	310
$Bu_2Sn(SCH_2CONHCH_2Bu)_2$	I	*	310
$(n-C_5H_{11})_2Sn\{S-n-C_{12}H_{25}\}_2$	I	**	429, 680
$(n-C_5H_{11})_2Sn\{SCH(CO_2C_{10}H_{11})CH_2CO_2C_{10}H_{21}\}_2$	I	**	805
$cyclo-C_6H_{11})_2Sn[SCH_2CH_2CHMeCH_2CMe_3]_2$	I	**	805
$n-C_8H_{17})_2Sn(SCH_2CO_2Bu)_2$	Ia	Liquid	273, 336
$n-C_{12}H_{25})_2Sn(S-n-C_5H_{11})_2$	I	*	429, 680
$n-C_{12}H_{25})_2Sn(S-n-C_{12}H_{25})_2$	I	*	429, 680
$n-C_{12}H_{25})_2Sn(SCH_2Ph)_2$	I	*	429, 680
$n-C_{12}H_{25})_2Sn(SCH_2CO_2Bu)_2$		**	805
$n-C_{12}H_{25})_2Sn(SCH_2CH_2O_2CC_{11}H_{23})_2$	I	*	750
$n-C_{12}H_{25})_2Sn(SCH_2CH_2CH_2CO_2Bu)_2$	II		308

	Prepn. Method		Ref.
$(n-C_{12}H_{25})_2Sn(SCHPhCH_2O_2CPr)_2$	I	*	750
$\{(n-C_{12}H_{25})_2Sn[SC_6H_4CH_2O_2C(CH_2)_4-]_2$	I	*	750
$Ph_2Sn(SCH_2CH_2O_2CC_{11}H_{23})_2$	I	*	750
$Ph_2Sn(SCHPhCH_2O_2CPr)_2$	I	*	750
$Ph_2Sn(S-n-C_{12}H_{25})_2$	I	*	429, 680
	II	*	311
	I	**	429, 680
$Ph_2Sn(SCH_2Ph)_2$			805
$Ph_2Sn(SCH_2CO_2-n-C_{12}H_{25})_2$	I	*	750
$Ph_2Sn[SC_6H_4CH_2O_2C(CH_2)_4-]_2$		**	805
$(o-MeC_6H_4)_2Sn[SCH(CO_2Bu)CH_2CO_2Bu]_2$		**	805
$(PhCH_2)Sn[SCH_2CH_2CO_2-1-C_8H_{17}]_2$		**	805
$EtPhSn(SCH_2CO_2-1-C_8H_{17})_2$		**	805

TABLE II

$R_2Sn(SR')_2$	Prepn. Method	Properties	Ref.
$Bu_2Sn[SCH_2CH_2OB(o\text{-dihydroabietyl})_2]_2$	I	*	749
$Bu_2Sn[SCH_2CH_2OB(OCH_2CHEtBu)_2]_2$	I	*	749
$Bu_2Sn[SCH_2CH_2OB(OCH_2CHBu-n-C_6H_{13})_2]_2$	I	*	749
$Bu_2Sn(SCH_2CH_2OBOCH_2CH_2O)_2$	I	*	749
$(Me_2Sn)_3[(p-SC_6H_4O)_3B]_2$	I	(1)	748
$(Bu_2Sn)_3[(SCH_2CH_2O)_3B]_2$	I	(1)	748
$(n-C_{12}H_{25})_2Sn)_3[(SCH_2CH_2O)_3B]_2$	I	(1)	748
$(Ph_2Sn)_3(SCH_2CH_2O)_3B]_2$	I	(1)	748
$(Ph_2Sn)_3(SCHPhCH_2O)_3B]_2$	I		748

(1) The compounds are useful as accelerator and oxidants for rubber and oil additives.
 * Useful as stabilizer for vinyl halide polymers.
** Useful as antioxidant for vulcanized rubber.

ORGANOTIN THIOCYANATES, XANTHATES, DITHIOCARBAMATES, AND DITHIOPHOSPHATES

The compounds are prepared by reacting organotin halides with alkali metal thiocyanates, xanthates, dithiocarbamates, and dithiophosphates, respectively.

Compound	Prepared from	Properties	Ref.
$Me_2Sn(CSN)_2$	Me_2SnCl_2 + NaSCN in Me_2CO	White needles, m. 198.6-199.4° (decompn.)	486
$(CH_2{:}CH)_2Sn(SCN)_2$	$(CH_2{:}CH)_2SnCl_2$ + NaSCN	m. 163.6-167° (decompn.)	487
$Bu_2Sn(SCN)_2$	Bu_2SnCl_2 + NaSCN	White needles, m. 144-145°	486
$Bu_2Sn[SC(S)OCHMe_2]_2$	Bu_2SnCl_2 + Me_2CHOCS_2Na	Reddish solid [1]	597
$Bu_2Sn[SC(S)OBu]_2$	Bu_2SnCl_2 + $KSC(S)OBu$	[2]	705, 909
$(n\text{-}C_{12}H_{25})_2Sn[SC(S)OCHMe_2]_2$	$(n\text{-}C_{12}H_{25})_2SnCl_2$ + Me_2CHOCS_2K	Yellow oil [2]	909
$Bu_2Sn[SC(S)N(n\text{-}C_5H_{11})_2]_2$	Bu_2SnCl_2 + $KSC(S)N(n\text{-}C_5H_{11})_2$	[2]	705, 909
$(n\text{-}C_{12}H_{25})_2Sn[SC(S)NBu_2]_2$	$(n\text{-}C_{12}H_{25})_2SnCl_2$ + $KSC(S)NBu_2$	[2]	909
$Bu_2Sn[SP(S)(OC_6H_{10}Me)_2]_2$	Bu_2SnCl_2 + $KSP(S)(OC_6H_{10}Me)_2$	[2]	909
$Bu_2Sn[SP(S)(OC_6H_4C_6H_{13})_2]_2$	Bu_2SnCl_2 + $KSP(S)(OC_6H_4C_6H_{13})_2$	[2]	909
$(n\text{-}C_{12}H_{25})_2Sn[SP(S)(OCHMe_2)_2]_2$	$(n\text{-}C_{12}H_{25})_2SnCl_2$ + $KSP(S)(OCHMe_2)_2$	[2]	909
$(Me_2CH)_3SnSC(S)OBu$	$(Me_2CH)_3SnCl$ + $KSC(S)OBu$	Reddish oil [1]	597

[1] Useful as stabilizer for vinyl chloride-acetate polymers (597).
[2] Useful as corrosion-stabilizer for lubricating oils (909).

BIS(TRIORGANOTIN) SULFIDES

The following compounds were prepared by reacting triorganotin halides with sodium sulfide or hydrogen sulfide, by reacting bis-(triorganotin) oxides with hydrogen sulfide, and by treating triorganotin hydrides with thiols in a molar ratio of 2:1.

$(R_3Sn)_2S$	Prepared from	Properties	Ref.
$(Me_3Sn)_2S$	$Me_3SnI + Na_2S$	b. $118°/18mm$	204
$(Et_3Sn)_2S$	$(Et_3Sn)_2O + H_2S$	b. $133-137°/1mm$; d_{25} 1.429; n^{20}_D 1.5468	18, 465
	$Et_3SnI + Na_2S$	b. $187-188°/20mm$ (1,2,3)	204
$(Pr_3Sn)_2S$	$Pr_3SnI + Na_2S$	b. $215-219°/19mm$	204
$(Ph_3Sn)_2S$	$Ph_3SnH + CH_2:CHCH_2SH$ (2:1)	m. $141.5-143°$	732
$[(p\text{-}EtO_2CC_6H_4)_3Sn]_2S$	$(p\text{-}EtO_2CC_6H_4)_3SnCl + H_2S$	m. $132-133°$	382

(1) Rxns. with Ag salt → the corresponding Et_3Sn salts (18).
(2) Rxns. with Ag_2O → $(Et_3Sn)_2O$ (18).
(3) Rxns. with metal halides → Et_3Sn halides (19).

TRIORGANOTIN MERCAPTIDES

Trialkyl-, triaryl-, and triaralkyl-organothiotin compounds, R_3SnSR', are formed:

1. By reacting triorganotin hydroxides with thiols in solvents, with azeotropic removal of water (Method I).
2. By reacting triorganotin halides with thiols in the presence of Na_2CO_3 or NaOH (Method II).
3. By reacting bis(triorganotin) oxides with thiols (Method III).
4. By treating tetraorganotin compounds of the general formulae R_4Sn or R_3SnR' with thiols at elevated temperature (Method IV).
5. By transetherification of triorganotin organomercaptides, R_3SnSR', with other thiols (Method V).
6. By reacting triorganotin carboxylates with thiols (Method VI).

TRIORGANOTIN MERCAPTIDES

R_3SnSR'	Prepared from	Properties	Ref.
$Me_3Sn-n-C_{12}H_{25}$	$Me_3SnOH + n-C_{12}H_{25}SH$	Liquid (5)	429, 680
	$Me_3SnCl + n-C_{12}H_{25}SH$		429, 680
Et_3SnSMe	$(Et_3Sn)_2O + MeSH$	b. 224°; d^{20}_4 1.319; n^{20}_D 1.529	18
Et_3SnSEt	$(Et_3Sn)_2O + EtSH$	b. 125-126°/12mm; b. 240°; d_{25} 1.278; n^{20}_D 1.5150	465
$Et_3SnSCH_2CO_2Et$	$(Et_3Sn)_2O + HSCH_2CO_2Et$	b. 107-109°/1.5mm; n^{25}_D 1.5099-1.5101	488
$Et_3SnSCH_2CHMe_2$	$(Et_3Sn)_2O + i-BuSH$	b. 86-88°/1mm; d_{25} 1.244; n^{20}_D 1.5122	465
		b. 84-86°/1mm; b. 244° (5, 6)	466
$Et_3SnSCH_2CH_2CHMe_2$	$Et_3SnSEt + i-BuSH$	b. 95-97°/1mm; 6.285°;	465
	$(Et_3Sn)_2O + i-C_5H_{11}SH$	d^{20}_D 1.5060; d_{25} 1.188	465, 466
$Et_3SnSCHMe_2$	$Et_3SnSCHMe_2 + i-C_5H_{11}SH$	b. 78-79°/1mm, b. 256°;	465
	$(Et_3Sn)_2O + Me_2CHSH$	n^{20}_D 1.5132 (7)	466
	$Et_3SnSEt + Me_2CHSH$	b. 78-80°/1mm	467
$Et_3SnS-n-C_6H_{13}$	$Et_4Sn + n-C_6H_{13}SH$ (excess)	b. 126-127°/1mm; n^{20}_D 1.5032	467
$Et_3SnS-n-C_7H_{15}$	$Et_4Sn + n-C_7H_{15}SH$	b. 134-135°/1mm; n^{20}_D 1.5006	467
Et_3SnSBu	$(Et_3Sn)_2O + BuSH$	b. 88-91°/1mm; b. 248°; n^{20}_D 1.5123	465
$Et_3SnSCMe_3$	$(Et_3Sn)_2O + Me_3CSH$	b. 84-86°/1mm; b. 254°; n^{20}_D 1.5051	465
$Et_3SnSCH_2CO_2(CH_2)_4-CHMe_2$		Antioxidant for vulcanized rubber	805
Et_3SnSPh	$Et_4Sn + PhSH$	b. 138-140°/1mm; n^{20}_D 1.5828	467
Et_3SnSCH_2Ph	$Et_4Sn + PhCH_2SH$	b. 138-140°/1mm; n^{20}_D 1.5675	467
	$(Et_3Sn)_2O + PhCH_2SH$	b. 136-138°/1mm; b. 260°; n^{20}_D 1.5682	465
$Et_3SnSC_6H_4Me-o$	$(Et_3Sn)_2O + o-MeC_6H_4SH$	b. 132-136°/1mm; b. 292°; n^{20}_D 1.5740	465
	$Et_4Sn + o-MeC_6H_4SH$	b. 132-136°/1mm; n^{20}_D 1.5720	467
$Et_3SnSC_6H_4Me-m$	$(Et_3Sn)_2O + m-MeC_6H_4SH$	b. 132-134°/1mm; b. 288°; n^{20}_D 1.5705	465

Compound	Preparation / Properties	Ref.
$Et_3SnSC_6H_4Me-p$	$(Et_3Sn)_2O + p-MeC_6H_4SH$	b. 128-130°/1mm; n^{20}_D 1.5712; b. 300°;	465
		b. 125-126°/1mm; (5)	466
	$Et_3SnS(CH_2)_2CHMe_2 + p-MeC_6H_4SH$		
$Et_3SnS-2-C_{10}H_7$	$Et_4Sn + p-MeC_6H_4SH$	b. 129-132°/1mm	467
$Pr_3SnS-n-C_7H_{15}$	$Et_4Sn + 2-C_{10}H_7SH$	b. 189-190°/1mm; n^{20}_D 1.5308	467
$Pr_3SnS-n-C_{10}H_{21}$	$Pr_4Sn + n-C_7H_{15}SH$	b. 158-160°/1mm; n^{20}_D 1.4981	468
	$Pr_4Sn + n-C_{10}H_{21}SH$	b. 180-183°/1mm; n^{20}_D 1.4998	468
Pr_3SnSPh	$(Pr_3Sn)_2O + PhSH$	(1-4)	
$Pr_3SnSC_6H_4Me-p$	$(Pr_3Sn)_2O + p-MeC_6H_4SH$	b. 157-159°/1mm; n^{20}_D 1.5626	466
	$Pr_3SnOCOEt + p-MeC_6H_4SH$	b. 157-159°/1mm; n^{20}_D 1.5516	466
Pr_3SnSCH_2Ph	$Pr_4Sn + p-MeC_6H_4SH$	b. 159-160°/1mm; n^{20}_D 1.5602	468
$1-Pr_3SnS-n-C_7H_{15}$	$Pr_4Sn + PhCH_2SH$	b. 165-167°/1mm; n^{20}_D 1.5558	468
$1-Pr_3SnS-n-C_{10}H_{21}$	$1-Pr_4Sn + n-C_7H_{15}SH$	b. 155-157°/1mm; n^{20}_D 1.5045	468
$1-Pr_3SnSPh$	$1-Pr_4Sn + n-C_{10}H_{21}SH$	b. 192-195°/1mm; n^{20}_D 1.5010	468
$1-Pr_3SnSC_6H_4Me-p$	$1-Pr_4Sn + PhSH$	b. 138-139°/1mm; n^{20}_D 1.5676	468
$1-Pr_3SnSCH_2Ph$	$1-Pr_4Sn + p-MeC_6H_4SH$	b. 157-158°/1mm; n^{20}_D 1.5648	468
$Bu_3SnS-n-C_{12}H_{25}$	$1-Pr_4Sn + PhCH_2SH$	b. 167-170°/1mm; n^{20}_D 1.5497	468
	$Bu_3SnOH + n-C_{12}H_{25}SH$	(8)	429, 680
	$Bu_3SnCl + n-C_{12}H_{25}SH$	Antioxidant for vulcanized rubber	429, 680
$Bu_3SnSCH_2CO_2-i-C_8H_{17}$		(8)	805
$Bu_3SnSCH_2CO_2-n-C_{10}H_{21}$	$Bu_3SnCl + HSCH_2CO_2-n-C_{10}H_{21}$		559
$Bu_3SnSCH_2CONH-n-C_5H_{11}$	$Bu_3SnCl + HSCH_2CH_2CONH-n-C_5H_{11}$	Viscous oil	310

(1) Rxn. with Ag salts → the corresponding Et_3Sn salts (18).
(2) Rxn. with $Ag_2O → (Et_3Sn)_2O$ (18).
(3) Rxn. with $MeSi(COCCF_3)_3 → Et_3SnOCOCF_3$ (19).
(4) Rxn. with $Me_3SiOAc → Et_3SnOAc$ (19).
(5) Rxn. with 100% $H_2SO_4 → (Et_3Sn)_2SO_4$ (466).
(6) Rxn. with PhOH → Et_3SnSPh (466).
(7) Rxn. with AcOH → Et_3SnOAc (466).
(8) The compounds are useful as stabilizers for vinyl halide polymers (429, 598, 680).

TRIORGANOTIN MERCAPTIDES (Cont'd.)

R_3SnSR'	Prepared from	Properties	Ref.
$Ph_3SnS-n-C_{12}H_{25}$	$Ph_3SnOH + n-C_{12}H_{25}SH$	(8)	429
	$Ph_3SnCl + n-C_{12}H_{25}SH$		429
$Ph_3SnSCH_2CO_2-n-C_{10}H_{21}$	$Ph_3SnCl + HSCH_2CO_2-n-C_{10}H_{21}$	Liquid (8)	559
$[(PhCH_2)_3SnSCH_2CO_2-CH_2CH_2-]_2$	$(PhCH_2)_3SnCl +$ $(HSCH_2CH_2CO_2CH_2CH_2-)_2$		702
$Ph_3SnSCH_2CO_2CH_2Ph$		(8)	307
$Ph_3SnSCH(CO_2Bu)CH_2CO_2Bu$	$Ph_3SnBr + HSCH_2CO_2CH_2Ph$	(8) Antioxidant for vulcanized rubber	805
$Ph_3SnSC(S)S-n-C_{12}H_{25}$	$Ph_3SnI + n-C_{12}H_{25}SC(S)SK$	Anticorrosion agent for lubricating oil	704
$[(PhCH_2)_3SnS(CH_2)_4OCO-(CH_2)_3]_2CH_2$	$(PhCH_2)_3SnCl + [HS(CH_2)_4OCO-(CH_2)_3]_2CH_2$	(8)	598

(8) The compounds are useful as stabilizers for vinyl halide polymers (429, 598, 680).

7. ORGANOTIN GRIGNARD COMPOUND

(TRIMETHYLSTANNYL)METHYLMAGNESIUM CHLORIDE Me_3SnCH_2MgCl
Prepn.: By reacting Me_3SnCH_2Cl with Mg in Et_2O (772).
Rxns. with: $MeSiCl_3 \rightarrow Me_3SnCH_2SiMeCl_2$ (484).
$MeSi(OMe)_3 \rightarrow Me_3SnCH_2SiMe(OMe)_2$ (484).
$BF_3 \rightarrow (Me_3SnCH_2)_3B$ (772).

8. ORGANOTIN DERIVATIVES OF BORON AND ALUMINUM

TRIS(TRIMETHYLSTANNYLMETHYL)BORANE $(Me_3SnCH_2)_3B$
Prepn.: By reacting Me_3SnCH_2Cl with Mg in Et_2O and treating the Grignard with BF_3; 58% yield (772).
Props.: b. 90-91°/0.15mm (772).
Rxn. with: NaOH in 95% EtOH + $H_2O_2 \rightarrow Me_2SnO + Me_3SnCH_2OH$ (772).

DIBUTYLTIN BISDI(2-ETHYLHEXYL)BORATE $Bu_2Sn[OB(OCH_2CHEtBu)_2]_2$
Prepn.: By refluxing Bu_2SnO with $B(OH)_3$ and 2-ethylhexanol in MePh with azeotropic distn. of H_2O (912).
Props.: Clear colorless liquid (912).
Use: Stabilizer for Cl-contg. polymers, polymerization accelerator, antioxidant and lubricating oil additive (912).

ORGANOTIN-ALUMINUM COMPOUNDS

The following complex compounds, having the formula $R_{4-x}Sn[HAl(OEt)_3]_x$, where x= 1-3, were prepd. from $R_{4-x}SnCl_x$ and $NaHAl(OEt)_3$ in Et_2O: $Ph_3SnHAl(OEt)_3$, $Ph_2Sn[HAl(OEt)_3]_2$, $PhSn[HAl(OEt)_3]_3$, and $(C_6H_{11})_3SnHAl(OEt)_3$ (475).

9. ORGANOTIN DERIVATIVES OF SILICON AND GERMANIUM

Organotin derivatives of silicon are prepared:

1. By reacting organotin halides with organosilylalkyl-magnesium halides or with organosilylaryllithium compounds.
2. By reacting trimethylstannylmethylmagnesium chloride with organosilyl halides.
3. By reacting tetraorganotin compounds containing at least one vinyl group with silanes containing one hydrogen on the silicon atom.
4. By reacting lithium or potassium triorganostannides with organosilyl halides.

(Triphenylgermyl)triphenyltin was prepared by reacting potassium triphenylgermanide with triphenyltin chloride.

TRIS[(TRIMETHYLSILYL)METHYL]TIN IODIDE $(Me_3SiCH_2)_3SnI$
Prepn.: By heating $(Me_3SiCH_2)_4Sn$ with I_2 (1:1) at 175°; 76.5% yield (493).
Props.: b. 105°/0.3mm; $n^{25}D$ 1.5237 (493).

DIMETHYLBIS[(TRIMETHYLSILYL)METHYL]TIN $Me_2Sn(CH_2SiMe_3)_2$
Prepn.: By refluxing Me_2SnCl_2 with Me_3SiCH_2MgCl (1.25-3.1) in C_4H_8O; 94% yield (493).
Props.: b. 55°/0.7mm; $n^{25}D$ 1.4702; d_{25} 1.073 (493).
Rxns. with: I_2 (1:1) in xylene at 175° → $(Me_3SiCH_2)_2MeSnI$ (493).
Br_2 (1:1) in CCl_4 at r.t. → $(Me_3SiCH_2)Me_2SnBr$ and $(Me_3SiCH_2)_2MeSnBr$ (493).
HBr (1:1) in $CHCl_3$ at -78° to r.t. → $(Me_3SiCH_2)Me_2SnBr$ and $(Me_3SiCH_2)_2MeSnBr$ (493).
$HgBr_2$ → $(Me_3SiCH_2)_2MeSnBr$ + $MeHgBr$ (493).

DIMETHYLBIS(DIMETHYLSILYLMETHYL)TIN $Me_2Sn(CH_2SiHMe_2)_2$
Prepn.: By reacting Me_2SnCl_2 with Me_2HSiCH_2MgCl in Et_2O under reflux; 82% yield (372a).
Props.: b. 101°/20mm; $n^{25}D$ 1.4743; d^{25}_4 1.108 (372a).
Rxn. with: H_2O in EtOH in the presence of KOH → $Me_2SiCH_2SnMe_2CH_2SiMe_2O$ (372a).

DIMETHYLBIS[3-(DIMETHYLSILYL)PROPYL]TIN $Me_2Sn[(CH_2)_3SiHMe_2]_2$
Prepn.: By reacting Me_2SnCl_2 with $Me_2HSi(CH_2)_3MgCl$ in Et_2O under
   reflux; 75.5% yield (372a).
Props.: b. 146°/15mm; $n^{25}D$ 1.4730; $d^{25}4$ 1.052; IR spectrum (372a).
Rxn. with: H_2O in EtOH in the presence of KOH followed by
   cracking at 250° → $Me_2Si(CH_2)_3SnMe_2(CH_2)_3SiMe_2O$ (372a).

BIS[(TRIMETHYLSILYL)METHYL]DIVINYLTIN $(CH_2:CH)_2Sn(CH_2SiMe_3)_2$
Prepn.: By refluxing $(CH_2:CH)_2SnBr_2$ with Me_3SiCH_2MgBr
   (0.83:3.5) in C_4H_8O; 79.5% yield (493).
Props.: b. 93-94°/3.4mm; $n^{25}D$ 1.4826; d_{25} 1.078 (493).

DIBUTYLBIS(DIMETHYLSILYLMETHYL)TIN $Bu_2Sn(CH_2SiHMe_2)_2$
Prepn.: By reacting Bu_2SnCl_2 with Me_2HSiCH_2MgCl in Et_2O under
   reflux; 66.6% yield (372a).
Props.: b. 130°/5mm; $n^{25}D$ 1.4810; $d^{25}4$ 1.043 (372a).
Rxns. with: EtOH in the presence of NaOEt → $[Me_2(EtO)SiCH_2]_2$-
   $SnBu_2$ (372a).

DIBUTYLBIS(ETHOXYDIMETHYLSILYLMETHYL)TIN $Bu_2Sn[CH_2Si(OEt)Me_2]_2$
Prepn.: By refluxing $(Me_2HSiCH_2)_2SnBu_2$ in EtOH in the presence of
   NaOEt; 83% yield (372a).
Props.: b. 186°/15mm; $n^{25}D$ 1.4655; $d^{25}4$ 1.056 (372a).
Rxn. with: H_2O in EtOH in the presence of KOH →
   $Me_2SiCH_2SnBu_2CH_2SiMe_2O$ (372a).

DIBUTYLBIS[(TRIMETHYLSILYL)METHYL]TIN $Bu_2Sn(CH_2SiMe_3)_2$
Prepn.: By refluxing Bu_2SnCl_2 with Me_3SiCH_2MgCl (1.65:4.0) in
   C_4H_8O; 91% yield (493).
Props.: b. 98°/0.45mm; $n^{25}D$ 1.4777; d_{25} 1.027 (493).
Rxns. with: I_2 (1:1) at 175° → $(Me_3SiCH_2)_2SnBuI$ (493).
Br_2 (1:1) in CCl_4 at r.t. → $(Me_3SiCH_2)_2BuSnBr$ and (Me_3SiCH_2)-
   Bu_2SnBr (493).
HBr (1:1) in $CHCl_3$ at -78° to r.t. → $(Me_3SiCH_2)_2BuSnBr$ and
   $(Me_3SiCH_2)Bu_2SnBr$ (493).
CF_3CO_2H (1:1) at r.t. to 80° → $(Me_3SiCH_2)Bu_2SnO_2CCF_3$ (493).

BIS[(TRIMETHYLSILYL)METHYL]DIPHENYLTIN $Ph_2Sn(CH_2SiMe_3)_2$
Prepn.: By refluxing Ph_2SnCl_2 with Me_3SiCH_2MgCl (2.5:6.2) in
   C_4H_8O; 88.4% yield (493).
Props.: b. 137°/0.35mm; $n^{25}D$ 1.5499; d_{25} 1.149 (493).
Rxns. with: I_2 (1:2) in C_6H_6 → $(Me_3SiCH_2)_2SnI_2$ (493).
HBr (1:2) in CH_2Cl_2 at -78° → $(Me_3SiCH_2)_2SnBr_2$ (493).

DIBUTYL[(TRIMETHYLSILYL)METHYL]TIN TRIFLUOROACETATE
$Bu_2(Me_3SiCH_2)SnOCOCF_3$

Prepn.: By reacting $(Me_3SiCH_2)_2SnBu_2$ with CF_3CO_2H (1:1) at r.t. to 80°; 54.7% yield (493).
Props.: m. 62.0-63.4° (493).

TRIMETHYL[(TRIMETHYLSILYL)METHYL]TIN $Me_3SnCH_2SiMe_3$
Prepn.: By refluxing Me_3SnBr with Me_3SiCH_2MgCl (1.5:2.05) in C_4H_8O; 86.2% yield (493).
Props.: b. 64-65°/24mm; $n^{25}D$ 1.4594; d_{25} 1.136 (493).

DICHLOROMETHYL(TRIMETHYLSTANNYL)METHYLSILANE $Me_3SnCH_2SiMeCl_2$
Prepn.: By reacting Me_3SnCH_2MgCl with $MeSiCl_3$ (1:2) in Et_2O; 50% yield (484).
Props.: b. 58-59°/3.8-4mm; $n^{25}D$ 1.4824; d_{25} 1.415 (484).

3-METHOXY-3,5,5-TRIMETHYL-2-OXA-3-SILA-5-STANNAHEXANE
$Me_2SnCH_2SiMe(OMe)_2$

Prepn.: By reacting Me_3SnCH_2MgCl with $MeSi(OMe)_3$ in Et_2O; 74.3% yield (484).
Props.: b. 77.5-81°/18mm; $n^{25}D$ 1.4523; d_{25} 1.248 (484).
Rxns. with: H_2O (acidified H_2O-Et_2O mixt.) → $[Me_3SnCH_2Si(Me)O]_n$ (494).

[2-(TRICHLOROSILYL)ETHYL]TRIETHYLTIN $Et_3SnCH_2CH_2SiCl_3$
Prepn.: By heating $Et_3SnCH:CH_2$ with $HSiCl_3$ in a sealed tube in the presence of Bz_2O_2 (491).
Props.: b. 85°/0.6mm (491).
Rxns. with: NaOMe in Et_2O → $Et_3SnCH_2CH_2Si(OMe)_3$ (491).

TRIETHYL[2-(TRIMETHYLOXYSILYL)ETHYL]TIN $Et_3SnCH_2CH_2Si(OMe)_3$
Prepn.: By refluxing $Et_3SnCH_2CH_2SiCl_3$ with NaOMe in Et_2O (491).
Props.: b. 78°/0.4mm; $n^{25}D$ 1.4638; d_{25} 1.209 (491).

TRIBUTYL(DIMETHYLSILYLMETHYL)TIN $Bu_3SnCH_2SiHMe_2$
Prepn.: By refluxing Me_2HSiCH_2MgCl with Bu_3SnCl in Et_2O; 99% yield (372a).
Props.: b. 133°/5mm; $n^{25}D$ 1.4764; $d^{25}4$ 1.047 (372a).
Rxns. with: EtOH in the presence of NaOEt → $Me_2(EtO)SiCH_2SnBu_3$ (372a).
H_2O in EtOH in the presence of KOH → $(Bu_3SnCH_2SiMe_2)_2O$ (372a).
$Me_2(CH_2:CH)SiCl$ in the presence of H_2PtCl_6 in 1-PrOH at 90-150° → $Bu_3SnCH_2SiMe_2H$ and $Bu_3SnCH_2SiMe_2CH_2CH_2SiMe_2Cl$ (372a).

TRIBUTYL(ETHOXYDIMETHYLSILYLMETHYL)TIN $Bu_3SnCH_2Si(OEt)Me_2$
Prepn.: By refluxing $Me_2HSiCH_2SnBu_3$ in EtOH in the presence of NaOEt; 97% yield (372a).
Props.: b. 147°/5mm; n^{25}_D 1.4682; d^{25}_4 1.053 (372a).

2-CHLORO-2,5,5-TRIMETHYL-7,7-DIBUTYL-2,5,7-DISILASTANNAHENDECANE
$Bu_3SnCH_2SiMe_2CH_2CH_2SiMe_2Cl$
Prepn.: By refluxing $Bu_3SnCH_2SiHMe_2$ with $CH_2:SiMe_2Cl$ in i-PrOH in the presence of $H_2PtCl_6 \cdot 6 H_2O$; 32.1% yield (372a).
Props.: b. 180°/30mm; n^{25}_D 1.4836; d^{25}_4 1.066 (372a).
Rxn. with: $H_2O \rightarrow (Bu_3SnCH_2SiMe_2CH_2CH_2SiMe_2)_2O$ (372a).

5,5,11,11-TETRABUTYL-7,7,9,9-TETRAMETHYL-8,7,9,5,11-OXADISILADISTANNAPENTADECANE $(Bu_3SnCH_2SiMe_2)_2O$
Prepn.: By refluxing $Bu_3SnCH_2SiHMe_2$ with H_2O in EtOH in the presence of KOH and removing the excess H_2O and EtOH by azeotropic distn. with C_6H_6; 40.5% yield (372a).
Props.: b. 240°/0.1mm; n^{25}_D 1.4852; d^{25}_4 1.102; IR spectrum (372a).

5,5,17,17-TETRABUTYL-7,7,10,10,12,12,15,15-OCTAMETHYL-11,7,10,12,15,5,17-OXATETRASILADISTANNAHENEICOSANE
$(Bu_3SnCH_2SiMe_2CH_2CH_2SiMe_2)_2O$
Prepn.: By shaking $Bu_3SnCH_2SiMe_2CH_2CH_2SiMe_2Cl$ with H_2O and neutralizing the rxn. mixt. with 5% NaOH (372a).
Props.: n^{25}_D 1.4850; d^{25}_4 1.045 (372a).

TRIPHENYL(TRIPHENYLSILYL)TIN $Ph_3SnSiPh_3$
Prepn.: By reacting $LiSnPh_3$ or $KSnPh_3$ with Ph_3SiCl in Et_2O under reflux (163, 168, 184).
Props.: m. 289-291° (163); m. 296-298° (168).

TETRAKIS(TRIMETHYLSILYL)METHYLTIN $(Me_3SiCH_2)_4Sn$
Prepn.: By adding $SnCl_4$ dissolved in C_6H_6 to Me_3SiCH_2MgCl in C_4H_8O and heating under reflux; 83.5% yield (493).
Props.: b. 94°/0.2mm; n^{25}_D 1.4839; d_{25} 1.018 (493).
Rxns. with: I_2 (1:1) at 175° $\rightarrow (Me_3SiCH_2)_3SnI$ (493).

TETRAKIS[p-(TRIMETHYLSILYL)PHENYL]TIN $(Me_3SiC_6H_4)_4Sn$
Prepn.: By reacting $Me_3SiC_6H_4$-p-Li with $SnCl_4$ (4.5:1) in Et_2O at 0-25°; 58% yield (370).
Props.: m. 343-345° (370).

2,2,4,4,6,6-HEXAMETHYL-1,2,6,4-OXADISILASTANNANE
$Me_2SiCH_2SnMe_2CH_2SiMe_2O$

Prepn.: By refluxing $(Me_2HSiCH_2)_2SnMe_2$ with H_2O in EtOH in the presence of KOH; 86.6% yield (372a).
Props.: b. 95°/21mm; $n^{25}D$ 1.4743; d^{25}_4 1.203 (372a).

2,2,6,6,10,10-HEXAMETHYL-1,2,10,6-OXADISILASTANNECANE
$Me_2Si(CH_2)_3SnMe_2(CH_2)_3SiMe_2O$

Prepn.: By refluxing $[Me_2HSi(CH_2)_3]_2SnMe_2$ in EtOH in the presence of H_2O and KOH, followed by thermal cracking at 250°; 20.4% yield (372a).
Props.: b. 148°/16mm; $n^{25}D$ 1.4840; d^{25}_4 1.147; IR spectrum (372a).

4,4-DIBUTYL-2,2,6,6-TETRAMETHYL-1,2,6,4-OXADISILASTANNANE
$Me_2SiCH_2SnBu_2CH_2SiMe_2O$

Prepn.: By adding Bu_2SnCl_2 to $(ClMgCH_2SiMe_2)_2O$ in Et_2O under reflux; washing with H_2O and 1% NaOH, adding powdered KOH, removing the solvent, and cracking the residue at 220°/5mm; 37% yield (372a).
By refluxing $[Me_2(EtO)SiCH_2]_2SnBu_2$ with H_2O in EtOH in the presence of KOH; 92% yield (372a).
Props.: b. 170°/25mm; $n^{25}D$ 1.4800; d^{25}_4 1.113 and 1.116; IR spectrum (372a).

(TRIMETHYLSTANNYLMETHL)METHYLSILOXANE POLYMER
$[(Me_3SnCH_2)MeSiO]_x$

Prepn.: By hydrolyzing $(Me_3SnCH_2)MeSi(OMe)_2$ in a mixt. of H_2O and Et_2O contg. HCl (494).
Props.: $n^{25}D$ 1.4993; d_{25} 1.448 (494).

BIS[(TRIMETHYLSILYL)METHYL]TIN OXIDE POLYMER $[(Me_3SiCH_2)_2SnO]_x$
Prepn.: By hydrolyzing $(Me_3SiCH_2)_2SnBr_2$ with aq. KOH (1:2) (493).
Props.: White, amorphous solid, m. 145-160° (493).

(DIETHYLSTANNYLOXY)POLYDIORGANOSILOXANE POLYMER
$[(-OR_2Si-)_n-OSnEt_2-]_x$

Prepn.: By hydrolyzing a mixture of diethyltin dichloride with organosilanes, such as Me_2SiCl_2, Et_2SiCl_2, and $PhSiCl_3$, in 10% aqueous NH_3-MePh mixture at 50-55° (27a).
Props.: Clear, yellow liquids sol. in C_6H_6, MePh, Et_2O, Me_2CO, and EtOH; at elevated temps. they become infusible and insoluble (27a).

For other silicon-containing organotin compounds see p. 146, 154, 220, 242, 243.

(TRIPHENYLGERMYL)TRIPHENYLTIN $Ph_3SnGePh_3$
Prepn.: By reacting Ph_3GeK with Ph_3SnCl; 60% yield (183).
Props.: m. 284-286° (183).
Rxns. with: I → Ph_3GeI (183).
PhLi → mixt. of prods. (183).
BuLi followed by CO_2 and acidification → Ph_3GeCO_2H and $(Ph_3Ge)_2$ (183).

10. ORGANOTIN-NITROGEN COMPOUNDS

N-TRIETHYLSTANNYLPHTHALIMIDE $Et_3SnN(CO)_2C_6H_4$
Prepn.: By refluxing Et_3SnOH with phthalimide in C_6H_6; 43% yield (268).
Props. m. 71-73° (268).

N,N'-(DIBUTYLSTANNYLENE)DISUCCINIMIDE $Bu_2Sn(N(COCH_2)_2)_2$
Prepn.: By reacting $Bu_2Sn(OMe)_2$ with succinimide under N at up to 145°; ~theoret. yield (349).
Props.: Waxy solid, sol. in MePh (349).
Use: Stabilizer for vinyl chloride polymers (349).

2,2'-(DIBUTYLSTANNYLENE)BIS(3-PROPYLOXAZIRIDINE) ? or
BUTYRALDOXIME O,O'-DIBUTYLTIN DERIV. ?
$Bu_2Sn(NCH(O)CH_2Et)_2$ or $Bu_2Sn(ON:CHPr)_2$
Prepn.: By heating $Bu_2Sn(OMe)_2$ with butyraldoxime (1:2) at 110-120° (349).
Props.: Liquid, d_{20} 1.2120; $n^{20}D$ 1.4998 (349).
Use: Stabilizer for vinyl chloride polymers (349).

ORGANOTIN SULFONAMIDES
Organotin sulfonamides were prepared by reacting:

1. Dialkyltin dihalides with $RSO_2NR'Na$ (R= alkyl or aryl; R'= H or alkyl) in the presence or absence of a solvent (Method I).
2. Dialkoxydialkyltin compounds with $RSO_2NR'Na$ at 130-140° in a nitrogen atmosphere (Method II).
3. Trialkyltin halides with $RSO_2NR'Na$ in the presence or absence of a solvent (Method III).
4. Trialkyltin hydroxides with RSO_2NH_2 in EtOH under reduced pressure (Method IV).

ORGANOTIN SULFONAMIDES

Compound	Prepared from	Method	Solvent	Properties	Ref.
$Bu_2Sn(NHSO_2\text{-}C_6H_{13})_2$	Bu_2SnCl_2 + $C_6H_{13}SO_2NHNa$	I		*	342, 346
$Bu_2Sn(NHSO_2\text{-}Ph)_2$	Bu_2SnCl_2 + $PhSO_2NHNa$	I		m. 135-137°*	342, 346
$Bu_2Sn(EtNSO_2\text{-}Ph)_2$	Bu_2SnCl_2 + $PhSO_2NEtNa$	I		*	346
$Bu_2Sn(BuN\text{-}SO_2Ph)_2$	Bu_2SnCl_2 + $PhSO_2NBuNa$	I	xylene	Clear, yellowish oil*	342, 346
$Bu_2Sn(NHSO_2\text{-}C_6H_4Me\text{-}p)_2$	Bu_2SnCl_2 + $p\text{-}MeC_6H_4SO_2\text{-}NHNa$	I	xylene	*	342, 346
$Bu_2Sn(OMe)\text{-}(NBuSO_2Ph)$	$Bu_2Sn(OMe)_2$ + $PhSO_2NHBu$	II		Liquid*	349
$Bu_3Sn(EtNSO_2\text{-}C_6H_4Me\text{-}p)$	Bu_3SnCl + $p\text{-}Me\text{-}C_6H_4SO_2NEtNa$	III		b. 200°/1.5mm*	342, 346
$Et_3SnNHSO_2Me$	Et_3SnCl + $MeSO_2NHNa$	III	EtOH	Crystals, m. 38°	268
$Et_3SnNHSO_2\text{-}C_6H_4Me\text{-}p$	Et_3SnOH + $p\text{-}Me\text{-}C_6H_4SO_2NH_2$	IV	EtOH	m. 69.5-71°	268
Ph_3SnNHO_2SPh				**	338

* Useful as stabilizer for vinyl halide polymers.
** Sternutatory effect (338).

11. ORGANOTIN PHOSPHONIUM COMPLEX SALTS

DIMETHYLSTANNYLENEBIS(METHYLENETRIPHENYLPHOSPHONIUM) COMPLEX SALTS

$[Me_2Sn(CH_2PPh_3)_2][Me_2SnBr_4]$, m. ~135° (decompn.), prepared by reacting Me_2SnBr_2 with Ph_3PCH_2 (663a).

$[Me_2Sn(CH_2PPh_3)_2][Cr(NH_3)_2(SCN)_4]_2$, m. 115-119°, prepared from the preceding salt by precipitating the phosphonium cation with $[Cr(NH_3)_2(SCN)_4]^-$ (663a).

$[Me_2Sn(CH_2PPh_3)][BPh_4]_2$, m. 78-81°, prepared from the first salt by precipitating the phosphonium cation with $[BPh_4]^-$ (663a).

$[Me_2Sn(CH_2PPh_3)_2][HgBr_2Cl_2]$, m. 139-140°, prepared by reacting Me_2SnCl_2 with Ph_3PCH_2 and pptg. the salt with $HgBr_2$ (663a).

[(TRIMETHYLSTANNYL)METHYL]TRIPHENYLPHOSPHONIUM COMPLEX SALTS

[Ph$_3$PCH$_2$SnMe$_3$][Me$_3$SnBr$_2$]

Prepn.: By reacting Me$_3$SnBr with Ph$_3$PCH$_2$ in Et$_2$O (663a).
Props.: White solid, m. ~152° (decompn.); The phosphonium compd. was also pptd. as [Ph$_3$PCH$_2$SnMe$_3$][Cr(NH$_3$)$_2$(SCN)$_4$], m. 123-125° (663a).

12. ORGANOTIN-METAL CARBONYLS AND -METAL CARBONYL PHOSPHINE COMPLEXES

Organotin-metal tetracarbonyl compounds of the general formulae R$_2$SnFe(CO)$_4$, R$_2$SnCo(CO)$_4$]$_2$, (R$_3$Sn)$_2$Fe(CO)$_4$ and R$_3$SnCo(CO)$_4$ are formed by reacting organotin halides (R$_2$SnCl$_2$ or R$_3$SnCl) with K$_2$Fe(CO)$_4$ or with KCo(CO)$_4$ or by reacting triorganotin hydroxides with Ca[HFe(CO)$_4$]$_2$.

Reactions of the compounds with triphenylphosphine lead to the substitution of one mole CO in the complex compounds by one mole of the phosphine.

ORGANOTIN-METAL CARBONYLS AND THEIR COMPLEXES

Compound	Prepared from	Solvent	Properties	Ref.
$Bu_2Sn[Fe(CO)_4]$	$Bu_2SnCl_2 + K_2[Fe(CO)_4]$		Light-yellow, shiny leaflets	227
$(Ph_3Sn)_2[Fe(CO)_4]$	$Ph_3SnOH + Ca[Fe(CO)_4]$		Colorless crystals, sol. in org. solvents, sinter at 126-130°	225 226 227
$Me_2Sn[Co(CO)_4]_2$	$Me_2SnCl_2 + NaCo(CO)_4$	H_2O	Yellow-brown crystals turning purple in air	226 227
$[Me_2SnO(H_2O)]\cdot Me_2Sn-[Co(CO)_4]_2$ (?)	$Me_2SnCl_2 + KCo(CO)_4$ in a soln. contg. KCN		Yellow solid stable in air	227
$Bu_2Sn[Co(CO)_4]_2$	$Bu_2SnCl_2 + NaCo(CO)_4$	MeOH	Oily liquid	227
$Bu_3Sn[Co(CO)_4]$	$Bu_3SnCl + NaCo(CO)_4$	MeOH	Yellow-brown oil (1,2)	226, 227
$Bu_2Sn[Fe(CO)_3PPh_3]$	$Bu_2Sn[Fe(CO)_4] + Ph_3P$	p-xylene	Reddish-brown, insol. solid	227
$Me_2Sn[Co(CO)_3PPh_3]_2$	$Me_2Sn[Co(CO)_4]_2 + Ph_3P$	p-xylene	Red-brown crystals, stable in air	226, 227
$Bu_2Sn[Co(CO)_3PPh_3]_2$	$Bu_2Sn[Co(CO)_4]_2 + Ph_3P$	Et_2O	Crystals stable in air	226, 227
$Bu_3Sn[Co(CO)_3PPh_3]$	$Bu_3Sn[Co(CO)_4] + Ph_3P$	C_6H_6	Reddish-brown crystals, (stable) slightly sol. in C_6H_6 (3)	226

(1) Rxn. with HCl → evolution of CO and H (227).
(2) Rxn. with PBr_3 → brown-black hydroscopic substance slightly sol. in Me_2CO, MeOH, and Et_2O (227).
(3) Sol. in C_5H_5N with evolution of a gas (227).

13. ORGANOTIN HETEROCYCLES

For the organotin heterocycles of divalent tin see "Organic Derivatives of Divalent Tin."

5,5-DICHLORO-10,11-DIHYDROBENZO[b,f]STANNIEPIN

Prepn.: By reacting (o-LiC$_6$H$_4$CH$_2$-)$_2$ with SnCl$_4$ in Et$_2$O-C$_6$H$_6$ under reflux, hydrolyzing the prod., and treating it with SnCl$_4$ on steam bath and then at 150°; 37.5% yield (303).
By chlorinating 10,11-dihydrodibenzo[b,f]stannoepin in CH$_2$Cl$_2$; 81% yield (303).
Props.: m. 106-106.5°; IR spectrum (303).
Rxns. with: MeMgBr → 5,5-dimethyl deriv. (303).
LiAlH$_4$ in Et$_2$O → 10,11-dihydrodibenzo[b,f]stannoepin (303).
NaOH in EtOH-H$_2$O → 5-oxo deriv. (303).

5,5-DIMETHYL-10,11-DIHYDRODIBENZO[b,f]STANNIEPIN

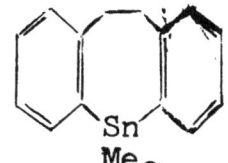

Prepn.: By reacting 5,5-dichloro-10,11-dihydrodibenzo-[b,f]stanniepin with MeMgBr in Et$_2$O (303).
Props.: b. 130-135°/0.2mm; n^{25}D 1.6130; IR spectrum (303).
Rxn. with: Br$_2$ in CCl$_4$ → (o-BrC$_6$H$_4$CH$_2$-)$_2$ (303).

5,5-DIPHENYL-10,11-DIHYDRODIBENZO[b,f]STANNIEPIN

Prepn.: By reacting (o-LiC$_6$H$_4$CH$_2$-)$_2$ with Ph$_2$SnCl$_2$ in Et$_2$O (303).
By reacting 5,5-dichloro-10,11-dihydrodibenzo[b,f]stanniepin with PhLi in Et$_2$O under reflux (303).
Props.: m. 136-137°, resolidifies and remelts at 146-147°; IR spectrum (303).

5-OXO-10,11-DIHYDRODIBENZO[b,f]STANNIEPIN (?)

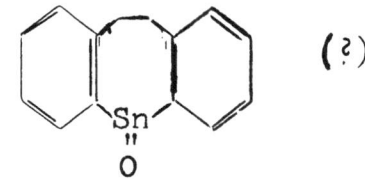

Prepn.: By treating 5,5-dichloro-10,11-dihydrodibenzo[b,f]stanniepin in dioxane with NaOH in 50% aq. EtOH; 86.5% yield (303).
Props.: IR spectrum (303).

5-THIO-10,11-DIHYDRODIBENZO[b,f]STANNIEPIN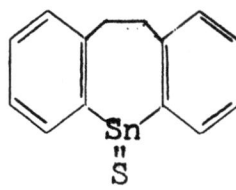

Prepn.: By treating 10,11-dihydrodibenzo[b,f]stannoepin with
S dissolved in MePh; 60% yield (303).
Props.: m. 274-275°; IR spectrum (303).

2,2-DIBUTYL-1,3,2-OXATHIASTANNOLANE $\overline{Bu_2SnSCH_2CH_2O}$
Prepn.: By refluxing Bu_2SnO with $HSCH_2CH_2OH$ (1:1) in MePh
with azeotropic removal of H_2O (431).
Use: Stabilizer for vinyl halide polymers (431).

2,2,4,4,6,6-HEXACHLORO-1,2,4,6-OXATRISTANNACYCLOHEXANE
ETHERATE (?) $\overline{SnCl_2CH_2SnCl_2CH_2SnCl_2O} \cdot Et_2O$
Prepn.: By reacting $SnCl_2$ with CH_2N_2 in Et_2O (574).

8,8-DIBUTYL-3-METHYL-5,11-DIOXO-1,4,7,9,8-DIOXADITHIASTANNA-
CYCLOHENDECANE $Bu_2SnSCH_2CO_2CH_2CHMeOCOCH_2S$
Prepn.: By heating Bu_2SnO with $HSCH_2CO_2CH_2CHMeOCOCH_2SH$ in
C_6H_6 with azeotropic removal of H_2O (702).
Use: Stabilizer for halogenated polymers (702).

DIBUTYLTIN OXYSULFIDES $Bu_2SnSBu_2SnSBu_2SnO$ or
$Bu_2SnOBu_2SnOBu_2SnS$ (?)
Prepn.: By passing H_2S through a suspension of Bu_2SnO in
toluene (806).
Use: Antioxidant for vulcanized rubber (806).

2,2,4,4,6,6-HEXAMETHYL-1,2,6,4-OXADISILASTANNANE
$\overline{Me_2SiCH_2SnMe_2CH_2SiMe_2O}$
Prepn.: By refluxing $(Me_2HSiCH_2)_2SnMe_2$ with H_2O in EtOH in the
presence of KOH; 86.6% yield (372a).
Props.: b. 95°/21mm; $n^{25}D$ 1.4743; d^{25}_4 1.203 (372a).

4,4-DIBUTYL-2,2,6,6-TETRAMETHYL-1,2,6,4-OXADISILASTANNANE
$\overline{Me_2SiCH_2SnBu_2CH_2SiMe_2O}$
Prepn.: By adding Bu_2SnCl_2 to $(ClMgCH_2SiMe_2)_2O$ in Et_2O under
reflux, washing the prod. with H_2O and 1% NaOH, adding powdered
KOH, and cracking the residue, after removal of the solvent,
at 220°/5mm; 37% yield (372a).
By refluxing $[Me_2(OEt)SiCH_2]_2SnBu_2$ with H_2O in the presence of KOH;
92% yield (372a).
Props.: b. 170°/25mm; $n^{25}D$ 1.4800; d^{25}_4 1.113 and 1.116;
IR spectrum (372a).

2,2,6,6,10,10-HEXAMETHYL-1,2,10,6-OXADISILASTANNACANE

$Me_2Si(CH_2)_3SnMe_2(CH_2)_3SiMe_2O$

Prepn.: By refluxing $[Me_2HSi(CH_2)_3]_2SnMe_2$ in EtOH in the presence of H_2O and KOH, followed by cracking at 250°; 20.4% yield (372a).
Props.: b. 148°/16mm; $n^{25}D$ 1.4840; d^{25}_4 1.147; IR spectrum (372a).

14. HEXAORGANODITIN COMPOUNDS AND TETRAORGANODITIN DIHALIDES AND DINITRATES

Hexaorganoditin compounds are prepared:

1. By reacting triorganotin halides or hydroxides with ammonia, alkali metals, or alkali metal triorganostannides.
2. By reacting alkali metal triorganostannides with halogens or with organic halides.
3. By treating sodium tetraorganodistannides with organohalides.
4. By reacting sodium-tin alloys with organomercury halides.
5. By reacting tin tetrahalides with organolithium compounds.
6. By reacting tin dihalides with organomagnesium halides or organolithium compounds.

Tetraorganoditin dihalides are prepared by treating diorganotin dihalides with organic bases, ammonia, hydrazine, or alkali metals.

Tetraorganoditin dinitrates are prepared by treating tetraorganoditin dihalides with silver nitrate.

HEXAMETHYLDITIN $(Me_3Sn)_2$

Prepn.: By treating Me_3SnCl with an alkali metal in NH_3 (63).
Props.: m. 23; b. 182°/756mm (63); Magnetic measurements indicate the dimeric structure (379).
Rxns. with: Alkali metals → cleavage of the Sn-Sn bond (63).
Na-K alloy in $(-CH_2OMe)_2$, followed by $(p-MeC_6H_4)_3SnBr$ → $(p-MeC_6H_4)_3SnSnMe_3$ (63).
Br_2 and I_2 at 25°, heat of rxn. was detd. (411).

HEXAETHYLDITIN $(Et_3Sn)_2$

Prepn.: By reacting Et_3Sn halides or Et_3SnOH with one atom equiv. of Na in liq. NH_3 (195, 201, 203).
By reacting Et_2SnNa (?) or $(Et_2SnNa)_2$ with ethyl halides (201, 203).
By digesting Et_3SnCl with Na in $1-C_5H_{11}OEt$ at $130°$; 70% yield (283, 387).

Props.: The compound was designated "triethyltin radical" by Harada (195, 201, 203), who claimed the existence of two isomers identified by the difference in the rate of oxidation. The isomer prepd. by the first method, b. $152-154°/17mm$, yielded about three times less Et_2SnO than the second isomer, b. $153-155°/17mm$, which was prepd. by the second method (201). The difference in the reactivity was ascribed to the presence of impurities (202); b. $160°/23mm$ (283, 387); Magnetic susceptibility and parachor (258).

Rxns. with: Na (1:1) in $NH_3 \rightarrow Et_3SnNa$ (195).
$HgCl_2$ (1:1) $\rightarrow Et_3SnCl$ (283, 387).
$PhHgCl \rightarrow PhSnEt_3$ and Et_3SnCl (283, 387).
R_2Hg (1:1) $\rightarrow RSnEt_3$ (283, 387).

Addn. Compd.: $Et_3Sn \cdot PhN:NC(S)NHNHPh$, prepd. in $CHCl_3$ (1), which in $CHCl_3$ soln. is converted by sunlight to $Et_2Sn \cdot$ dithizone complex (2).

HEXAPROPYLDITIN $(Pr_3Sn)_2$

Props.: Magnetic susceptibility and parachor (258).

HEXABUTYLDITIN $(Bu_3Sn)_2$

Prepn.: By-prod. in the rxn. of Bu_2SnCl_2 with BuCl and Na (265)
Rxns. with: Cl_2 at r.t. $\rightarrow Bu_3SnCl$ (265).

HEXAPHENYLDITIN $(Ph_3Sn)_2$

Prepn.: By reacting Ph_3SnLi with Ph_3SnCl in Et_2O under reflux or with Ph_3SnBr in Et_2O at r.t. (571).
By reacting Ph_3SnLi with I_2 in Et_2O (170).
By reacting Ph_3SnLi with Ph_2CO at $-10°$; 14.8% yield (170).
By shaking Ph_3SnLi with H_2O in Et_2O; the prod. contains also Ph_2SnO (571).
By reacting $SnCl_4$ with PhLi in Et_2O at $-10°$; 10.9% yield (170).
By passing CO_2 over a soln. of Ph_3SnLi in Et_2O at $-10°$ (170).
By heating Ph_2Sn with PhN_2CPh_3 in C_6H_6 or CCl_4 under CO_2 (432).
By reacting Ph_3SnBr with Li in NH_3 (Ph_3SnH is formed as by-prod.) (571).
By reacting Ph_3SnH with RNH_2 or $CH_2:CHCOR$ (2:1) (732).
By electrolysis of Ph_3SnCl in MeOH; cathodic prod. (438).
By-prod. in the rxn. of Ph_3SnLi with aryl halides in Et_2O (167).
By-prod. in the rxn. of Na_2S with PhHgCl in xylene (382).

HEXAPHENYLDITIN $(Ph_3Sn)_2$ (Cont'd.)
Props.: m. 229-231° (163); m. 226-228° (170); m. 229° (432); m. 231-232° (382); m. 229.5-231° (571).
Rxns. with: $KMnO_4$ in Me_2CO followed by $H_2O \rightarrow Ph_3SnOH$ (36).
PhN(NO)Ac in $CCl_4 \rightarrow Ph_3SnCl$ (432).
PhLi $\rightarrow Ph_4Sn$ (584).

HEXA-p-CHLOROPHENYLDITIN $[(p-ClC_6H_4)_3Sn]_2$
Prepn.: By reacting Na_2Sn with $p-ClC_6H_4HgCl$ in xylene under reflux (382).
Props.: m. 224-226° (382).
Rxn. with: I_2 in $C_6H_6 \rightarrow (p-ClC_6H_4)_3SnI$ (382).

HEXA-o-TOLYLDITIN $[(o-MeC_6H_4)_3Sn]_2$
Prepn.: By reacting $SnCl_4$ with $o-MeC_6H_4Li$ (1:3) in Et_2O at -10° and refluxing the resulting $(o-MeC_6H_4)_3SnLi$ with $o-MeC_6H_4I$; 44.6% yield (170).
By reacting $(o-MeC_6H_4)_3SnCl$ with Na in C_6H_6-xylene under reflux; 62.1% yield (170).
By refluxing $(o-MeC_6H_4)_3SnLi$ with $PhCH_2Cl$; 37.6% yield (170).
Props.: m. 298-300° (170); m. 208-212° (380); Magnetic measurements in C_6H_6 indicate the distannane structure (380).

HEXA-p-TOLYLDITIN $[(p-MeC_6H_4)_3Sn]_2$
Prepn.: By treating $(p-MeC_6H_4)_3SnBr$ with Na-K alloy (63).
Props.: m. 251-252° (63).

HEXA(2-BIPHENYLYL)DITIN $[(2-PhC_6H_4)_3Sn]_2$
Prepn.: By reacting $2-PhC_6H_4MgBr$ or $2-PhC_6H_4Li$ with $SnCl_2$ in C_4H_8O (41).
By reacting $(2-PhC_6H_4)_3SnBr$ with Na in xylene (41).
Props.: The product occurs in two forms: a low-melting form prepd. by the first method, m. 170°; and a high-melting form prepd. from $(2-PhC_6H_4)_3SnBr$, m. 288-289° (41).

1,1,1-TRIMETHYL-2,2,2-TRI-p-TOLYLDITIN $(p-MeC_6H_4)_3SnSnMe_3$
Prepn.: By treating $(Me_3Sn)_2$ with Na-K in $(-CH_2OMe)_2$ and adding the resulting green soln. to $(p-MeC_6H_4)_3SnBr$ (63).
Props.: m. 139.5-141° (63).

TETRAORGANODITIN DIHALIDES AND DINITRATES

Compound	Prepared from	Yield %	Properties	Ref.
$(Et_2SnCl)_2$	$Et_2SnCl_2 + Et_3N$ (1:1)	77	White crystals discoloring at 120° (1)	249
$(Pr_2SnCl)_2$	$Pr_2SnCl_2 + H_2NNH_2$ (1:1)	93	White crystals, m. 120.5-121.5°	249
$(Bu_2SnBr)_2$	$Bu_2SnBr_2 + Et_3N$ (1:1)	61	White crystals, m. 103-104°	249
$(Bu_2SnCl)_2$	$Bu_2SnCl_2 + Na$ (1:1)	91	m. 115-116°	248
	Bu_2SnCl_2 or $Bu_2SnClBr$ + organic bases	96	m. 110-112 (2,3,4)	249
$(n-C_5H_{11})_2SnCl_2$	$(n-C_5H_{11})_2SnCl_2 + H_2NNH_2$ (1:1)	92.8	White crystals, m. 96.5-97.5°	249
$(Ph_2SnCl)_2$	$Ph_2SnCl_2 + H_2NNH_2$ (1:1)	20	White crystals, m. 185-187°	249
$(Bu_2SnNO_3)_2$	$(Bu_2SnCl)_2 + AgNO_3$	50	m. 225°	248

(1) Sublimes at 160° with decompn., yielding Et_2SnCl_2 and a white residue, m. 240° (249).
(2) Rxn. with Br_2 in $CCl_4 \rightarrow Bu_2SnClBr$ (248, 249).
(3) Rxn. with RMgBr $\rightarrow Bu_2SnR_2$ (248).
(4) Rxn. with $AgNO_3 \rightarrow (Bu_2SnNO_3)_2$ (248).

15. BRIDGED BIS(TRIORGANOTIN) COMPOUNDS

Acetylene-bridged bis(triorganotin) compounds are prepared:

1. By reacting triorganotin halides with ethylnylene-bis(magnesium bromide) or with sodium acetylide.
2. By reacting sodium triorganostannides with diiodoacetylene.

Other bridged bis(triorganotin) compounds are prepared:

1. By reacting triorganotin hydrides with acetylenic compounds or with compounds containing two double bonds, e.g. allene, diallyl ether, acrylic anhydride, and divinylbenzene in molar ratio of 2:1.
2. By reacting triorganotin halides with α-ω-alkylene-bis(magnesium bromide) or with arylidenedilithium compounds.
3. By reacting haloaryltriaryltin compounds with aryl-lithium derivative.

ACETYLENE-BRIDGED BIS(TRIORGANOTIN) COMPOUNDS

Bridged compound	Prepared from	Yield %	Properties	Ref.
$Et_3SnC\!:\!CSnEt_3$	$Et_3SnCl + (:CMgBr)_2$		m. 52°; b. 123°/0.05mm; b. 294° d_{20} 1.3411; n^{20}_D 1.5089; soly. data; (1:4)	50
	$Et_3SnCl + NaC\!:\!CH$ (1:1)	69		50
$Ph_3SnC\!:\!CSnPh_3$	$Ph_3SnCl + (:CMgBr)_2$	90	m. 153°; soly. data (4)	50
	$Ph_3SnCl + NaC\!:\!CH$	69		50
	$Ph_3SnNa + C_2I_2$			50
$(p\text{-}ClC_6H_4)_3SnC\!:\!CSn(C_6H_4Cl\text{-}p)_3$	$(p\text{-}ClC_6H_4)_3SnBr + (:CMgBr)_2$		m. 191° (5)	214
$(PhCH_2)_3SnC\!:\!CSn(CH_2Ph)_3$	$(PhCH_2)_3SnCl + (:CMgBr)_2$		m. 94° (5)	214
$(o\text{-}MeC_6H_4)_3SnC\!:\!CSn(C_6H_4Me\text{-}o)_3$	$(o\text{-}MeC_6H_4)_3SnCl + (:CMgBr)_2$		m. 143° (5)	214
$(p\text{-}MeC_6H_4)_3SnC\!:\!CSn(C_6H_4Me\text{-}p)_3$	$(p\text{-}MeC_6H_4)_3SnCl + (:CMgBr)_2$		m. 149° (5)	214

1. Rxn. with $I_2 \rightarrow Ph_3SnI + C_2I$ (50).
2. Rxn. with ammoniacal Cu^+ solns. $\rightarrow Cu_2C_2$ (50).
3. Rxn. with $SnCl_4 \rightarrow$ white polymeric solid which evolves C_2H_2 on contact with H_2O and explodes when touched in the dry state (50).
4. Rxn. with $AgNO_3 \rightarrow Ag_2C_2$ (50).
5. Rxn. with alkali $\rightarrow$ triorganotin hydroxide and C_2H_2 (214).

OTHER BRIDGED BIS(TRIORGANOTIN) COMPOUNDS

Bridged Compound	Prepared from	Yield %	Properties	Ref.
$Pr_3SnCH_2CH(CO_2Me)SnPr_3$	$Pr_3SnH + CH:CCO_2Me$	69	b. 120-123°/0.0002mm	273a
$(Pr_3SnCH_2CH_2CH_2-)_2$			Magnetic susceptibility and parachor	258
$(Ph_3SnCH_2-)_2$	$Ph_3SnH + CH:CH$	42	White solid, m. 206-207°	273a
$(Ph_3SnCH_2)_2CH_2$	$Ph_3SnH + CH_2:C:CH_2$	18	m. 182-185°	273a
$(Ph_3SnCH_2CH_2-)$	$Ph_3SnCl + (CH_2CH_2MgBr)_2$		m. 149-150.5°	595
$Ph_3SnC_6H_4SnPh_3$	$Ph_3SnCl + p-LiC_6H_4Li$	67	m. 289-292°	595
$p-(Ph_3SnCH_2CH_2)_2C_6H_4$	$Ph_3SnH + p-CH_2:CHC_6H_4CH:CH_2$	40	m. 136-142°	273a
$(Ph_3SnC_6H_4-)_2$	$Ph_3SnCl + (p-LiC_6H_4-)_2$		m. 235-236°	595
$Ph_3SnCH_2CH(CO_2Me)SnPh_3$	$Ph_3SnH + CH:CCO_2Me$	52	m. 111-112°	273a
$Ph_3SnCH_2CH(Ph)SnPh_3$	$Ph_3SnH + PhC:CH$	49	m. 139-140°	273a, 276
$(Ph_3SnCH_2CHMeCH_2-)_2$	$Ph_3SnH + (CH_2:CMeCH_2-)_2$	30	m. 93-94°	273a
$(Ph_3SnCH_2CH_2CO_2CH_2-)_2$	$Ph_3SnH + (CH_2:CHCO_2CH_2-)_2$	91	m. 93-96°	273a
$(Ph_3SnCH_2CH_2CH_2)_2O$	$Ph_3SnH + (CH_2:CHCH_2)_2O$	33	m. 105-108°	273a
$(Ph_3SnCH_2CH_2CO)_2O$	$Ph_3SnH + (CH_2:CHCO)_2O$	60	m. 182-186°	273a
$[p-(Ph_3Sn)C_6H_4-]_2$	$p-IC_6H_4SnPh_3 + PhLi$	84.5	m. 303°	584
	$Ph_3SnCl + p-BrC_6H_4MgBr$	by-prod.		583
	$Ph_3SnC_6H_4Br-p + Mg$	by-prod.	m. 298°	585

16. MISCELLANEOUS ORGANOTIN COMPOUNDS INCLUDING POLYMERS

TRIETHYLTIN CYANIDE   Et$_3$SnCN
Prepn.: By reacting (Et$_3$Sn)$_2$O with anhyd. HCN in CCl$_4$ at 0° (18).
By refluxing Et$_3$SnSMe, or (Et$_3$Sn)$_2$S, or Et$_3$Sn halides with AgCN (18).
By refluxing (Et$_3$Sn)$_2$O with Et$_3$SiCN (19).
By mixing an ether soln. of Et$_3$SnBr with aq. KCN; 80% yield (268).
Props.: Cotton-like substance, m. 153°; b. 249° (decompn.); sublimes at 8 mm (18); Needles shaped crystals, m. 163.5-164° (268).
Rxns. with: Ag salts → the corresponding Et$_3$Sn salts (18).
Ag$_2$O → (Et$_3$Sn)$_2$O (18).
Use: Constituent of antifouling marine paints (549).

TRIPHENYLTIN CYANIDE   Ph$_3$SnCN
Prepn.: By reacting Ph$_3$SnCl with NaCN in liquid NH$_3$ (50).
By passing anhyd. HCN into Ph$_3$SnOH and washing the ppt. with anhyd. HCN; 95% yield (592).
Props.: m. 255-256° (592).
Rxn. with: H$_2$O → Ph$_3$SnOH + HCN (592).

DIMETHYLTIN FERROCYANIDE-DIMETHYLTIN OXIDE COMPLEX (?)
                                        (Me$_2$Sn)$_2$Fe(CN)$_6$·Me$_2$SnO
Prepn.: By metathetic reaction of Me$_2$SnCl$_2$ with K$_4$Fe(CN)$_6$ in aq. soln. (455).
Props.: White ppt., which becomes yellow on drying, darkens at 310° (455).

DIMETHYLTIN FERRICYANIDE   (Me$_2$Sn)$_3$[Fe(CN)$_6$]$_2$
Prepn.: By metathetic reaction of Me$_2$SnCl$_2$ with K$_3$Fe(CN)$_6$ in aq. soln. (455).
Props.: Organo-brown solid, decomp. at 335-360°, slightly sol. in H$_2$O (455).
Rxns. with: H$_2$O$_2$ → green solid (455).

TRIETHYLTIN ISOCYANATE   Et$_3$Sn(NCO)
Prepn.: By refluxing Et$_3$SnBr, Et$_3$SnSMe, or (Et$_3$Sn)$_2$S with AgNCO in MePh (18).
By refluxing (Et$_3$Sn)$_2$O with MeOSi(NCO)$_3$; 90% yield (19).
Props.: m. 48°; b. 239°; b. 162-163°/80mm; b. 120-121°/11mm (18); m. 53°; b. 160°/70mm (19).
Rxns. with: AgX (X= AcO or F) → Et$_3$SnX (18).

TRIETHYLTIN ISOTHIOCYANATE $Et_3Sn(NCS)$

Prepn.: By refluxing Et_3SnSMe, $(Et_3Sn)_2S$, or Et_3SnX (X= I, Br, or CN) with AgNCS in MePh (18).
By reacting $(Et_3Sn)_2O$ with HNCS in Et_2O; 70% yield (18).
By refluxing $(Et_3Sn)_2O$ with $Me_3Si(NCS)$ or $PO(NCS)_3$; 86% yield (19).
Props.: m. 33°; b. 130°/1mm; b. 282° (decompn.); b. 180°/9mm (18); n^{20}_D 1.582 (19).
Rxns. with: Ag salts → the corresponding Et_3Sn salts (18).
Ag_2O → $(Et_3Sn)_2O$ (18).

ETHOXYDIETHYLTIN FLUORIDE $EtOSnEt_2F$

Prepn.: By passing AcF vapor into a soln. of $Et_2Sn(OEt)_2$ (1:1) in Et_2O (577).
Props.: White crystalline powder, m. 90-91°; sol. in Et_2O and C_6H_6; decomp. on storing (577).

TRIS(TRIMETHYLTIN)OXONIUM BROMIDE $(Me_3Sn)_3OBr$

Prepn.: By reacting $(Me_3Sn)_2O$ with Me_3SnBr in C_6H_6 or petr. ether or in the absence of solvent (200, 203).
Props.: m. 88° (decompn.) (200).

TRIS(TRIMETHYLTIN)OXONIUM IODIDE $(Me_3Sn)_3OI$

Prepn.: By reacting $(Me_3Sn)_2O$ with Me_3SnI (excess) in C_6H_6 or petr. ether or in the absence of solvent (200, 203).
Props.: Crystalline substance (200).
Rxns. with: H_2O (by dissolving in moist C_6H_6) → $(Me_3SnOH)_2 \cdot Me_3SnI$ (200).
Na in NH_3 in the presence of H_2O → Me_3SnNa (200).

POLYMERIC ORGANOTIN OXYCHLORIDE $ClSnR_2[OSnR_2]_xCl$

The compounds of the above formula in which R is an alkyl, aryl, or aralkyl group and x is a whole number are useful as antioxidants for vulcanized rubber (791).

METHYLENE-TIN CHLORIDE POLYMER $-SnCl_2(SnCl_2CH_2)_n-$

Prepn.: By reacting $SnCl_2$ with CH_2N_2 in C_6H_6 (574).

ORGANOTIN COMPOUNDS $R'OSn(R_2)O]_nSnR_2OR'$

Compounds of the general formula $R'OSnR_2O]_nSnR_2OR'$, where R is an alkyl, aryl, or aralkyl group, R' is an acyl, alkyl or aryl group, and n is an integer number, are useful as antioxidants for rubber.

The compounds can be prepared by reacting R_2SnO with R'OH or R_2SnCl_2 with R'OH in the presence of basic substances (789a).

OXASTANNA COMPOUNDS FROM ORGANOTIN OXIDES AND CARBOXYLIC ACID ESTERS (?)

$$\left[-O-\underset{R}{\overset{OR'}{C}}-O-\underset{R''}{\overset{R''}{Sn}}-O-\underset{R''}{\overset{R''}{Sn}}- \right]_n$$

Compounds of the above structure (useful as stabilizers for polyvinyl halide resin) were prepared by heating R''_2SnO with RCO_2R' (2:1) at $200°$. The following compounds were reacted (242a):

$Bu_2SnO + PrCO_2Bu$
$Bu_2SnO + AcOC_6H_{13}$
$Bu_2SnO + (:CHCO_2Et)_2$ (both stereomers)
$Bu_2SnO + AcOBu$
$Bu_2SnO + AcOEt$
$Bu_2SnO + o-C_6H_4(CO_2Ph)_2$
$Bu_2SnO + C_{11}H_{23}CO_2Me$
$Bu_2SnO + (:CHCO_2CH_2CHEtBu)_2$
$Bu_2SnO + PrCO_2Et$
$Bu_2SnO + MeCH:CHCO_2Bu$
$Bu_2SnO + Me(CH:CH)_2CO_2Et$
$Me_2SnO + C_{17}H_{35}CO_2Ph$
$Me_2SnO + o-C_6H_4(CO_2CH_2CHEtBu)_2$
$Ph_2SnO + PrCO_2Bu$

For other organotin polymers see the section on organotin-silicon compounds, p. 236.

STABILIZERS FOR HALOGENATED SYNTHETIC POLYMERS

Tin compounds containing Sn-S and Sn-O bonds, useful as stabilizers for halogenated synthetic resins, were prepared by reacting an alkyl, aryl, or aralkyl alc., a dibasic org. acid or its anhydride, a dialkyl-, diaryl-, or diaralkyltin oxide and a mercaptoacid ester (803).

ORGANOTIN COMPOUNDS

Anthelmintic props. (683).
Toxicity of triethyltin compounds (700).

17. ORGANIC DERIVATIVES OF DIVALENT TIN

Diorganotin compounds are prepared:

1. From diorganotin dihalides by reducing them with sodium or with lithium aluminum hydride.
2. By reacting sodium-tin alloy with alkyl halides.
3. By reacting tin dihalides with alkyl- or aryl-alkali or alkaline earth compounds or with the corresponding Grignard compounds.
4. By reacting diorganotin dihalides with sodium diorganostannides.

DIETHYLTIN Et_2Sn

Prepn.: By reacting Et_2SnBr_2 with Na (195).
By reacting Et_2SnNa_2 with Et_3SnBr_2 (195).
By reacting Na_2Sn with EtBr in NH_3 (210).
Props.: Orange-colored subst., decomp. >150° yielding $(Et_3Sn)_2$ + Sn (195, 203).
Rxns. with: Et_2SnBr_2 and Na in $NH_3 \rightarrow (Et_2SnNa)_2$ (195).
H_2O in $CHCl_3 \rightarrow H(Et_2SnO)_3OH \cdot SnEt_2Cl_2$ (210).
EtBr in the presence of $H_2O \rightarrow H(Et_2SnO)_3OH \cdot Et_2SnBr_2$ (210).
MeBr, followed by hydrolysis $\rightarrow Me_3SnOH$ (210).
$HgCl_2 \rightarrow Et_2SnCl_2$ (283, 387).
$Ph_2Hg \rightarrow Et_2SnPh_2$ (283, 387).

BISCYCLOPENTADIENYLTIN $(C_5H_5)_2Sn$

Prepn.: By reacting C_5H_5Li with $SnCl_2$ in $HCONH_2$ under exclusion of air and moisture (129).
By reacting C_5H_5M (M= alk. or alk. earth metal) with anhyd. $SnCl_2$ in liq. NH_3 or other N base; 52% yield (602).
Props.: Colorless crystals, m. 104-105°, sol. in org. solvents (129); Turns yellow on exposure to air; upon heating passes various stages of decompn.; at 213° turns black (602); Magnetic data (130); Dipole moment (561); IR spectrum (602, 657).

DIPHENYLTIN Ph_2Sn

Prepn.: By reacting anhyd. $SnCl_2$ with PhLi (1:2) in Et_2O; 73% yield (571).
Props.: Golden-yellow ppt. (571); Magnetic data (238).
Rxns. with: PhN_2CPh_3 in $C_6H_6 \rightarrow (Ph_3Sn)_2$ (432).
PhN(NO)Ac in $CCl_4 \rightarrow Ph_4Sn$ (432).
PhLi (1:1) in $C_6H_6-Et_2O \rightarrow Ph_3SnLi$ (571).
Ph_3CNa (1:1) in $Et_2O \rightarrow Ph_2(Ph_3C)SnNa \cdot Et_2O$ (571).

DI(9-PHENANTHRYL)TIN $(C_{14}H_9)_2Sn$

Prepn.: By reacting $C_{14}H_9MgBr$ with $SnCl_2$ in $Et_2O-C_6H_6$ or in C_4H_8O; 69% yield (42).
Props.: Yellow to orange-yellow powder, m. 179-180°; soly. data (42).

BIS(α-THIENYL)TIN $(α-C_4H_3S)_2Sn$

Use: Corrosion inhibiter for dielectrics (122).

10,11-DIHYDRODIBENZO[b,f]STANNOEPIN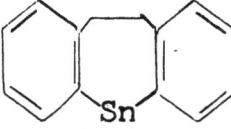

Prepn.: By reducing 5,5-dichloro-10,11-dihydrodibenzo[b,f]-stanniepin with LiAlH$_4$ in Et$_2$O (303).
Props.: m. 60-160°; IR spectrum (303).
Rxns. with: Cl in CH$_2$Cl$_2$ → 5,5-dichloro-10,11-dihydrodibenzo[b,f]-stanniepin (303).
S in MePh → 5-thio deriv. (303).

C. LEAD

1. TETRAORGANOLEAD COMPOUNDS

SYMMETRICAL, NON-FUNCTIONAL TETRAORGANOLEAD COMPOUNDS

Symmetrical tetraorganolead compounds are prepared:

1. By reacting lead, lead alloys, or mixtures of lead and other metals with organohalides in the presence or absence of various catalyzing or activating substances. The method has been extensively studied with respect to the preparation of tetraethyllead.
2. By reacting lead dihalides with organomagnesium halides.
3. By reacting lead amalgam or lead powder with organolithium compounds.
4. By subjecting hexaorganodilead compounds to a thermal treatment in the presence or absence of any catalysts; or to a treatment with magnesium and magnesium iodide or with iodine; or to a treatment with organohalides of organolithium compounds.
5. By treating diorganolead dihalides with four equivalents of lithium and reacting the reaction mixture with the corresponding organic halides.
6. By displacement of the aryl groups from tetraaryllead compounds by alkyl groups, using alkyllithium compounds.
7. By reacting aryldiazonium fluoroborates with lead or lead alloys.

TETRAMETHYLLEAD Me_4Pb

Prepn.: By reacting a mixt. of an Na-Pb alloy and Mg with MeCl at $45-85°/75$ psi, then indroducing a mixt. of MeCl and Et_2O, and continuing the rxn. under the same conditions (81). A ternary alloy Na-Pb-Mg may be employed as a source of Pb (82, 83).

By reacting an Na-Pb alloy with MeI in the presence of C_5H_5N; 50% yield (216).

By reacting Mg_2Pb (18-22% Mg) with MeBr (excess) at 80-120° under pressure in the presence of aliphatic and aromatic amines, C_5H_5N, or tetraalkyl- or arylammonium iodides (496).

By reacting PbI_2 with MeMgCl or MeMgI at r.t. (155, 646).

By reacting $PbCl_2$ with MeMgI (2:4) at 100°; 90% yield (646).

By reacting powd. Pb having a non-oxidized surface (Pb left from the rxn. of Na-Pb in the prepn. of Et_4Pb) with MeCl or MeBr in the presence of I_2 or iodides of Pb, Na, K, Me, Et, Hg or Cu (410).

TETRAMETHYLLEAD   Me$_4$Pb (Cont'd.)

Prepn. (Cont'd.): By reacting PbI$_2$ with MeLi and MeI in Et$_2$O; theoret. yield (155).
By treating (Me$_3$Pb)$_2$ with a silica-type catalyst at 0-110° (339).
Props.: m. 242.92°K (529); b. 101°/760mm (260); log P= 6.9381-1378.7/(t° + 230) (70); Viscosity data (677); Mol. structure, Pb-C distance, and symmetry (60, 572); Bond length (511); Mass spectrum (111a, 112, 426, 733); Thermodynamic data (529, 661a); Magnetic susceptibility and parachor (259, 261); Potential barrier value (136); Force consts. (774); Polarity (464); Ionization potential (135a); Raman spectrum (193a, 236, 330, 510, 558); IR spectrum (330, 505); Thermal decompn. yields the Me· radicals (105); Mechanism of pyrolysis (335); Kinetics of the thermal decompn. (557a); Thermal decompn. at 1500-1700° yields C$_2$H$_6$ in the quantities proportional to the Me· radical (701); Photolysis yields C$_2$H$_6$, C$_2$H$_4$, C$_2$H$_2$ and traces of CH$_4$; absorption spectrum of the photolysis prods. (98, 99, 101); Catalytic toxicity to Pt black (873); Isotopic analysis (623).

Rxns. with: I$_2$ in Et$_2$O at -60° → Me$_3$PbI (68).
Br$_2$ in CHCl$_3$ → Me$_2$PbBr$_2$ (155).
HCl at r.t. → Me$_3$PbCl (79, 216).
R$_4$Pb → redistribution rxn. (69, 69a, 71, 73, 910).
Et$_3$PbBr → 17.6% Me$_4$Pb, 44% Me$_3$EtPb, 22% Me$_2$Et$_2$Pb, and 15.8% MeEt$_3$Pb + Et$_4$Pb (74).
Et$_2$Hg → Hg has a greater affinity than Pb for the Me group (76, 77).
Carboxylic acids → Me$_3$Pb carboxylates (79, 216).
Sulfonic acids → Me$_3$Pb sulfonates (216).
CBr$_4$ or C$_2$Br$_6$ in the presence of O$_2$ → Me$_2$PbBr$_2$ and COBr$_2$ (220).
Et$_4$Pb + C$_2$H$_4$Br$_2$ + AlCl$_3$ → mixt. of Pb alkyls and alkylene-dilead alkyls (568).
Al(BH$_4$)$_3$ → MeAl(BH$_4$)$_2$ with intermediate formation of Me$_3$PbBH$_4$ (?) (675).
AuCl$_3$ (3:1) → 3Me$_3$PbCl + Au + 3Me· (536).
AuCl$_3$ (2:1) → Me$_3$PbCl + Me$_2$PbCl$_2$ + Au + 3Me· (536).
AuCl$_3$ (3:2) → 3PbCl$_2$ + 2Au + 6Me· (536).
AgNO$_3$ or Cu(NO$_3$)$_2$·3 H$_2$O at -70° → AgMe and CuMe, respectively (48, 105, 480, 481, 482) and Me$_3$PbNO$_3$, C$_2$H$_6$, C$_2$H$_4$, and CH$_4$ (152).
O in a diffusion flame supported by H, CH$_4$, CO, or MeCHO → PbO with an intermediate formation (?) of Me$_2$PbO$_2$ (653).

Use: Catalyst for polymerization of esters (628).
Catalyst for polymerization of olefins (862, 887).
In conjunction with Cu^{++} as a polymerization catalyst (48).

TETRAETHYLLEAD Et_4Pb

Prepn.: By reacting a Na-Pb alloy with EtCl at 50-70°/75 psi and treating the unreacted Pb in the rxn. mixt. with EtMgCl in Et_2O; 83.5% yield (84, 605). In a similar process effected at 50-100° the unreacted Pb is treated with EtCl premixed with EtLi; 66.2% yield (85). A yield of 76.4% was obtd. by reacting Na-Pb with EtCl at 75°/75 psi under N and treating the unreacted Pb with EtI in the presence of Et_2Zn and Et_2O (86). Activation energy and kinetics of the rxn. of Na-Pb with EtCl were investigated (724).

By treating Na-Pb with EtCl under pressure at a temp. above the normal b.p. of EtCl and recycling the unreacted EtCl (815).

By adding molten Na-Pb alloy at 400-425° to liquid EtCl through an orifice or spray nozzle (709).

By reacting a Na-Pb alloy, immediately after its solidification, with EtCl at 85°; 68% yield (540).

By reacting Na-Pb shot with liquid EtCl at 55-60° under a moderate pressure in a turbulent flow (526) or at 70-120° in a countercurrent flow (458).

By treating Na-Pb alloy flakes with liquid EtCl or Et_2SO_4 in a vertical cylinder 5-30 min. at 70-100° (606).

By reacting Na-Pb alloy with EtCl injected at 60°/500-1500 psi (525).

By reacting Na-Pb alloy with a large excess EtCl at 85°/110 psi in a continuous operation; 93.7% yield (384).

By initiating the rxn. of Na-Pb with a small amt. EtCl at 35-50° under pressure and then adding the bulk of EtCl and maintaining the rxn. temp. by cooling (603).

By reacting finely divided Pb with Et_2SO_4 or $(EtO)_3PO$ at 120-200° in the presence of a catalyst (PbI_2); up to 45% yield (302, 504).

By reacting Na-Pb with EtCl at 100°/200 psi under N; an automatic, continuous process (687).

By reacting Pb or Na-Pb with EtCl at 50-70° under pressure in the presence of an addn. compd. of an alkali metal and a polycyclic aromatic hydrocarbon, an ether, or aromatic amine (498).

By reacting a Na-Pb alloy with EtCl at 85°/autogeneous pressure in the presence of carboxylic acid anhydrides (99a).

By treating a Na-Pb alloy with EtCl in the presence of aldehydes, ketones, or carboxylic acid esters (53, 96a, 229, 230, 500). A yield of 89.46% was obtd. at 80-120°/107-225 psi in the presence of Me_2CO and C_2H_2 (809). The process can be carried out continuously (45). Methods were developed for the dissipation of the heat of rxn. (110, 908) and a convenient working-up procedure (56). A yield of 85% was obtd. by conducting the rxn. at 70-80° in the presence of I_2 or iodine compounds (145). The agglomeration of the solid by-prods. can be prevented by adding oil-sol. surfactants (674).

TETRAETHYLLEAD   Et$_4$Pb (Cont'd.)

Prepn. (Cont'd.): Activation of the rxn. of a Na-Pb alloy with EtCl by acetals such as CH$_2$(OMe)$_2$, MeCH(OEt)$_2$, and ClCH$_2$CH(OEt)$_2$ gave a yield of 88.4% (416).

By reacting a Na-Pb alloy with EtCl in the presence of amides (AcNH$_2$ or BzNH$_2$); up to 89% yield (231).

By reacting a Na-Pb alloy with EtCl or EtBr in an HCl or HBr atm. at 0.5-3.0 atm. pressure (46, 462).

By reacting a Na-Pb alloy with EtCl in an H halide atm. in the presence of Br$_2$ (461a).

By reacting a Na-Pb alloy with EtBr in the presence of C$_5$H$_5$N on a steam bath (469).

By reacting Pb, a Na-Pb mixt., or Na-Pb alloy with EtI in liquid NH$_3$ at -30°; up to 95% yield (291).

By reacting Na-Pb with EtCl at 50° in an autoclave, raising the temp. to 70° and the pressure to 75 psi, then adding EtCl, Et$_2$Cd and Et$_2$O, and completing the rxn. at 80° (629a).

By reacting a Na-Pb alloy contg. 1% graphite with EtCl; the presence of C prevents agglomeration of Pb (670a).

By reacting a mixt. of a Na-Pb alloy and Mg with a small quantity EtCl at 45-85°/75 psi., then adding the remaining EtCl mixed with Et$_2$O, and completing the rxn. under the same conditions. Any (Et$_3$Pb)$_2$, if formed, can be converted to Et$_4$Pb by heating at 85°; 76% yield (81). A ternary alloy of Na, Pb, and Mg, contg. up to 22.5% Na and 2.5-25% Mg may be used as the source of Pb (82). The ratio of the constituents in the ternary alloy may range from 0.5 to 15% Na and from 3 to 19.5% Mg, the total amt. of Na and Mg being between 5 and 20% (83). The presence of Mg shortens the induction period of the rxn. with EtCl, and optimum results were obtd. with 0.1 wt. % Mg (778).

By reacting a Na-Pb alloy contg. 0.15-0.2% Mg with Et halide at 75° in an autoclave (604).

By reacting a Na-Pb or K-Na-Pb melt with EtCl vapors (kinetics data) (509).

By reacting a ternary alloy of K, Na, and Pb (5-35% Na and 1-5% K) in the liquid phase at a temp. below 100° under pressure in the presence of O- or N-contg. compounds; 97% yield (501). Kinetics data (723a).

By reacting Mg$_2$Pb with excess EtCl in a sealed tube at 80° in the presence of Et$_2$O; 32% yield (266).

By reacting Mg$_2$Pb with Et halides at 80-120° under pressure in the presence of aliphatic ethers, tert-aliphatic or aromatic amines, C$_5$H$_5$N, or tetraalkyl- or arylammonium iodide (495, 496, 497).

By reacting active Na-Pb alloy flakes contg. Cu at 85° in the presence of Me$_2$CO (541).

By reacting a Ca-Pb alloy (Ca:Pb = 1-1.5:1) with EtCl at 50-70° in an autoclave in the presence of carbonyl compounds; 40% yield (301). If the molar ratio of Ca:Pb = 2.2:1, the yield is lower (499).

By reacting a Na-Pb-Hg alloy with EtCl at 35-50°/2.5-3.5 atm. (914).
By reacting EtCl with Na-Pb and Zn or with Na-Pb and EtZnCl at 50-200°/0-50 atm (751).
By reacting Pb having non-oxidized surface (recovered from the rxns. of Na-Pb with EtCl) with EtCl or EtBr at 100-130° in the presence of I_2 or iodides such as PbI_2, NaI, KI, MeI, EtI, HgI_2 and Cu_2I_2 (410).
By reacting $PbCl_2$ with EtMgBr (155) in an H halide atm. (462).
By electrolyzing Et_3Al with a Pb anode; theoret yield (589).
By treating $(Et_3Pb)_2$ with EtI or EtBr at 60-70°, 98% yield (300).
By heating $(Et_3Pb)_2$ with activated C at 80-95° (659).
By treating $(Et_3Pb)_2$ with a silica-type catalyst (339).
By-prod. in the rxn. of Ph_2Pb dihalides with 4 equiv. Li followed by EtBr (29).
Et_4Pb is recovered from the rxn. mixts. contg. Et_4Pb, EtCl, Pb, and NaCl by vaporization followed by steam distn. (658).
Et_4Pb purified by vacuum distn. can be stored under a layer of p-$(PhCH_2NH)C_6H_4OH$ dissolved in $(HOCH_2CH_2)_3N$ (462).
Props.: Polymorphic crystals melting at six different temps. in the range 135.6-141.5°K (527, 528); Eight melting points were established in the range 135.6-142.9°K (529); log P= 9.428-2938/T (63a); log P= 8.1547-2184.6/(t + 230) (70); Viscosity data (677); Stability (757); Mass spectrum (111a, 426); Magnetic susceptibility and parachor (259, 261); Mol. wt. measurements in various solvents indicate monomeric structure (785); IR spectrum (236); IR spectrum of $(C_2D_5)_4Pb$ (236, 622); UV spectrum (439); Raman spectrum (193a, 236); Raman spectrum of $(C_2D_5)_4Pb$ (236); Ionization potential (135a). Thermal decompn. and halflife of the Et radical (810).
Rxns. with: Me_4Pb, Pr_4Pb, and mixts. of Me_4Pb and Pr_4Pb → redistribution of the alkyl groups (69, 71, 72, 73, 910).
Me_4Pb in the presence of $AlCl_3$, $FeCl_3$, $HgCl_2$, or $ZrCl_4$ → redistribution of the alkyl groups (69a).
Me_4Pb + $C_2H_4Br_2$ + $AlCl_3$ → a complex mixt. of alkyllead compds. and alkylenebis(trialkyllead) (568).
Me_2Hg → Hg shows a greater affinity than Pb for the Me group (76, 77).
Na in NH_3-Et_2O → Et_3PbNa (55, 171) and Et_2PbNa_2 (55).
PhLi (1:1) in Et_2O, the Ph group does not displace the Et group (150)
$NaNH_2$ in NH_3 followed by RCl → Et_3PbR (156).
Carboxylic acids in the presence of SiO_2 gel → Et_3Pb carboxylates (6, 169, 216) and Et_2Pb dicarboxylates (338).
Carboxylic acids in the presence or absence of SiO_2 gel → Et_2Pb dicarboxylates (217, 388).
Nitrophenols over SiO_2 gel → the corresponding Et_3Pb phenoxides (5).
CBr_4 or C_2Br_6 + O_2 → Et_2PbBr_2 and $COBr_2$ (220).
MeCOSH → $Et_2Pb(OCOMe)_2$, $Et_3PbSCOMe$, and $Pb(SCOMe)_2$ (217).
CH_2FCO_2H in Et_2O → $Et_3PbOCOCH_2F$ (471).
CCl_3CHO at 160-170° under N, followed by hydrolysis with AcOH → Et_3PbCl (157a).
Pb^{210} → radioactive Et_4Pb (142).

TETRAETHYLLEAD Et_4Pb (Cont'd.)

Rxns. with (Cont'd.): $Pb(OAc)_4$ (3:1) → Et_3PbOAc (289, 290).
Org. compds. contg. -SH and -OH groups → cleavage of the Et group (146a, 385).
C^{14}-labeled C_6H_6 in UV light → C^{14} exchange between the reactants (294).
$AgNO_3$ or $Cu(NO_3)_2 \cdot 3\ H_2O$ at $-80°$ → EtAg or EtCu and Et_3PbNO_3 (48, 142, 152, 479, 480, 481).
$HgCl_2$ in EtOH → Et_3PbCl (361).
HgO + H_3PO_4 (1:1:2) → Et_2Hg and $Et_2Pb[OP(O)(OH)_2]_2$ (613a).
HgO + H_3PO_4 (1:2:2) → $(EtHg)_2HPO_4 \cdot 2\ Et_2PbHPO_4$ (?) (613a).
$Hg(OAc)_2$ → $Pb(OAc)_4$ and EtHgAc (655).
$AuCl_3$ (3:1) → 3 Et_3PbCl + Au + 3 Et· (536).
$AuCl_3$ (2:1) → $EtPbCl$ + Et_2PbCl_2 + Au + 3 Et· (536).
$AuCl_3$ (3:2) → 3 $PbCl_2$ + Au + 6 Et· (536).
$AlCl_3$ or $EtAlBr_2$-Et_2AlBr → activated alkylaluminum compd. useful as catalyst for the polymerization of olefins (656).
$AlCl_3$ in $CHCl_3$ under N → $PbCl_2$ (362). A rxn. mechanism proposed (146).
$SnCl_4$ (2.5:1) → Et_2SnCl_2 (useful in deleading gasoline) (74a).
$SnCl_4$ (1:1) → Et_2PbCl_2 + Et_2SnCl_2 (279).
$C(NO_2)_4$ 65:177 → explosive mixt. (811).
NO_2 (1:4) at -50 to $+50°$ → $Et_2Pb(NO_3)_2$ (754).
PBr_3 → $EtPBr_2$ and other prods. (C_4H_{10}, EtBr) (764).
PCl_3 → $EtPCl_2$ (278).
$EtAsCl_2$ (1:2) → Et_2PbCl_2 + $2Et_2AsCl$ (278).
$AsCl_3$ (3:1) at $100°$ → $EtAsCl_2$ (278).
$AsCl_3$ (1:2) → Et_2PbCl_2 + $2EtAsCl_2$ (278).
$BiCl_3$ → Et_3Bi, Et_2BiCl (146), Et_3PbCl and $PbCl_2$ (363).
$SbCl_3$ (1:3) → $EtSbCl_2$ (278).
S and S compds. in fuels decrease the antiknock powder of Et_4Pb (332).
SO_2 in Et_2O contg. H_2O → Et_2PbSO_3 (217, 338).
H_2SO_3 → Et_2PbSO_3 (169).
Aq. HCl at r.t. → 86% Et_3PbCl (79, 216).
Satd. soln. of HCl or HI in CCl_4 → $PbCl_2$ and PbI_2, respectively (141).
Dry HCl in MePh at $90°$ → Et_2PbCl_2 (217, 469).
I_2 in Et_2O → Et_3PbI (79).
UF_4 at $170°$ → black solid (642).
$FeCl_3$ → Et_3PbCl, Et_2PbCl_2, and $FeCl_2$ (146).
$FeCl_3$ in $CHCl_3$ under N, followed by a treatment with H_2O and HCl → $PbCl_2$ (362).
$FeCl_2$, used for the removal of Et_4Pb from gasoline (825).
H_2PtCl_6 → Pt (146).

Use: 1. **As fuel additive**

Antiknock agent (46, 47, 49, 106, 114, 115, 119, 332, 333, 366, 398, 421, 436, 552, 582, 611, 640, 644, 645, 651, 671, 698, 712, 735, 755, 767, 794, 795, 812, 813, 819, 827, 835, 836, 846, 853, 857, 859, 860, 861, 870, 876, 894, 902).

Antiknock mechanism (106, 436, 556, 640, 654, 755, 756, 758, 795, 838, 861, 878).

Effect on the octane no. in various fuels (421, 552, 735, 822, 830, 840, 898).

Effect on the octane no. of gasoline in the presence of PhNHMe and BuOH (444).

Antiknock effect in fuels in the presence of the S compds. (364, 744, 849, 868, 869, 899).

Fuel additive in conjunction with Sn compds. and halogenated hydrocarbons (232).

Fuel additive in conjunction with P compds. (621, 696).

Aviation gasoline additive in conjunction with aromatic compds. of As, Sb, or Bi and $C_2H_4Br_2$ (580).

Effect on the ignition of isoöctane and triptane (257, 878).

Effect on the ignition of gasoline (401, 534, 893).

Effect on the precombustion of $n\text{-}C_7H_{16}$ (366, 371, 858, 877, 878).

Effect on the ignition temp. of various hydrocarbons (844, 892).

Effect on the combustion of $n\text{-}C_7H_{16}$ in the presence of halogenated hydrocarbons (374, 377).

Antiknock effect on the combustion of $n\text{-}C_6H_{14}$ (859).

Accelerating effect on the EtOH ignition (401, 415).

Effect on the combustion of $CH_4 + O_2$ in the presence of N_2 (418).

Antiknock effect on CO in the H-CO mixts. (615).

Inhibiting effect on the vapor-phase slow oxidation of $i\text{-}Pr_2O$ and on the ignition of Et_2O (88).

Inhibiting effect on the burning rate of the mixts. of $C(NO_2)_4$ and a fuel of the general formula C_nH_{2n-6} (194).

Inhibiting effect on the oxidation of HCHO (89).

Promoting effect on the peroxide formation in aviation gasoline (557).

Effect on the decompn. of peroxides during combustion of hydrocarbons (838).

Effect on the concentration of aldehydes and peroxides in fuels (567, 659).

Effect on the oxidation of pure hydrocarbons (299, 762, 763, 775, 776).

Effect on the oxidation of heptane and isoöctane (736, 878).

Effect on the decomposition of Me_3COOH at 170° (652).

Effect on the oxidation of MeOH (879).

Inhibiting effect on the C deposition in engines (828).

Effect on the air requirements for the suppression of the C deposit formation (784).

Retardation of the Et_4Pb dissocn. to $Et_3Pb\cdot$ and $Et\cdot$ during a liquid-phase oxidation of the aviation gasoline (763a).

Decomposition of Et_4Pb in a combustion chamber (693) in fuels (444a).

TETRAETHYLLEAD Et_4Pb (Cont'd.)

Use (Cont'd.): Decomposition of Et_4Pb in gasoline at $100°/20-60$ atm. → mainly PbO; $100°$/various pressures → $PbBr_2$; $100°/60$ atm. → Et_3PbBr (814).
Thermal decomposition in hydrocarbons at $200-300°$ (105a, 444a).
Composition of the combustion prods. of a mixt. of Et_4Pb, C_8H_{18}, and air (102, 105a).
Effect upon the flame propagation in H_2-air; CH_4-air; $n-C_4H_8$-air; and C_3H_8-air mixts. (837, 891).
Stabilization of leaded fuels (557, 671, 672).
Effect on the formation of the detonation wave (890).

2. As catalyst
In the sulfochlorination of aliphatic and aromatic hydrocarbons with $SO_2 + Cl_2$ (670, 907).
For the chlorination of paraffinic hydrocarbons (900).
For the condensation of arylalkanes (792).
For the addition of H_2S and mercaptans to olefins (841).
For the oxidation of dialkylbenzenes by air (852).
For telomerization of C_2H_4 with HCl (848).
For alkylation of hydrocarbons (905).
For the polymerization of rosins (824).
For the copolymerization of C_4H_6 with $PhCH:CH_2$ (654a).
For the polymerization of olefins (842).
In conjunction with Cu^{++} for the polymerization rxns. (48).
In conjunction with Me_2CHCN for the polymerization of C_2H_4 (668).
In conjunction with $AlCl_3$ or $EtAlBr_2 \cdot Et_2AlBr$ for the polymerization of olefins (913).

3. As filler for Geiger counters (895).
4. As additive to the lubricating oils contg. S (673).

Physiological props.: Toxicity (660, 831, 865, 872, 882, 897).
Toxicity and symptoms (686, 713, 714, 715, 716, 717, 718, 719, 720, 721, 739, 742, 766, 797, 818, 820, 821, 823, 829, 839, 845, 854, 864, 874, 884, 885, 886, 888).
Toxicity, antidotes, and therapy (624, 626, 770, 793, 866, 881).
Effect on the central nervous system (662a).
Concentration in the animal body (783).
Inhibition of cholinesterase (845).
Penetration through the skin (843, 863).
Absorption through the lungs (863, 880).
Legal limits for the Et_4Pb concentration in air (617).

TETRAPROPYLLEAD Pr_4Pb

Prepn.: Two-step process: a mixt. of NaPb alloy and Mg is reacted with PrCl under pressure (75 psi) at $45-85°$, then PrCl mixed with Et_2O is introduced, and the rxn. if contd. under the same condns. (81). Ternary alloys of Na-Pb-Mg contg. up to 22.5% Na and 2.5-25% Mg may be used as the source of Pb (82). Ternary alloys of Na-Pb-Mg contg. 0.5-15 Na and 3-19.5% Mg with the sum of Na and Mg between 5 and 20% may be used as the source of Pb (83).

By treating (Pr$_3$Pb)$_2$ with a silica-type catalyst at 0-110° (339).
By reacting Na-Pb alloy (78% Pb) with PrI in the presence of C$_5$H$_5$N
  under N on a steam bath; 45% yield (469).
Props.: b. 80°/0.4mm (469); Viscosity data (677); Mass spectrum
  (426); Magnetic susceptibility and parachor (259, 261).
Rxns. with: Et$_4$Pb, redistribution rxn. (73).
Me$_4$Pb + Et$_4$Pb, redistribution rxn. (73).
HCl gas in petr. ether → Pr$_3$PbCl (469).
Use: Catalyst in the prepn. of polyesters (628).

TETRAISOPROPYLLEAD   i-Pr$_4$Pb
Prepn.: By reacting Mg$_2$Pb (18-22% Mg) with excess i-PrBr at
  80-120°/14 psi in the presence of aliphatic ethers, tert.
  aliphatic and aromatic amines, C$_5$H$_5$N, or tetraalkyl- or aryl-
  ammonium iodides (495, 496, 497).

TETRABUTYLLEAD   Bu$_4$Pb
Prepn.: By reacting Ph$_4$Pb with BuLi (1:4) in Et$_2$O (149).
By treating (Bu$_3$Pb)$_2$ with a silica-type catalyst at 0-110° (339).
By reacting Mg$_2$Pb with BuI at 80-120°/elevated pressure in the
  presence of aliphatic ethers, tert. aliphatic or aromatic
  amines, or tetraalkyl- or aryl-ammonium iodides (495).
Props.: Viscosity data (677). Magnetic susceptibility and
parachlor (259, 261).
Rxns. with: FeCl$_3$ in CHCl$_3$ → Bu$_3$PbCl (362).
HCl gas → Bu$_3$PbCl (469).
Use: Catalyst in the prepn. of polyesters (628).

TETRAISOBUTYLLEAD   i-Bu$_4$Pb
Prepn.: By heating i-BuBr with Pb-Na alloy (20% Na) in the
  presence of pyridine (216).
By reacting Mg$_2$Pb (18-22% Mg) with excess i-BuBr at 80°/6 psi
  in the presence of aliphatic ethers, tert. aliphatic or
  aromatic amines, pyridin, or tetraalkyl- or aryl-ammonium
  iodides (495).
Props.: Oily liquid, decompg. during distn. in vacuo (216).
Mass spectrum (426).
Rxns. with: HCl gas in light petr. → i-Bu$_3$PbCl (216).
Carboxylic acids → i-Bu$_3$Pb carboxylates (216).
Sulfonic acids → i-Bu$_3$Pb sulfonates (216).

TETRAPENTYLLEAD   (n-C$_5$H$_{11}$)$_4$Pb
Prepn.: By reacting C$_5$H$_{11}$Cl with Pb (4:3) (618).
Props.: Viscosity data (677); Magnetic susceptibility (261).
Use: Catalysts for the prepn. of polyesters (628).
Antiknock agent (816).

TETRACYCLOHEXYLLEAD $(n-C_6H_{11})_4Pb$
Use: Catalyst for polymerization of vinyl compd. (690).

TETRAHEPTYLLEAD $(n-C_7H_{15})_4Pb$
Props.: Viscosity data (677); Magnetic susceptibility and parachor (259).

TETRAKIS(TETRADECYL)LEAD $(n-C_{14}H_{29})_4Pb$
Prepn.: By reacting $(n-C_{14}H_{29})_3PbCl$ with $n-C_{14}H_{29}MgBr$ (369).
Props.: m. 31° (369).

TETRACETYLLEAD $(n-C_{16}H_{33})_4Pb$
Prepn.: By reacting $(n-C_{16}H_{33})_3PbCl$ with $n-C_{16}H_{33}MgBr$ (369).
Props.: m. 42° (369).

TETRAPHENYLLEAD Ph_4Pb
Prepn.: By reacting $PbCl_2$ with PhLi at -5°; rxn. mechanism:
$PbCl_2 + 3PhLi \to Ph_3PbLi$; $2Ph_3Pb + PhLi \to Ph_4Pb + Ph_3PbLi$ (54, 306).
By pyrolyzing $(Ph_3Pb)_2$ at 100° in a sealed tube or at 60-65° in C_5H_5N; 93% yield (147).
By reacting $(Ph_3Pb)_2$ with a mixt. of Mg and MgI_2; 56% yield (147).
By treating Ph_2PbF_2 with Ca in liq. NH_3 (160).
By adding PhLi (3) to a mixt. of $PbCl_2$ (1) and PhI (1) in Et_2O and heating the mixt. under reflux for 2 hrs.; 80% yield (166, 306, 535).
By adding PhMgBr (3) to $PbCl_2$ (1) and PhI (1) in Et_2O and refluxing the rxn. mixt. for 5 hrs.; 37% yield (166, 306).
By reacting $PbCl_2$ with PhMgBr; the prod. contains Ph_4Pb and $(Ph_3Pb)_2$ (329).
By disproportionation of Ph_2PbCl_2 or $Ph_2Pb(OAc)_2$ in the presence of $H_2NNH_2 \cdot H_2SO_4$ of Na_2CO_3 in EtOH (383).
By reacting PhN_2BF_4 with Pb and Na-Pb (10% Na) in Me_2CO at 4-6°; 30% yield (389, 393).
By refluxing PhMgBr and $PbCl_2$ (1.5:0.65) in Et_2O-MePh; 82-83% yield along with a small amt. of Ph_3PbBr and Ph_2 (483).
By reacting PhLi and Pb powder in Et_2O under N; 27% yield (537); 5.2% yield (538).
By refluxing Pb powder with PhBr and Li in Et_2O; 23% yield (538).
By refluxing PhLi with Pb-Hg in Et_2O; 8.5% yield (538).
By-prod. in the rxn. of PhLi with a double amalgam (Pb-Sn-Hg 1:1:2 atom ratio) in Et_2O under reflux; 2.6% yield along with 46.5% Ph_4Sn (538).
Props.: m. 224-225° (147); m. 227-229° (166); m. 226° (389, 538); m. 226-227° (537); d. 1.745 (144); Soly. in C_6H_6 (723); Crystal structure (234, 335, 587, 588); Lattice consts. and symmetry (144, 587, 588); Interatomic distances (235); UV spectrum (373); IR spectrum (579); Raman spectrum (579); Diffraction data (137); Diamagnetic anomaly (406); Cocrystallization with Ph_4Sn and Ph_4Si (723).

Rxns. with: Alc. HCl followed by alc. KOH → Ph$_3$PbOH (35).
I$_2$ (1:1) in CHCl$_3$ → Ph$_3$PbI (43, 147).
AlCl$_3$ → Ph$_3$PbCl + PhAlCl$_2$; Ph$_4$Pb + PhAlCl$_2$ → Ph$_3$PbCl + Ph$_2$AlCl;
Ph$_3$PbCl + AlCl$_3$ → Ph$_2$PbCl$_2$ + PhAlCl$_2$ (146).
AgNO$_3$ or Cu(NO$_3$)$_2$ → Ph$_3$PbNO$_3$, Ag or Cu and Ph· (152, 524).
N$_2$O$_4$ in CHCl$_3$ → Ph$_2$Pb(NO$_3$)$_2$ (156, 372).
Na in NH$_3$-Et$_2$O → Ph$_3$PbNa (55, 171).
BuLi (1:4) → Bu$_4$Pb + 4PhLi (149).
Alkyl-Li → displacement of the Ph group (150).
H$_2$ at 200°/60 atm. → Pb + C$_6$H$_6$ (143).
Tetraalkyl- or tetraaryllead → redistribution rxn. (73).
Carboxylic acids → Ph$_3$Pb carboxylates (132).
Pb(OAc)$_4$ (1:1) → Ph$_2$Pb(OAc)$_2$ (289, 290).
Carboxylic acids at 50-100° → Ph$_2$Pb dicarboxylates (295, 296, 402, 404).
AcOH under reflux followed by H$_2$S → Ph$_2$PbS (329).
AcCl in the presence of AlCl$_3$ → Ph$_3$PbCl + PhAc (355).
HSO$_3$Cl → Pb$_2$PbCl$_2$ (372).
H$_2$SO$_4$ → prod. sol. in H$_2$O (372).
CCl$_4$ in the presence of Bz$_2$O$_2$ under reflux → 60.8% Ph$_2$PbCl$_2$ and 5.2% Ph$_2$Pb(OBz)$_2$ (399); the rxn. does not occur in the presence of Ac$_2$O$_2$ (400).
Hg(OAc)$_2$ in CHCl$_3$ → Ph$_2$Sn(OAc)$_2$ (404).
HgO + RCO$_2$H → Ph$_2$Sn(OOCR)$_2$ (404).
CoCl$_2$ or FeCl$_3$ in EtOCH$_2$CH$_2$OH → Ph$_2$PbCl$_2$ (433).
NiCl$_2$ or CuCl$_2$ → PbCl$_2$ (433).
MnCl$_2$ → Ph$_3$PbCl (433).
HNO$_3$ followed by Na halides → Ph$_2$Pb dihalides (483).
Li in Et$_2$O → Ph$_3$PbLi + PhLi (537).
o-HSC$_6$H$_4$CO$_2$H in EtOH at 100° → C$_6$H$_4$SPb(O)OCO (689).
N$_2$O$_3$ (1:4) → Ph$_2$Pb(NO$_3$)$_2$ + 2PhN$_2$NO$_3$ (350a).
Phenols → cleavage of the Ph groups (296a, 689a).
Me$_3$NHCl in EtOH at 130° → Ph$_3$PbCl, at 150° → Ph$_2$PbCl$_2$ (297a).
p-HOC$_6$H$_4$NO$_2$ at 150° → C$_6$H$_4$(OH)NO$_2$·2 Pb(OC$_6$H$_4$NO$_2$)$_2$ (689a).
p-MeC$_6$H$_4$OH at 50° → (MeC$_6$H$_4$O)$_2$Pb·2 H$_2$O (689a).
2,4,6-Cl$_3$C$_6$H$_2$OH → PbCl$_2$ (689a).
PhSH at 130-150° → (PhS)$_2$Pb (298a).
p-MeC$_6$H$_4$SH at 150° → (p-MeC$_6$H$_4$S)$_2$Pb (298a).
Use: Catalyst for sulfochlorination of non-aromatic hydrocarbons with SO$_2$ and Cl$_2$ (670).
Catalyst for the polymerization of vinyl monomers (690).
Curing agent for resinous siloxanes (697).
Stabilizer for vinyl halide polymers (578).

TETRA-m-TOLYLLEAD   (m-MeC$_6$H$_4$)$_4$Pb
Prepn.: By pyrolyzing [(m-MeC$_6$H$_4$)$_3$Pb]$_2$ in xylene; 90% yield (43, 147).
Props.: m. 122-123° (43, 147); Crystal structure and lattice consts. (356, 548).
Rxn. with: Br$_2$ (1:1) in C$_5$H$_5$N → (m-MeC$_6$H$_4$)$_3$PbBr (143, 147).

TETRA-o-TOLYLLEAD $(o\text{-MeC}_6H_4)_4Pb$

Prepn.: By pyrolyzing $[(o\text{-MeC}_6H_4)_3Pb]_2$ in xylene; 37% yield (147).
Props.: m. 198-200° (147).

TETRA-p-TOLYLLEAD $(p\text{-MeC}_6H_4)_4Pb$

Prepn.: By pyrolyzing $[(p\text{-MeC}_6H_4)_3Pb]_2$ in xylene; 49% yield (147).
By reacting $[(p\text{-MeC}_6H_4)_3Pb]_2$ with $Mg + MgI_2$; 49% yield (147).
By reacting Na-Pb alloy (10% Na) with $p\text{-MeC}_6H_4N_2BF_4$ in Me_2CO at 4°; 15.6% yield (393).
By refluxing Pb powder with $p\text{-MeC}_6H_4Br$ and Li in Et_2O; 10% yield (538).
Props.: m. 240° (393); m. 240-241° (538).

TETRABENZYLLEAD $(PhCH_2)_4Pb$

Prepn.: By reacting $PbCl_2$ with $PhCH_2MgCl$ (1:4.4) in Et_2O; 41% yield (37).
Props.: Large yellow crystals or shiny, yellow prisms, m. 65.5-66.5°; sol. in org. solvents, sensitive to air (37).

SYMMETRICAL, FUNCTIONALLY SUBSTITUTED TETRAORGANOLEAD COMPOUNDS

Tetravinyllead was prepared by reacting K_2PbCl_6, $(NH_4)_2PbCl_6$, $PbCl_4$, or $PbCl_2$ with vinylmagnesium bromide or chloride in tetrahydrofuran.

Attempts were also made to prepared tetraallyllead from lead dichloride and allylmagnesium bromide, but the compound was not isolated in the pure state.

Tetrakis(haloaralkyl)lead compounds were prepared by reacting lead dichloride with haloaralkylmagnesium halides in ether.

Tetrakis(alkoxyaryl)lead compounds were prepared from hexakis-(alkoxyaryl)dilead compounds by pyrolysis in xylene or by treatment with a mixture of magnesium and magnesium iodide and with iodine in chloroform.

Tetrakis(aminoaryl)lead compounds were prepared by reacting lead dichloride with aminoaryllithium in a ratio of 1:3 and treating the resulting lithium tris(aminoaryl)plumbide with the corresponding aminoaryl iodide.

TETRAVINYLLEAD $(CH_2:CH)_4Pb$

Prepn.: By reacting K_2PbCl_6 with $CH_2:CHMgBr$ in C_4H_8O; 69.5% yield (622).
By reacting $PbCl_4$ with $CH_2:CHMgBr$ in C_4H_8O (680a).
By reacting $(NH_4)_2PbCl_6$ or $PbCl_4$ with $CH_2:CHMgBr$ in C_4H_8O at $-20°$ and completing the rxn. under reflux; up to 17.2% yield (710).
By reacting $PbCl_2$ with $CH_2:CHMgCl$ in C_4H_8O; 85% yield (711).
Props.: b. $50°/4mm$; IR spectrum (622); b. $74-77°/12mm$ (710); b. $69-70°/11mm$; $d^{20}D$ 1.5462; d_{20} 1.7882; stable in H_2O (711).
Rxns. with: Cl_2 or $HCl \rightarrow (CH_2:CH)_2PbCl_2$ and $(CH_2:CH)_3PbCl$ (680a).
$AcOH \rightarrow (CH_2:CH)_3PbOAc$ (680a).
$CCl_3CO_2H \rightarrow (CH_2:CH)_3PbO_2CCCl_3$ (680a).
MX_3 (M = P, As, Sb, Bi; X = Cl or Br) in CCl_4 under reflux $\rightarrow$ $(CH_2:CH)_2PbX_2$ (711).

TETRAALLYLLEAD (?) $(CH_2:CHCH_2)_4Pb$

Prepn.: By reacting powdered $PbCl_2$ with $CH_2:CHCH_2MgBr$ under reflux; the change in color of the rxn. mist. is explained as the formation of diallyl and triallyllead; however, none of the compds. was isolated (554).

TETRAKIS(HALOARALKYL)LEAD COMPOUNDS

The following compounds were prepared from $PbCl_2$ and $RMgBr$ in Et_2O.

R_4Pb	Molar ratio $PbCl_2:RMgBr$	Yield %	Properties	Ref.
$(o-FC_6H_4CH_2)_4Pb$	1:4.3	30	Yellow needles, m. 74-75°; magnetic data (1)	37
$(o-ClC_6H_4CH_2)_4Pb$	1:4.3	33	Yellow prisms, m. 99.5°(1)	37
$(o-BrC_6H_4CH_2)_4Pb$	1:4.3	18	Yellow prisms, m. 122°(1,2)	37

(1) The compounds are very sensitive to air.
(2) Reactions with $I_2 \rightarrow o-BrC_6H_4CH_2I$ and PbI; with $KMnO_4 \rightarrow o-BrC_6H_4CO_2H$.

TETRAKIS(ALKOXYARYL)LEAD COMPOUNDS

Compound	Prepared from	Yield %	Properties	Ref.
$(o\text{-MeOC}_6H_4)_4Pb$	$[(o\text{-MeOC}_6H_4)_3Pb]_2$ pyrolysis in xylene	53	m. 148-149°	43, 147
$(p\text{-MeOC}_6H_4)_4Pb$	$[(p\text{-MeOC}_6H_4)_3Pb]_2$ pyrolysis in xylene	82	m. 145-146° (1)	43, 147, 148
	$[(p\text{-MeOC}_6H_4)_3Pb]_2$ + Mg + MgI$_2$	58		43, 147
	$[(p\text{-MeOC}_6H_4)_3Pb]_2$ + I$_2$ in CHCl$_3$	50.6		43, 147
$(o\text{-EtOC}_6H_4)_4Pb$	$[(o\text{-EtOC}_6H_4)_3Pb]_2$ pyrolysis in xylene	63	m. 219-200°	43, 147
$(p\text{-EtOC}_6H_4)_4Pb$	$[(p\text{-EtOC}_6H_4)_3Pb]_2$ pyrolysis in xylene	91	m. 110°	43, 147
	$[(p\text{-EtOC}_6H_4)_3Pb]_2$ + Mg + MgI	56.7		147

(1) Rxn. with HCl in CHCl$_3$ → $(p\text{-MeOC}_6H_4)_3PbCl$ + $(p\text{-MeOC}_6H_4)_2PbCl_2$ (148).

TETRAKIS(p-DIMETHYLAMINOPHENYL)LEAD $(p\text{-Me}_2NC_6H_4)_4Pb$

Prepn.: By reacting PbCl$_2$ with $p\text{-Me}_2NC_6H_4Li$ (1:3) in Et$_2$O at -10° and treating the resulting $(p\text{-Me}_2NC_6H_4)_3PbLi$ with $p\text{-Me}_2NC_6H_4I$ in Et$_2$O under reflux; 63% yield (166, 535).
Props.: White crystals, m. 187-189° (166).
Salt: Tetramethiodide, $Pb(C_6H_4NMe_3I)_4$, white needles, m. 187-189°, prepared by warming $(p\text{-Me}_2NC_6H_4)_4Pb$ with MeI (1:4) in MeOH (165).

ASYMMETRICAL, NON-FUNCTIONAL TETRAORGANOLEAD COMPOUNDS

Tetraorganolead compounds of the general formula R$_3$PbR' are prepared:

1. By reacting triorganolead halides, R$_3$PbX, with organomagnesium halides, R'MgX, or with organolithium compounds, R'Li.
2. By reacting lithium or sodium triorganoplumbides, R$_3$PbM, with hydrocarbyl halides, RX. The alkali metal triorganoplumbides are prepared by cleaving hexaorganodilead compounds with alkali metals in liquid ammonia or by treating lead dichloride with organolithium compounds in a molar ration of 1:3.

Tetraorganolead compounds of the general formula $R_2PbR'_2$, are prepared:

1. By reacting diorganolead dihalides, R_2PbX_2, with organomagnesium halides, $R'MgX$, or with organolithium compounds, $R'Li$.
2. By treating diorganolead dihalides, R_2PbX_2, with lithium in a molar ratio of 1:4 and treating the resulting dilithium diorganoplumbides with organic halides.
3. By treating powdered lead with an equimolar mixture of an organolithium compound, RLi, and a different organic halide, $R'X$.

Tetraorganolead compounds of the general formula $R_2R'R''Pb$ are prepared by reacting sodium triorganoplumbides of the general formula $R_2R'PbNa$ with organic halides, $R''X$.

ASYMMETRICAL TETRAORGANOLEAD COMPOUNDS R₃PbR'

R_3PbR'	Prepared from	Yield %	Properties	Ref.
Me_3PbEt			$\log P = 7.2760 - 1602.5/(t + 230)$ (1,2)	70
$Me_3PbCHMe_2$			Redistribution rxns. with Me_2Pb-$(CHMe_2)_2$ and with Me_2Pb-$(CH_2CHMe_2)_2$	73 426
Me_3PbBu			Mass spectrum	68
$Me_3PbCHMeEt$	Me_3PbBr or Me_3PbI + $EtCH(Me)MgBr$	50	b. 59°/13mm; d^{20}_4 1.7011; n^{20}_D 1.5133	68
			Mass spectrum	426
Me_3PbCMe_3	Me_3PbI + Me_3CMgCl	88	b. 47-47.2°/13mm; m. 5.7°; d^{20}_4 1.6615; n^{20}_D 1.5089; camphor-like odor; cubic crystal structure (3)	68
			Mass spectrum	426
Me_3PbPh			UV spectrum	59
Et_3PbMe			$\log P = 7.8768 - 2000.5/(t + 230)$ (4,5)	70
			Raman and IR spectra	236
			Mass spectrum	426
$Et_3PbCHMeEt$	Et_3PbNa + $EtMeCHBr$	82	b. 108-109°/15mm; d^{20}_4 1.5318	171
Et_3PbPh	Et_3PbNa + PhI or $PhBr$ (7)	77	b. 127-131°/7mm; b. 137-140°/13mm; d^{21}_4 1.5915; n^{20}_D 1.5532, 1.5752 (6)	171, 353
			Dipole moment	59
			UV spectrum	160
Et_3PbCH_2Pb-$(cyclo$-$C_6H_{11})_3$-$PbCH_2Ph$	Et_3PbLi (or Ca) + $PhCH_2Cl$ $(C_6H_{11})_3PbNa$ + $PhCH_2Cl$	40 39	b. 147-151°/13mm Yellow needles, m. 228° (decompn.)	43, 147
Ph_3PbEt	Ph_3PbNa + $EtBr$ Ph_3PbLi + $EtBr$ (5)	83 45	White crystals, m. 39° m. 48-50° (8)	134, 171 166, 535
Ph_3PbPr	Ph_3PbLi + $PrBr$ $PbCl_2$ + $PhLi$ (1:3) + $PrBr$	58 63	m. 68-69° m. (9)	166 535

Compound	Preparation	Yield (%)	Properties	Refs.
Ph_3PbCH_2Ph	$Ph_3PbLi + PhCH_2Cl$	69	m. 91-93°	54, 166, 535
	$(Ph_3Pb)_2 + Li$ (1:2) + $PhCH_2Cl$	72	m. 91-92°	160, 306
	$(Ph_3Pb)_2 + Na$ (1:2) + $PhCH_2Cl$	80		147, 171
$Ph_3PbCHPh_2$	$Ph_3PbNa + Ph_2CHCl$	54	m. 122°	43, 147
Ph_3PbCPh_3 (?)	$Ph_3PbCl + Ph_3CMgCl$		Yellow crystals, m. 196-197°; dissoc. in C_6H_6 soln.	43
$(p\text{-}MeC_6H_4)_3PbCH_2Ph$	$(p\text{-}MeC_6H_4)_3PbNa + PhCH_2Cl$	53	m. 81-82°	43, 147
	$PbCl_2 + p\text{-}MeC_6H_4Li$ (1:3), followed by $PhCH_2Cl$	65	m. 81-83°	166, 535
$Ph_3Pb(3\text{-indenyl})$	$Ph_3PbCl + 3\text{-}C_9H_7Na$	75	m. 122° (10)	614
$Ph_3Pb(9\text{-fluorenyl})$	$Ph_3PbCl + 9\text{-}C_{13}H_9Li$	75	m. 118-120° (11)	614
$Ph_3Pb\text{-}(9\text{-phenanthryl})$	$Ph_3PbCl + 9\text{-}C_{14}H_9MgBr$	57	m. 169-171°	160, 306
$Ph_3Pb(7\text{-benz[a]-anthracenyl})$	$Ph_3PbCl + 7\text{-}C_{18}H_{11}Li$	52	m. 204-205° (12)	160, 306

1) Raman spectrum (236); Mass spectrum (426).
2) Rxns. with: $Et_3PbMe \to$ redistribution of the radicals (73); with $HCl \to Me_2EtPbCl$ (79).
3) Redistribution rxn. in the presence of $AlCl_3$ (73).
4) Rxn. with $Me_3PbEt \to$ redistribution of the alkyl groups (73).
5) Rxn. with HCl at r.t. → oily prod. (79).
6) Rxn. with Na in $NH_3 \to Et_2PhPbNa + C_2H_6 + NaNH_2$ (55).
7) By-prod. in the rxn. of Ph_2PbX_2 with Li (1:4) followed by EtBr (29).
8) Rxn. with HCl in petr. ether → $EtPh_2PbCl$ (160).
9) Useful as stabilizer for vinyl resins (578).
10) Rxn. with aq. alc. → hydrolysis (614).
11) Rxn. with HCl in $C_6H_6 \to Ph_2(C_{13}H_9)PbCl$ (614).
12) Rxn. with HCl in $CHCl_3 \to Ph_2PbCl_2$ (160, 306).

ASYMMETRICAL TETRAORGANOLEAD COMPOUNDS $R_2PbR'_2$

$R_2PbR'_2$	Prepared from	Yield %	Properties	Ref.
Me_2PbEt_2			$\log_{10} P = 7.5903 - 1810.7/(t + 230)$ (1)	70
			Raman spectrum	236
			Mass spectrum	426
			(2)	73
			(2)	73
$Me_2Pb(CHMe_2)_2$				
$Me_2Pb(CH_2CHMe_2)_2$				
$Me_2Pb(CHMeEt)_2$	Me_3PbI or Me_3PbBr + MeEtCH$_2$MgBr	26		68
Et_2PbPr_2	Et_2PbCl_2 + PrMgBr		b. 70-72°/4mm (3, 4)	217
Et_2PbPh_2	Ph_2PbX_2 + Li (1:4) + EtBr	60	b. 158-161°/2mm; n^{20}_D 1.6150; d^{20}_{20} 1.6401 (5)	29
			(6)	535
$(cyclo-C_6H_{11})_2$-$PbPh_2$	Pb powder + PhLi + EtI			160, 306
$Ph_2Pb(C_6H_4Me-p)_2$			(7)	149
$Ph_2Pb(3-indenyl)_2$	Ph_2PbCl_2 + 3-C_9H_7Li	30	m. 107-110° (8)	614
Ph_2Pb-(9-fluorenyl)$_2$	Ph_2PbO_2 + 9-$C_{13}H_9Li$		m. 138-140°	614
Ph_2Pb-(9-phenanthryl)$_2$	Ph_2PbBr_2 + 9-$C_{14}H_9Li$	68	m. 208-210° (9)	160, 306
$Ph_2Pb(-7-benz[a]-$anthracenyl)$_2$	Ph_2PbCl_2 + 7-$C_{18}H_{11}Li$	3.6	m. 295-296°	160, 306

(1) Rxn. with HCl → oily prod. (79).
(2) Rxn. with $Me_3PbCHMe_2$ → redistribution of the alkyl groups (73).
(3) Rxn. with carboxylic acids → Et_2PrPb carboxylates (217, 338).
(4) Rxn. with dry HCl in petr. ether → $EtPr_2PbCl$ (217, 338).
(5) Rxn. with dry HCl in petr. ether → $Et_2PhPbCl$ (306).
(6) Rxn. with HCl in petr. ether → (cyclo-C_6H_{11})$_2$PhPbCl (160, 306).
(7) Rxn. with BuLi in Et_2O → cleavage of the aryl groups, mainly MeC_6H_4 (149).
(8) Rxn. with aq. alc. →hydrolysis (614).
(9) Rxn. with HCl in $CHCl_3$ → Ph_2PbCl_2 (160, 306).

ASYMMETRICAL TETRAORGANOLEAD COMPOUNDS $R_2R'R''Pb$

$R_2R'R''Pb$	Prepared from	Properties	Ref.
$Et_2MeBuPb$	$Et_2MePbNa + BuCl$	b. $67°/5mm$; n^{20}_D 1.5125; d^{20}_4 1.5817	55
$Et_2Me(EtMeCH)Pb$	$Et_2MePbNa + EtMeCHCl$	b. $87°/10mm$; n^{20}_D 1.5180; d^{20}_4 1.5892	55
$Et_2MePhPb$	$Et_2MePbNa + PhCl$	b. $132°/15mm$; d^{20}_4 1.7025	55

ASYMMETRICAL, HOMOFUNCTIONAL TETRAORGANOLEAD COMPOUNDS

Tetraorganolead compounds containing at least one alkenyl group bound to the metal were prepared:

1. By reacting organolead halides with alkenylmagnesium halides or with alkenyllithium compounds.
2. By reacting sodium triorganoplumbides with alkenyl halides.

Tetraorganolead compounds containing at least one halogen-substituted radical were prepared:

1. By reacting organolead halides with diazomethane in the presence of copper bronze to yield the corresponding chloromethyl derivatives or with haloarylmagnesium halides to yield the corresponding haloaryl derivatives.
2. By reacting lithium triorganoplumbides with ω-chloroalkyl bromides to yield the corresponding ω-chloroalkyl derivatives or with bromoaralkyl chlorides to yield the corresponding bromoaralkyl derivatives.

Asymmetrical tetraorganolead compounds containing a basic nitrogen group were prepared:

1. By reacting lithium or sodium triorganoplumbides with aminoalkyl halides.
2. By reacting triorganolead halides with aminoalkyllithium derivatives.
3. By reacting triorganolead halides with aminoarylmagnesium halides or with aminoaryllithium compounds.
4. Triphenyl(2-pyridyl)lead was prepared by reacting diphenyl-(2-pyridyl)lead iodide with phenylmagnesium bromide.

Tetraorganolead compounds containing one radical with a carboxy or carbalkoxy group were prepared:

1. By oxidizing tetraorganolead compounds containing one alcoholic hydroxy group with potassium permanganate. The resulting carboxylic acid was converted to the methyl ester by treating it with diazomethane.
2. By pyrolyzing triorganolead carbalkoxycarboxylates.

Tetraorganolead compounds containing one alcoholic or phenolic hydroxy group or the corresponding ether group were prepared by reacting triorganolead halides, e.g. Ph_3PbCl, with hydroxyaryl-, hydroxyalkylaryl-, or alkoxyaryl-lithium or -magnesium halides. Tris(alkoxyaryl)aralkyllead compounds were prepared from tris(alkoxyaryl)lead halides and aralkylmagnesium halides or from sodium tris(alkoxyaryl)plumbides and aralkyl halides.

COMPOUNDS CONTAINING OLEFINIC GROUPS

Compound	Prepared from	Yield %	Properties	Ref.
$(CH_2:CH)_2PbEt_2$	Et_2PbCl_2 + $CH_2:CHMgBr$	74.8	b. 30°/1mm; IR spectrum	622
$(CH_2:CH)_2PbPh_2$	Ph_2PbCl_2 + $CH_2:CHMgBr$			680a
$CH_2:CHCH_2PbEt_3$	Et_3PbNa + $CH_2:CHCH_2Cl$	72	b. 85-86°/8mm; n^{22}_D 1.5410; d^{23}_4 1.6056	171
$Me_2C:CHPbEt_3$	Et_3PbCl + $Me_2C:CHLi$		b. 99-101°/12mm; thermodynamic data; IR spectrum (1,2)	191
$CH_2:CHPbPh_3$	Ph_3PbCl + $CH_2:CHMgBr$			191, 680a
$CH_2:CHCH_2PbPh_3$	Ph_3PbCl + $CH_2:CHCH_2MgBr$	73	Decomp. 165-175°	178

(1) Rxn. with Br_2 in $CHCl_3$ → Et_3PbBr + $Me_2CBrCHBr$ (191).
(2) Rxn. with $AgNO_3$ → $Me_2C:CH_2$, $(Me_2C:CH)_2$, and Et_3PbNO_3 (191).

HALOGEN-SUBSTITUTED TETRAORGANOLEAD COMPOUNDS

Compound	Prepared from	Yield %	Properties	Ref.
$ClCH_2PbEt_3$	Et_3PbCl + CH_2N_2 (over Cu)		b. 65.5-66.5°/3mm; d_{20} 1.7917; n^{20}_D 1.5434	576
$p\text{-}BrC_6H_4PbEt_3$	Et_3PbCl + $p\text{-}BrC_6H_4MgBr$	30	b. 143°/0.002mm; d^{20}_{27} 1.8586; n^{20}_D 1.5968 (1)	156, 372
$Cl(CH_2)_3PbPh_3$ (?)	Ph_3PbLi + $Br(CH_2)_3Cl$		m. 68-71°	165
$p\text{-}ClC_6H_4PbPh_3$			(2)	150
$p\text{-}BrC_6H_4PbPh_3$	Ph_3PbCl + $p\text{-}BrC_6H_4MgBr$	43	m. 121°	583
$p\text{-}BrC_6H_4CH_2PbPh_3$	Ph_3PbLi + $p\text{-}BrC_6H_4CH_2Cl$	67	Yellow crystals, m. 66-68°	165, 535
$(ClCH_2)_2PbEt_2$	Et_2PbCl_2 + CH_2N_2 (over Cu)		b. 96°/2mm; d_{20} 1.9890	576
$(p\text{-}ClC_6H_4)_2PbPh_2$			(3)	150

(1) Rxn. with Cu-Mg → $Et_3PbC_6H_4MgBr$ (372).
(2) Rxn. with BuLi (1:1) in Et_2O, followed by CO_2 → 24% BzOH + 76% $p\text{-}ClC_6H_4CO_2H$ (150).
(3) Rxn. with BuLi (1:1) in Et_2O, followed by CO_2 → 2% BzOH + 98% $p\text{-}ClC_6H_4CO_2H$ (150).

TETRAORGANOLEAD COMPOUNDS CONTAINING BASIC NITROGEN GROUPS

Compound	Prepared from	Yield %	Properties	Ref.
$Et_2NCH_2CH_2PbEt_3$	$Et_3PbNa + Et_2NCH_2CH_2Cl$	29	b. 85-90°/0.01mm; d^{25}_{25} 1.459; n^{20}_D 1.5235	156
$Et_2N(CH_2)_3PbEt_3$	$Et_3PbNa + Et_2N(CH_2)_3Cl$	54	b. 90°/0.001mm; d^{28}_{28} 1.432; n^{20}_D 1.5229	156
$Et_2NCH(Et)PbEt_3$	$Et_3PbNa + Et_2NCH(Et)Cl$			372
$p-Me_2NC_6H_4PbEt_3$	$Et_3PbCl + p-Me_2NC_6H_4Li$		b. 130°; n^{25}_D 1.5442; n^{25}_D 1.4982 (1,2)	533
$Et_2N(CH_2)_3PbPh_3$	$Ph_3PbLi + Et_2N(CH_2)_3Cl$	91	Oily liquid (3,4,5)	165, 535
$o-H_2NC_6H_4PbPh_3$	$Ph_3PbCl + o-H_2NC_6H_4MgBr$ or with $o-H_2NC_6H_4Li$	60	m. 164-165° (6,7)	159, 533
$p-H_2NC_6H_4PbPh_3$	$Ph_3PbCl + p-H_2NC_6H_4MgBr$ or $p-H_2NC_6H_4Li$	66	m. 166-167° (7,8,9) m. 172° (10)	151 533
$p-MeNHC_6H_4PbPh_3$	$Ph_3PbCl + p-MeNHC_6H_4Li$	10	m. 97-98°	159, 533
$o-Me_2NC_6H_4PbPh_3$	$Ph_3PbCl + o-Me_2NC_6H_4Li$	50	m. 101-102° (11)	159, 533
2-Pyridyl-$PbPh_3$	$Ph_2(2-C_5H_4N)PbI + PhMgBr$	67.1	m. 220°	161
$p-[o-C_6H_4(CO)_2N]-C_6H_4PbEt_3$	$Et_3PbCl + p-[o-C_6H_4(CO)_2N]-C_6H_4Li$		Light yellow, glassy solid	533

(1) Rxn. with dry HCl in $CHCl_3 \to Et_3SnCl + PhNMe_2$ (533).
(2) Rxn. with $p-O_2NC_6H_4N_2Cl \to Et_3PbCl + p-(p-Me_2NC_6H_4N_2)C_6H_4NO_2$ (533).
(3) Salts: methiodide, m. 153-155° (165, 535); methosulfate, m. 137-138° (from C_6H_6) and 241-243° (from EtOH) (165), inhibits growth of microörganism (799); hydrochloride, m. 240° (165).
(4) Rxn. with HCl in $C_6H_6 \to PbCl_2$ (535).
(5) Rxn. with 5% aq. HCl $\to Ph_2Pb(Et_2NCH_2CH_2CH_2)PbCl \cdot HCl$ (535).
(6) Rxn. with $NaNO_2$, followed by $2-C_{10}H_7OH \to o-(2-HOC_{10}H_6N_2)C_6H_4PbPh_3$ (159, 533).
(7) Rxn. with HCl in $CHCl_3 \to Ph_3PbCl + PhNH_2$ (533).
(8) Rxn. with $NaNO_2$, followed by $2-C_{10}H_7OH \to p-(2-HOC_{10}H_6N_2)C_6H_4PbPh_3$ (151, 159, 533).
(9) Rxn. with NO_2^- in $H_2O \to p-HOC_6H_4PbPh_3$ (?) (533).
(10) Rxn. with HCl in $CHCl_3 \to Ph_3PbCl + PhNHMe$ (533).
(11) Rxns. with $RN_2Cl \to 2-Me_2N-5-RN_2-C_6H_3PbPh_3$ (159, 533).

CARBOXY- AND CARBALKOXY-SUBSTITUTED TETRAORGANOLEAD COMPOUNDS

Compound	Prepared from	Yield %	Properties	Ref.
p-$HO_2CC_6H_4PbPh_3$	p-$HOCH_2C_6H_4PbPh_3$ + $KMnO_4$	25	m. 256-258° (1,2)	156, 372
p-$MeO_2CC_6H_4PbPh_3$	p-$HO_2CC_6H_4PbPh_3$ + CH_2N_2	79	m. 125-127°	156, 372
$PhCH_2CH(CO_2Et)PbPh_3$	$Ph_3PbOCOCH(CO_2Et)$-CH_2Ph by pyrolysis at 128-132°/25mm	61	m. 82-84°	285
$EtO_2CCH_2PbPh_3$	$Ph_3PbOCOCH_2CO_2Et$ by pyrolysis at 160-165°/15mm	40	m. 59-60°	285

(1) Na and K salts are insol. in H_2O (156, 372).
(2) Rxn. with CH_2N_2 → $Ph_3PbC_6H_4CO_2Me$ (156, 372).

HYDROXY-SUBSTITUTED TETRAORGANOLEAD COMPOUNDS AND THEIR DERIVATIVES

Compound	Prepared from	Yield %	Properties (1)	Ref.
o-HOC$_6$H$_4$PbPh$_3$	Ph$_3$PbCl + o-HOC$_6$H$_4$MgBr or o-HOC$_6$H$_4$Li	70	m. 216-218°	159 160, 306, 533
m-HOCH$_2$C$_6$H$_4$PbPh$_3$	Ph$_3$PbCl + m-HOCH$_2$C$_6$H$_4$Li	41	m. 113-114° (2)	156, 372
o-HOCH$_2$C$_6$H$_4$PbPh$_3$	Ph$_3$PbCl + o-HOCH$_2$C$_6$H$_4$Li	71	m. 134-136° (3)	156, 372
p-HOCH$_2$C$_6$H$_4$PbPh$_3$	Ph$_3$PbCl + p-HOCH$_2$C$_6$H$_4$Li	63	m. 98-100° (4,5,6)	156, 372
m-HOCH$_2$CH$_2$C$_6$H$_4$PbPh$_3$	Ph$_3$PbCl + m-HOCH$_2$CH$_2$C$_6$H$_4$Li	57	m. 87-88°	156
p-HOCH$_2$CH$_2$C$_6$H$_4$PbPh$_3$	Ph$_3$PbCl + p-HOCH$_2$CH$_2$C$_6$H$_4$Li		m. 87-88°	372
p-(MeCHOH)C$_6$H$_4$PbPh$_3$	Ph$_3$PbCl + p-(MeCHOH)C$_6$H$_4$Li	52	m. 68-70°	156, 372
o-MeOC$_6$H$_4$PbPh$_3$	Ph$_3$PbCl + o-MeOC$_6$H$_4$MgBr	81	m. 128-129°	156, 372
p-MeOC$_6$H$_4$PbPh$_3$	Ph$_3$PbCl + p-MeOC$_6$H$_4$MgBr	86	m. 150-151° (decompn.)	156, 372
p-MeOC$_6$H$_4$PbPh$_3$	Ph$_3$PbCl + p-MeOC$_6$H$_4$Li		m. 152° (8,9)	148
(o-MeOC$_6$H$_4$)$_3$PbCH$_2$Ph	(o-MeOC$_6$H$_4$)$_3$PbNa + PhCH$_2$Cl	40	m. 80-81°	43, 147
(p-EtOC$_6$H$_4$)$_3$PbCH$_2$Ph	(p-EtOC$_6$H$_4$)$_3$PbNa + PhCH$_2$Cl	82	m. 76-77°	43, 147
(o-MeOC$_6$H$_4$)$_3$PbCPh$_3$	(o-MeOC$_6$H$_4$)$_3$PbI + Ph$_3$CMgCl		m. 145-146°	43

(1) Rxns. with RN$_2$Cl → 5-RN$_2$-2-HO-C$_6$H$_3$PbPh$_3$ (159, 533).
(2) Rxns. with KMnO$_4$ → mixt. of prods. (372).
(3) Rxn. with KMnO$_4$ → Ph$_2$PbC$_6$H$_4$COO (156, 372).
(4) Rxn. with Br$_2$ in C$_5$H$_5$N → p-HOCH$_2$C$_6$H$_4$Br + Ph$_3$PbBr (156).
(5) Rxn. with HCl in CHCl$_3$ → Ph$_3$PbCl + Ph$_2$PbCl$_2$ (156).
(6) Rxn. with KMnO$_4$ → Ph$_3$PbC$_6$H$_4$CO$_2$H (156, 372).
(7) Rxn. with RN$_2$Cl → 2-MeO-5-RN$_2$-C$_6$H$_3$PbPh$_3$ (533).
(8) Rxn. with HCl in CHCl$_3$ → Ph$_2$PbCl$_2$ + Ph$_3$PbCl (148).
(9) Rxn. with RN$_2$Cl → 4-MeO-5-RN$_2$-C$_6$H$_3$PbPh$_3$ (533).

ASYMMETRICAL HETEROFUNCTIONAL TETRAORGANOLEAD COMPOUNDS

Heterofunctional tetraorganolead compounds are prepared:

1. By converting the amino-substituted tetraorganolead compounds to the corresponding diazonium salts and reacting them with functionally-substituted aromatic compounds.
2. By reacting functionally-substituted tetraorganolead compounds with substituted aryldiazonium salts.
3. By reacting functionally-substituted organolead halides with heterocyclylmagnesium halides.

ASYMMETRICAL HETEROFUNCTIONAL TETRAORGANOLEAD COMPOUNDS

Compound	Prepared from	Yield %	Properties	Ref.
p-(2-HOC$_{10}$H$_6$N$_2$)C$_6$H$_4$PbPh$_3$	p-H$_2$NC$_6$H$_4$PbPh$_3$ + NaNO$_2$ + β-naphthol	35	m. 135° (decompn.)	151, 159, 533
2-HO-5-(p-ClC$_6$H$_4$N$_2$)-C$_6$H$_3$PbPh$_3$	2-HOC$_6$H$_4$PbPh$_3$ + p-ClC$_6$H$_4$N$_2$Cl	45		159, 533
2-HO-5-(p-BrC$_6$H$_4$N$_2$)-C$_6$H$_3$PbPh$_3$	2-HOC$_6$H$_4$PbPh$_3$ + p-BrC$_6$H$_4$N$_2$Cl	48		159, 533
2-HO-5-(p-IC$_6$H$_4$N$_2$)-C$_6$H$_3$PbPh$_3$	2-HOC$_6$H$_4$PbPh$_3$ + p-IC$_6$H$_4$N$_2$Cl	50		159, 533
2-HO-5-(p-HO$_2$CC$_6$H$_4$N$_2$)-C$_6$H$_3$PbPh$_3$	2-HOC$_6$H$_4$PbPh$_3$ + p-HO$_2$CC$_6$H$_4$-N$_2$Cl	15		159, 533
(Ph$_3$PbC$_6$H$_3$-2-OH-5-C$_6$H$_4$-p)$_2$	2-HOC$_6$H$_4$PbPh$_3$ + (p-C$_6$H$_4$N$_2$Cl)$_2$	86	m. 180° (1)	533
2-HO-3,5-(p-O$_2$NC$_6$H$_4$N$_2$)$_2$-C$_6$H$_2$PbPh$_3$	2-HOC$_6$H$_4$PbPh$_3$ + p-O$_2$NC$_6$H$_4$N$_2$Cl		Red-colored dyestuff	533
4-MeO-5-(p-O$_2$NC$_6$H$_4$N$_2$)-C$_6$H$_3$PbPh$_3$	p-O$_2$NC$_6$H$_4$N$_2$Cl + p-MeOC$_6$H$_4$PbPh$_3$		Red-colored dyestuff	533
2-MeO-5-(p-O$_2$NC$_6$H$_4$N$_2$)-C$_6$H$_3$PbPh$_3$	2-MeOC$_6$H$_4$PbPh$_3$ + p-O$_2$NC$_6$H$_4$N$_2$Cl			533
o-(2-HOC$_{10}$H$_6$N$_2$)C$_6$H$_4$PbPh$_3$	o-H$_2$NC$_6$H$_4$PbPh$_3$ + NaNO$_2$ + β-naphthol	70	Red-colored dyestuff (2)	159, 533
2-Me$_2$N-5-(p-O$_2$NC$_6$H$_4$N$_2$)-C$_6$H$_3$PbPh$_3$	o-Me$_2$NC$_6$H$_4$PbPh$_3$ + p-O$_2$NC$_6$H$_4$N$_2$Cl			533
2-Me$_2$N-5-(p-ClC$_6$H$_4$N$_2$)-C$_6$H$_3$PbPh$_3$	o-Me$_2$NC$_6$H$_4$PbPh$_3$ + p-ClC$_6$H$_4$N$_2$Cl			533
2-Me$_2$N-5-(p-BrC$_6$H$_4$N$_2$)-C$_6$H$_3$PbPh$_3$	o-Me$_2$NC$_6$H$_4$PbPh$_3$ + p-BrC$_6$H$_4$N$_2$Cl			533
2-Me$_2$N-5-(p-IC$_6$H$_4$N$_2$)-C$_6$H$_3$PbPh$_3$	o-Me$_2$NC$_6$H$_4$PbPh$_3$ + p-IC$_6$H$_4$N$_2$Cl			533
2-Me$_2$N-5-(p-HO$_2$CC$_6$H$_4$N$_2$)-C$_6$H$_3$PbPh$_3$	o-Me$_2$NC$_6$H$_4$PbPh$_3$ + p-HO$_2$CC$_6$H$_4$N$_2$Cl			533
(p-MeOC$_6$H$_4$)$_3$PbC$_4$H$_3$O	(p-MeOC$_6$H$_4$)$_3$PbCl + 2-furyl-MgI	84	m. 83°	148
(p-MeOC$_6$H$_4$)$_2$Pb(C$_4$H$_3$O)$_2$	(p-MeOC$_6$H$_4$)$_2$PbCl$_2$ + 2-furyl-MgI	55.9	m. 72-73°	148

(1) Rxn. with SnCl$_2$ in HCl → 2,4-(H$_2$N)$_2$C$_6$H$_3$OH (159).
(2) Rxn. with SnCl$_2$ in AcOH → p-Me$_2$NC$_6$H$_4$NHAc (159, 533).

TETRAHETEROCYCLYLLEAD COMPOUND

TETRATHIENYLLEAD $(C_4H_3S)_4Pb$
Rxns. with: Me_2CHCO_2H at 60-70° → $(C_4H_3S)_2Pb(O_2CCHMe_2)_2$ (737).

2. ALKALI METAL TRIORGANOPLUMBIDES

Sodium triorganoplumbides are prepared by reacting tetraorganolead or hexaorganodilead compounds, or triorganolead halides with metallic sodium in liquid ammonia or ammonia-ether mixture. Lithium triorganoplumbide is formed by reacting lead dichloride with organolithium compounds in a molar ratio of 1:3.

SODIUM TRIETHYLPLUMBIDE $NaPbEt_3$
Prepn.: By treating Et_4Pb with Na in liq. NH_3 or Et_2O-NH_3 (55, 171).
Rxns. with: $p-MeC_6H_4SO_3CH_2CH_2Cl$ → $C_{12}H_{30}OPb_2$ (169).
RCl → Et_3PbR (55, 171, 372).

SODIUM TRICYCLOHEXYLPLUMBIDE $NaPb(C_6H_{11})_3$
Prepn.: By shaking $[(C_6H_{11})_3Pb]_2$ with Na in Et_2O under N (221).
Rxns. with: Hg — $(C_6H_{11})_3Pb$ + NaHg (221).
I_2 → $(C_6H_{11})_2PbI_2$ + PbI_2 (221).

SODIUM TRIPHENYLPLUMBIDE $NaPbPh_3$
Prepn.: By reacting Ph_4Pb with Na in NH_3-Et_2O (55, 171).
By reacting $(Ph_3Pb)_2$ with Na in NH_3 (134).
By reducing Ph_3PbCl with Na in liq. NH_3 (147).
Props.: Cream-colored solid stable at r.t.; exists in the ionic form in liquid NH_3 (134).
Rxns. with: $C_2H_4Br_2$ → $(Ph_3Pb)_2$ (55).
RBr → Ph_3PbR (134, 147, 171).
NH_4Br → Ph_3PbNH_4 (?) (134).
CH_2Cl_2 → $(Ph_3Pb)_2$ (?) (134).

LITHIUM TRIPHENYLPLUMBIDE $LiPbPh_3$
Prepn.: By reacting $PbCl_2$ with PhLi (1:3) in Et_2O at -10° (165, 166)
Props.: Grayish white solid insol. in Et_2O (535); The compd. dissoc. into Ph_2Pb + PhLi (28, 535).
Rxns. with: R halides → Ph_3PbR (165, 166, 306, 535).
CO_2, followed by hydrolysis → $PhCO_2H$ (166).
$CH_2(CH_2Br)_2$ → $(Ph_3PbCH_2)_2CH_2$ (535).
$X(CH_2)_nX$ (X = halogen) → $Ph_3Pb(CH_2)_nPbPh_3$ (535).

3. ORGANOLEAD HYDRIDES

The first organolead hydride, trimethyllead hydride (Me_3PbH), was prepared by reducing trimethyllead chloride (Me_3PbCl) with potassium hydroborate (KBH_4) in liquid ammonia at $-33°$. The compound is a colorless solid melting at about $-100°$ and decomposing above this temperature to a red solid stable at $-78°$, probably $Me_3PbPbMe_2H$, and CH_4. The reaction of trimethyllead hydride with hydrogen chloride (1:1) yields trimethyllead chloride and hydrogen (647).

4. ORGANOLEAD HALIDES

HOMOGENEOUS TRIORGANOLEAD HALIDES

The compounds are prepared:

1. By treating homogeneous tetraorganolead compounds with: 1) hydrogen halides in a molar ratio of 1:1; 2) halogens in a molar ratio of 1:1; 3) $HgCl_2$, $FeCl_3$, or $MnCl_2$; 4) CCl_3CHO at $165-170°$ and hydrolyzing the reaction product; or 5) $Me_3N \cdot HCl$.
2. By treating homogeneous hexaorganodilead compounds with iodine or magnesium iodide.
3. By treating heterogeneous tetraorganolead compounds, R_3PbR', with hydrogen halides in a molar ratio of 1:1.
4. By treating diorganolead dihalides with lead or lead-sodium alloy at elevated temperature in a solvent.
5. By reacting lead dichloride with organomagnesium halides.
6. By decomposing aryldiazonium chloride-lead tetrachloride complex with copper powder.
7. By pyrolyzing diorganolead dihalides.

TRIMETHYLLEAD CHLORIDE Me_3PbCl
Prepn.: By treating Me_4Pb with HCl at r.t. in Et_2O; 60% yield (79, 216).
Props.: Colorless needles, m. $190°$ (216); Data for the dielec. const., density, and polarization (325); Dipole moment (521); Sternutatory effect (338).
Rxns. with: $Et_3PbCl \rightarrow$ 8.6% Me_3PbCl, 34.9% $Me_2EtPbCl$, 41.5% $MeEt_2PbCl$ and 15% Et_3PbCl (74).
KBH_4 (1:1) at $-33°$ in liq. $NH_3 \rightarrow Me_3PbH$, BH_3, and KCl (647).

TRIMETHYLLEAD IODIDE Me_3PbI

Prepn.: By treating Me_4Pb with I_2 at $-60°$ in Et_2O; 60% yield (68).
Rxns. with: $s-BuMgBr \to s-BuPbMe_3$, $s-Bu_2PbMe_2$ and Me_4Pb (68).
$t-BuMgCl \to t-BuPbMe_3$ (68).
$Na \to (Me_3Pb)_2$ (68).

TRIETHYLLEAD CHLORIDE Et_3PbCl

Prepn.: By treating Et_4Pb with HCl in Et_2O at r.t.; 96% yield (79, 216, 378).
By treating Et_4Pb with $HgCl_2$ in Et_2O; 27% yield (361).
By reacting Et_4Pb with CCl_3CHO at $165-170°$ under N and hydrolyzing the rxn. mixt. with $AcOH$ (157a).
Props.: m. $165-167°$ (361); Soly. data; decomp. during distn. with steam according to the equations: $3Et_3PbCl = 2Et_4Pb + PbCl_2 + EtCl$; $2Et_3PbCl = Et_4Pb + PbCl_2 + 2Et$ (80); UV spectrum (439); Dipole moment (353, 520, 521); Data for dielec. const., density, and polarization (325, 440); Sternutatory effect (338).
Rxns. with: Radioactive $Et_4Pb \to$ **radical exchange (redistribution)** (72).
$Me_3PbCl \to$ radical redistribution (74).
Wet $Ag_2O \to Et_3PbOH$ (79, 216).
$KCN \to Et_3PbCN$ (79, 216).
$Na_2S \to (Et_3Pb)_2S$ (79).
RMg halides $\to Et_3PbR$ (156).
$RLi \to Et_3PbR$ (191, 533).
$KCNS \to Et_3PbCNS$ (216).
$KCNSe \to Et_3PbCNSe$ (216).
$KCNO \to Et_3PbCNO$ (216)
Aq. alkali $\to Et_3PbOH$ (469).
Li or Ca in NH_3, followed by $RCl \to Et_3PbR$ (160).
EtONa in EtOH $\to Et_3PbOEt$ (169).
$ClCH_2CO_2Ag \to Et_3PbOCOCH_2Cl$ (216).
$CH_2N_2 \to Et_3PbCH_2Cl$ along with Et_4Pb (576).
$MeCHN_2 \to Et_4Pb$ and $PbCl_2$ ($Et_3PbCHClMe$ is instable) (576)
$RSO_2NR'Na \to Et_3PbNR'SO_2R$ (338, 472).
$Ag_2PO_3F \to (Et_3Pb)_2POF$ (470).
2,3-Dihydro-1,4-phthalazinedione $\to$ N-Et_3Pb-derivative (327).
Polarographic reduction (765).

TRIETHYLLEAD BROMIDE Et_3PbBr

Prepn.: By bubbling HBr through a soln. of Et_4Pb in Et_2O; 68.6% yield; in petr. ether the yield reaches 84% (147).
Props.: m. $105°$ (147); Soly. data (80); Decomp. during steam distn. according to the equations: $3Et_3PbBr = 2Et_4Pb + PbBr_2 + EtBr$; $2Et_3PbBr = Et_4Pb + PbBr_2 + 2Et$ (80); Dipole moment (521); Data for dielec. const., density, and polarization (325).
Rxns. with: $Me_4Pb \to$ redistribution of the alkyl groups (74).
Mg in $Et_2O \to Et_4Pb$, EtMgBr, and Pb (147).
Ag_2O in EtOH $\to Et_3PbOH$ (328).
$(EtO)_3P \to$ no Arbuzov rxn. (30).
Electrochemical reduction $\to (Et_3Pb)_2$ (104).

TRIBUTYLLEAD CHLORIDE Bu_3PbCl

Prepn.: By pyrolyzing Bu_2PbCl_2 at $130°$ (124).
By treating Bu_4Pb and $FeCl_3$ in $CHCl_3$; 36% yield (362)
By treating Bu_4Pb with dry HCl in petr. ether; 77% yield (469).
Props.: White crystals, m. 106.5-$108.5°$ (469); b. 147-$149°/5$-$7mm$ (362); Sternutatory effect (338).
Rxns. with: $Ag_2O \rightarrow Bu_3PbOH$ (469).
Polarographic reduction (765).

TRICYCLOHEXYLLEAD IODIDE $(C_6H_{11})_3PbI$

Prepn.: By reacting $[(C_6H_{11})_3Pb]_2$ with I_2 in $CHCl_3$ (43).
Props.: m. 91-$92°$ (222).
Rxns. with: 30% KOH $\rightarrow (C_6H_{11})_3PbOH$ (222).
$CaFe(CO_4H)_2 \rightarrow [(C_6H_{11})_3Pb]_2Fe(CO)_4$ (222).
HC:CNa in liq. $NH_3 \rightarrow [(C_6H_{11})_3PbC]_2$ (665).

TRIPHENYLLEAD CHLORIDE Ph_3PbCl

Prepn.: By treating Ph_4Pb or an aryltriphenyllead with dry HCl in $CHCl_3$; 75% yield (134, 156).
By decomposing $(PhN_2Cl)_2 \cdot PbCl_4$ with Cu powder in Et_2O at 18-$20°$ (282, 386).
By reacting Ph_2PbCl_2 with NaPb or Pb in xylene at 150-$155°$; 22.5% yield (383).
By heating Ph_4Pb with $MnCl_2$ in $EtOCH_2CH_2OH$; 51% yield (433).
By reacting Ph_4Pb with $Me_3N \cdot HCl$ at $130°$ in EtOH (297a).
Props.: m. $205°$ (386); Dipole moment (353, 520, 521); Conductance in non-aqueous solvents (788, 789); Data for the dielec. const., density, and polarization (325).
Rxns. with: $(:CMgBr)_2 \rightarrow Ph_3PbC:CPbPh_3$ (50).
NH_3 at $-33.4° \rightarrow$ several solid phases (133).
$Na_4Pb_9 \rightarrow (Ph_3Pb)_2$ (134).
Na in liq. $NH_3 \rightarrow Ph_3PbNa$ (147).
RLi $\rightarrow Ph_3PbR$ (148, 156, 159, 160, 306, 372, 533, 614).
o-$LiOC_6H_4Li$, followed by hydrolysis $\rightarrow (Ph_3Pb)_2CO_3$, which may be due to hydrolysis of $Ph_3PbOC_6H_4Li$ and rxn. with CO_2 (533).
$Ph_3CMgCl \rightarrow Ph_3PbCPh_3$ (?) (43).
RMg halides $\rightarrow Ph_3PbR$ (151, 156, 159, 178, 372, 583, 680a).
p-$BrC_6H_4MgBr \rightarrow Ph_4Pb$ and p-$BrC_6H_4PbPh_3$ (583).
$Et_2Ba \rightarrow Ph_3PbEt$ along with Ph_4Pb (152).
$RCO_2K \rightarrow Ph_3PbOCOR$ (285).
N_2O_3 (1:2) $\rightarrow PhN_2NO_3$ (350a).
Polarographic reduction (765).
$Ph_3SiONa \rightarrow Ph_3PbOSiPh_3$ (535).

TRIPHENYLLEAD BROMIDE Ph_3PbBr

Prepn.: By-prod. in the rxn. of $PhMgBr$ with $PbCl_2$ in Et_2O-MePh (483).

Props.: Dipole moment (353, 521); Data for the dielec. const., density, and polarization (325).

Rxns. with: $(:CMgBr)_2 \rightarrow Ph_3PbC:CPbPh_3$ (50).
$NaC:CH \rightarrow Ph_3PbC:CPbPh_3$ (50).
$RMgX \rightarrow Ph_3PbR$ (160).
$CaFe(CO)_4H_2 \rightarrow [Ph_3Pb]_2Fe(CO)_4$ (222).
$CaFe(CO)_4H_2 \rightarrow [Ph_2PbFe(CO)_4]_2$ (225).

TRIPHENYLLEAD IODIDE Ph_3PbI

Prepn.: By reacting Ph_4Pb with I dissolved in $CHCl_3$; 97% yield (134, 147).
By reacting $(Ph_3Pb)_2$ with MgI_2 in Et_2O-C_6H_6 at r.t.; 57% yield (43, 147).

Props.: White crystals, m. 142° (134); m. 138-139° (147); Dipole moment (521); Data for the dielec. const., density, and polarization (325); Polarographic studies (550).

Rxns. with: $(EtO)_3P \rightarrow Ph_4Pb$ (30).
Na (1:1) $\rightarrow (Ph_3Pb)_2$ (134).
Na in liq. $NH_3 \rightarrow (Ph_3Pb)_2$ (147).

OTHER HOMOGENEOUS TRIORGANOLEAD HALIDES

R_3PbX	Prepared from	Yield %	Properties	Ref.
Et_3PbI	$Et_4Pb + I_2$ in Et_2O at $-65°$	73	Yellow liquid, freezing at 19-20°	79
				680a
$(CH_2:CH)_3PbCl$	$(CH_2:CH)_4Pb + HCl$			469
Pr_3PbCl	$Pr_4Pb + HCl$	89	m. 133-134° (1,2)	104
Pr_3PbBr			(3)	104
Bu_3PbBr			(4)	
$(Me_2CHCH_2)_3PbCl$	$(Me_2CHCH_2)_4Pb + HCl$		Darkens at 145°, decomp. at 164°; Sternutatory effect	216
				338
n-$C_{12}H_{25})_3PbCl$	$PbCl_2 + n$-$C_{12}H_{25}MgBr$		m. 63.5° (5,6)	369
n-$C_{14}H_{29})_3PbCl$	$PbCl_2 + n$-$C_{14}H_{29}MgBr$		m. 74-75° (7)	369
n-$C_{16}H_{33})_3PbCl$	$PbCl_2 + n$-$C_{16}H_{33}MgCl$		m. 79-80° (8)	369
n-$C_{18}H_{37})_3PbCl$	$PbCl_2 + n$-$C_{18}H_{37}MgBr$		m. 82-83°	369
m-$MeC_6H_4)_3PbBr$	$(m$-$MeC_6H_4)_4Pb + Br_2$	78	m. 146-147°	43, 147
m-$MeC_6H_4)_3PbI$			(9)	147
p-$MeC_6H_4)_3PbI$			(10)	147
o-$MeOC_6H_4)_3PbCl$	$(p$-$MeC_6H_4)_3PbI_2 + MgI_2$	70	m. 167-171°	383
	$(o$-$MeOC_6H_4)_2PbCl_2 + Pb$	47		369
o-$MeOC_6H_4)_3PbI$	$(o$-$MeOC_6H_4)_2PbI_2 + I_2$	41	m. 122-123° (11)	43, 148
p-$MeOC_6H_4)_3PbCl$	$(p$-$MeOC_6H_4)_4Pb + HCl$	90	m. 152-153° (decompn.)	148
	$(p$-$MeOC_6H_4)_3PbC:CHCH:CHO + HCl$	43	(12)	148
$(p$-$EtOC_6H_4)_3PbI$	$[(p$-$EtOC_6H_4)_3Pb]_2 + I_2$	45.5	m. 99-100°	43, 147
$2,4,6$-$Me_3C_6H_2)_3PbI$	$[Me_3C_6H_2)_3Pb]_2 + I_2$		m. 200-201° (13)	43
$(PhCH_2CH_2)_3PbBr$				665

(1) Rxn. with $RSO_2NR'Na \rightarrow Pr_3PbNR'SO_2R$ (472).
(2) Polarographic reduction $(Pr_3Pb)_2$ (765).
(3) Electrochemical redn. $\rightarrow (Pr_3Pb)_2$ (104).
(4) Electrochemical redn. $\rightarrow (Bu_3Pb)_2$ (104).
(5) Rxn. with $AgNO_3$ in EtOH $\rightarrow (n$-$C_{12}H_{25})_3PbNO_3$ (369).
(6) Rxn. with AgOAc in EtOH $\rightarrow (n$-$C_{12}H_{25})_3PbOAc$ (369).
(7) Rxn. with n-$C_{14}H_{29}MgBr \rightarrow (n$-$C_{14}H_{29})_4Pb$ (369).
(8) Rxn. with n-$C_{16}H_{33}MgBr \rightarrow (n$-$C_{16}H_{33})_4Pb$ (369).
(9) Rxn. with Na in liq. $NH_3 \rightarrow [(m$-$MeC_6H_4)_3Pb]_2$ (147).
(10) Rxn. with HC≡CNa in liq. $NH_3 \rightarrow [(p$-$MeC_6H_4)_3Pb]_2$ (665).
(11) Rxn. with $Ph_3CMgCl \rightarrow (o$-$MeOC_6H_4)_3PbCPh_3$ (43).
(12) Rxn. with 2-furyl-MgI $\rightarrow (p$-$MeOC_6H_4)_3PbC:CHCH:CHO$ (148).
(13) Rxn. with HC≡CNa $\rightarrow [(PhCH_2CH_2)_3PbC:]_2$ (665).

HETEROGENEOUS TRIORGANOLEAD HALIDES

Triorganolead halides containing mixed organic groups are prepared:

1. By treating heterogeneous tetraorganolead compounds with hydrogen halides. In tetraorganolead compounds containing mixed alkyl substituents the longer chain substituent is preferentially cleaved off. In compounds containing aliphatic and aromatic groups, the aromatic groups are preferentially cleaved off.
2. By reacting diorganolead dihalides with Grignard compounds, derived from organic compounds different from the organic groups of the organolead derivative, in a molar ratio of 1:1.
3. By hydrolyzing inner salts of triorganolead compounds.

Compound	Prepared from	Yield %	Properties	Ref.
$Me_2EtPbCl$	$Me_3PbEt + HCl$		(1)	79
$Et_2PhPbCl$	$Et_2PbPh_2 + HCl$	81	Sinters at $142°$, decomp. $146-147°$	306
$(cyclo-C_6H_{11})_2$-$PhPbCl$	$(cyclo-C_6H_{11})_2PbPh_2 + HCl$		m. $195°$, decomp. $205°$ (2)	160, 306
$Ph_2EtPbCl$	$Ph_3PbEt + HCl$	81	Sinters at $142°$, decomp. $146-147°$ (3)	160
$Ph_2(2-C_5H_4N)PbI$	$Ph_2PbI_2 + 2-C_5H_4NMgBr$	63	Yellow crystals, m. $137-140$ (4)	161
$Ph_2[Et_2N(CH_2)_3]$-$PbCl \cdot HCl$	$Ph_3Pb(CH_2)_3NEt_2 + 5\%$ HCl		White crystals, decompg. without melting	535
$Ph_2(o-HO_2CC_6H_4)$-$PbCl$	$Ph_2\overline{PbC_6H_4COO} + HCl$		m. $210-220$ (5)	372
$Ph_2(o-MeO_2CC_6H_4)PbCl$	$Ph_2(o-HO_2CC_6H_4)PbCl + CH_2N_2$		m. $170-171°$	372
$Ph_2(9$-fluorenyl$)PbCl$	$Ph_3Pb(9-C_{13}H_9) + HCl$		m. $160°$ (decompn.)	614

(1) At $60°$ in $Me_2CO \rightarrow Me_3PbCl$, $Me_2EtPbCl$, $Et_2MePbCl$, and Et_3PbCl (74).
(2) Rxn. with Na in liq. $NH_3 \rightarrow$ dark-red, unidentified compd. (160, 306)
(3) Rxn. with Na in liq. $NH_3 \rightarrow$ instable compd. (160).
(4) Rxn. with PhMgBr $\rightarrow Ph_3Pb(2-C_5H_4N)$ (161).
(5) Rxn. with $CH_2N_2 \rightarrow Ph_2(o-MeO_2CC_6H_4)PbCl$ (372).

DIORGANOLEAD DIHALIDES

Diorganolead dihalides are prepared:

1. By treating tetraorganolead compounds with: 1) halogens in a molar ratio of 1:2 in a suitable solvent; 2) hydrogen halides in a molar ratio of 1:2; 3) halides of P, As, Sb, Bi, Fe, and Co; 4) chlorosulfonic acid; 5) carbon tetrachloride in the presence of benzoic anhydride; 6) trimethylammonium hydrochloride; or 7) nitric acid and reacting the resulting diorganolead dinitrates with hydrogen or sodium halides.
2. By reacting hexaorganodilead compounds with hydrogen halides or with halogens.
3. By decomposing aryldiazonium halide-lead dihalide complexes, $RN_2Cl \cdot PbCl_2$, in the presence of copper.
4. By treating diorganolead dicarboxylates with hydrogen halides.

Diorganolead difluorides are prepared by treating diorganolead dichlorides with potassium fluoride.

DIETHYLLEAD DICHLORIDE Et_2PbCl_2

Prepn.: By treating Et_4Pb with dry HCl in MePh at 90°; 80% yield (217, 469).

Props.: m. 220° (217); Soly. data (80, 217); Decomp. during steam distn. according to the equations: $2Et_2PbCl_2 = Et_3PbCl + PbCl_2 + EtCl$; $Et_2PbCl_2 = PbCl_2 + 2Et$ (80); Dipole moment (520, 521); Data for the dielec. const., density, and polarization (325); IR spectrum (622).

Rxns. with: RMg halides → Et_2PbR_2 (217, 622).
Ag_2O → $Et_2Pb(OH)_2$ (79).
$ClCH_2CO_2Ag$ in H_2O, followed by H_2S → $Et_2Pb(OCOCH_2Cl)_2$ and Et_2PbS (217).
$ClCH_2CO_2Ag$ in hot $EtOH-H_2O$ → $Pb(OCOCH_2Cl)_2$, $Et_3PbOCOCH_2Cl$, and $AgCl$ (217).
Aq. alkali → $Et_2Pb(OH)_2$ (217).
$SnCl_2$ → Et_2SnCl_2 and $PbCl_2$ (289, 290).
KCN in H_2O → $Et_2Pb(CN)_2$ (503).
CH_2N_2 → $Et_2Pb(CH_2Cl)_2$ (576).
Polarographic reduction (765).

BIS(2-CHLOROVINYL)LEAD DICHLORIDE (ClCH:CH)$_2$PbCl$_2$

Prepn.: By treating (ClCH:CH)$_2$Pb(OAc)$_2$ with 6% HCl in EtOH; 70% yield (390).

Props.: Solid, decomp. 163-167° (390).

Rxns. with: Hg in MeOH → ClCH:CHHgCl (390).
SnCl$_2$ in EtOH → (ClCH:CH)$_2$SnCl + PbCl$_2$ (390).
PhMgBr in Et$_2$O → C$_2$H$_4$ and Ph$_4$Pb (390).
H$_2$NNH$_2$·H$_2$SO$_4$ or H$_2$NNH$_2$·2HCl → C$_2$H$_2$ (390).
H$_2$O → evolution of AcOH (390).

DIPHENYLLEAD DICHLORIDE Ph$_2$PbCl$_2$

Prepn.: By reacting Ph$_4$Pb with: 1) AlCl$_3$ (146); 2) HCl, 98% yield (483); 3) ClSO$_3$H (372); 4) CoCl$_2$ or FeCl$_3$ in EtOCH$_2$CH$_2$OH, 62-86% yield (433); 5) CCl$_4$ in the presence of Bz$_2$O$_2$ (399); or 6) Me$_3$N·HCl at 150° in EtOH (297a).
By reacting Ph$_3$PbR with HCl in CHCl$_3$; 5% yield (156, 160).
By reacting (Ph$_3$Pb)$_2$ with HCl in CHCl$_3$; 16.5% yield (43, 147).
By decomposing PhN$_2$Cl·PbCl$_2$ with Cu powder in Me$_2$CO at 50-55° (282); or in Et$_2$O (386).
By treating Ph$_2$Pb(NO$_3$)$_2$ with concd. HCl; 93% yield (483).

Rxns. with: RLi (1:2) → Ph$_2$PbR$_2$ (160, 306, 614).
Alkali metals (1:2) in liq. NH$_3$ → black soln. contg. (Ph$_3$Pb)$_2$ and inorg. Pb (29).
Ca or Li in liq. NH$_3$ → (Ph$_3$Pb)$_2$ and Ph$_4$Pb (306).
Li (1:4) in liq. NH$_3$, followed by EtBr → Ph$_2$PbEt$_2$, Ph$_3$PbEt, PhPbEt$_3$, and Et$_4$Pb (28).
Hg in Me$_2$CO → PhHgCl (289, 290).
H$_2$ in the presence of a catalyst → inorg. Pb compds. and C$_6$H$_6$ (28).
SnCl$_2$ → Ph$_2$SnCl$_2$ and PbCl$_2$ (289, 290).
Hydrazine sulfate + Na$_2$CO$_3$ in EtOH → Ph$_4$Pb (383).
NaPb → Ph$_3$PbCl (383).
FeCl$_2$ → PbCl$_2$ (433).
KF → Ph$_2$PbF$_2$ (483).
CH$_2$:CHMgBr → Ph$_2$Pb(CH:CH$_2$)$_2$ (680a).

OTHER DIORGANOLEAD DIHALIDES

R_2PbX_2	Prepared from	Yield %	Properties	Ref.
Me_2PbBr_2	$Me_4Pb + Br_2$ in $CHCl_3$	92	(1,2)	155
Me_2PbCl_2			Soly. data (3)	80
Et_2PbBr_2			m. 300°	711
$(CH_2{:}CH)_2PbCl_2$	$(CH_2{:}CH)_4Pb + MCl_3$ (M = P, As, Sb, Bi) in CCl_4			680a
	$(CH_2{:}CH)_4Pb + Cl_2$			
$(CH_2{:}CH)_2PbBr_2$	$(CH_2{:}CH)_4Pb + MBr_3$ (M = P, As, Sb, or Bi) in CCl_4		m. 300°	711
Pr_2PbCl_2	$Pr_4Pb + HCl$ in Et_2O	82	White powder, m. 245° (4)	469
Bu_2PbCl_2			At 130° → Bu_3PbCl, $PbCl_2$, $BuCl$	124
Ph_2PbF_2	$Ph_2PbCl_2 + KF$	92	m. 300° (5,6)	483
Ph_2PbBr_2	$Ph_4Pb + HNO_3 + NaBr$	96	(6,7)	483
Ph_2PbI_2	$Ph_4Pb + HNO_3 + NaI$	98	(8,9)	483
$(o\text{-}MeOC_6H_4)_2PbCl_2$	$(o\text{-}MeOC_6H_4)_2Pb(OAc)_2 + HCl$	96	m. 187-188.5° (10)	383
$(p\text{-}MeOC_6H_4)_2PbCl_2$	$(p\text{-}MeOC_6H_4)_4Pb + HCl$ in C_6H_6	98.6		148
	$(p\text{-}MeOC_6H_4)_3PbC{:}CHCH{:}CHO + HCl$ in $CHCl_3$	57.3		148
	$(p\text{-}MeOC_6H_4)_2Pb(\delta{:}CHCH{:}CHO)_2 + HCl$ in C_6H_6	87.4		148
$(p\text{-}MeOC_6H_4)_2PbI_2$	$(p\text{-}MeOC_6H_4)_3PbI_2 + Cl_2$ in $CHCl_3$		m. 122-123° (11)	43
$(m\text{-}O_2NC_6H_4)_2PbCl_2$	$(p\text{-}MeOC_6H_4)_3PbI_2 + I_2$ in $CHCl_3$	7.1	Orange needles, m. 122-123°	43, 147
$(m\text{-}O_2NC_6H_4)_2PbBr_2$	$(m\text{-}O_2NC_6H_4)_2Pb(NO_3)_2 + NaCl$	97	m. 285-288°, sublimes at 250° (12,13)	160, 306
$(m\text{-}O_2NC_6H_4)_2PbI_2$	$(m\text{-}O_2NC_6H_4)_2Pb(NO_3)_2 + NaI$			286
$(p\text{-}EtO_2CC_6H_4)_2PbCl_2$	$(EtO_2CC_6H_4)_2Pb(OAc)_2 + alc.$ HCl	97	Yellow crystals, decomp. 135° m. 270°	160, 306 383

(1) Rxn. with: $SnCl_2 → Me_2SnCl_2$ (289, 290).
(2) Rxn. with $NaOH → Me_2PbO$ (474).
(3) Decomp. during steam distn. according to the equations: $Et_2PbBr_2 = PbBr_2 + 2Et$; $2Et_2PbBr_2 = Et_3PbBr + PbBr_2 + EtBr$; $3Et_2PbBr_2 = 2PbBr_2 + Et_4Pb + EtBr$ (80).
(4) Rxn. with moist Ag_2O, followed by $AcOH → Pr_3PbOAc + Pb(OAc)_2$ (217, 469).
(5) Rxn. with Ca or Li in liq. $NH_3 → Ph_4Pb + (Ph_3Pb)_2$ (160, 306).
(6) Rxns. with RMg halides → Ph_2PbR_2 (483).

(7) Rxn. with allyl-MgBr → Ph$_2$Pb(CH$_2$CH:CH$_2$)$_2$ (?), which decomp. to Ph$_3$PbCH$_2$CH:CH$_2$ (178).
(8) Rxn. with 2-pyridyl-MgBr → Ph$_2$(2-C$_5$H$_4$N)PbI (161).
(9) Rxn. with H$_2$ in the presence of catalysts → Ph$_2$ and inorg. Pb compds. (28).
(10) Rxn. with Pb in xylene under reflux → (o-MeOC$_6$H$_4$)$_3$PbCl (383).
(11) Rxn. with 2-furyl-MgI → (p-MeOC$_6$H$_4$)$_2$Pb(C$_4$H$_3$O)$_2$ (148).
(12) Rxns. with: Fe, TiCl$_3$, or SnCl$_2$ in dil. HCl → PhNH$_2$ (286).
(13) Rxn. with (NH$_4$)$_2$S → PhS (286).

ORGANOLEAD TRIHALIDES

Alkyllead triiodides were prepared by reacting scandium trichloroplumbide, ScPbCl$_3$, with alkyl iodides in the presence of iodine.

RPbI$_3$	Prepared from	Properties	Ref.
EtPbI$_3$	ScPbCl$_3$ + EtI	Brown crystals, decompg. at 90° (1,2)	314
PrPbI$_3$	ScPbCl$_3$ + PrI		314
BuPbI$_3$	ScPbCl$_3$ + BuI		314

(1) Rxn. with C$_5$H$_5$N at 0° → EtPbI$_3$·2 C$_5$H$_5$N, orange solid, instable at elevated temp. (314).
(2) Rxn. with H$_2$O → EtPb(OH)$_3$ (314).

5. ORGANOLEAD-OXYGEN COMPOUNDS

ORGANOPLUMBONIC ACIDS

The following organoplumbonic acids were prepared by reacting alkyl iodides with $PbO \cdot H_2O$ in aqueous sodium hydroxide solution at $5°$: $MePb(OH)_3$, $EtPb(OH)_3$, $PrPb(OH)_3$, $Me_2CHPb(OH)_3$, $BuPb(OH)_3$, $CH_2:CHCH_2Pb(OH)_3$, and $PhCH_2Pb(OH)_3$. The reaction is catalyzed by iodine or sodium iodide. The compounds, when moist, decompose at $35°$ into lead oxide and the corresponding alcohols according to the equation: $RPb(OH)_3 = PbO + ROH + H_2O$. $PhCH_2Pb(OH)_3$ disproportionates and forms $(PhCH_2)_2Pb$ (314a).

The following three arylplumbonic acids are prepared by hydrolyzing organolead tricarboxylates with diluted ammonia in ethanol or methanol.

BENZENEPLUMBONIC ACID $PhPbO_2H$
Prepn.: By hydrolyzing $PhPb(OCOCHMe_2)_3$ or $PhPb(OAc)_3$ with NH_4OH in EtOH; 88.5% yield (287, 403).
Props.: Yellowish, amorphous powder, burning with smoky flame, sol. in cold dil. HCl and 40-50% NaOH; in storage loses H_2O and approaches the composition $(PhPbO)_2PbO$ (287, 403).
Rxns. with: Carboxylic acids → PhPb tricarboxylates (287, 403).

p-TOLUENEPLUMBONIC ACID $p\text{-}MeC_6H_4PbO_2H$
Prepn.: By hydrolyzing $p\text{-}MeC_6H_4Pb(OAc)_3$ with 5% NH_4OH in EtOH (288, 403).
Props.: Yellowish, amorphous solid, sol. in dil. HCl and org. acids, difficultly sol. in 40-50% NaOH (288, 403).
Rxns. with: Concd. HCl → MePh (288, 403).
Carboxylic acids → $p\text{-}MeC_6H_4Pb$ tricarboxylates (288).

1-NAPHTHALENEPLUMBONIC ACID $1\text{-}C_{10}H_7PbO_2H$
Prepn.: By hydrolyzing $C_{10}H_7Pb(OAc)_3$ with NH_4OH in aq. MeOH; 62.6% yield (334).
Props.: Yellow, amorphous solid, which ignites in flame and is almost insol. in 50% KOH and concd. HCl (334).
Rxns. with: Carboxylic acids → $1\text{-}C_{10}H_7Pb$ tricarboxylates (334).

ORGANOLEAD TRICARBOXYLATES

Monoorganolead tricarboxylates are prepared:

1. By cleaving one organic group from diorganolead dicarboxylates with mercury dicarboxylates or with mercury oxide in a carboxylic acid.
2. By treating lead tetracarboxylates with diorganomercury compounds.
3. By reacting organoplumbonic acids with carboxylic acids.
4. By transesterifying organolead tricarboxylates with other carboxylic acids.

PHENYLLEAD TRIACETATE $PhPb(OAc)_3$

__Prepn.__: By dearylation of $Ph_2Pb(OAc)_2$ with $Hg(OAc)_2$ in AcOH and treating the rxn. mixt. with dild. HCl in EtOH (to convert PhHgOAc into PhHgCl); 79% yield (287, 402).
By reacting $Pb(OAc)_4$ with Ph_2Hg in $CHCl_3$ (405).
__Props.__: m. 101-102°; insol. in hydrocarbons (287, 402, 405).
__Rxns. with__: Aq. NaOH → $PhPbO_2H$ (287, 402).
NH_4OH in EtOH → $PhPbO_2H$ (287, 402).
I_2 in $CHCl_3$ → PbI_2 (287).
H_2O → $PhPbO_2H$ (287).

PHENYLLEAD TRIISOBUTYRATE $PhPb(OCOCHMe_2)_3$

__Prepn.__: By dearylation of $Ph_2Pb(OCOCHMe_2)_2$ with HgO in warm Me_2CHCO_2H; 48.8% yield (287).
By dearylation of $Ph_2Pb(OCOCHMe_2)_2$ with $Hg(OCOCHMe_2)_2$ in Me_2CHCO_2H; 50% yield (402).
By triturating $PhPbO_2H$ with Me_2CHCO_2H in C_6H_6 (403).
By arylating $Pb(OCOCHMe_2)_4$ with Ph_2Hg in $CHCl_3$; 50% yield (405).
__Props.__: m. 77-78° (287, 402, 403, 405).
__Rxns. with__: BzOH → $PhPb(OBz)_3$ (287, 402).
NH_4OH in EtOH → $PhPbO_2H$ (287, 403).
I_2 in $CHCl_3$ → PbI_2 (287).
H_2O → $PhPbO_2H$ (287).
HgO in Me_2CHCO_2H → $Pb(OCOCHMe_2)_4$ (287).

p-TOLYLLEAD TRIACETATE $p-MeC_6H_4Pb(OAc)_3$

__Prepn.__: By dearylation of $(p-MeC_6H_4)_2Pb(OAc)_2$ with $Hg(OAc)_2$ in AcOH; 67% yield (288, 402).
By arylating $Pb(OAc)_4$ with $(p-MeC_6H_4)_2Hg$ in $CHCl_3$; 66.6% yield (405).
__Props.__: m. 86-88° (from EtOH); crystallizes from C_6H_6 with one mole of the solvent, which can be removed in vacuo at r.t. (288); m. 87° (405).
__Rxns. with__: $CH_2:CMeCO_2H$ → $p-MeC_6H_4Pb(OCOCMe:CH_2)_3$ (288, 402).
Dild. alkali or ammonia → $p-MeC_6H_4PbO_2H$ (288, 402, 403).

OTHER ORGANOLEAD TRICARBOXYLATES

$RPb(OCOR')_3$	Prepared from	Yield %	Properties	Ref.
$PhPb(OBz)_3$	$PhPb(OCOCHMe_2)_3$ + BzOH	41.5	m. 149.5-151°	287, 288, 402
$p\text{-}MeC_6H_4Pb(OCOCMe:CH_2)_3$	$p\text{-}MeC_6H_4Pb(OAc)_3$ + CH_2:CMe-CO_2H	42	Solid, decomp. 120° (1)	288, 402, 403
	$p\text{-}MeC_6H_4PbO_2H$ + CH_2:CMeCO$_2$H			
$1\text{-}C_{10}H_7Pb(OAc)_3$	$Pb(OAc)_4$ + $(1\text{-}C_{10}H_7)_2Hg$	55	m. 168-169° (2,3)	334
$1\text{-}C_{10}H_7Pb(OCOCHMe_2)_3$	$Pb(OCOCHMe_2)_4$ + $(1\text{-}C_{10}H_7)_2Hg$	46	m. 99.5-101°	334
$1\text{-}C_{10}H_7Pb(OBz)_3$	$1\text{-}C_{10}H_7Pb(OAc)_3$ + BzOH	40.5	m. 173-174°	334

(1) Rxn. with dil. alk. or $NH_4OH \rightarrow p\text{-}MeC_6H_4PbO_2H$ (402).
(2) Rxn. with NH_4OH in MeOH $\rightarrow 1\text{-}C_{10}H_7PbO_2H$ (334).
(3) Rxn. with BzOH $\rightarrow 1\text{-}C_{10}H_7Pb(OBz)_3$ (334).

DIORGANOLEAD OXIDES

Diorganolead oxides are prepared by hydrolyzing diorganolead dicarboxylates, dihalides, or dinitrates and dehydrating the intermediate dihydroxides.

R_2PbO	Prepared from	Properties	Ref.
Me_2PbO	Me_2PbCl_2 + NaOH		474
Et_2PbO	$Et_2Pb(OH)_2$ over H_2SO_4		217
Ph_2PbO		Useful as stabilizer for vinyl resins (1,2)	578
$(p\text{-}MeC_6H_4)_2PbO$	$(p\text{-}MeC_6H_4)_2Pb(NO_3)_2$ + KOH		288
	$(p\text{-}MeC_6H_4)_2Pb(OAc)_2$ + KOH	(3,4)	288
$(O_2NC_6H_4)_2PbO$	$Ph_2Pb(OAc)_2$ + HNO_3, followed by dil. NaOH	(5)	474

(1) Rxn. with Na_2 pyrocatecholdisulfonate → Na diphenyllead pyrocatecholdisulfonate (474).
(2) Rxn. with alkali metals in C_5H_8O → complex compd. sol. with brown color (669).
(3) Rxn. with EtOAc in AcOH → $(p\text{-}MeC_6H_4)_2Pb(OAc)_2$ (288).
(4) Rxn. with RCO_2H → $(p\text{-}MeC_6H_4)_2Pb(OCOR)_2$ (402).
(5) Rxn. with $TiCl_3$ in concd. HCl-MeOH → $(H_2NC_6H_4)_2Pb(OH)_2$ (474).

DIORGANOLEAD DICARBOXYLATES

Diorganolead dicarboxylates are formed:

1. By reacting lead tetracarboxylates with diorganomercury compounds in $CHCl_3$ in a molar ratio of 1:2.
2. By treating diorganolead dihydroxide with carbon dioxide and treating the resulting bicarbonate with Al_2O_3; diorganolead carbonate is formed.
3. By reacting diorganolead oxides or dihydroxides with carboxylic acids.
4. By treating diorganolead dihalides with silver carboxylates.
5. By reacting diorganolead sulfides or sulfites with carboxylic acids.
6. By transesterifying diorganolead dicarboxylates with other carboxylic acids.
7. By dealkylating trialkyllead acetates with mercury diacetate.
8. By treating triorganolead hydroxides with equimolar quantities of carboxylic acids.

9. By reacting homogeneous tetraorganolead compounds with carboxylic acids in the presence or absence of catalysts (SiO_2).
10. By reacting homogeneous tetraorganolead compounds with equimolar amounts of lead tetracarboxylates.
11. By treating homogeneous tetraorganolead compounds with mercury dicaroboxylates or with mercuric oxide in carboxylic acids.
12. By treating hexaorganodilead compounds with carboxylic acids.
13. By reacting lead tetracarboxylates with thiophene. Dithienyllead diisobutyrate was prepared by this method.

DIETHYLLEAD CARBONATE Et_2PbCO_3
Prepn.: By passing CO_2 through $Et_2Pb(OH)$ and converting the resulting bicarbonate to the carbonate in the presence of Al_2O_3 (79).
Props.: Soly. data (80).
Rxn. with: H_2O according to the equations: $Et_2PbCO_3 + H_2O = Et_2Pb(OH)_2 + CO_2$; and $2Et_2PbCO_3 + H_2O = Et_3PbOH + PbCO_3 + CO_2 + EtOH$ (80).

DIETHYLLEAD DIACETATE $Et_2Pb(OAc)_2$
Prepn.: By reacting $Pb(OAc)_4$ with Et_2Hg (1:2) in $CHCl_3$ at r.t.; 91% yield (383).
By reacting Et_3PbOAc with $Hg(OAc)_2$ (402).
Props.: m. 200-201° (383); m. 130° (402).
Rxn. with: $ClCH_2CO_2H$ in EtOH → $Et_2Pb(O_2CCH_2Cl)_2$ (402).

DIETHYLLEAD BIS(CHLOROACETATE) $Et_2Pb(OCOCH_2Cl)_2$
Prepn.: By treating Et_2PbS with a concd. aq. soln. of $ClCH_2CO_2H$ (217).
By reacting Et_2PbSO_3 with $ClCH_2CO_2H$ (338).
By grinding $ClCH_2CO_2Ag$ together with Et_2PbCl_2 in H_2O to a paste and treating it with H_2S; Et_2PbS is formed as by-prod. (217).
By treating $Et_2Pb(OAc)_2$ with $ClCH_2CO_2H$ in EtOH; 41% yield (402).
Props.: m. 180° (217); m. 176-177° (402); Disproportionates in aq. solns. to $Et_3PbOCOCH_2Cl$, $Pb(OCOCH_2Cl)_2$, and $ClCH_2CO_2Et$ (217, 338).

BIS(2-CHLOROVINYL)LEAD DIACETATE $(ClCH:CH)_2Pb(OAc)_2$
Prepn.: By reacting $Pb(OAc)_4$ with $(trans-ClCH:CH)_2Hg$ in $CHCl_3$; 72% yield (390).
Props.: Solid, at 115-130° evolves C_2H_2, at 140° turns yellow and at 185° black; soly. data (390).
Rxns. with: HCl in EtOH → $(ClCH:CH)_2PbCl_2$ (390).
BzOH in EtOH → $(ClCH:CH)_2Pb(OBz)_2$ (390).
H_2SO_4 in EtOH → $(ClCH:CH)_2PbSO_4$ (390).
Addn. compd.: $(ClCH:CH)_2(AcO)_2Pb \cdot NH_4OAc$, decomp. 177-178° (390).

BIS(2-CHLOROVINYL)LEAD DIBENZOATE (ClCH:CH)$_2$Pb(OBz)$_2$
Prepn.: By treating (ClCH:CH)$_2$Pb(OAc)$_2$ with BzOH in EtOH; 72% yield (390).
Props.: Solid, decomp. 204-204.5°; soly. data; above 155° evolves C$_2$H$_2$ (390).
Rxns. with: 30% KOH → C$_2$H$_2$ evolution (390).
Aq. KI → C$_2$H$_2$ evolution (390).

DI-2-THIENYLLEAD DIISOBUTYRATE (C$_4$H$_3$S)$_2$Pb(O$_2$CCHMe$_2$)$_2$
Prepn.: By reacting (Me$_2$CHCO$_2$)$_4$Pb with thiophene at r.t. for several days, then treating the rxn. mixt. with EtOH for two days (737).
By reacting (C$_4$H$_3$S)$_4$Pb with Me$_2$CHCO$_2$H at 60-70°; 67% yield (737).
Props.: Decomp. 192° (737).
Rxn. with: ClCH$_2$CO$_2$H in C$_6$H$_6$ → (C$_4$H$_3$S)$_2$Pb(O$_2$CCH$_2$Cl)$_2$ (737).

DIPHENYLLEAD DIACETATE Ph$_2$Pb(OAc)$_2$
Prepn.: By refluxing Ph$_4$Pb with Pb(OAc)$_4$ (4:5.5 by wt.) in CHCl$_3$ with a trace of AcOH; 46% yield (289, 290).
By heating Ph$_4$Pb with AcOH at 70-100°; 98% yield (295, 329, 402).
By reacting Pb(OAc)$_4$ with Ph$_2$Hg (1:2) in CHCl$_3$; 31.3% yield (383).
By reacting Ph$_4$Pb with Hg(OAc)$_2$ in CHCl$_3$ contg. AcOH; 62% yield (404).
Props.: m. 200-201° (289, 290, 383); m. 200-201.5° (402, 404); Mono or dihydrate, m. 194-195° (295).
Rxns. with: Hg(OAc)$_2$ in AcOH → PhPb(OAc)$_3$ + PhHgOAc (287, 402).
H$_2$S → Ph$_2$PbS (329).
HNO$_3$, followed by NaOH → (O$_2$NC$_6$H$_4$)$_2$PbO (474).
Stearic acid at 120-140° → Ph$_2$Pb distearate (404).
Hydrazine sulfate + Na$_2$CO$_3$ in EtOH under reflux → Ph$_4$Pb (383).
NaCa pyrocatecholdisulfonate → Ph$_2$Pb-pyrocatecholdisulfonic acid NaCa salt (474).

DIPHENYLLEAD DIISOBUTYRATE Ph$_2$Pb(OCOCHMe$_2$)$_2$
Prepn.: By heating Ph$_4$Pb with Me$_2$CHCO$_2$H; 84% yield (402, 404).
By treating Ph$_4$Pb with HgO in Me$_2$CHCO$_2$H (404).
Props.: m. 204-205° (402, 404).
Rxns. with: HgO in Me$_2$CHCO$_2$H → PhPb(OCOCHMe$_2$)$_3$ (287).
Hg(OCOCHMe$_2$)$_2$ → PhPb(OCOCHMe$_2$)$_3$ (402).
RCO$_2$H → Ph$_2$Pb(OCOR)$_2$ (transesterification) (402, 404).

DIPHENYLLEAD DIBENZOATE Ph$_2$Pb(OBz)$_2$
Prepn.: By reacting Ph$_4$Pb with BzOH at 160° (296).
By reacting Ph$_2$Pb(OCOCHMe$_2$)$_2$ with BzOH in EtOH; 83% yield (402, 404).
By-prod. in the rxn. of Ph$_4$Pb with CCl$_4$ in the presence of Bz$_2$O$_2$ (399).
Props.: m. 231-232° (296); m. 228-229° (399); m. 240° (402, 404).

OTHER DIORGANOLEAD DICARBOXYLATES

$R_2Pb(OCOR)_2$	Prepared from	Yield %	Properties	Ref.
$Et_2Pb(OCOCCl_3)_2$	$Et_2PbSO_3 + Cl_3CCO_2H$		m. 151° (decompn.)	217
$Et_2Pb(OBz)_2$	$Et_4Pb + BzOH$ over SiO_2		m. 168°; on heating decomp.	217
	$Et_2PbSO_3 + BzOH$	100	With explosion (1)	217
$Et_2Pb(OCOC_6H_4Cl-p)_2$	$Et_2PbOH + p-ClC_6H_4CO_2H$ (1:1)	24	m. 185°	169
$Et_2Pb(OCOC_6H_4Br-m)_2$	$Et_4Pb + m-BrC_6H_4CO_2H$ (1:2)	9.8	m. 178-179° (decompn.)	169
$Et_2Pb(OCOC_6H_4Me-p)_2$	$Et_3PbOH + p-MeC_6H_4CO_2H$ (1:1)	30	m. 186° (decompn.)	169
$Et_2Pb(OCOC_6H_4NO_2-m)_2$	$Et_4Pb + m-O_2NC_6H_4CO_2H$ (1:2)	77	m. 179-180° (decompn.)	169
$Et_2Pb(OCOC_6H_4NHBu)_2$	$Et_3PbOH + o-(BuNH)C_6H_4-CO_2H$ (1:1)	32.7	m. 169-169.5° (decompn.)	169
Et_2Pb dinicotinate	Et_3PbOH + nicotinic acid (1:1)	26.6	m. 143° (decompn.)	169
$Pr_2Pb(OAc)_2$	$Pr_2Pb(OH)_2$ + aq. AcOH		m. 122°	469
$Ph_2Pb(OOCH)_2$	$Ph_4Pb + HCOOH$ at 50°	76	Doesn't melt up to 230°	295
$Ph_2Pb(OCOCH_2Cl)_2$	$Ph_2Pb(OCOCHMe_2)_2 + ClCH_2CO_2H$	77.5	Decomp. 185-205°	402, 404
$Ph_2Pb(OCOEt)_2$	$Ph_4Pb + EtCO_2H$ at 100°	80	m. 170-172°	296
			m. 183.5-184.5°	402, 404
$Ph_2Pb(OCOCHOHMe)_2$	$Ph_4Pb + MeCHOHCO_2H$	81	m. 212-215° (decompn.)	296
$Ph_2Pb(OCOPr)_2$	$Ph_4Pb + PrCO_2H$	72	m. 132-134°	296
$Ph_2Pb(OCOMe:CH_2)_2$	$Ph_2Pb(OCOEt)_2$ or $Ph_2Pb(OCOCHMe_2)_2 + CH_2:CMe-CO_2H$	96	m. 205-212° (decompn.)	402, 404
$Ph_2Pb(OCOCHOHEt)_2$	$Ph_4Pb + EtCHOHCO_2H$	80	m. 198-201° (decompn.)	296
$Ph_2Pb(OCOCH_2CHMe_2)_2$	$Ph_4Pb + Me_2CHCH_2CO_2H$		m. 166-168°	296
$Ph_2Pb(OCO(CH_2)_4Me)_2$	$Ph_4Pb + n-C_5H_{11}CO_2H$	54	Yellow, viscous oil	296
$Ph_2Pb(OCO-n-C_{17}H_{35})_2$	$Ph_2Pb(OCOCHMe_2)_2 + n-C_{17}H_{35}CO_2H$	64	m. 100-101°	402, 404
	$Ph_2Pb(OAc)_2 + n-C_{17}H_{35}CO_2H$			404
	$Ph_2Pb(OAc)_2 + n-C_{17}H_{35}CO_2H$			404

$(p-MeC_6H_4)_2Pb(OAc)_2$	$(p-MeC_6H_4)_2PbO$ + AcOH	64	m. 209-210° (2)	288, 402
$(p-MeC_6H_4)_2Pb(OCOCHMe_2)_2$	$[(p-MeC_6H_4)_3Pb]_2$ + Me_2-CHCO$_2$H		m. 202-203° (3)	288
	$(p-MeC_6H_4)_2PbO$ + Me_2-CHCO$_2$H	48.3		402
$(p-EtO_2CC_6H_4)_2Pb(OAc)_2$	$Pb(OAc)_4$ + $(p-EtO_2CC_6H_4)_2Hg$ (1:2)	51	m. 207-208° (4)	383
$(o-MeOC_6H_4)_2Pb(OAc)_2$	$Pb(OAc)_4$ + $(o-MeOC_6H_4)_2Hg$	63	m. 191-193° (5)	383
$(2-C_{10}H_7)_2Pb(OAc)_2$	$Pb(OAc)_4$ + $(2-C_{10}H_7)_2Hg$	31	m. 236-236.5°	383

(1) Rxn. with dry HCl → Et_2PbCl_2 (217).
(2) Rxn. with $Hg(OAc)_2$ in AcOH, followed by EtOH-HCl → $p-MeC_6H_4Pb(OAc)_3$ + $p-MeC_6H_4HgCl$ (288, 402).
(3) Rxn. with aq. NaOH → $(p-MeC_6H_4)_2PbO$ (288).
(4) Rxn. with HCl in EtOH → $(p-EtO_2CC_6H_4)_2PbCl_2$ (383).
(5) Rxn. with HCl in EtOH → $(o-MeOC_6H_4)_2PbCl_2$ (383).

DI- and TRIORGANOLEAD HYDROXIDES

Organolead hydroxides are prepared by hydrolyzing organolead halides with wet silver oxide or aqueous alkali or by hydrating organolead oxides.

DIETHYLLEAD DIHYDROXIDE $Et_2Pb(OH)_2$

Prepn.: By treating Et_2PbCl_2 with wet Ag_2O; 84% yield (79). By shaking Et_2PbCl_2 with 30% aq. NaOH in Et_2O (217).
Props.: Soly. data; decomp. during steam distn. according to the following equation: $Et_2Pb(OH)_2 = Pb(OH)_2 + 2Et$ (80); Forms hexahydrate which loses H_2O at ~120° and blackens > 210°; over $H_2SO_4 \to Et_2PbO$ (217).
Rxns. with: $CO_2 \to Et_2Pb(CO_3H)_2$, which on standing over activated Al_2O_3 yields Et_2PbCO_3 (79).
$HCl \to Et_2PbCl_2$ (217).

BIS(AMINOPHENYL)LEAD DIHYDROXIDE $(H_2NC_6H_4)_2Pb(OH)_2$

Prepn.: By reducing $(O_2NC_6H_4)_2PbO$ with $TiCl_3$ in concd. HCl-MeOH (474).
Rxn. with: Na_2 pyrocatecholdisulfonate $\to Na_2$ bis(aminophenyl)lead pyrocatecholdisulfonate (474).

TRIMETHYLLEAD HYDROXIDE Me_3PbOH

Rxn. with: $[Na(OH)_x]_2Fe(CO)_4 \to CH_4$ (223).

TRIETHYLLEAD HYDROXIDE Et_3PbOH

Prepn.: By treating Et_3PbCl or Et_3PbBr with wet Ag_2O; 93% yield (79, 328).
By shaking Et_3PbCl with aq. NaOH in C_6H_6 or Et_2O (79, 469).
Props.: Soly. data; decomp. during steam distn. according to the equations: $2Et_3PbOH = Et_4Pb + Pb(OH)_2 + 2Et$: $2Et_3Pb(OH)_2 + Et_4Pb$ (80).
Rxns. with: $RCO_2H \to Et_3PbOCOR$ and $Et_2Pb(OCOR)_2$ (169, 216, 217, 469).
$HCO_2H \to Pb(O_2CH)_2$ (169).
$RSH \to Et_3PbSR$ (216, 338).
Phthalimide $\to Et_3PbN(CO)_2C_6H_4$ (216, 338).
$Fe(CO)_5 \to$ iron carbonyl complex (219).
$Ca[Fe(CO)_4H]_2 \to Et_2PbFe(CO)_4$ (219).
$RSO_2NHR \to Et_3PbNHSO_2R$ (328).

TRIPROPYLLEAD HYDROXIDE Pr_3PbOH
Prepn.: By shaking Pr_3PbCl with moist Ag_2O in EtOH (216, 469).
Rxns. with: Phthalimide → $Pr_3PbN(CO)_2C_6H_4$ (216).
$Ca[Fe(CO)_4H]_2$ → $[Pr_2PbFe(CO)_4]_2$ (222).
RSO_2NH_2 → Pr_3PbNHO_2SR (338, 472).
RCO_2H → Pr_3PbO_2CR (469).
KCN → Pr_3PbCN (469).

TRIBUTYLLEAD HYDROXIDE Bu_3PbOH
Prepn.: By shaking an alc. soln. of Bu_3PbCl with moist Ag_2O (469).

TRICYCLOHEXYLLEAD HYDROXIDE $(C_6H_{11})_3PbOH$
Prepn.: By reacting $[(C_6H_{11})_3Pb]_2$ with ether soln. of $FeCl_3$ + HCl and treating the resulting chloride with alc. KOH under reflux; 99% yield (35).
By reacting $(C_6H_{11})_3PbI$ with 30% KOH (222).
Props.: White flocs sensitive to light (35).
Rxn. with: $Ca[Fe(CO)_4H]_2$ → $[(C_6H_{11})_3Pb]_2Fe(CO)_4$ (222).

TRIPHENYLLEAD HYDROXIDE Ph_3PbOH
Prepn.: By reacting Ph_4Pb with alc. HCl and hydrolyzing the resulting halide with alc. KOH; 76% yield (35).
By oxidation of $(Ph_3Pb)_2$ with $KMnO_4$ in Me_2CO to $(Ph_3Pb)_2O$, followed by hydration of the oxide (35).
Props.: Dipole moment (353).
Rxns. with: Maleic anhydride → $Ph_3PbO_2CCH:CHCO_2H$ and $(Ph_3PbO_2CCH:)_2$ (160, 306).
Cond. HNO_3 → Ph_3PbNO_3 (160).
$Fe(CO)_5$ → dark-gray, insol. prod. (219).
$Ca[Fe(CO)_4H]_2$ → $(Ph_3Pb)_2Fe(CO)_4$ (222).
$NaCo(CO)_4$ → $Ph_3PbCo(CO)_4$ (225).
Use: Stabilizer for vinyl resins (578).

DI- and TRIORGANOLEAD ALKOXIDES AND ARYLOXIDES

Organolead alkoxides are formed by reacting organolead halides with sodium alkoxides.

Organolead aryloxides are formed by reacting tetraorganolead compounds with phenols in the presence or absence of silica gel.

DIETHYLBIS(p-NITROPHENOXY)LEAD $Et_2Pb(OC_6H_4NO_2-p)_2$

Prepn.: By heating Et_4Pb with $p-O_2NC_6H_4OH$ (1:1) in xylene in the presence of silica gel (469).
Props.: Yellow solid; does not melt but burns at high temp. (469).

TRIETHYLLEAD PICRATE $Et_3PbOC_6H_2(NO_2)_3$

Prepn.: By heating Et_4Pb with picric acid in C_6H_6 at 60° (5).
Props.: Mild explosive (5).

ETHOXYTRIETHYLLEAD Et_3PbOEt

Prepn.: By reacting Et_3PbCl with EtONa in EtOH; 94.5% yield (169).
Props.: Instable, colorless oil (169).

DI- and TRIORGANOLEAD SULFONATES

Diorganolead diarylsulfonates are prepared:

1. By reacting diorganolead sulfites with arylsulfonic acids in water.
2. By reacting diorganolead oxides with arylenedisulfonic acid di-sodium salts.

Triorganolead alkyl- and arylsulfonates are prepared:

1. By reacting triorganolead hydroxides with alkyl- or arylsulfonic acids.
2. By reacting tetraorganolead compounds with alkyl- or arylsulfonic acids in the presence of silica gel.
3. By reacting triorganolead carboxylates with alkali metal salts of arylsulfonic acids.

DIORGANOLEAD DISULFONATES

Compound	Prepared from	Yield %	Properties	Ref.
$Et_2Pb(O_3SC_6H_4Me-o)_2$	Et_2PbSO_3 + $o-MeC_6H_4SO_3H$	60	Colorless, rhombohedral crystals	217
$Et_2Pb(O_3SC_6H_4Me-p)_2$	Et_2PbSO_3 + $p-MeC_6H_4SO_3H$	60	Long needles without m.p.	217
$Ph_2Pb(O_3S)_2C_6H_2(ONa)_2$	Ph_2PbO + sodium pyrocatechol-disulfonate		Sol. in H_2O (1)	474

(1) Rxn. with H_2S → white ppt. (474).

TRIORGANOLEAD SULFONATES AND SULFINATES

R_3PbO_3SR'	Prepared from	Yield %	Properties	Ref.
$Et_3PbO_3SC_6H_4Me-o$	$Et_4Pb + o-MeC_6H_4SO_3H$		m. 189°	216
$Et_3PbO_3SC_6H_4Me-p$	$Et_4Pb + p-MeC_6H_4SO_3H$		m. 170°	216
	$Et_3PbOAc + p-MeC_6H_4SO_3Na$	67.8	m. 109°	169
$Et_3PbO_3S-1-C_{10}H_6-4-NH_2$	$Et_3PbOAc + 4$-aminonaphthalenesulfonic acid salt	20	m. 238-240° (decompn.)	169
$Et_3PbO_3S-2-C_{10}H_7$	$Et_4Pb + 2-C_{10}H_7SO_3H$		m. 152°	216
$Et_3PbO_3SC_6H_3-5-Me-2-NH_2$	$Et_3PbOAc + 2$-amino-5-toluenesulfonic acid salt	22	m. 210° (decompn.)	169
$Et_3PbO_3SC_6H_4NH_2-p \cdot H_2O$	$Et_3PbOAc + p-H_2NC_6H_4SO_3Na$	46	m. 84-86°	169
$Et_3PbO_2SC_6H_4Me-p$	$Et_3PbOAc + p-MeC_6H_4SO_2Na$	44	m. 86-88°	169
$Et_3PbO_2S-cyclo-C_6H_{11}$	$Et_3PbOAc + cyclo-C_6H_{11}SO_2Na$	40	m. 132-134°	169
$Pr_3PbO_3SC_6H_4Me-o$	$Pr_3PbOH + o-MeC_6H_4SO_3H$		m. 86-87°*	469
	$Pr_4Pb + o-MeC_6H_4SO_3H$		m. 87°	216
$Pr_3PbO_3SC_6H_4Me-p$	$Pr_3PbOH + p-MeC_6H_4SO_3H$		m. 82-83°	469
	$Pr_4Pb + p-MeC_6H_4SO_3H$		m. 73-74.5°*	216
$Bu_3PbO_3SC_6H_4Me-p$	$Bu_3PbOH + p-MeC_6H_4SO_3H$		m. 81-82°	469
$Bu_3PbO_3S-2-C_{10}H_7$	$Bu_3PbOH + 2-C_{10}H_7SO_3H$		m. 68°	469
$(Me_2CHCH_2)_3PbO_3SC_6H_4Me-p$	$(Me_2CHCH_2)_4Pb + p-MeC_6H_4SO_3H$		Darkens at 176-180°, sinters at 190-195°*	216
$(Me_2CHCH_2)_3PbO_3SC_6H_4Me-o$	$(Me_2CHCH_2)_4Pb + o-MeC_6H_4SO_3H$		Darkens at 166°, sublimes on decomp. 172°	216
Et_3Pb-camphorsulfonate	Et_3PbOH or $Et_4P + d-10$-camphorsulfonic acid	86.3	m. 172°	169

* Sternutatory effect (216).

TRIORGANOLEAD CARBOXYLATES

Triorganolead carboxylates are formed:

1. By reacting tetraorganolead compounds with equivalent amounts of carboxylic acids in alcohol in the presence or absence of silica gel.
2. By reacting triorganolead hydroxides with carboxylic acids or acid anhydrides in alcohol.
3. By reacting triorganolead halides with silver or potassium carboxylates.
4. By treating triorganolead acetates with alkali carboxylates in water.
5. By treating lead tetraacetate with three mole equivalents of tetraorganolead compounds.
6. By passing carbon dioxide through triorganolead hydroxides. The reaction yields triorganolead carbonates.

TRIETHYLLEAD CARBONATE $(Et_3Pb)_2CO_3$

Prepn.: By passing CO_2 through triorganolead hydroxide.
Props.: Soly. data; decomp. during steam distn. according to the equations: $(Et_3Pb)_2CO_3 = Et_4Pb + Et_2PbCO_3$; $(Et_3Pb)_2CO_3 + H_2O = Et_3PbOH + CO_2$; and $(Et_3Pb)_2CO_3 + H_2O = Et_4Pb + Et_2Pb(OH)_2 + CO_2$ (80).

TRIORGANOLEAD CARBOXYLATES

$R_3PbOCOR$	Prepared from	Yield %	Properties	Ref.
$Me_3PbOOCH$	$Me_4Pb + HCOOH$ (reflux)		m. 113°	216
Me_3PbOAc	$Me_4Pb + AcOH$	84	White crystals, m. 183-184°	79
			m. 194°	216
$Me_3PbOCOCH_2Cl$	$Me_3PbCl + ClCH_2CO_2Ag$ in H_2O		White needles, m. 169°*	216
	$Me_4Pb + ClCH_2CO_2H$ in Et_2O		m. 167°	216
$Me_3PbOCOCCl_3$	$Me_4Pb + Cl_3CCO_2H$ in Et_2O		Darkens at 180°	216
$Me_3PbOCOCH_2CHMe_2$	$Me_4Pb + Me_2CHCH_2CO_2H$		m. 160°*	216
Et_3PbOAc	$Et_4Pb + Pb(OAc)_4$ (3:1) in $CHCl_3$		m. 160-160.5°* (1,2)	289, 290, 469
$Et_3PbOCOCH_2F$	$Et_4Pb + FCH_2CO_2H$		m. 180.5° (decompn.)*(3)	471
$Et_3PbOCOCH_2Cl$	$Et_3PbCl + ClCH_2CO_2Ag$	79	m. 147°	217
$Et_3PbOCOCCl_3$	$Et_3PbOH + Cl_3CCO_2H$		m. 141°	216
	$Et_4Pb + Cl_3CCO_2H$			216
$Et_3PbOCOCH_2Br$	$Et_3PbOH + BrCH_2CO_2H$		m. 120°	216
$Et_3PbOCOEt$	$Et_4Pb + EtCO_2H$ (SiO_2)		m. 141°	216
$Et_3PbOCOCH_2CH_2Cl$	$Et_3PbOH + ClCH_2CH_2CO_2H$		Colorless prisms, m. 106°*	216
$Et_3PbOCOOEt$	$Et_3PbOH + EtOCO_2H$ in EtOH		Colorless needles, m. 55° (4)	216
$Et_3PbOCOCH:CH_2$	$Et_3PbOH + CH_2:CHCO_2H$		Sinters at 120°*	216
$Et_3PbOCOCH:CHMe$	$Et_3PbOH + MeCH:CHCO_2H$	98.5	Needles, m. 135-136°	216
$Et_3PbOCOCH_2CHMe_2$			*	338
$Et_3PbOCO(CH_2)_4Me$	$Et_4Pb + Me(CH_2)_4CO_2H$ (SiO_2)		White crystals, m. 94.5-95.3° (5)	6
Et_3PbSAc	$Et_3PbOH + AcSH$ in H_2O		White needles, m. 45°	217

* The compounds have sternutatory effect (338).
(1) Rxn. with carboxylic acids → Et_3Pb carboxylates (169).
(2) Rxn. with $Hg(OAc)_2$ → $Et_2Pb(OAc)_2$ (402).
(3) Convulsant effect upon subcutaneous administration (337).
(4) Decomp. at 140° with effervescence and solidifies (216).
(5) Useful as rust and emulsion inhibitor for lubricating oil (6).

TRIORGANOLEAD CARBOXYLATES (Cont'd.)

$R_3PbOCOR$	Prepared from	Yield %	Properties	Ref.
$Et_3PbOCOCH:CHCO_2C_6H_{13}$			(6;7)	817
$Et_3PbOCOCH:CHCO_2CH_2CHEtBu$			(6)	817
$Et_3PbOCOCCl:CHMe$	$Et_3PbOH + MeCH:CClCO_2H$	75	m. 153-155°	169
$Et_3PbOCOC:CHCH:CHO$	$Et_3PbOH + $ 2-furoic acid	55	m. 156-157° (decompn.)	169
$(Et_3PbOCOCH:)_2$	$Et_3PbOH + $ maleic acid	83	Decomp. 165°	169
$(Et_3PbOCOCH_2CH_2)_2$	$Et_3PbOH + (CH_2CH_2CO_2H)_2$	83	Chars at 190°	169
Et_3PbOBz	$BzOH$ in $EtOH$		m. 126°	217
$Et_3PbO_2CC_6H_4NH_2$-o	$Et_3PbOH + $ o-$H_2NC_6H_4CO_2H$	63	m. 96°	469
$Et_3PbO_2CC_6H_4NO_2$-p	$Et_3PbOH + $ p-$O_2NC_6H_4CO_2H$	29.7	Needles, m. 167-168.5°	169
$Et_3PbO_2CC_6H_4NO_2$-m	$Et_3PbOH + $ m-$O_2NC_6H_4CO_2H$	11.4	m. 172-173° (decompn.)	169
$Et_3PbO_2CC_6H_4NHMe$-o	$Et_3PbOH + $ o-$MeNHC_6H_4CO_2H$	36	m. 132.7 (decompn.)	169
$Et_3PbO_2CC_6H_4NO_2$-o	$Et_3PbOH + $ o-$O_2NC_6H_4CO_2H$	81.6	m. 142-143	169
$Et_3PbO_2CC_6H_4NHPh$-o	$Et_3PbOH + $ o-$PhNHC_6H_4CO_2H$	57	m. 124.5-125°	169
$Et_3PbO_2CC_6H_4OMe$-o	$Et_3PbOH + $ o-$MeOC_6H_4CO_2H$	80	m. 97-98°	169
$Et_3PbO_2CC_6H_4Cl$-p	$Et_3PbOH + $ p-$ClC_6H_4CO_2H$	62.5	m. 123-124°	169
$Et_3PbO_2CC_6H_4Br$-m	$Et_3PbOH + $ m-$BrC_6H_4CO_2H$	30	m. 113-114°	169
$Et_3PbO_2CC_6H_4Br$-o	$Et_3PbOH + $ o-$BrC_6H_4CO_2H$	45	m. 134-135°	169
$Et_3PbO_2CC_6H_4Br$-p	$Et_3PbOH + $ p-$BrC_6H_4CO_2H$	53.3	m. 127-128°	169
$Et_3PbO_2CC_6H_4I$-m	$Et_3PbOH + $ m-$IC_6H_4CO_2H$	50.4	m. 135-136°	169
$Et_3PbO_2CC_6H_4I$-o	$Et_3PbOH + $ o-$IC_6H_4CO_2H$	28	m. 138.5-139°	169
$Et_3PbO_2CC_6H_4I$-p	$Et_3PbOH + $ p-$IC_6H_4CO_2H$	52	m. 129.5-130.5°	169
$Et_3PbO_2CC_6H_4OH$-o	$Et_3PbOH + $ o-$HOC_6H_4CO_2H$	19	m. 74.5-75.5°	169
$Et_3PbO_2CCH_2Ph$	$Et_3PbOH + PhCH_2CO_2H$	90	m. 96-97°	169
$Et_3PbO_2CC:CPh$	$Et_3PbOH + PhC:CCO_2H$	84	m. 149-150° (decompn.)	169
$Et_3PbO_2CCHPh_2$	$Et_3PbOH + Ph_2CHCO_2H$	87	m. 164-165°	169
$Et_3PbO_2CCH:CHPh$	$Et_3PbOH + PhCH:CHCO_2H$	79	m. 122-123° (decompn.)	169
$Et_3PbO_2CCH:CHCH_2COPh$	$Et_3PbOH + BzCH_2CH:CHCO_2H$	50	m. 139-141° (decompn.)	169
$Et_3PbO_2CCH:CHCH_2C:CHCH:CHO$	$Et_3PbOH + $ 2-furyl-CH_2-$CH:CHCO_2H$		m. 132-133° (decompn.)	169
Et_3PbO_2C-9-fluorenyl	$Et_3PbOH + $ 9-fluorenyl-CO_2H	23.7	Darkens at 195-196°, decomp. 208°	169
Et_3PbO_2C-9-ethylcarbazole	$Et_3PbOH + $ 9-ethylcarbazole-carboxylic acid	76	m. 195° (decompn.)	169
Et_3PbO_2C-2-lepidine	$Et_3PbOH + $ 2-lepidine-carboxylic acid	86	m. 153-155°, resolidified and decomp. at 197-199°	169

$Et_3PbO_2C(CH_2)_2-2-C_{10}H_7$	$Et_3PbOH + 2-C_{10}H_7(CH_2)_2CO_2H$	78.8	$m.$ 134.5°	169
Et_3Pb-styphnate			(8)	64
$Et_3PbO_2CCPh_3$	$Et_3PbOH + Ph_3CCO_2H$		m. 134-136°	169
$(Et_3Pb)_2$ camphorate	Et_3PbOH + camphoric acid	71	Darkens at 175-190°	169
$(Et_3Pb)_3$ citrate	Et_3PbOH + citric acid	92		169
$(CH_2:CH)_3PbOAc$	$(CH_2:CH)_4Pb + AcOH$			680a
$(CH_2:CH)_3PbO_2CCCl_3$	$(CH_2:CH)_4Pb + Cl_3CCO_2H$			680a
Pr_3PbOAc	$Pr_4Pb + AcOH (SiO_2)$		m. 128°	216
$Pr_3PbOCOCH_2Cl$	$Pr_4Pb + ClCH_2CO_2H (SiO_2)$		m. 109-110°	216
	$Pr_3PbOH + ClCH_2CO_2H$			469
$Pr_3PbOCOCH_2Br$	$Pr_3PbOH + BrCH_2CO_2H$		m. 93-94°*	469
$Pr_3PbOCOCH_2I$	$Pr_3PbOH + ICH_2CO_2H$		m. 88-89°*	469
$Pr_3PbOCOCCl_3$	$Pr_3PbOH + Cl_3CCO_2H$		m. 139-140°*	469
$Pr_3PbOCOCH:CH_2$	$Pr_3PbOH + CH_2CHCO_2H$		m. 123°*	469
$Pr_3PbOCOCH_2CH_2Cl$	$Pr_3PbOH + ClCH_2CH_2CO_2H$		m. 99-100°*	469
$Pr_3PbOCOCH:CHMe$	$Pr_3PbOH + MeCH:CHCO_2H$		m. 135°	469
$Pr_3PbOCOPr$	$Pr_3PbOH + PrCO_2H$		m. 105-106°**	469
$Pr_3PbOCOCH_2CHMe_2$	$Pr_3PbOH + Me_2CHCH_2CO_2H$		m. 110-111°**	469
$Pr_3PbO_2CC_6H_4NH_2$-o	$Pr_3PbOH + o-H_2NC_6H_4CO_2H$		m. 57-58°	469
$Pr_3PbSCSOEt$	$Pr_3PbCl + KSCSOEt$		m. 57.5°	469
$Pr_3PbOCOEt$	$Pr_3PbOH + EtCO_2H$		m. 121-122°	469
Bu_3PbOAc	$Bu_3PbOH + AcOH$		m. 86°*	469
$Bu_3PbOCOCH_2Cl$	$Bu_3PbOH + ClCH_2CO_2H$		60°	469
$Bu_3PbOCOCCl_3$	$Bu_3PbOH + Cl_3CCO_2H$		m. 119°*	469
$Bu_3PbOCOCH_2Br$	$Bu_3PbOH + BrCH_2CO_2H$		m. 54-55°*	469
$Bu_3PbOCOCH_2I$	$Bu_3PbOH + ICH_2CO_2H$		m. 83°*	469
$Bu_3PbOCOEt$	$Bu_3PbOH + EtCO_2H$		m. 79-80°*	469
$Bu_3PbOCOCH_2CH_2Cl$	$Bu_3PbOH + ClCH_2CH_2CO_2H$		m. 65-66°	469
$Bu_3PbOCOCH:CHMe$	$Bu_3PbOH + MeCH:CHCO_2H$		m. 119°	469
$(Bu_3PbCH:)_2$			(9)	817
$(Me_2CHCH_2)_3PbOAc$	$(Me_2CHCH_2)_4Pb + AcOH$		m. 117°*	216

(6) Useful as stabilizer for vinyl halides resins (817).
(7) In conjunction with Bu_2SnPh_2 useful as stabilizer for vinyl halide polymers (752).
(8) Useful as component in electric blasting caps (64).
(9) Useful as stabilizer for vinyl resins (817).
* Sternutatory effect (216, 338).

TRIORGANOLEAD CARBOXYLATES (Cont'd.)

$R_3PbOCOR$	Prepared from	Yield %	Properties	Ref.
$\{Me_2CHCH_2\}_3PbOCOEt$	$(Me_2CHCH_2)_4Pb + EtCO_2H$		m. 118°	216
$\{Me_2CHCH_2\}_3PbOCOPr$	$(Me_2CHCH_2)_4Pb + PrCO_2H$		Colorless needles, m. 119°*	216, 369
$(n-C_{12}H_{25})_3PbOAc$	$(n-C_{12}H_{25})_3PbCl + AgOAc$		m. 59°	160, 306
$Ph_3PbOCH:CHCO_2H$	$Ph_3PbOH + (:CHCO)_2O$ (1:1)	90	m. 207°	285
$Ph_3PbOCOCH_2CO_2Et$	$Ph_3PbCl + KO_2CCH_2CO_2Et$	63.6		
$Ph_3PbOCOCH(CH_2Ph)CO_2Et$	$Ph_3PbCl + KO_2CCH(CH_2Ph)CO_2Et$	34.6	m. 131-132°; at 128-132°/25mm → $Ph_3PbCH(CH_2Ph)CO_2Et$	285
$(Ph_3PbOCOCH:)_2$	$Ph_3PbOH + (:CHCO)_2O$ (2:1)	82	m. 198-199°	160, 306
$Ph_3PbOCOC_{11}H_{23}$	$Ph_4Pb + n-C_{11}H_{23}CO_2H$		m. 91°	132
$Ph_3PbOCOC_{13}H_{27}$	$Ph_4Pb + n-C_{13}H_{27}CO_2H$		m. 102-103°	132
$Ph_3PbOCOC_{15}H_{31}$	$Ph_4Pb + n-C_{15}H_{31}CO_2H$		m. 110°	132
$Ph_3PbOCOC_{17}H_{35}$	$Ph_4Pb + n-C_{17}H_{35}CO_2H$		m. 112°	132

$R_2R'PbOCOR''$				
$Et_2PrPbOCOCH_2Cl$	$Et_2PbPr_2 + ClCH_2CO_2H$		m. 112°	217, 338
$Et_2PrPbOCOEt$	$Et_2PbPr_2 + EtCO_2H$		m. 86°	217, 338
$Et_2BuPbOCOCH_2Cl$	$Et_2PbBu_2 + ClCH_2CO_2H$		m. 82-83°	217

* Sternutatory effect (216).

6. ORGANOLEAD-SULFUR COMPOUNDS

ORGANOLEAD SULFIDES, MERCAPTIDES, AND THIOCARBOXYLATES

Diorganolead sulfides are prepared by reacting diorganolead dicarboxylates with hydrogen sulfide in diluted acetic acid.

Bis(triorganolead) sulfides are prepared by reacting triorganolead halides with sodium sulfide in water at $0°$.

Triorganolead mercaptides are prepared by reacting triorganolead hydroxides with mercaptans or thiophenols.

Diorganolead bisthiocarboxylates are prepared by reacting diorganolead sulfites with thiocarboxylic acids or by cleaving tetraorganolead compounds with thiocarboxylic acids at elevated temperatures.

Triorganolead thiocarboxylates are prepared by reacting triorganolead hydroxides with thiocarboxylic acids.

DIPHENYLLEAD SULFIDE Ph_2PbS
Prepn.: By passing H_2S through a cooled soln. of $Ph_2Pb(OAc)_2$ in dild. AcOH (329).
Props.: Pale yellow prisms, m. $130°$; mol. wt. detd. cryoscopically in C_6H_6 indicates a polymer form (329).

BIS(TRIETHYLLEAD)SULFIDE $(Et_3Pb)_2S$
Prepn.: By adding dropwise $Na_2S \cdot 9H_2O$ in H_2O to Et_3PbCl in H_2O at $0°$; 86% yield (79).
Props.: Yellow-green liquid, f.p. $-45.1°$; $n^{20}D$ 1.6249; d^{20}_4 2.05 (79).
Rxn. with: Air $\rightarrow (Et_3Pb)_2SO_4$ (79).

TRIETHYLLEAD ETHYL SULFIDE Et_3PbSEt
Prepn.: By reacting Et_3PbOH with $EtSH$ in Et_2O (216, 338).
Props.: Colorless liquid, b. $76-78°/0.075mm$ (216); Sternutatory effect (216, 338).

TRIETHYLLEAD PHENYL SULFIDE Et_3PbSPh
Prepn.: By shaking Et_3PbOH with PhSH in Et_2O (216, 338).
Props.: Pale-yellow oil, decomp. during distn. forming $(PhS)_2Pb$ (216); Sternutatory effect (338).
Rxn. with: Dry HCl in $Et_2O \to Et_3PbCl$ + PhSH (216).

DIETHYLLEAD BIS(THIOACETATE) $Et_2Pb(SAc)_2$
Prepn.: By reacting Et_2PbSO_3 with AcSH in the cold (217).
By heating Et_4Pb with AcSH at 100° (217).
Props.: Yellow, rectangular tables, m. 84.5-85°; decomp. 160° (217).
Rxn. with: Concd. HCl $\to Et_2PbCl_2$ (217).

TRIETHYLLEAD THIOACETATE Et_3PbSAc
Prepn.: By reacting triethyllead hydroxide with thioacetic acid in water (217).
Props.: White needles, m. 45° (217).

7. ORGANOLEAD-AMIDES AND -IMIDES

ORGANOLEAD-IMIDES

N-Triorganoplumbyl-arylenedicarbimides of the general formula $R_3PbNCORCO$ are prepared by reacting triorganolead hydroxides or chlorides with arylenedicarbimides or hydrazides. The compounds have sternutatory effect.

Compound	Prepared from	Solvent	Yield %	Properties	Ref.
$Et_3Pb\overline{NCOC_6H_4\text{-}o\text{-}CO}$	Et_3PbOH + o-$C_6H_4(CO)_2NH$	EtOH	90	m. 131°; fungicide	216, 328
$Et_3Pb\overline{NCOC_6Cl_4\text{-}o\text{-}CO}$	Et_3PbOH + o-$C_6Cl_4(CO)_2NH$	EtOH	83		328
$Pr_3Pb\overline{NCOC_6H_4\text{-}o\text{-}CO}$	Pr_3PbOH + o-$C_6H_4(CO)_2NH$	EtOH			328
$Et_3Pb\overline{NCOC_6H_4\text{-}o\text{-}CONH}$	Et_3PbCl + Na-$\overline{NCOC_6H_4\text{-}o\text{-}CONH}$	H_2O	59.5	Fungicide, bactericide, insecticide	327

ORGANOLEAD-SULFONAMIDES AND SULFOCARBIMIDES

N-Triorganoplumbyl-sulfonamides and -carbosulfonimides of the general formula $R_3PbNR'SO_2R''$ are prepared by reacting triorganolead chlorides, R_3PbCl, with sulfonamides, $R''SO_2NHR'$, or triorganolead hydroxides, R_3PbOH, with sodium sulfonamides, $R''SO_2NH'Na$ in benzene or ethanol. The compounds have a sternutatory effect.

Compound	Prepared from	Solvent	Yield %	Properties	Ref.
$Et_3PbNHSO_2Me$	$Et_3PbCl + MeSO_2NH_2$	C_6H_6	100	m. 97°*	472
$Et_3PbNHSO_2Ph$	$Et_3PbCl + PhSO_2NHNa$	EtOH		m. 132°*	472
$Et_3PbNHSO_2C_6H_4NH_2$-p	$Et_3PbOH + p-H_2NC_6H_4SO_2NH_2$	EtOH		m. 171°*	472
$Et_3PbNHSO_2C_6H_4Me$-o	$Et_3PbOH + o-MeC_6H_4SO_2NH_2$	EtOH		m. 133°	472
$Et_3PbNHSO_2C_6H_4Me$-p	$Et_3PbCl + p-MeC_6H_4SO_2NHNa$	EtOH	73	m. 127°, extremely potent sternutator (1)	338, 472
$Et_3PbN(Ph)SO_2CH:CH_2$	$Et_3PbOH + p-MeC_6H_4SO_2NH_2$	EtOH	80	m. 116°*	472
$[Et_3PbN(Ph)SO_2]_2CH_2$	$Et_3PbCl + CH_2CHSO_2N(Ph)Na$	EtOH	55	Decompn. 180°*	472
$Et_3PbN(Ph)SO_2C_6H_4Me$-p	$Et_3PbOH + CH_2(SO_2NHPh)_2$	EtOH		m. 134°	472
$Et_3PbN(Ph)SO_2C_6H_4Me$-p	$Et_3PbOH + p-MeC_6H_4SO_2NHPh$	EtOH		m. 123°	472
$Et_3PbN(C_6H_4Cl)SO_2C_6H_4Me$	$Et_3PbOH + p-MeC_6H_4SO_2NHC_6H_4Cl$-p	EtOH		m. 117°	472
$Et_3PbN(C_6H_4Br)SO_2C_6H_4Me$	$Et_3PbOH + p-MeC_6H_4SO_2NHC_6H_4Br$-p	EtOH		m. 135°	472
$Et_3PbNCOC_6H_4$-o-SO_2	$Et_3PbOH + o-COC_6H_4SO_2NH$	EtOH		m. 100-101°*	472
$Pr_3PbNHSO_2C_6H_4Me$-p	$Pr_3PbCl + p-MeC_6H_4SO_2NHNa$	EtOH	73	m. 96°*	338, 472
$Pr_3PbNHSO_2Ph$	$Pr_3PbOH + PhSO_2NH_2$	C_6H_6		m. 67°**	472
$Pr_3PbNHSO_2Me$	$Pr_3PbOH + MeSO_2NH_2$	EtOH		m. 101°*	472
$Pr_3PbNHSO_2C_6H_4NH_2$-p	$Pr_3PbOH + p-H_2NC_6H_4SO_2NH_2$	EtOH		m. 104°	472
$Pr_3PbN(Ph)SO_2C_6H_4Me$-p	$Pr_3PbOH + p-MeC_6H_4SO_2NHPh$	EtOH		m. 130°*	472
$Pr_3PbNCOC_6H_4$-o-SO_2	$Pr_3PbOH + o-COC_6H_4SO_2NH$	EtOH			472
$Et_2PrPbNHSO_2C_6H_4Me$-p	$Et_2PrPbCl + p-MeC_6H_4SO_2NHNa$	Et_2O		m. 94.5°	217, 338
$Et_2PrPbNHSO_2C_6H_4Me$-p	$Et_2PrPbOH + p-MeC_6H_4SO_2NH_2$				217

(1) Rxn. with dry HCl in $Et_2O \rightarrow Et_2PbCl_2 + Et_3PbCl$ (472).
* Sternutatory effect (338).
** Most potent sternutator (338).

8. ORGANOLEAD CYANIDES, CYANATES, THIOCYANATES, SELENOCYANATES, NITRATES, SULFATES, SULFITES, SELENITES, AND PHOSPHONATE

Organolead cyanides, cyanates, thiocyanates, and selenocyanates are prepared by reacting organolead halides, carboxylates, or hydroxides with alkali metal cyanides, cyanates, thiocyanates, and selenocyanates, repectively. Triethyllead cyanide is also formed from diethyllead dicyanide at 120-260°.

Diorganolead dinitrates are prepared by reacting tetraorganolead or hexaorganodilead compounds with nitric acid or by treating tetraorganolead compounds with dinitrogen trioxide or tetroxide. Triorganolead nitrates are prepared by treating tetraorganolead compounds or triorganolead halides with silver or cupric nitrate, or by reacting triorganolead hydroxides with nitric acid.

Diorganolead sulfites are prepared by treating tetraorganolead compounds with sulfur dioxide in ether in the presence of water.

Diorganolead selenite is prepared by treating tetraorganolead compounds with selenous acid.

Diorganolead sulfate is formed by treating diorganolead dicarboxylates with sulfuric acid.

Triorganolead sulfates are formed by the air oxidation of bis(triorganolead) sulfide.

Bis(triethyllead) fluorophosphonate is prepared by refluxing triethyllead chloride with silver fluorophosphonate in benzene.

ORGANOLEAD SALTS

Compound	Prepared from	Yield %	Properties	Ref.
$Et_2Pb(CN)_2$	Et_2PbCl_2 + KCN		Stable at r.t.; at 120-260° disproportionates → Et_3PbCN + $Pb(CN)_2$ (1)	502, 503
Et_3PbCN	Et_3PbCl + KCN	77	(2,3)	79, 216
	$Et_2Pb(CN)_2$ at 120-260°	63	m. 189°	502
Et_3PbCNO	Et_3PbCl + KCNO		Colorless crystals turning yellow at 230°; shrinking at 183.5°	216
Et_3PbCNS	Et_3PbCl + KCNS	60	White crystals, m. 35°	216
	Et_3PbOAc + KCNS	93	m. 26.5-27°	169
$Et_3PbCNSe$	Et_3PbCl + KCNSe		m. 29.5-30.5°	216
	Et_3PbOAc + KCNSe		m. 33-34°	169
Pr_3PbCN	Pr_3PbOH + KCN		White needles, m. 135°	469
$Et_2Pb(NO_3)_2$	Et_4Pb + N_2O_4 (1:4)			754
$Ph_2Pb(NO_3)_2$	Ph_4Pb + N_2O_4 in $CHCl_3$			156, 372
	Ph_4Pb + concd. HNO_3		(4,5)	483
	Ph_4Pb + N_2O_3 (1:4)		Solid (6)	350a
$(p-MeC_6H_4)_2Pb(NO_3)_2$	$(p-MeC_6H_4)_3Pb$ + HNO_3			402
$(m-O_2NC_6H_4)_2Pb(NO_3)_2$	$Ph_2Pb(NO_3)_2$ + HNO_3 at -50°		(7)	306
Me_3PbNO_3	Me_4Pb + $AgNO_3$ or $Cu(NO_3)_2 \cdot 3H_2O$ at -70° in EtOH	82.7		152
Et_3PbNO_3	Et_4Pb + $AgNO_3$ or $Cu(NO_3)_2 \cdot 3H_2O$	75.5		152
	Et_3PbCl + $AgNO_3$			191
$(n-C_{12}H_{25})_3PbNO_3$	$(n-C_{12}H_{25})_3PbCl$ + $AgNO_3$		m. 44-45°	369

(1) Fungicidal properties (503).
(2) Sternutatory effect (338).
(3) Useful in compounding antifouling marine paints (549).
(4) Rxn. with HNO_3 → $(m-O_2NC_6H_4)_2Pb(NO_3)_2$ (306).
(5) Rxns. with NaBr, NaI, or HCl → Ph_2Pb dihalides (483).
(6) Rxn. with alc. alkali → $(p-MeC_6H_4)_2PbO$ (402).
(7) Rxns. with NaCl or NaI → $(m-O_2NC_6H_4)_2Pb$ dihalides (160, 306).

ORGANOLEAD SALTS (Cont'd.)

Compound	Prepared from	Yield %	Properties	Ref.
Ph_3PbNO_3	$Ph_4Pb + AgNO_3$ or $Cu(NO_3)_2 \cdot 3H_2O$ in EtOH under reflux	86.5	(8)	152, 524
	$Ph_3PbOH + HNO_3$		Needles, sintering at 220-225°	160
	$Ph_3PbNO_3 + HNO_3$			160
$(m-O_2NC_6H_4)_3PbNO_3$				217, 338
Et_2PbSO_3	$Et_4Pb + SO_2 + H_2O$	99	White, amorph. powder (9-11)	169
Et_2PbSeO_3	$Et_4Pb + H_2SeO_3$	15.5		169
$(ClCH:CH)_2Pb(OAc)_2$	$(ClCH:CH)_2Pb(OAc)_2 + H_2SO_4$	66		390
$(Et_3Pb)_2SO_4$	$(Et_3Pb)_2S$ air oxidation	55	Darkens at 171-172°; insol. (12)	79
$(Et_3PbO)_2P(O)F$	$Et_3PbCl + Ag_2PO_3F$		m. 260°	470

(8) Rxn. with $HNO_3 \rightarrow (m-O_2NC_6H_4)_3PbNO_3$ (160).
(9) Rxn. with dil. $HCl \rightarrow Et_2PbCl_2$ (217).
(10) Rxns. with carboxylic acids $\rightarrow Et_2Pb$ dicarboxylates (217, 338).
(11) Rxns. with $RSO_3H \rightarrow Et_3Pb(O_3SR)_2$ (217).
(12) Useful in compounding antifouling marine paints (549).

9. HEXAORGANODILEAD COMPOUNDS

According to the molecular weight and magnetic measurement, triorganolead compounds reported in the older literature are dimeric. In very diluted solutions they are extensively dissociated.

Hexaorganodilead compounds are prepared:

1. By reacting triorganolead halides with metallic sodium.
2. By reacting lead dichloride with organomagnesium halides or with organolithium compounds.
3. By electrolytic reduction of triorganolead halides.
4. By treating diorganolead dihalides with calcium or lithium.

Alkyllithium compounds displace aryl groups in hexaaryldilead compounds and yield the corresponding hexaalkyldilead compounds. Hexaethyldilead, which is formed as a by-product in the preparation of tetraethyllead from lead or lead alloys and ethyl chloride, disproportionates at elevated temperatures to tetraethyllead.

HEXAMETHYLDILEAD $(Me_3Pb)_2$

Prepn.: By treating Me_3PbI with Na in liq. NH_3; 7% yield (68).
By reacting $PbCl_2$ with MeMgI; 61% yield (68).
By electrolytic reduction of Me_3PbCl dissolved in an amt. of 2 N aq. KOH required for the formation of Me_3PbOH (218).
Props.: Yellow plates, m. 37-38°, d. 2.38; decomp. slowly in darkness, rapidly in diffused daylight; crystallizes in hexagonal system (68); Mol. structure and bond length data (513).
Rxns. with: Et_6Pb_2 at elevated temp. → 18% Me_4Pb, 15% Me_3EtPb, 23% Me_2Et_2Pb, 31% $MeEt_3Pb$, and 13% Et_4Pb (78).
CH_2Br_2 in the presence of $AlCl_3$ → $(Me_3Pb)_2CH_2$ (568).
Silica type catalyst at 0-110° → Me_4Pb (339).

HEXAETHYLDILEAD $(Et_3Pb)_2$

Prepn.: By reducing Et_3PbCl with Na in liq. NH_3; 69% yield (147).
By electrolytic reduction of Et_3PbCl dissolved in an amt. of 2 N aq. KOH required for an intermediate formation of Et_3PbOH (218).
By-prod. in the prepn. of Et_4Pb from Pb or PbNa and EtCl (659).
Props.: Yellow liquid decomp. on standing even in a sealed container (147); On heating at 80-95° in the presence of activated C → Et_4Pb (659).
Rxns. with: $(Me_3Pb)_2$ at elevated temp. → 18% Me_4Pb, 15% Me_3EtPb, 23% Me_3Et_2Pb, 31% $MeEt_3Pb$, and 13% Et_4Pb (78).
EtI or EtBr → Et_4Pb (300)!
$Mg + MgI_2$ → Et_4Pb + EtMgI (147).
Silica type catalyst at 0-110° → Et_4Pb (339).

HEXAPROPYLDILEAD $(Pr_3Pb)_2$

<u>Use</u>: Stabilizer for vinyl resins (578).

HEXABUTYLDILEAD $(Bu_3Pb)_2$

<u>Prepn.</u>: By reacting $(Ph_3Pb)_2$ or $[(p-MeC_6H_4)_3Pb]_2$ with BuLi (1:6) in Et_2O (149).

HEXACYCLOHEXYLDILEAD $[(C_6H_{11})_3Pb]_2$

<u>Prepn.</u>: By reacting $PbCl_2$ with $C_6H_{11}Mg$ halides at elevated temp.; 64% yield (147).

<u>Props.</u>: Decompn. 196°; instable in light; at 190° in xylene in a sealed tube → deposition of Pb (147); Mol. wt. detn. indicates the dimeric form (351); Magnetic measurements of 0.5-1.0% solns. indicate that dissocn. to $(C_6H_{11})_3Pb\cdot$ amts. to 10% (238); Dimeric form confirmed by magnetic measurements (379).

<u>Rxns. with</u>: Na in NH_3, followed by $PhCH_2Cl$ → $(C_6H_{11})_3PbCH_2Ph$ (43, 147).
$FeCl_3$-HCl in Et_2O, followed by alc. KOH → $(C_6H_{11})_3PbOH$ (35).
$Mg + MgI_2$ → $(C_6H_{11})_3PbI$ (147).
I_2 in $CHCl_3$ → $(C_6H_{11})_3PbI$ (43).
O_2 in CCl_4 → $(C_6H_{11})_3PbCl + (C_6H_{11})_2PbCl_2 + CO_2 + C_6H_{11}Cl$ (220).
CBr_4 → explosive formation of $(C_6H_{11})_2PbBr_2$ (220).
Na in Et_2O under N → $(C_6H_{11})_3PbNa$ (221).

HEXAPHENYLDILEAD $(Ph_3Pb)_2$

<u>Prepn.</u>: By reacting Ph_3PbI with Na (1:1) in liq. NH_3; 90% yield (134, 147).
By reacting Ph_3PbNa with $C_2H_4Br_2$ (55).
By reacting Ph_3PbCl with Na_4Pb_9 in liq. NH_3; 90% yield (134).
By treating Ph_2PbF_2 or Ph_2PbCl_2 with Ca or Li in liq. NH_3 (160, 306).
By refluxing $PbCl_2$ with PhLi (1:2) in MePh; 79% yield (166, 535).
By-prod. in the rxn. of $PbCl_2$ with PhMgBr (329).

<u>Props.</u>: Stable in darkness; decompn. starts at 170°, melts completely at 225° (m. of Ph_4Pb); insol. in H_2O; sol. in org. solvents (134, 329); Mol. wt. detn. indicates the dimeric form (325, 329, 351); Magnetic susceptibility and dissocn. data (422); Data for dielec. const., density, and polarization (325); Pyrolysis in xylene → Ph_4Pb (147).

<u>Rxns. with</u>: $KMnO_4$, followed by hydrolysis → Ph_3PbOH (35).
$AlCl_3$ → $PbCl_2$, Ph_2PbCl_2, Ph_4Pb, and Ph_3PbCl (146).
Na (1:2) in NH_3, followed by $PhCH_2Cl$ in Et_2O → Ph_3PbCH_2Pb (43, 147).
$Mg + MgI_2$ → Ph_4Pb (147).
MgI_2 in Et_2O-C_6H_6 → Ph_3PbI (147).
HCl in $CHCl_3$ → $PbCl_2$ and Ph_2PbCl_2 (43, 147).
BuLi (1:6) → $(Bu_3Pb)_2$ + PhLi (149).
$(:CHCO)_2O$ → $CH(PbPh_3)CO_2O$ (160, 306).
Li in NH_3, followed by $PhCH_2Cl$ → Ph_3PbCH_2Ph (160, 306).
Alkali or alkaline earth metals in liq. NH_3 → the corresponding metal triphenylplumbides (306).

HEXA-m-TOLYLDILEAD $[(m\text{-MeC}_6H_4)_3Pb]_2$
Prepn.: By reacting $PbCl_2$ with $m\text{-MeC}_6H_4Mg$ halides (147).
By reducing $(m\text{-MeC}_6H_4)_3PbI$ with Na in liq. NH_3; 70% yield (147).
Props.: m. 109°; pyrolysis → $(m\text{-MeC}_6H_4)_4Pb$ (43, 147).

HEXA-o-TOLYLDILEAD $[(o\text{-MeC}_6H_4)_3Pb]_2$
Rxn. with: $AlCl_3$ → $(o\text{-MeC}_6H_4)_4Pb$, $PbCl_2$, and unchanged starting matl. (146).

HEXA-p-TOLYLDILEAD $[(p\text{-MeC}_6H_4)_3Pb]_2$
Props.: Disproportionates in xylene under reflux to $(p\text{-MeC}_6H_4)_4Pb$ (147).
Rxns. with: $Mg + MgI_2$ → $(p\text{-MeC}_6H_4)_4Pb$ (147).
MgI_2 → $(p\text{-MeC}_6H_4)_3PbI$ (147).
Na in NH_3 followed by $PhCH_2Cl$ → $(p\text{-MeC}_6H_4)_3PbCH_2Ph$ (43).
BuLi (1:6) → $(Bu_3Pb)_2$ (149).
Concd. HNO_3 → $(p\text{-MeC}_6H_4)_2PbO$ (288).
Me_2CHCO_2H → $(p\text{-MeC}_6H_4)_2Pb(OCOCHMe_2)_2$ (288).
HNO_3 under reflux → $(p\text{-MeC}_6H_4)_2Pb(NO_3)_2$ (402).

HEXA-o-METHOXYPHENYLDILEAD $[(o\text{-MeOC}_6H_4)_3Pb]_2$
Prepn.: By reacting $PbCl_2$ with $o\text{-MeOC}_6H_4Mg$ halides; 59% yield (147).
Props.: m. 198-201° (decompn.) (43, 147); Pyrolysis → $(o\text{-MeOC}_6H_4)_4Pb$ + Pb (43, 147).
Rxns. with: Na in NH_3, followed by $PhCH_2Cl$ → $(o\text{-MeC}_6H_4)_3PbCH_2Ph$ (43, 147).
$Mg + MgI_2$ → $(o\text{-MeC}_6H_4)_4Pb$ (43, 147).
I_2 in $CHCl_3$ → $(o\text{-MeC}_6H_4)_3PbI$ (43, 147).
Na followed by NH_4Br → PhOMe, $[(p\text{-MeOC}_6H_4)_3Pb]_2$, and $(p\text{-MeOC}_6H_4)_4Pb$ (147)

HEXA-p-METHOXYPHENYLDILEAD $[(p\text{-MeOC}_6H_4)_3Pb]_2$
Prepn.: By reacting $PbCl_2$ with $p\text{-MeOC}_6H_4Mg$ halides; 65.4% yield (147, 148).
Props.: m. 198-200° (decompn.); disproportionates in xylene under reflux → $(p\text{-MeOC}_6H_4)_4Pb$ + Pb (43; 147, 148).
Rxns. with: I_2 in $CHCl_3$ → $(p\text{-MeOC}_6H_4)_4Pb$ and $(p\text{-MeOC}_6H_4)_2PbI_2$ (43, 147).
HCl in $CHCl_3$ → $(p\text{-MeOC}_6H_4)_3PbCl$ and $(p\text{-MeOC}_6H_4)_2PbCl_2$ (148).

HEXA-o-ETHOXYPHENYLDILEAD $[(o\text{-EtOC}_6H_4)_3Pb]_2$
Prepn.: By reacting $EtOC_6H_4Mg$ halides with $PbCl_2$; 23.4% yield (147).
Props.: m. 170-171° (decompn.) (43, 147); pyrolysis → $(o\text{-EtOC}_6H_4)_4Pb$ (43, 147).

HEXA-p-ETHOXYPHENYLDILEAD $[(p\text{-}EtOC_6H_4)_3Pb]_2$
Prepn.: By reacting $EtOC_6H_4Mg$ halides with $PbCl_2$; 58.4% yield (147).
Props.: m. 178-179° (decompn.) (43, 147); pyrolysis → $(p\text{-}EtOC_6H_4)_4Pb$ (43, 147).
Rxns. with: Na in liq. NH_3 followed by $PhCH_2Cl$ → $(p\text{-}EtOC_6H_4)_3PbCH_2Ph$ (43, 147).
I_2 in C_5H_5N → $(p\text{-}EtOC_6H_4)_3PbI$ (43, 147).

HEXAMENTYLDILEAD $[(2,4,6\text{-}Me_3C_6H_2)_3Pb]_2$
Prepn.: By reacting $PbCl_2$ with $Me_3C_6H_2Mg$ halides; 12.3% yield (147, 322).
Props.: Yellow, crystalline powder, decomp. 325° (322); m. 325° (43); Disproportionates on heating in xylene in a sealed tube at 190° → $(Me_3C_6H_2)_4Pb$ and Pb (43, 147); Electron paramagnetic resonance (381).
Rxns. with: Mg + MgI_2 → $(Me_3C_6H_2)_3PbI$ (147).
I_2 in C_5H_5N → $(Me_3C_6H_2)_3PbI$ (43).

HEXA(o-CHLOROBENZYL)DILEAD $[(o\text{-}ClC_6H_4CH_2)_3Pb]_2$
Prepn.: By reacting $PbCl_2$ with $o\text{-}ClC_6H_4CH_2MgBr$ (3:2) in $Et_2O\text{-}C_6H_6$; 22% yield (37).
Props.: Light red crystals, decompn. 170°; magnetic measurements indicate dimeric structure (37).

HEXA(o-BROMOBENZYL)DILEAD $[(o\text{-}BrC_6H_4CH_2)_3Pb]_2$
Prepn.: By reacting $PbCl_2$ with $o\text{-}BrC_6H_4CH_2MgBr$ (3:2) in $Et_2O\text{-}C_6H_6$; 14% yield (37).
Props.: Deep red crystals; decomp. 170°; magnetic measurements indicate dimeric structure (37).

HEXA-1-NAPHTHYLDILEAD $[(1\text{-}C_{10}H_7)_3Pb]_2$
Prepn.: By reacting $PbCl_2$ with $1\text{-}C_{10}H_7MgBr$ or $1\text{-}C_{10}H_7Li$; 6-10% yield (147).
Props.: m. 268-269°; pyrolysis → $(C_{10}H_7)_4Pb$ + Pb (43).

10. BRIDGED BIS(TRIORGANOLEAD) COMPOUNDS

Acetylene-bridged bis(triorganolead) compounds, $R_3PbC\!:\!CPbR_3$, are prepared by reacting triorganolead halides with sodium acetylide in liquid ammonia or with ethynylenebis(magnesium bromide) in ether or chloroform.

Methylene- and polymethylene-bridged organolead compounds are formed by reacting lithium triorganoplumbides, tetraorganolead, or hexaorganodilead compounds with methylene or polymethylene dihalides.

p-Bis(triphenylplumbyl)benzene is formed by reacting triphenyllead chloride with p-dilithiobenzene in ether.

Bis(triphenylplumbyl)succinic anhydride is formed by reacting hexaphenyldilead with maleic anhydride in chloroform in a sealed tube at a temperature below 160°.

ACETYLENE-BRIDGED TRIORGANOLEAD COMPOUNDS

$[R_3PbC\!:\!]_2$	Prepared from	Properties	Ref.
$[Et_3PbC\!:\!]_2$	$Et_3PbBr + NaC\!:\!CH$	b. 135-140°/0.05mm; $d_{20.5}$ 1.904; $n^{20.5}D$ 1.567; Soly. data (1,2)	50
$[(cyclo\text{-}C_6H_{11})_3PbC\!:\!]_2$	$(C_6H_{11})_3PbI + NaC\!:\!CH$	m. 136°, sensitive to moisture	665
$[Ph_3PbC\!:\!]_2$	Ph_3Pb halide + $(\!:\!CMgBr)_2$	m. 138.5°; soly. data (3) (1)	50 50
$[(p\text{-}MeC_6H_4)_3PbC\!:\!]_2$	$Ph_3PbBr + NaC\!:\!CH$ $(p\text{-}MeC_6H_4)_3PbI + NaC\!:\!CH$	m. 130° (3)	665
$[(o\text{-}MeC_6H_4)_3PbC\!:\!]_2$	$(o\text{-}MeC_6H_4)_3PbBr + NaC\!:\!CH$	m. 121° (3)	665
$[(PhCH_2CH_2)_3PbC\!:\!]_2$	$(PhCH_2CH_2)_3PbBr + NaC\!:\!CH$	m. 62°, sensitive to moisture (3)	665

(1) Rxns. with $AgNO_3 \rightarrow Ag_2C_2$ (50).
(2) Rxns. with ammoniacal $Cu^+ \rightarrow Cu_2C_2$ (50).
(3) Rxns. with alkali $\rightarrow R_3PbOH + HC\!:\!CH$ (665).

ALKYLENE- and BENZENE-BRIDGED TRIORGANOLEAD COMPOUNDS

$R_3PbXPbR_3$	Prepared from	Yield %	Properties	Ref.
$(Me_3Pb)_2CH_2$	$(Me_3Pb)_2 + CH_2Br_2/AlCl_3$		Antiknock agent	568
$Me_3PbCH_2CH_2PbMe_2(CH_2-CH_2Br)$	$Me_4Pb + C_2H_4Br_2/AlCl_3$		Antiknock agent	568
$(Ph_3PbCH_2)_2CH_2$	$Ph_3PbLi + Br(CH_2)_3Cl$ or $Br(CH_2)_3Br$	58	m. 94-95°	165, 535
$(Ph_3PbCH_2)_2$	$Ph_3PbLi + Br(CH_2)_4Cl$	46	m. 134-136°	165, 535
$Ph_3PbC_6H_4PbPh_3$	$Ph_3PbCl + p-C_6H_4Li_2$	9	m. 285-288°	156, 372
$[-CH(PbPh_3)CO]_2O$	$(Ph_3Pb)_2 + (:CHCO)_2O$	81	Grayish-brown crystals darkening at 265° (1)	160, 306

(1) Rxn. with 10% NaOH – di-sodium salt $C_{40}H_{32}Na_2O_4Pb_2$ (160, 306).

11. ORGANOLEAD-METAL CARBONYL COMPOUNDS

Organolead-iron and cobalt carbonyl compounds of the general formulae $[R_2PbFe(CO)_4]_2$, $(R_3Pb)_2Fe(CO)_4$, and $Ph_3PbCo(CO)_4$ are formed by reacting triorganolead hydroxides or halides with calcium iron carbonyl hydride or sodium cobalt tetracarbonyl.

Compound	Prepared from	Properties	Ref.
$[Me_2PbFe(CO)_4]_2$	$Me_3PbOH + Ca[Fe(CO)_4H]_2$	Red-orange substance, turns black at 120-130°	222
$[Et_2PbFe(CO)_4]_2$	$Et_3PbOH + Ca\{Fe(CO)_4H\}_2$	Red plates or needles (1)	219
$[Pr_2PbFe(CO)_4]_2$	$Pr_3PbOH + Ca\{Fe(CO)_4H\}_2$	Red-orange substance, m. 98° (2)	222
$[Bu_2PbFe(CO)_4]_2$	$Bu_3PbOH + Ca\{Fe(CO)_4H\}_2$	m. 87°	222
$[Ph_2PbFe(CO)_4]_2$	$Ph_3PbBr + Ca\{Fe(CO)_4H\}_2$	Red crystals, decomp. 152°	225
$[(cyclo-C_6H_{11})_3Pb]_2Fe(CO)_4$	$(cyclo-C_6H_{11})_3PbOH$ or $(cyclo-C_6H_{11})_3PbI + Ca\{Fe(CO)_4H\}_2$	Decomp. 140°	222
$(Ph_3Pb)_2Fe(CO)_4$	Ph_3PbOH or $Ph_3PbBr + Ca\{Fe(CO)_4H\}_2$	Yellow-orange crystals, decomp. 135-140° (3,4)	222
$Ph_3Pb[Co(CO)_4]$	$Ph_3PbOH + NaCo(CO)_4$	Solid, decompn. 98°	225

(1) The compd. burns luminously. At 110° slowly darkens. At 138° gives off a yellow vapor and forms an oily prod.; deflagrates when heated under N; very sol. in org. solvents; decomp. on standing in air and light (219). Mol. wt. (222).
(2) Rxn. mechanism of its formation (223).
(3) Rxn. with Hg halides → $HgFe(CO)_4$ and Ph_3Pb halide (224).
(4) Rxn. with phenanthroline → evolution of one CO/mole of the complex and formation of a brick-red, crystalline substance (225).

12. MISCELLANEOUS ORGANOLEAD COMPOUNDS

Mixed alkyllead compounds containing Me_3EtPb, Me_2Et_2Pb, $MeEt_3Pb$, and Et_4Pb were prepared by reacting EtCl and Me halide with a NaPb alloy at 80-135° in the presence of $AlCl_3$ (911).

A mixture of organotin and lead compounds was prepared by treating an alloy of Pb, Sn, and an alkali metal with an alkyl, aryl, or aralkyl halide (871).

Alkyllead compounds were prepared by reacting alkyl radicals formed by thermal decomposition of gasoline with solid or vaporized lead (883, 895, 896).

Organolead compounds were prepared by reacting Grignard compounds of chlorinated hydrocarbon oils with Pb salts (889).

(o-CARBOXYPHENYL)DIPHENYLLEAD CHLORIDE $Ph_2Pb(C_6H_4CO_2H-o)Cl$
Prepn.: By treating $Ph_2Pb(OH)C_6H_4CO_2H-o$ with 6 N HCl in EtOH; 87% yield (156).
Props.: m. 210-220°; Me ester, m. 170-171°; evolves HCl in cold H_2SO_4 (156).

o-CARBOXYPHENYLDIPHENYLLEAD HYDROXIDE $Ph_2Pb(OH)C_6H_4CO_2H-o$
Prepn.: By treating $Ph_3PbC_6H_4CH_2OH-o$ with $KMnO_4$ in Me_2CO; 35% yield (156).
Props.: m. 300-305° (156).
Rxns. with: HCl in EtOH → $Ph_2Pb(C_6H_4-o-CO_2H)Cl$ (156).

TRIPHENYL(TRIPHENYLSILYLOXY)LEAD $Ph_3SiOPbPh_3$
Prepn.: By reacting Ph_3PbCl with Ph_3SiONa; 56% yield (535).
Props.: White crystals, m. 123-124° (535).

(o-CARBOXYPHENYL)DIPHENYLLEAD INNER SALT $\overline{Ph_2PbC_6H_4COO}$
Prepn.: By oxidizing $Ph_3PbC_6H_4CH_2OH$ with $KMnO_4$ in Me_2CO (372).
Props.: m. 300-305° (372).
Rxn. with: HCl in EtOH → $Ph_2(o-HO_2CC_6H_4)PbCl$ (372).

13. ORGANIC DERIVATIVES OF DIVALENT LEAD

Diorganolead compounds are prepared:

1. By reacting lead dihalides or dinitrate with organo-alkali metal derivatives.
2. By treating tetraorganolead compounds with four equivalents of sodium in an ether-ammonia mixture.
3. By reacting lead diethoxide with malonic acid esters.

DIETHYLLEAD Et_2Pb
Prepn.: By cleaving the Et groups from Et_4Pb by Na in Et_2O-NH_3 (55).

BIS[DI(CARBETHOXY)METHYL]LEAD $Pb[CH(CO_2Et)_2]_2$
Prepn.: By reacting $Pb(OEt)_2 \cdot 2EtOH$ with $CH_2(CO_2Et)_2$ in Et_2O (570).
Rxn. with: $BzH \rightarrow PhCHCH(CO_2Et)_{2\ 2}$ (570).

BISCYCLOPENTADIENYLLEAD $(C_5H_4)_2Pb$
Prepn.: By reacting C_5H_5Na with $Pb(NO_3)_2$ in Me_2NCHO (128).
By reacting C_5H_5M (M = alk. or alk. earth metal) with $PbCl_2$ in liq. NH_3 or other N base; 55% yield (602).
Props.: Yellow, diamagnetic crystals insol. in H_2O, sol. in org. solvents; dipole moment 1.63D (128, 561); m. 132-135° (602); Magnetic data (130); IR spectrum (602, 657); X-ray diffraction data (602).

DI-9-FLUORENYLLEAD $(C_{13}H_9)_2Pb$
Prepn.: By reacting $PbCl_2$ with 9-fluorenyllithium (1:2) in Et_2O at -15° (614).
Props.: m. 238-240° (614).

DI(p-DIMETHYLAMINOPHENYL)LEAD (?) $(Me_2NC_6H_4)_2Pb$
Prepn.: By reacting $p-Me_2NC_6H_4Li$ with $PbCl_2$ (3:1) in Et_2O under reflux and pouring the Et_2O sol. fraction into H_2O (166).
Props.: Orange-red ppt. (166).

m,α,α-TRINITROTOLUENE LEAD DERIVATIVE $[m-O_2NC_6H_4C(NO_2)_2]_2Pb$
Props.: Orange-yellow, explosive ppt. (376).

III. BIBLIOGRAPHY TO DATA SECTION

1. Aldridge, W. N. and Cremer, J. E., Analyst. $\underline{82}$, 37-43 (1957); CA $\underline{51}$, 9417.
2. ----- and -----, Nature $\underline{178}$, 1306-7 (1956); CA $\underline{51}$, 4826.
3. -----, Biochem. J. $\underline{69}$, 367-76 (1958); CA $\underline{52}$, 18834.
4. ----- and Cremer, J. E., ibid., $\underline{61}$, 406-417 (1955).
5. Altamura, M. S., U.S. 2,171,423, Aug. 29, 1939; CA $\underline{34}$, 116.
6. -----, U.S. 2,493,213, Jan. 3, 1950; CA $\underline{44}$, 2229.
7. Anderson, H. H., J. Am. Chem. Soc. $\underline{71}$, 1799-1801 (1949); CA $\underline{43}$, 6536.
8. -----, ibid., $\underline{72}$, 194-6 (1950); CA $\underline{44}$, 5254.
9. -----, ibid., 2089-90; CA $\underline{44}$, 7226.
10. -----, ibid., $\underline{73}$, 5439-40 (1951); CA $\underline{46}$, 10029.
11. -----, ibid., 5440-1; CA $\underline{51}$, 10030.
12. -----, ibid., 5798-9; CA $\underline{46}$, 6083.
13. -----, ibid., 5800-2; CA $\underline{46}$, 5528.
14. -----, ibid., $\underline{74}$, 1421-3 (1952); CA $\underline{46}$, 6538.
15. -----, ibid., 2370-1; CA $\underline{48}$, 2574.
16. -----, ibid., 2371-2; CA $\underline{48}$, 2574.
17. -----, ibid., $\underline{75}$, 814-16 (1953); CA $\underline{48}$, 1949.
17a. -----, ibid., 1576-8.
18. ----- and Vasta, J. A., J. Org. Chem. $\underline{19}$, 1300-5 (1954); CA $\underline{50}$, 164.
19. -----, ibid., 1766-9; CA $\underline{50}$, 856.
20. -----, ibid., $\underline{20}$, 536-41 (1955); CA $\underline{50}$, 3996.
21. -----, ibid., 900-9; CA $\underline{50}$, 5518.
22. -----, ibid., $\underline{21}$, 869-70 (1956); CA $\underline{52}$, 1940.
23. -----, J. Am. Chem. Soc. $\underline{78}$, 1692-4 (1956); CA $\underline{50}$, 13732.
24. -----, ibid., $\underline{79}$, 326-8 (1957); CA $\underline{51}$, 5613.
25. -----, ibid., 4913-15; CA $\underline{52}$, 949.
26. -----, J. Org. Chem. $\underline{22}$, 147-8 (1957); CA $\underline{51}$, 14546.
27. Andrews, T. M., Bower, F. A., LaLiberte, B. R., and Montermoso, J. C., J. Am. Chem. Soc. $\underline{80}$, 4102-4 (1958); CA $\underline{53}$, 3040.
27a. Andrianov, K. A., Ganina, T. N., and Khrustaleva, E. N., Izvest. Akad. Nauk S.S.S.R; Otdel. Khim. Nauk $\underline{1956}$, 798-804; CA $\underline{51}$, 3487.
28. d'Ans, J., Zimmer, H., Endrulat, E., and Lübke, K., Naturwissenschaften $\underline{39}$, 450 (1952); CA $\underline{47}$, 8674.
29. Apperson, L. D., Iowa State Coll. J. Sci. $\underline{16}$, 7-9 (1941); CA $\underline{36}$, 4476.
30. Arbuzov, B. A. and Pudovik, A. N., Zhur. Obshchei Khim. $\underline{17}$, 2158-65 (1947); CA $\underline{42}$, 4522.
31. ----- and Grechkin, N. P., ibid., 2166-77; CA $\underline{42}$, 4522.
32. ----- and -----, ibid., $\underline{20}$, 107-15 (1950); CA $\underline{44}$, 5832.

33. Arbuzov, B. A. and Grechkin, N. P., Izvest. Akad. Nauk S.S.S.R., Otdel. Khim. Nauk. 1956, 440-2; CA 50, 16661.
34. Arntzen, C. E., Iowa State Coll. J. Sci. 18, 6-9 (1943); CA 38, 61.
34a. Bachman, G. B., Carlson, C. L., and Robinson, M., J. Am. Chem. Soc. 73, 1964-5 (1951); CA 46, 2512.
35. Bähr, G., Z. anorg. Chem. 253, 330-6 (1947); CA 44, 1921.
36. -----, ibid., 256, 107-12 (1948); CA 44, 1436.
37. ----- and Zoche, G., Chem. Ber. 88, 542-50 (1955); CA 50, 7734.
38. ----- and -----, ibid., 1450-54; CA 50, 10669.
39. ----- and Gelius, R., ibid., 91, 812-18 (1958); CA 52, 20004.
40. ----- and -----, ibid., 818-24; CA 52, 20004.
41. ----- and -----, ibid., 825-9; CA 52, 20004.
42. ----- and -----, ibid., 829-33; CA 52, 20005.
43. Bailie, J. C., Iowa State Coll. J. Sci. 14, 8-10 (1939); CA 34, 6241.
44. Balandin, A. A., Klabunovskii, E. I., Kozina, M. P. and Ul'yanova, O. D., Izvest. Akad. Nauk S.S.S.R., Otdel. Khim. Nauk 1957, 12-17; CA 52, 8714.
45. Baldwin, R. H. and Bley, R. E., U.S. 2,678,907; May 18, 1954; CA 48, 9402.
46. Bartholomew, E., U.S. 2,398,281, Apr. 9, 1946; CA 40, 3889.
47. -----, U.S. 2,398,282, Apr. 9, 1946; CA 40, 3889.
48. Bawn, C. E. H., and Whitby, F. J., Discussions Faraday Soc. 1947, No. 2, 228-36; CA 43, 5732.
49. Beatty, H. A. and Lovell, W. G., Ind. Eng. Chem. 41, 886-8 (1949); CA 43, 5578.
50. Beermann, C. and Hartmann, H., Z. anorg. u. allgem. Chem. 276, 20-32 (1954); CA 49, 6087.
51. Benkeser, R. A., DeBoer, C. E., Robinson, R. E., and Sauve, D. M., J. Am. Chem. Soc. 78, 682-6 (1956); CA 50, 12871.
52. Best, C. E., U.S. 2,713,585, July 19, 1955; CA 50, 5725.
53. Beste, G. W., Tanner, H. M., and Shapiro, H., U.S. 2,653,159, Sept. 22, 1953; CA 48, 10057.
54. Bindschadler, E. and Gilman, H., Proc. Iowa Acad. Sci. 48, 273-4 (1941); CA 36, 1595.
55. -----, Iowa State Coll. J. Sci. 16, 33-6 (1941); CA 36, 4476.
56. Blitzer, S. M. and Brown, O. M., U.S. 2,622,093, Dec. 16, 1952; CA 47, 2971.
57. Bobashinskaya, C. S. and Kocheshkov, K. A., J. Gen. Chem. (U.S.S.R.) 8, 1850-6 (1938); CA 33, 5820.
58. Booth, R. B. and Kraus, C. A., J. Am. Chem. Soc., 74, 1415-17 (1952); CA 46, 6536.
59. Bowden, K. and Brande, E. A., J. Chem. Soc. 1952, 1068-77; CA 46, 8960.
59a. Brainina, E. M., and Freidlina, R. Kh., Bull. acad. sci. U.S.S.R., Classe sci. chim. 1947, 623-30; CA 42, 5863.
60. Brockway, L. O. and Jenkins, H. O., J. Am. Chem. Soc. 58, 2036-44 (1936).

61. Brook, A. G. and Gilman, H., J. Am. Chem. Soc. <u>76</u>, 77-80 (1954); CA <u>49</u>, 2344.
62. -----, ibid., <u>77</u>, 4827-9 (1955); CA <u>50</u>, 6347.
63. Brown, M. P., and Fowles, G. W. A., J. Chem. Soc. <u>1958</u>, 2811-14; CA <u>53</u>, 205.
63a. Buckler, E. J. and Norrish, R. G. W., ibid., <u>1936</u>, 1567-9; CA <u>31</u>, 921.
64. Burrows, L. A., Filbert, W. F., and Reid, E. E., U.S. 2,105,635, Jan. 18, 1938; CA <u>32</u>, 2357.
65. Burschkies, K., Ber. <u>69B</u>, 1143-6 (1936); CA <u>30</u>, 5189.
66. Burt, S. L., U.S. 2,583,084, Jan. 22, 1952; CA <u>47</u>, 146.
67. Caldwell, J. R., U.S. 2,720,507, Oct. 11, 1955; CA <u>50</u>, 2205.
68. Calingaert, G. and Soroos, H., J. Org. Chem. <u>2</u>, 535-9 (1938); CA <u>32</u>, 6619.
69. ----- and Beatty, H. A., J. Am. Chem. Soc. <u>61</u>, 2748-54 (1939); CA <u>34</u>, 373.
69a. ----- and -----, U.S. 2,270,108, Jan. 13, 1942; CA <u>36</u>, 3190.
70. -----, ----- and Neal, H. R., J. Am. Chem. Soc. <u>61</u>, 2755-8 (1939); CA <u>34</u>, 373.
71. ----- and Soroos, H., ibid., 2758-60; CA <u>34</u>, 373.
72. -----, Beatty, H. A., and Hess, L., ibid., 3300-1; CA <u>34</u>, 980.
73. -----, ----- and Soroos, H., ibid., <u>62</u>, 1099-104 (1940); CA <u>34</u>, 6219.
74. -----, Soroos, H., and Shapiro, H., ibid., 1104-7; CA <u>34</u>, 6219.
74a. -----, -----, and -----, U.S. 2,390,988, Dec. 18, 1945; CA <u>40</u>, 1025.
75. -----, ----- and Hnizda, V., J. Am. Chem. Soc. <u>62</u>, 1107-10 (1940); CA <u>34</u>, 6219.
76. -----, -----, and Thomson, G. W., ibid., 1542-5; CA <u>34</u>, 6219.
77. -----, -----, and Shapiro, H., ibid., <u>63</u>, 947-8 (1941); CA <u>35</u>, 3600.
78. -----, -----, and -----, ibid., <u>64</u>, 462-3 (1942); CA <u>36</u>, 1899.
79. -----, Dykstra, F. J., and Shapiro, H., ibid., <u>67</u>, 190-2 (1945); CA <u>39</u>, 1840.
80. -----, Shapiro, H., Dykstra, F. J., and Hess, L., ibid., <u>70</u>, 3902-6 (1948); CA <u>43</u>, 2111.
81. ----- and -----, U.S. 2,535,190; Dec. 26, 1950; Brit. 673,997, June 18, 1952; CA <u>45</u>, 3864.
82. ----- and -----, U.S. 2,535,191; Dec. 26, 1950; Ger. 875,355; CA <u>45</u>, 3864.
83. ----- and -----, U.S. 2,535,192; Dec. 26, 1950; CA <u>45</u>, 3865.
84. ----- and -----, U.S. 2,535,193; Dec. 26, 1950; CA <u>45</u>, 3865.
85. ----- and -----, U.S. 2,558,207; June 26, 1951; CA <u>46</u>, 131.
86. ----- and -----, U.S. 2,562,856; July 31, 1951; CA <u>46</u>, 1581.

87. Caujolle, F., Lesbre, M., Meynier, D., and Blaigot, A., Compt. rend. 240, 1732-4 (1955); CA 49, 11875.
88. Chamberlain, G. H. N. and Walsh, A. D., Proc. Roy. Soc. (London) A 215, 175-86 (1952); CA 47, 4588.
89. ----- and -----, Inst. intern. chim. Solvay, 8e Conseil chim., Univ. Bruselles, Mécanisme de oxydation, Rapp. et disc. 1950, 155-64, 165-175; CA 47, 9599.
90. Chatt, J. and Williams, A. A., J. Chem. Soc. 1954, 4403-11; CA 49, 5081.
91. ----- and -----, Silicium, Schwefel, Phosphate, Colloq. Sek. Anorg. Chem. Intern. Union Reine Angew. Chem. Münster 1954, 60-3 (Pub. 1955); CA 51, 13796.
92. Church, J. M., Johnson, E. W., and Ramsden, H. E., U.S. 2,591,675, Apr. 8, 1952; Brit. 718,393, Nov. 10, 1954; CA 48, 5207.
93. -----, ----- and -----, U.S. 2,593,267, Apr. 15, 1952; CA 48, 6460.
94. -----, U.S. 2,630,436, Mar. 3, 1953; CA 48, 1420.
95. -----, Ramsden, H. E., Hirschland, H., and Buchanan, H. W., U.S. 2,630,442, Mar. 3, 1953; CA 48, 1420.
96. -----, -----, ----- and -----, U.S. 2,743,257, Apr. 24, 1956; CA 50, 11056.
96a. Clem, W. J. and Plunkett, R. J., U.S. 2,515,821, July 18, 1950; CA 44, 9978.
97. Cleverdon, D., Staudinger, J. J. P., Faulkner, D., and Milner, J. N., U.S. 2,623,892, Dec. 30, 1952; Brit. 692,554; CA 47, 3036.
98. Clouston, J. G. and Cook, C. L., Nature 179, 1240-1 (1957); CA 51, 17468.
99. ----- and -----, Trans. Faraday Soc. 54, 1001-1007 (1958).
99a. Clem, W. J. and Podolsky, H., U.S. 2,426,598, Sept. 2, 1947; CA 42, 203.
100. Colaitis, D. and Lesbre, M., Bull. soc. chim. France 1952, 1069-72; CA 47, 3754.
100a. Cook, B. E., Ph. D. Thesis, Univ. of Michigan, 1949.
101. Cook, C. L. and Clouston, J. G., Nature 177, 1178-9 (1956); CA 50, 14370.
102. Cook, E. B., Smith, R. W., Jr., and Brinkley, S. R., Jr., U.S. Bur. Mines, Rept. Invest. No. 4947, 15 pp. (1953); CA 47, 6127.
103. Costa, G., Gazz. chim. ital. 80, 42-62 (1950); CA 44, 9826, and Ann. chim. (Rome) 41, 207-20 (1951); CA 45, 10094.
104. -----, Ann. chim. (Rome) 40, 541-58 (1950); CA 45, 6507.
105. -----, Camus, A. M., and Pauluzzi, E., Gazz. chim. ital. 86, 997-1013 (1956); CA 53, 1122.
105a. Cramer, P. L., J. Am. Chem. Soc. 60, 1406-10 (1938); CA 32, 5772.
106. ----- and Campbell, J. M., Ind. Eng. Chem. 41, 893-7 (1949); CA 43, 5576.

107. Cremer, J. E., Biochem. J. 67, 87-96 (1957).
108. -----, ibid., 68, 685-92 (1958).
109. Cresswell, W. T., Leicester, J., and Vogel, A. I., Chemistry & Industry 1953, 19.
110. Daudt, H. W., U.S. 2,091,114, Aug. 24, 1937; CA 31, 7446.
111. Davies, A. G. and Hall, C. D., Chemistry & Industry 1958, 1695.
111a. Dibeler, V. H. and Mohler, F. L., J. Research Natl. Bur. Standards 47, 337-42 (1951).
112. -----, J. Research Natl. Bur. Standards 49, 235-9 (1952); CA 47, 4191.
113. Dillard, C. R., McNeill, E. H., Simmons, D. E. and Yeldell, J. B., J. Am. Chem. Soc. 80, 3607-8 (1958); CA 53, 2076.
114. Downs, D., Walsh, A. D., and Wheeler, R. W., Trans. Royal Soc. (London) A 243, 299-358 (1951); CA 46, 244.
115. -----, Street, J. C., and Wheeler, R. W., Fuel 32, 279-309 (1953); CA 47, 9599.
116. Eberly, K. C., U.S. 2,560,034, July 10, 1951; CA 46, 7583.
117. -----, Smith, G. E., Jr., and Albert, H. E., U.S. 2,560,042, July 10, 1951; CA 46, 1581.
118. ----- and Best, C. E., U.S. 2,626,955, Jan. 27, 1953; CA 47, 11224.
119. Edgar, G., Advances Chem. Ser. No. 5, 221-34 (1951); CA 45, 10559.
120. Edgell, W. F., and Ward, C. H., J. Am. Chem. Soc. 76, 1169 (1954); CA 49, 1543.
121. Egorov, Yu. P., Izvest. Akad. Nauk S.S.S.R., Otdel. Khim. Nauk 1957, 124; CA 51, 7861.
122. Ellenburg, A. M., U.S. 2,573,894, Nov. 6, 1951; CA 46, 1428.
123. Eskin, I. T., Nesmeyanov, A. N., and Kocheshkov, K. A., J. Gen. Chem. U.S.S.R. 8, 35-41 (1938); CA 32, 5386.
124. Evans, D. P., J. Chem. Soc. 1938, 1466; CA 33, 129.
125. Faulkner, D. and Milne, J. N., U.S. 2,583,419, Jan. 22, 1952; CA 47, 146; also Brit. 692,556, June 10, 1953.
126. Finholt, A. E., Bond, A. C., Jr., Wilzbach, K. E., and Schlesinger, H. I., J. Am. Chem. Soc. 69, 2692-6 (1947); CA 42, 1144.
126a. -----, ----- and Schlesinger, H. I., ibid., 1199-1203; CA 41, 4396.
127. Fischer, A. K., West, R. C., and Rochow, E. G., ibid., 76, 5878 (1954); CA 49, 2242.
128. Fischer, E. O. and Grubert, H., Z. anorg. u. allgem. Chem. 286, 237-42 (1956); CA 51, 2448.
129. ----- and -----, Z. Naturforsch. 11b, 423-4 (1956); CA 51, 17560.
130. ----- and Piesbergen, U., ibid., 758-9; (1956); CA 51, 12579.
131. Flood, E. A., Godfrey, K. L., and Foster, L. S., Inorganic Synthesis. Vol. III, 64-7; CA 49, 5181.

132. Ford, G. M., Iowa State Coll. J. Sci. 12, 121-2 (1937); CA 32, 4923.
133. Foster, L. S., Gruntfest, I. J., and Fluck, L. A., J. Am. Chem. Soc. 61, 1687-90 (1939); CA 33, 6812.
134. -----, Dix, W. M., and Gruntfest, I. J., ibid., 1685-7; CA 33, 6812.
135. Fox, V. W., Hedricks, J. G., and Ratti, H. J., Ind. Eng. Chem. 41, 1774 (1949); CA 43, 8737.
135a. Fraser, R. G. J. and Jewitt, T. N., Proc. Roy. Soc. (London) A 160, 563-74 (1937); CA 31, 7744.
136. French, F. A. and Rasmussen, R. S., J. Chem. Phys. 14, 389-94 (1946); CA 40, 4603.
136a. Freidlina, R. Kh., Kochetkov, A. K., and Nesmeyanov, N. A., Izvest. Akad. Nauk. S.S.S.R., Otdel Khim. Nauk. 1948, 445-50; CA 44, 10569.
137. Frevel, L. K., Rinn, H. W., and Anderson, H. C., Ind. Eng. Chem., Anal. Ed. 18, 83-93 (1946); CA 40, 2049.
138. Fuchs, R. and Gilman, H., J. Org. Chem. 22, 1009 (1957); CA 52, 5323.
139. -----, Moore, L. O., Miles, D., and Gilman, H., ibid., 21, 1113-17 (1956); CA 52, 295.
140. ----- and Gilman, H., ibid., 23, 911-13 (1958); CA 53, 1121.
141. Garcia Escolar, L. and Martinez, E. N., Anales, real. soc. españ. fis. y quim. (Madrid) 54B, 441-50 (1958); CA 53, 6908.
142. Genta, V. and Ansaloni, A., Gazz. chim. ital. 84, 921-6 (1954); CA 49, 5993.
143. Gershbein, L. L. and Ipatieff, V. N., J. Am. Chem. Soc. 74, 1540-2 (1952); CA 48, 143.
144. Giacomello, G., Gazz. chim. ital. 68, 422-8 (1938); CA 32, 9046.
145. Gilbert, O. G., U.S. 2,621,199, Dec. 9, 1952; CA 47, 9996.
146. Gilman, H. and Apperson, L. D., J. Org. Chem. 4, 162-8 (1939); CA 33, 5819.
146a. ----- and Nelson, J. F., J. Am. Chem. Soc. 59, 935-7 (1937).
147. ----- and Bailie, J. C., ibid., 61, 731-8 (1938); CA 33, 3346.
148. ----- and Towne, E. B., ibid., 739-41; CA 33, 3346.
149. ----- and Moore, F. W., ibid., 62, 3206-8 (1940); CA 35, 85.
150. ----- and -----, and Jones, R. G., ibid., 63, 2482-5 (1941); CA 35, 7369.
151. ----- and Stuckwisch, ibid., 64, 1007-8 (1942); CA 36, 3160.
152. ----- and Woods, L. A., ibid., 65, 435-7 (1943); CA 37, 2355.
153. -----, Haubein, A. H., O'Donnell, G., and Woods, L. A., ibid., 67, 922-6 (1945); CA 39, 3267.
154. ----- and Melvin, H. W., Jr., ibid., 71, 4050-1 (1949); CA 44, 10655.
155. ----- and Jones, R. G., ibid., 72, 1760-1 (1950); CA 44, 7226.

156. Gilman, H. and Melstrom, D. S., J. Am. Chem. Soc. $\underline{72}$, 2953-8; CA $\underline{44}$, 9930.
157. ----- and Arntzen, C. E., J. Org. Chem. $\underline{15}$, 994-1002 (1950); CA $\underline{45}$, 2890.
157a. ----- and Abbott, R. K., ibid., $\underline{8}$, 224-9 (1943); CA $\underline{37}$, 5017.
158. ----- and Arntzen, C. E., J. Am. Chem. Soc. $\underline{72}$, 3823 (1950); CA $\underline{45}$, 1056.
159. ----- and Stuckwisch, C. G., ibid., 4553-5; CA $\underline{45}$, 3343.
160. ----- and Leeper, R. W., J. Org. Chem. $\underline{16}$, 466-75 (1951); CA $\underline{46}$, 1479.
161. -----, Gregory, W. A., and Spatz, S. M., ibid., 1788-91; CA $\underline{46}$, 9100.
162. ----- and Goreau, Th. N., ibid., $\underline{17}$, 1470-4 (1952); CA $\underline{47}$, 8674.
163. ----- and Rosenberg, S. D., J. Am. Chem. Soc. $\underline{74}$, 531-2 (1952); CA $\underline{47}$, 8674.
164. ----- and -----, ibid., 5580-2; CA $\underline{48}$, 1948.
165. ----- and Summers, L., ibid., 5924-7; CA $\underline{47}$, 12283.
166. -----, ----- and Leeper, R. W., J. Org. Chem. $\underline{17}$, 630-40 (1952); CA $\underline{47}$, 1584.
167. ----- and Rosenberg, S. D., ibid., $\underline{18}$, 680-5 (1953); CA $\underline{48}$, 6981.
168. ----- and Wu, T. C., ibid., 753-64; CA $\underline{48}$, 6982.
169. -----, Spatz, S. M., and Kolbezen, M. J., ibid., 1341-51; CA $\underline{48}$, 11304.
170. ----- and Rosenberg, S. D., ibid., 1554-60; CA $\underline{48}$, 13648.
171. ----- and Bindschadler, E., ibid., 1675-8; CA $\underline{49}$, 5275.
172. ----- and Rosenberg, S. D., J. Am. Chem. Soc. $\underline{75}$, 2507-9 (1953); CA $\underline{49}$, 6090.
173. ----- and -----, ibid., 3592-3; CA $\underline{49}$, 5346.
174. ----- and Wu, T. C., J. Am. Chem. Soc. $\underline{77}$, 3228-30 (1955); CA $\underline{50}$, 3274.
175. ----- and Gerow, C. W., ibid., 4675-6; CA $\underline{50}$, 6348.
176. ----- and -----, ibid., 5509-12; CA $\underline{50}$, 6351.
177. ----- and -----, ibid., 5740-2; CA $\underline{50}$, 8556.
178. ----- and Eisch, J., J. Org. Chem. $\underline{20}$, 763-9 (1955); CA $\underline{50}$, 8498.
179. ----- and Gerow, C. W., J. Am. Chem. Soc. $\underline{78}$, 5435-8 (1956); CA $\underline{51}$, 2604.
180. ----- and -----, ibid., 5823-4; CA $\underline{51}$, 2605.
181. ----- and -----, ibid., $\underline{79}$, 342-5 (1957); CA $\underline{51}$, 6533.
182. ----- and Gist, L. A., Jr., J. Org. Chem. $\underline{22}$, 250-5 (1957); CA $\underline{51}$, 16333.
183. ----- and Gerow, C. W., ibid., 334-6; CA $\underline{52}$, 2787.
184. ----- and Gist, L. A., Jr., ibid., 689-90; CA $\underline{52}$, 1940.
185. ----- and -----, ibid., 368-71; CA $\underline{52}$, 1088.
186. ----- and Gerow, C. W., ibid., $\underline{23}$, 1582-4 (1958); CA $\underline{53}$, 7066.

187. Gilman, H., Hughes, M. B., and Gerow, C. W., J. Org. Chem. 24, 352-6 (1959).
188. Gilta, G., Bull. soc. chim. Beg. 46, 263-74 (1937); CA 32, 851.
189. Gingold, K., Rochow, E. G., Seyferth, D., Smith, A. C., Jr., and West, R., J. Am. Chem. Soc. 74, 6306 (1952); CA 47, 11905.
190. Glarum, S. N. and Kraus, C. A., ibid., 72, 5398-5401 (1950); CA 45, 6154.
191. Glockling, I. F., J. Chem. Soc. 1955, 716-20; CA 49, 12277.
192. Gloskey, C. R., U.S. 2,718,522, Sept. 20, 1955; CA 50, 8709.
193. Gregory, W. A., U.S. 2,636,891, Apr. 28, 1953; CA 48, 3397.
193a. Hall, R. W., Ph. D. Thesis, Univ. of Oklahoma (1952).
194. Hannum, J. A., U.S. 2,559,071, July 3, 1951; CA 45, 9836.
195. Harada, T., Sci. Papers Inst. Phys. Chem. Research (Tokyo) 35, 290-329 (1939); CA 33, 5357.
196. -----, ibid., 36, 497-500 (1939); CA 34, 3674.
197. -----, ibid., 501-3; CA 34, 3674.
198. -----, ibid., 504-8; CA 34, 3675.
199. -----, ibid., 38, No. 999, 115-17 (1940); CA 35, 1027.
200. -----, Bull. Chem. Soc. Japan, 15, 455-8 (1940); CA 35, 2110.
201. -----, ibid., 481-3; CA 35, 3225.
202. -----, ibid., 16, 292 (1941); CA 36, 1899.
203. -----, Sci. Papers Inst. Phys. Chem. Research (Tokyo) 38, 146-66 (1940); CA 35, 2470.
204. -----, Bull. Chem. Soc. Japan, 17, 281-2 (1942); CA 41, 4444.
205. -----, ibid., 283-6; CA 41, 4444.
206. -----, Papers Inst. Phys. Chem. Research (Tokyo) 42, 57-8 (1947); CA 43, 7899.
207. ----- and Okuba, T., ibid., 59-61 (173-5 Engl. transl.); CA 43, 7900.
208. -----, ibid., 62-3 (176-7 Engl. transl.): CA 43, 7900.
209. -----, ibid., 64-66 (178-80 Engl. transl.): CA 43, 7900.
210. -----, J. Sci. Research Inst. (Tokyo) 43, 31-3 (91-3 Engl. transl.) (1948); CA 43, 4632.
211. -----, Repts. Sci. Research Inst. (Japan) 24, 177-81 (1948); CA 45, 2356.
211a. Harris, D. M., Nebergall, W. H., and Johnson, O. H., Inorg. Syntheses 5, 70-4.
212. Harris, J. O., U.S. 2,431,038, Nov. 18, 1947; CA 42, 1606.
213. Hartmann, H. and Ahrens, J. U., Angew. Chem. 70, 75 (1958); CA 52, 10922.
214. ----- and Honig, H., ibid., 69, 614 (1957).
215. ----- and -----, ibid., 70, 75 (1958); CA 52, 10922.
216. Heap, R. and Saunders, B. C., J. Chem. Soc. 1949, 2983-8; CA 44, 3880.
217. -----, ----- and Stacey, G. J., ibid., 1951, 658-64; CA 45, 8972.
218. Hein, F. and Klein, A., Ber. 71B, 2381-4 (1938); CA 33, 962.

219. Hein, F. and Pobloth, H., Z. anorg. allgem. Chem. 248, 84-104 (1941); CA 37, 2676.
220. -----, Nebe, E., and Reimann, W., ibid., 251, 125-60 (1943); CA 37, 5326.
221. ----- and -----, Ber. 75B, 1744-50 (1942); CA 38, 1213.
222. ----- and Heuser, E., Z. anorg. Chem. 254, 138-50 (1947); CA 43, 2111.
223. ----- and -----, ibid., 255, 125-40 (1947); CA 43, 4596.
224. ----- and Scheiter, H., ibid., 259, 183-96 (1949); CA 44, 1017.
225. -----, Kleinert, P., and Jehn, W., Naturwissenschaften 44, 34 (1957).
226. Hieber, W. and Breu, R., Angew. Chem. 68, 679-80 (1956); CA 51, 12732.
227. ----- and -----, Chem. Ber. 90, 1270-4 (1957).
228. Hill, J. W., U.S. 2,484,508, Oct. 11, 1949; CA 44, 5899.
229. Holbrook, G. E., U.S. 2,464,397, Mar. 15, 1949; CA 43, 4287.
230. -----, U.S. 2,464,398, Mar. 15, 1949; CA 43, 4287.
231. -----, U.S. 2,464,399, Mar. 15, 1949; CA 43, 4287.
232. Howell, Wm. C., U.S. 2,586,660, Feb. 19, 1952; CA 46, 4211.
233. Ismailzade, I. G. and Zhdanov, G. S., Zhur. Fiz. Khim. 27, 550-3 (1953); CA 48, 7969.
234. -----, Ibid., 26, 1139-43 (1952); CA 49, 2147.
235. ----- and Zhdanov, G. S., ibid., 1619-30; CA 49, 2811.
236. Jackson, J. A., Jr., Dissertation Abstr. 16, 357-8.
237. Jacobs, G., Compt. rend. 238, 1825-6 (1954); CA 48, 11231.
238. Jensen, K. A. and Clauson-Kaas, N., Z. anorg. allgem. Chem. 250, 277-86 (1943); CA 37, 5292.
239. Johnson, E. W. and Church, J. M., U.S. 2,570,686; Oct. 9, 1951; CA 46, 5084; Brit. 661,241, Nov. 21, 1951.
240. -----, U.S. 2,599,557, June 10, 1952; CA 47, 1728.
241. ----- and Church, J. M., U.S. 2,608,567, Aug. 26, 1952; CA 47, 8089.
242. ----- and -----, U.S. 2,672,471, Mar. 16, 1954; CA 49, 2483; Brit. 710,224, June 9, 1954.
242a. -----, Brit. 737,033, Sept. 21, 1955; U.S. 2,763,632, Sept. 18, 1956; CA 50, 13992.
243. Johnson, O. H. and Nebergall, W. H., J. Am. Chem. Soc. 70, 1706-7 (1948); CA 42, 6743.
244. ----- and -----, ibid., 71, 1720-2 (1949); CA 43, 6989.
245. ----- and Harris, D. M., ibid., 72, 5564-6 (1950); CA 45, 6166.
246. ----- and -----, ibid., 5566-8; CA 45, 6166.
247. ----- and Jones, L. V., J. Org. Chem. 17, 1172-6 (1952); CA 47, 4839.

248. Johnson, O. H. and Fritz, H. E., J. Org. Chem. 19, 74-6 (1954); CA 49, 3000.
249. -----, -----, Halvorson, D. O., and Evans, R. L., J. Am. Chem. Soc. 77, 5857-8 (1955); CA 50, 2348.
250. ----- and Schmall, E. A., ibid., 80, 2931-4 (1958); CA 52, 20003.
251. ----- and Holum, J. R., J. Org. Chem. 23, 738-40 (1958).
252. -----, Inorganic Syntheses, Vol. 5, 67.
253. ----- and Harris, D. M., ibid., 74-6.
254. -----, Nebergall, W. H., and Harris, D. M., ibid., 76-8.
255. Jones, L. V., Dissertation Abstr. 13, 308-9 (1953); CA 47, 8635.
256. Jones, W. J., Davies, W. C., Bowden, S. T., Edwards, C., Davis, V. E. and Thomas, L. H., J. Chem. Soc., 1947, 1446-50; CA 42, 4920.
257. Jovellanos, J. U., Taylor, E. S., Taylor, C. F., and Leary, W. A., Natl. Advisory Comm. for Aeronautics, Tech. Note No. 2127, 103 pp. (1950); CA 44, 8629.
258. Kadomtzeff, I., Compt. rend. 226, 407-9 (1948); CA 42, 5731.
259. -----, ibid., 661-3; CA 42, 5731.
260. -----, ibid., 230, 536-7 (1950); CA 44, 7104.
261. -----, Bull. soc. chim. France 1949, D 394-6; CA 44, 1296.
261a. Kaesz, H. D. and Stone, F. G. A., Spectrochimica Acta 1959, 360-77.
262. Karantassis, T. and Basileiados, K., Compt. rend. 205, 460-2 (1937); CA 31, 8505.
263. ----- and -----, ibid., 206, 842-4 (1938); CA 32, 4522.
264. Kerk, G. J. M. van der and Luijten, J. G. A., Organic Syntheses, Vol. 36, Wiley & Sons, Inc., 1956, pp. 86-9.
265. ----- and -----, J. Appl. Chem. (London) 4, 301-7 (1954); CA 49, 15728.
266. ----- and -----, ibid., 307-13; CA 49, 15729.
267. ----- and -----, ibid., 314-19; CA 49, 15729.
268. ----- and -----, ibid., 6, 49-55 (1956); CA 51, 12842.
269. ----- and -----, ibid., 56-60; CA 51, 12842.
270. ----- and -----, ibid., 93-6 (1956); CA 51, 1027.
271. -----, Noltes, J. G., and Luijten, J. G. A., ibid., 7, 356-65 (1957); CA 52, 290.
272. -----, -----, and -----, ibid., 366-9; CA 52, 291.
273. ----- and Luijten, J. G. A., ibid., 369-74; CA 52, 291.
273a. ----- and Noltes, J. G., ibid., 9, 106-113 (1959).
273b. ----- and -----, ibid., 113-120.
274. -----, Luijten, J. G. A., and Noltes, J. G., Chemistry & Industry 1956, 352; CA 51, 262.
275. -----, Noltes, J. G., and Luijten, J. G. A., ibid., 1958, 1290-1; CA 53, 8038.

276. Ker, G. J. M. van der, Luijten, J. G. A., and Noltes, J. G., Angew. Chem. 70, 298-306 (1958); CA 52, 18281.
277. Kerr, K. B., Poultry Sci. 31, 328-36 (1952); CA 46, 11557.
278. Kharasch, M. S., Jensen, F. V., and Weinhouse, S., J. Org. Chem. 14, 429-32 (1949); CA 43, 6971.
279. -----, U.S. 2,453,138, Nov. 9, 1948; CA 44, 2746.
280. Kocheshkov, K. A. and Nad, M. M., J. Gen. Chem. (U.S.S.R.) 5, 1158-67 (1935); CA 30, 1036.
281. -----, Nesmeyanov, A. N., and Klimova, V. A., ibid., 6, 167-71 (1936); CA 30, 4834.
282. -----, -----, and Gipp, N. K., ibid., 172-5.
282a. -----, Nad, M. M., and Aleksandrov, A. P., ibid., 1672-5; CA 31, 2590.
283. -----, Nesmayanov, A. N., and Pusyreva, V. P., Ber. 69B, 1639-42 (1936); CA 30, 6325.
284. -----, Uchenyie Zapriski (Sci. Repts. Moskow State-Univ.) 3, 297-303 (1934); CA 30, 8184.
285. ----- and Aleksandrov, A. P., J. Gen. Chem. (U.S.S.R.) 7, 93-6 (1937); CA 31, 4291.
286. ----- and Borodina, G. M., Bull. aca. sci. U.R.S.S., Classe sci. math. nat., Ser. chim. 1937, 569-75 (In English 576); CA 32, 2095.
287. ----- and Panov, E. M., Izvest. Akad. Nauk., S.S.S.R., Otdel. Khim. Nauk. 1955, 711-17; Bull Acad. Sci. U.S.S.R., Div. Chem. Sci. 1955, 633-8 (Engl. transl.); CA 50, 7075.
288. ----- and -----, Izvest. Akad. Nauk, S.S.S.R., Otdel. Khim. Nauk, 1955, 718-22; CA 50, 7076.
289. Kochetkov, A. K. and Freidlina, R. Kh., ibid., 1950, 203-8; CA 44, 9342.
290. ----- and -----, Uchenyie Zapiski Moskov. Gosudarst. Univ. M. V. Lomonosova No. 132, Org. Khim. No. 7, 144-50 (1950); CA 50, 7728.
291. Kolka, A. J. and Krohn, I. T., U.S. 2,621,200, Dec. 9, 1952; Brit. 719,913, Dec. 8, 1954; CA 47, 10550.
292. Korsching, H., Z. Naturforsch. 1, 219-21 (1946); CA 41, 1902.
293. Korshak, V. V., Polyakova, A. M., Petrov, A. D., and Mironov, V. F., Doklady Akad. Nauk S.S.S.R. 112, 436-8 (1957); CA 51, 13815.
294. Korshunov, I. A. and Orlova, A. A., Zhur. Obshchei Khim. 28, 45-47 (1958); CA 52, 11769.
295. Koton, M. M., J. Gen. Chem. (U.S.S.R.) 9, 2283-6 (1939); CA 34, 5049.
296. -----, ibid., 11, 376-8 (1941); CA 35, 5870.
296a. -----, ibid., 17, 1307-8 (1947); CA 42, 1903.
297. -----, ibid., 26, 3212-14 (1956); CA 51, 8031.
297a. -----, ibid., 18, 936-40 (1948); CA 43, 559.
298. ----- and Kiseleva, T. M., ibid., 27, 2553-8 (1957); CA 52, 7136.

298a. Koton, M. M., Moskvina, E. P., and Florinskii, F. S., J. Gen. Chem. (U.S.S.R.), 20, 2093-5 (1950); CA 45, 5644.
299. Krein, S. E., Izvest. Vsesoyuz. Teplotekh. Inst. 14, No. 3, 15-21 (1941); CA 38, 3462.
299a. Krespan, C. G. and Engelhardt, V. A., J. Org. Chem. 23, 1565 (1958).
300. Krohn, I. T. and Shapiro, H., U.S. 2,555,891, June 5, 1951; CA 46, 523.
301. ----- and -----, U.S. 2,594,183, Apr. 22, 1952; CA 47, 145.
302. -----, U.S. 2,727,053, Dec. 13, 1955; CA 50, 10761.
303. Kuivila, H. G. and Beumel, O. F., Jr., J. Am. Chem. Soc., 80, 3250-3 (1958); CA 53, 1376.
304. Langkammerer, C. M., U.S. 2,253,128, Aug. 19, 1941; CA 35, 8151.
305. Laubengayer, A. W. and Allen, B., Paper presented at the 117-th ACS meeting in Detroit, Apr. 16-20, 1950.
306. Leeper, R. W., Iowa State Coll. J. Sci. 18, 57-9 (1943); CA 38, 726.
307. Leistner, Wm. E. and Knoepke, O. H., U.S. 2,641,588, June 9, 1953; CA 48, 5207.
308. ----- and Hecker, A. C., U.S. 2,641,596, June 9, 1953; CA 48, 5208.
309. ----- and -----, U.S. 2,680,107, June 1, 1954; CA 49, 4713.
310. ----- and -----, U.S. 2,704,756, Mar. 22, 1955; CA 50, 2687; Brit. 737,508, Sept. 28, 1955.
311. ----- and Knoepke, O. H., U.S. 2,726,227, Dec. 6, 1955; CA 50, 6095.
312. ----- and -----, U.S. 2,726,254, Dec. 6, 1955; CA 50, 6095.
313. Lengel, J. H. and Dibeler, V. H., J. Am. Chem. Soc. 74, 2683-4 (1952).
314. Lesbre, M., Compt. rend. 204, 1822-4 (1937); CA 31, 5757.
314a. -----, ibid., 210, 535-6 (1940); CA 34, 4006.
315. ----- and Rouet, J., Bull. soc. chim. France (5) 18, 490 (1951).
316. ----- and Dupont, R., Compt. rend. 237, 1700-2 (1953) and Bull. soc. chim. France (5) 20, 796 (1953); CA 49, 1543.
317. ----- and Roques, G., Compt. rend. 78e Congr. socs. savantes Paris et depts., Sect. sci. 1953, 423-8; CA 49, 15768.
318. ----- and Dupont, R., ibid., 429-37; CA 49, 15729.
319. -----, Buisson, R., Luijten, J. G. A. and Kerk, G. J. M. van der, Rec. trav. chim. 74, 1056-61 (1955); CA 50, 8501.
320. -----, Mazerolles, P., and Voigt, D., Compt. rend. 240, 622-4 (1955); CA 49, 7903.

321. Lesbre, M. and Buisson, R., Bull soc. chim. France 1957, 1204-6; CA 52, 6165.
322. -----, Satgé, J., and Voigt, D., Compt. rend. 246, 594-6 (1958); CA 53, 270.
323. ----- and Mazerolles, P., ibid., 1708-11 (1958); CA 52, 17093.
324. ----- and Satgé, J., ibid., 47, 471-4 (1958); CA 53, 6996.
325. Lewis, G. L., Oesper, P. F., and Smyth, C. P., J. Am. Chem. Soc. 62, 3243-4 (1940); CA 35, 356.
326. Lide, D. R., Jr., J. Chem. Phys. 19, 1605-5 (1951); CA 46, 5972.
327. Ligett, W. B., Closson, R. D., and Wolf, C. N., U.S. 2,595,798, May 6, 1952; CA 46, 7701.
328. -----, ----- and -----, U.S. 2,640,006, May 26, 1953; CA 47, 8307.
329. Lile, W. J. and Menzies, R. J., J. Chem. Soc., 1950, 617-21; CA 44, 7178.
330. Lippincott, E. R. and Tobin, M. C., J. Am. Chem. Soc. 75, 4141-4 (1953).
331. -----, Mercier, Ph., and Tobin, M. C., J. Phys. Chem. 57, 939-42 (1953); CA 48, 2477.
332. Livingston, H. K., Ind. Eng. Chem. 41, 888-93 (1949); CA 43, 5579.
333. -----, ibid., 43, 663-71 (1951); CA 45, 4435.
334. Lodochnikova, V. I., Panov, E. M., and Kocheshkov, K. A., Izvest. Akad. Nauk. S.S.S.R., Otdel. Khim. Nauk. 1957, 1484-6; CA 52, 7245.
335. Long, L. H., J. Chem. Soc. 1956, 3410-16; CA 51, 8565.
336. Luijten, J. G. A. and Pazarro, A., Brit. Plast. 30, 183-6 (1957).
337. McCombie, H. and Saunders, B. C., Nature 158, 382-5 (1946); CA 41, 2807.
338. ----- and -----, ibid., 159, 491-4 (1947); CA 42, 1190.
339. McDyer, Th. W. and Closson, R. D., U.S. 2,571,987, Oct. 16, 1951; CA 46, 3556.
340. Mack, G. P., U.S. 2,604,460, July 22, 1952; CA 46, 10684.
341. ----- and Parker, E., U.S. 2,604,483, July 22, 1952; CA 47, 4358.
342. ----- and -----, U.S. 2,618,625, Nov. 18, 1952; CA 47, 1977.
343. ----- and -----, U.S. 2,626,953, Jan. 27, 1953; CA 47, 11224.
344. ----- and -----, U.S. 2,628,211, Feb. 10, 1953; CA 47, 5165.
345. ----- and Savarese, F. B., U.S. 2,631,990, Mar. 17, 1953; CA 47, 5718.
346. ----- and Parker, E., U.S. 2,634,281, Apr. 7, 1953; CA 48, 1420.

347. Mack, G. P., and Savarese, F. B., U.S. 2,684,973, July 27, 1954; CA 49, 10359.
348. ----- and Parker, E., U.S. 2,700,675, Jan. 25, 1955; CA 50, 397.
349. ----- and -----, U.S. 2,727,917, Dec. 20, 1955; Brit. 766,875; CA 50, 10761.
350. Maier, L., Seyferth, D., Stone, F. G. A., and Rochow, E. G., J. Am. Chem. Soc., 79, 5884-9 (1957); CA 52, 6167.
350a. Makarova, L. G. and Nesmeyanov, A. N., J. Gen. Chem. (U.S.S.R.) 9, 771-9 (1939); CA 34, 391.
351. Malatesta, L., Gazz. chim. ital. 73, 176-80 (1943); CA 38, 5128.
352. ----- and Pizzotti, R., ibid., 344-9 (1943); CA 41, 1137.
353. ----- and -----, ibid., 349-55; CA 41, 1137.
354. -----, Sacco, A. and Ormezzano, L., ibid., 80, 658-62 (1950); CA 46, 4473.
355. Malinovskii, M. S., Trudy Gor'kov. Gosudarst. Pedagog. Inst. 1940, No. 5, 39-41; Khim Reperat. Zhur. 4, No. 2, 45 (1941); CA 37, 3070.
356. Malý, J. and Teplý, J., Chem. Zvesti 7, 553-62 (1953); CA 48, 10401.
357. Manulkin, Z. M., J. Gen. Chem. U.S.S.R. 11, 386-91 (1941); CA 35, 5854.
358. -----, ibid., 13, 42-5 (1943); CA 38, 331.
359. -----, ibid., 46-50; CA 38, 332.
360. -----, ibid., 14, 1047-53 (1944); CA 41, 89.
361. -----, ibid., 16, 235-42 (1946); CA 41, 90.
362. -----, ibid., 18, 299-305 (1948); CA 42, 6742.
363. -----, ibid., 20, 2004-8 (1950); CA 45, 5611.
364. Mapstone, G. E. and Durham, J. S., J. Inst. Petroleum 37, 737-9 (1951); CA 46, 3249; see also Petroleum Refiner 31, No. 2, 132-3 (1952).
365. Mathis, R., Lesbre, M., and Sérée de Roch, I., Compt. rend. 243, 257-9 (1956); CA 50, 16386.
366. Mauss, F., Rev. inst. franç. pétrôle 10, 1126-33 (1955); CA 50, 7423.
367. Mazerolles, P. and Voigt, D., Compt. rend. 240, 2144-6 (1955); CA 49, 14401.
368. ----- and Lesbre, M., ibid., 248, 2018-20 (1959).
369. Meals, R. N., J. Org. Chem. 9, 211-18 (1944); CA 38, 4563.
370. Meen, R. H. and Gilman, H., ibid., 22, 564-5 (1957); CA 52, 1090.
371. Melby, A. O., J. Inst. Petroleum 38, 965-73 (1952); CA 47, 4068.
372. Melstrom, D. S., Iowa State Coll. J. Sci. 18, 65-7 (1943); CA 38, 727.
372a. Merker, R. L. and Scott, M. J., J. Am. Chem. Soc. 81, 975-8 (1959).
373. Milazzo, G., Gazz. chim. ital. 71, 73-81 (1941); CA 37, 1928.

374. Miyanishi, T. and Murata, K., Rept. Japan Assoc. Advancement Sci. 17, 199-204 (1943); CA 44, 2739.
375. Milligan, J. G. and Kraus, C. A., J. Am. Chem. Soc. 72, 5297-3000 (1950); CA 45, 4670.
376. Milone, M. and Massa, A., Gazz. chim. ital. 70, 196-201.
377. Miyanishi, M. and Murata, K., Sci. Papers Inst. Phys. Chem. Research (Tokyo) 41, 99-107 (1943); CA 41, 6696.
378. Monserrat, M. P., Combustibles (Zaragoza) 5, No. 25/6, 15-21, No. 27/8, 67-77 (1945).
379. Morris, H. and Selwood, P. W., J. Am. Chem. Soc. 63, 2509-10 (1941); CA 35, 7248.
380. -----, Byerly, W., and Selwood, P. W., ibid., 64, 1727-9 (1942); CA 36, 5409.
381. Müller, E., Günter, F., Scheffler, K., and Fettel, H., Chem. Ber. 91, 2888-9 (1958); CA 53, 4910.
382. Nad', M. M. and Kocheshkov, K. A., J. Gen. Chem. U.S.S.R. 8, 42-50 (1938); CA 32, 5387.
383. ----- and -----, ibid., 12, 409-13 (1942); CA 37, 3068.
384. Neher, C. M., Padgitt, F. L., and Weimer, P. E., U.S. 2,644,827, July 7, 1953; Brit. 712,644, July 28, 1954; CA 48, 5208.
384a. Neiman, M. B. and Shushunov, V. A., J. Phys. Chem. (U.S.S.R.) 22, 145-55 (1948); CA 42, 5315.
385. Nelson, J. F., Iowa State Coll. J. Sci. 12, 145-7 (1937); CA 32, 3756.
386. Nesmeyanov, A. N., Kocheshkov, K. A., Klimova, V. A., and Gipp, N. K., Ber. 68B, 1877-83 (1935); CA 30, 1679.
387. -----, ----- and Puzyreva, V. P., J. Gen. Chem. U.S.S.R. 7, 118-21 (1937); CA 31, 4290.
388. -----, Borisov, A. E., and Abramova, A. N., Izvest. Akad. Nauk. S.S.S.R., Otdel. Khim. Nauk 1946, 647-50; CA 42, 6316.
389. ----- and Kocheshkov, K. A., ibid., 1945, 522-4; CA 42, 5870.
390. -----, Freidlina, R. Kh., and Kochetkov, A., ibid., 1948, 127-33; CA 43, 1716.
391. -----, Borisov, A. E., and Abramova, A. N., ibid., 1949, 570-7; CA 44, 7759.
392. -----, and Makarova, L. G., Doklady Akad. Nauk S.S.S.R. 87, 421-2 (1952); CA 48, 623.
393. ----- and -----, Izvest. Akad. Nauk S.S.S.R., Otdel. Khim. Nauk. 1954, 380; CA 49, 5346.
394. -----, Borisov, A. E., and Vol'kenau, N. A., ibid., 1956, 162-71; CA 50, 13831.
395. -----, ----- and Novikova, N. V., Doklady Akad. Nauk S.S.S.R. 119, 504-5 (1958); CA 52, 17093.
396. -----, Emel'yanova, L. I., and Makarova, L. G., ibid., 122, 403-4 (1958); CA 53, 3115.
397. Oesper, P. F. and Smyth, C. P., J. Am. Chem. Soc. 64, 173-5 (1942); CA 36, 1222.

398. Ogilvie, J. D. B., Davis, S. G., Thompson, A. L., Grunmitt, W. T., and Winkler, G. A., Can. J. Research 26F, 246-63 (1948); CA 42, 7967.
399. Ol'dekop, Yu. A. and Sokolova, R. F., Zhur. Obshchei Khim. 23, 1159-62 (1953); CA 47, 12226.
400. ----- and Moryganov, B. N., ibid., 2020-3; CA 49, 3097.
401. Ono, S., Rev. Phys. Chem. Japan, Shinkichi Horiba Commem. Vol. 1946, 42-9; CA 44, 4661.
402. Panov, E. M. and Kocheshkov, K. A., Doklady Akad. Nauk S.S.S.R. 85, 1037-40 (1952); CA 47, 6365.
403. ----- and -----, ibid., 1293-5 (1952); CA 47, 6887.
404. ----- and -----, Zhur. Obshchei Khim. 25, 489-92 (1955); CA 50, 3271.
405. -----, Lodochnikova, V. I., and Kocheshkov, K. A., Doklady Akad. Nauk S.S.S.R. 111, 1042-4 (1956); CA 51, 9512.
406. Pascal, P., Compt. rend. 218, 57-9 (1944); CA 40, 1366.
407. -----, Pacault, A., and Tchakirian, A., ibid., 226, 849-51 (1948); CA 42, 4409.
408. Passino, H. J. and Mantell, R., U.S. 2,569,492, Oct. 2, 1951; CA 46, 3556.
409. Pavlovskaya, M. E. and Kocheshkov, K. A., Compt. rend. acad. sci. U.R.S.S. 49, 263-4 (1945); CA 40, 5696.
410. Pearsall, H. W., U.S. 2,414,058, Jan. 7, 1947; CA 41, 2430.
411. Pedley, J. B., Skinner, H. A. and Chernick, C. L., Trans. Faraday Soc. 53, 1612-17 (1957).
412. Petrov, A. D., Mironov, V. F., and Dolgii, I. E., Izvest. Akad. Nauk S.S.S.R., Otdel. Khim. Nauk 1956, 1146-8; CA 51, 4938.
413. ----- and -----, ibid., 1957, 1491-3; CA 52, 7136.
414. Pikina, E. I., Talalaeva, T. V., and Kocheshkov, K. A., J. Gen. Chem. U.S.S.R. 8, 1844-9 (1938); CA 33, 5839.
415. Pitesky, I. and Wiebe, R., Ind. Eng. Chem. 37, 577-9 (1945); CA 39, 3417.
416. Plunkert, R. J., U.S. 2,477,465, July 26, 1949; CA 43, 8398.
417. Polkinhorne, H. and Tapley, C. G., Brit. 736,822, Sept. 14, 1955; CA 50, 8725.
418. Polyakov, M. V. and Zhigailo, Ya. V., Doklady Akad. Nauk U.S.S.R. 73, 1229-30 (1950); CA 45, 1331.
419. Ponomarenko, V. A. and Vzenkova, G. Ya., Izvest. Akad. Nauk S.S.S.R., Otdel. Khim. Nauk 1957, 994-6; CA 52, 4473.
420. -----, ----- and Egorov, Yu. P., Doklady Akad. Nauk S.S.S.R. 122, 405-8 (1958); CA 53, 112.
421. Potter, R. I., Petroleum Refiner 27, No. 9, 497-504 (1948); CA 43, 383.
422. Preckel, R. and Selwood, P. W., J. Am. Chem. Soc., 62, 2765-6 (1940); CA 34, 7705.
423. Price, S. J. W. and Trotman-Dickenson, A. F., Trans. Faraday Soc. 54, 1630-7 (1958).

424. Price, S. J. W. and Trotman-Dickenson, A. F., J. Chem. Soc. 1958, 4205-7; CA 53, 2752.
425. Ptitsyna, O. A., Reutov, O. A., and Turchinskii, M. F., Doklady Akad. Nauk S.S.S.R. 114, 110-12 (1957); CA 52, 1090.
426. Quinn, E. I., Dibeler, V. H., and Mohler, F. L., J. Res. Nat. Bur. Stand. 57, 41-3 (1956); CA 51, 73.
427. Ramsden, H. E. and Davidson, H., Brit. 695,610, Aug. 12, 1953; U.S. 2,675,398, Apr. 13, 1954; CA 48, 10057.
428. -----, Brit. 701,714, Dec. 30, 1953; U.S. 2,675,397, Apr. 13, 1954; CA 49, 4011.
429. -----, Buchanan, H. W., Church, J. M., and Johnson, E. W., Brit. 719,421, Dec. 1, 1954; CA 50, 397.
430. ----- and Gloskey, C. R., U.S. 2,675,399, Apr. 13, 1954; Brit. 740,274, Nov. 9, 1955; CA 49, 5512.
431. -----, Brit. 743,313, Jan. 11, 1956; CA 50, 16829.
432. Razuvaev, G. A. and Fedotova, E. I., Zhur. Obshchei Khim. 21, 1118-22 (1951); CA 46, 5006.
433. -----, -----, Zaichenko, T. N., and Kul'vinskaya, N. A., Sbornik Statei Obshchei Khim. 2, 1514-16 (1953); CA 49, 5346.
434. ----- and Fetyukova, V., Zhur. Obshchei Khim. 21, 1010-15 (1951); CA 46, 1479.
435. -----, Akad. Nauk. S.S.S.R. Inst. Org. Khim. Sintezy Org. Soedinenii Sbornaik I, 41-2 (1950); CA 47, 8004.
436. Retailliau, E. R., Richards, H. A., Jr., and Jones, M. C. K., Am. Scientists 39, 656-71 (1951); CA 46, 8825.
437. Reutov, O. A., Ptitsyna, O. A., and Patrina, N. D., Zhur. Obshchei Khim. 28, 588-92 (1958); CA 52, 17151.
438. Riccoboni, L., Atti ist. Veneto sci. 96, II, 183-92 (1937); CA 33, 7208.
439. -----, Gazz. chim. ital. 71, 696-713 (1941); CA 36, 6902.
440. -----, ibid., 72, 47-62 (1942); CA 37, 574.
441. ----- and Popoff, P., Atti ist. veneto sci., Venezia 107, Pt. II., 123-48 (1949); CA 44, 6752.
442. Rieche, A. and Bertz, T., Angew. Chem. 70, 507 (1958); CA 53, 3040.
443. Riemschneider, R., Menge, K., and Klauge, P., Z. Naturforsch. 11b, 115-16 (1956); CA 50, 10589.
444. Roberti, G., Minervini, C., Pipparelli, E., and Semmola, E., Riv. ital. petrolio 14, No. 161, 10-12 (1946); CA 41, 4297.
444a. -----, Pipparelli, E., and Simonetti, A., Ricerca sci. 12, 347-52 (1941); CA 35, 5681.
445. Rochow, E. G., J. Am. Chem. Soc. 69, 1729-31 (1947); CA 41, 6187.

446. Rochow, E. G., U.S. 2,444,270, June 29, 1948; CA 42, 7318.
447. -----, J. Am. Chem. Soc. 70, 436-7 (1948); CA 42, 2231.
448. -----, ibid., 1801-2; CA 42, 6260.
449. -----, U.S. 2,451,871, Oct. 19, 1948; CA 43, 2631.
450. -----, U.S. 2,506,386, May 2, 1950; CA 44, 7344.
451. -----, J. Am. Chem. Soc. 72, 198 (1950); CA 45, 115.
452. ----- and Sindler, B. M., ibid., 1218-20; CA 44, 6038.
453. -----, Didtschenko, R., and West, R. C., Jr., ibid., 73, 5486-7 (1951); CA 47, 487.
454. ----- and Seyferth, D., ibid., 75, 2877-8 (1953); CA 47, 9836.
455. -----, ----- and Smith, A. C., Jr., ibid., 3099-3101; CA 47, 9837.
456. -----, U.S. 2,679,506, May 25, 1954; Brit. 713,130, Aug. 4, 1954; CA 49, 4705.
457. ----- and Allred, A. L., J. Am. Chem. Soc. 77, 4489-90 (1955); CA 49, 15593.
458. Rodekohr, H. M. and Blitzer, S. M., U.S. 2,574,759, Nov. 13, 1951; Brit. 685,230, Dec. 31, 1952; CA 46, 2288.
459. Rosenberg, S. D., Gibbons, A. J., Jr., and Ramsden, H. E., J. Am. Chem. Soc. 79, 2137-8 (1957); CA 51, 12846.
460. ----- and -----, ibid., 2138-40; CA 51, 12822.
460a. -----, Debreczeni, E., and Weinberg, E. L., ibid., 81, 972-5 (1959).
461. Ruddies, G. F., U.S. 2,415,444, Feb. 11, 1947; CA 41, 2886.
461a. -----, U.S. 2,277,781, Mar. 31, 1942; CA 36, 4832.
462. -----, U.S. 2,473,972, June 21, 1949; CA 44, 7344.
463. Saitow, A., Rochow, E. G., and Seyferth, D., J. Org. Chem. 23, 116-18 (1958); CA 52, 18195.
464. Sanderson, R. T., J. Am. Chem. Soc. 77, 4531-2 (1955).
465. Sasin, G. S., J. Org. Chem. 18, 1142-5 (1953); CA 48, 11305.
466. ----- and Sasin, R., ibid., 20, 387-90 (1955); CA 50, 4055.
467. Sasin, R. and Sasin, G. S., ibid., 770-3; CA 50, 8499.
468. Sasin, G. S., Borror, A. L., and Sasin, R., ibid., 23, 1366-7 (1958).
469. Saunders, B. C. and Stacey, G. J., J. Chem. Soc., 1949, 919-25; CA 44, 542.
470. -----, -----, Wild, F., and Winding, I. G. E., ibid., 1948, 699-703; CA 42, 6741.
471. ----- and -----, ibid., 1773-9; CA 43, 2932.
472. -----, ibid., 1950, 684-7; CA 44, 7793.
473. Schmall, E. A., Dissertation Abstr. 19, 667; CA 53, 3115.
474. Schmidt, H., Med. u. chem. Abhandl. med.-chem. Forschungsstätten I. G. Farbenind. 3, 418-28 (1936); CA 31, 5866.

474a. Schmitz-Dumont, O., Z. anorg. allgem. Chem. 248, 289-96 (1941); CA 37, 2721.
475. ----- and Bungard, G., Angew. Chem. 67, 208-9 (1955); CA 49, 8725.
476. Schwarz, R. and Schmeisser, M., Ber. 69B, 579-85 (1936); CA 30, 4419.
477. Scott, D. W., Good, W. D., and Waddington, G., J. Phys. Chem. 60, 1090-5 (1956); CA 51, 64.
478. Selwood, P. W., J. Am. Chem. Soc. 61, 3168-9 (1939); CA 34, 298.
479. Semerano, G., Riccobini, L., and Gallegari, F., Ber. 74B, 1297-1308 (1941); CA 36, 4802.
480. ----- and -----, ibid., 1089-99; CA 36, 4801.
481. ----- and -----, Ricerca sci. 11, 269-70 (1940); CA 37, 71.
482. ----- and -----, Z. physik. Chem. A189, 203-18 (1941); CA 37, 3043.
483. Setzer, W. C., Leeper, R. W., and Gilman, H., J. Am. Chem. Soc. 61, 1609-10 (1939); CA 33, 6264.
484. Seyferth, D. and Rochow, E. G., J. Org. Chem. 20, 250-6 (1955); CA 49, 13888.
485. ----- and -----, J. Am. Chem. Soc. 77, 907-8 (1955); CA 50, 1643.
486. ----- and -----, ibid., 77, 1302-4 (1955); CA 50, 1643.
487. ----- and Stone, F. G. A., ibid., 79, 515-17 (1957); CA 51, 9516.
488. -----, ibid., 2133-6; CA 51, 12018.
489. -----, J. Org. Chem. 22, 478 (1957); CA 52, 1057.
490. -----, J. Am. Chem. Soc. 79, 2738-40 (1957); CA 51, 14544.
491. -----, J. Org. Chem. 22, 1252-3 (1957); CA 52, 6167.
492. -----, ibid., 1599-1602; CA 52, 7136.
493. -----, J. Am. Chem. Soc. 79, 5881-4 (1957); CA 52, 6234.
494. ----- and Rochow, E. G., J. Polymer Sci. 18, 543-58 (1955); CA 52, 781.
495. Shapiro, H., U.S. 2,535,235, Dec. 26, 1950; Brit. 668,561; CA 45, 3865.
496. -----, U.S. 2,535,236, Dec. 26, 1950; CA 45, 3865.
497. -----, U.S. 2,535,237, Dec. 26, 1950; CA 45, 3865.
498. ----- and DeWitt, E. G., U.S. 2,575,323, Nov. 20, 1951; CA 46, 5073.
499. ----- and Krohn, I. T., U.S. 2,594,225, Apr. 22, 1952; CA 47, 146.
500. -----, U.S. 2,597,754, May 20, 1952; CA 47, 1183.
501. ----- and DeWitt, E. G., U.S. 2,635,106, Apr. 14, 1953; CA 48, 2762.
502. -----, Hnizda, V. F., and Calingaert, G., U.S. 2,674,519, Apr. 6, 1954; CA 48, 10308.
503. -----, -----, and -----, U.S. 2,674,610, Apr. 6, 1954; CA 49, 4706.

504. Shapiro, H. and Krohn, I. T., U.S. 2,688,628, Sept. 7, 1954; CA 49, 14797.
505. Sheline, R. K. and Pitzer, K. S., J. Chem. Phys. 18, 595-601 (1950); CA 44, 10398.
506. Shishido, K. and Kinukawa, J., Japan. 6626 ('53); CA 49, 9690.
507. ----- and -----, Japan. 6428 ('54); CA 50, 7856.
508. Shostakovskii, M. F., Kotrelev, V. N., Kochkin, D. A., Kuznetsova, G. I., Kalinina, S. P., and Borisenko, V. V., Zhur. Priklad. Khim. 31, 1434-6 (1958); CA 53, 3040.
509. Shushunov, V. A. and Boryshnikov, Yu. N., Zhur. Fiz. Khim. 27, 830-9 (1953); CA 49, 2838.
510. Siebert, H., Z. anorg. u. allgem. Chem. 263, 82-6 (1950); CA 45, 1423.
511. -----, ibid., 268, 177-90 (1952); CA 46, 10714.
512. Silbert, F. C. and Kirner, W. R., Ind. Eng. Chem., Anal. Ed. 8, 353 (1936).
513. Skinner, H. A. and Sutton, R. E., Trans. Faraday Soc. 36, 1209-12 (1940); CA 35, 2390.
514. ----- and -----, ibid., 40, 164-24 (1944); CA 38, 5701.
515. Skoldinov, A. P. and Kocheshkov, K. A., J. Gen. Chem. (U.S.S.R.) 12, 398-401 (1942); CA 37, 3064.
516. Smith, A. C., Jr., and Rochow, E. G., J. Am. Chem. Soc. 75, 4103-5 (1953); CA 49, 8096.
517. ----- and -----, ibid., 4105-6; CA 49, 8096.
518. Smith, F. A., U.S. 2,625,559, Jan. 13, 1953; CA 47, 11224.
519. Smith, F. B. and Kraus, C. A., J. Am. Chem. Soc. 74, 1418-20 (1952); CA 46, 10121.
520. Smyth, C. P., ibid., 63, 57-66 (1941); CA 35, 1281.
521. -----, J. Org. Chem. 6, 421-6 (1941); CA 35, 4647.
522. Solerio, A., Gazz. chim. ital. 81, 664-7 (1951); CA 46, 6083.
523. -----, ibid., 85, 61-8 (1955); CA 50, 3992.
524. Spice, J. E. and Twist, W., J. Chem. Soc. 1956, 3319-23.
525. Stanton, R., U.S. 2,572,887, Oct. 30, 1951; CA 46, 2565.
526. -----, U.S. 2,619,496, Nov. 25, 1952; CA 47, 3334.
527. Staveley, L. A. K., Paget, H. P., Goalby, B. B., and Warren, J. B., Nature 164, 787-8 (1949); CA 44, 1897.
528. -----, -----, -----, and -----, J. Chem. Soc. 1950, 2290-2301; CA 45, 7518.
529. -----, Warren, J. B., Paget, H. P., and Dowrick, D. J., J. Chem. Soc. 1954, 1992-2001; CA 48, 12537.
530. Stefl, E. P. and Best, C. E., U.S. 2,713,580, July 19, 1955; CA 50, 5762.
531. Stoner, H. B. and Threlfall, C. J., Biochem. J. 69, 376-85 (1958); CA 52, 18834.
532. -----, Barnes, J. M., and Duff, J. I., Brit. J. Pharmacol. 10, 16-25 (1955); CA 49, 11181.

533. Stuckwisch, C. G., Iowa State Coll. J. Sci. 18, 92-4 (1943); CA 38, 728.
534. Sturgis, B. M., Cantwell, E. N., Morris, W. E., and Schultz, D. L., Proc. Am. Inst. Petroleum 34, [III] 256-69 (1954); CA 50, 2963.
535. Summers, L., Iowa State Coll. J. Sci. 26, 292-4 (1952); CA 47, 8673.
536. Tagliavini, G., Schiavon, G., and Belluco, U., Gazz. chim. ital. 88, 746-54 (1958).
537. Talalaeva, T. V. and Kocheshkov, K. A., J. Gen. Chem. (U.S.S.R.) 8, 1831-8 (1938); CA 33, 5819.
538. ----- and -----, Ibid., 12, 403-8 (1942); CA 37, 3067.
539. -----, Zaitseva, N. A., and Kocheshkov, K. A., ibid., 16, 901-6 (1946); CA 41, 2014.
540. Tanner, H. M., U.S. 2,635,105, Apr. 14, 1953, Brit. 718,619, Nov. 17, 1954; CA 48, 2762.
541. -----, U.S. 2,635,107, Apr. 14, 1953; CA 48, 2762.
542. Tatlock, W. S. and Rochow, E. G., J. Org. Chem. 17, 1555-63 (1952); CA 48, 566.
543. Tchakirian, A. and Lewinsohn, M., Compt. rend. 201, 835-7 (1935); CA 30, 1682.
544. -----, Lesbre, M., and Lewinsohn, M., ibid., 202, 138-40 (1936); CA 30, 2549.
545. -----, Ann. chim. 12, 415-99 (1939); CA 34, 2271.
546. ----- and Bevillard, P., Bull. soc. chim. France 1950, 1300; CA 45, 6110.
547. Teal, G. K. and Kraus, C. A., J. Am. Chem. Soc. 72, 4706-9 (1950); CA 45, 3743.
548. Teplý, J., and Malý, J., Chem. Zvesti 7, 463-74 (1953); CA 48, 10427.
549. Tisdate, W. H., Brit. 578,312, June 23, 1943.
550. Toropova, V. F. and Saikina, M. K., Sbornik Statei Obshchei Khim., Akad. Nauk S.S.S.R. 1, 210-15 (1953); CA 48, 12579.
551. Trautman, C. E. and Ambrose, H. A., U.S. 2,416,360, Feb. 25, 1947; CA 42, 2760.
552. Veksler, Z. V. and Zabryanskii, E., Novosti Tekhniki 1936, No. 26, 18-19; Novosti Neftepererabotki 3, No. 17, 6-7 (1936); CA 31, 2404.
553. Vijayaraghavan, K. V., J. Indian Chem. Soc. 22, 135-40 (1945); CA 40, 2787.
554. -----, ibid., 227 (1945); CA 40, 4659.
555. Voigt, D., Lesbre, M., and Gallais, F., Compt. rend. 239, 1485-7 (1954); CA 49, 5276.
556. Walsh, A. D., J. Roy. Inst. Chem. 75, 323-6 (1951); CA 46, 2276.
557. Walters, E. L. and Busso, C. J., Ind. Eng. Chem. 41, 907-14 (1949); CA 43, 5578.

557a. Waring, C. E. and Horton, W. S., J. Am. Chem. Soc. 67, 540-7 (1945); CA 39, 2246.
558. Waters, D. N. and Woodward, L. A., Proc. Roy. Soc. (London) A246, 119-32 (1958); CA 52, 19514.
559. Weinberg, E. L. and Johnson, E. W., U.S. 2,648,650, Aug. 11, 1953; CA 48, 10056.
560. -----, U.S. 2,679,505, May 25, 1954; Brit. 709,594, May 26, 1954; CA 49, 4705.
561. Weiss, E., Z. anorg. u. allgem. Chem. 287, 236-41 (1956); CA 51, 7785.
562. West, R. and Rochow, E. Ge., J. Am. Chem. Soc. 74, 2490-1; CA 46, 8432.
563. -----, ibid., 4363-5; CA 47, 8673.
564. -----, Webster, M. H., and Wilkinson, G., ibid., 5794-5; CA 47, 10920.
565. -----, ibid., 75, 6080-1 (1953); CA 48, 13648.
566. -----, Hunt, H. R., Jr., and Whipple, R. O., ibid., 76, 310 (1954); CA 49, 160.
567. Wheeler, R. W., Downs, D., and Walsh, A. D., Nature 162, 893-4 (1948); CA 43, 2411.
568. Wiczer, S. B., U.S. 2,447,926, Aug. 24, 1948; CA 42, 7975.
569. Widmaier, O., Rev. inst. franç. petrole 4, 543-54 (1949); CA 44, 10293.
570. Wittig, G., Todt, U., and Nagel, K., Chem. Ber. 83, 110-19 (1950); CA 44, 5334.
571. -----, Meyer, F. J., and Lange, G., Ann. 571, 167-201 (1951); CA 45, 5556.
572. Wong, C. H. and Schomaker, V., J. Chem. Phys. 28, 1007-9 (1958); CA 52, 16001.
573. Worrall, D. E., J. Am. Chem. Soc. 62, 3267 (1940); CA 35, 87.
574. Yakubovich, A. Ya., Makarov, S. P., Ginsburg, V. A., Gavrilov, G. I., and Merkulova, E. N., Doklady Akad. Nauk S.S.S.R. 72, 69-72 (1950); CA 45, 2856.
575. -----, ----- and Gavrilov, G. I., Zhur. Obshchei Khim. 22, 1788-93 (1952); CA 47, 9257.
576. -----, Merkulava, E. N., Makarov, S. P., and Gavrilov, G. I., ibid., 2060-3; CA 47, 9257.
577. -----, Makarov, S. P., and Ginsburg, V. A., ibid., 28, 1036-38 (1958).
578. Yngve, V., Brit. 497,879, Dec. 30, 1938; CA 33, 3923.
579. Young, C. W., Koehler, J. S. and McKinney, D. S., J. Am. Chem. Soc. 69, 1410-15 (1947).
580. Yust, V. and Bame, J. L., U.S. 2,750,267, June 12, 1956; CA 50, 13428.
581. Zakharkin, L. I. and Okhlobystin, O. Yu., Doklady Akad. Nauk S.S.S.R. 116, 236-8 (1957); CA 52, 6167.

582. Zang, W. I. and Lovell, W. G., Ind. Eng. Chem. $\underline{43}$, 2826-33 (1951); CA $\underline{46}$, 2276.
583. Zasosov, V. A. and Kocheshkov, K. A., Sbornik Statei Obshchei Khim., Akad. Nauk S.S.S.R. $\underline{1}$, 278-84 (1953); CA $\underline{49}$, 912.
584. ----- and -----, ibid., 285-9; CA $\underline{49}$, 913.
585. ----- and -----, ibid., 290-8; CA $\underline{49}$, 913.
586. Zeman, W., Gadermann, E., and Hardebeck, K., Deut. Arch. klin. Med. $\underline{198}$, 713-21 (1951); CA $\underline{46}$, 6265.
587. Zhdanov, G. S. and Ismailzade, I. G., Doklady Akad. Nauk S.S.S.R. $\underline{68}$, 95-8 (1949); CA $\underline{43}$, 8764.
588. ----- and -----, Zhur. Fiz. Khim. $\underline{24}$, 1495-1501 (1950); CA $\underline{45}$, 4112.
589. Ziegler, K. and Lehmkuhl, H., Angew. Chem. $\underline{67}$, 424 (1955); CA $\underline{50}$, 6303.
590. Zietz, J. R., Blitzer, S. M., Redman, H. E., and Robinson, S. C., J. Org. Chem. $\underline{22}$, 60-2 (1957); CA $\underline{51}$, 11239.
591. Zimmer, H., Angew. Chem. $\underline{64}$, 620 (1952).
592. ----- and Lübke, K., Chem. Ber. $\underline{85}$, 1119-21 (1952); CA $\underline{47}$, 12280.
593. ----- and Sparmann, H. W., Naturwissenschaften $\underline{40}$, 220 (1953); CA $\underline{48}$, 12044.
594. ----- and -----, Chem. Ber. $\underline{87}$, 645-51 (1954); CA $\underline{49}$, 9589.
595. ----- and Moslé, H. G., ibid., 1255-7; CA $\underline{49}$, 13925.
596. ----- and Gold, H., ibid., $\underline{89}$, 712-14 (1956); CA $\underline{51}$, 403.
597. Argus Chemical Lab., Inc., Brit. 740,397, Nov. 9, 1955; U.S. 2,759,906, Aug. 21, 1956; CA $\underline{50}$, 13987.
598. -----, Brit. 743,304, Jan. 11, 1956; CA $\underline{50}$, 16828.
599. Bakelite Corp., Brit. 664,133, Jan. 2, 1952; CA $\underline{46}$, 11230.
600. British Thomson-Houston Co. Ltd., Brit. 626,398, July 14, 1949; CA $\underline{44}$, 2548.
601. -----, Brit. 654,571, June 20, 1951; CA $\underline{46}$, 4561.
602. California Research Corp., Brit. 774,529, May 8, 1957; CA $\underline{52}$, 2928.
603. E. I. duPont de Nemours & Co., Brit. 453,271, Sept. 1, 1936; CA $\underline{31}$, 1043.
604. -----, Ger. 660,442, May 25, 1938; CA $\underline{32}$, 6261.
605. Ethyl Corp., Brit. 673,440, June 4, 1952; CA $\underline{47}$, 3335.
606. -----, Brit. 707,075, Apr. 14, 1954; CA $\underline{49}$, 6990.
607. Firestone Tire & Rubber Co., Brit. 728,953, Apr. 27, 1955; U.S. 2,731,440, Jan. 17, 1956; CA $\underline{49}$, 13693.
608. -----, Brit. 728,954, Apr. 27, 1955; CA $\underline{49}$, 13693.
609. Metal & Thermit Corp., Brit. 739,883, Nov. 2, 1955; CA $\underline{50}$, 13986.
610. E. H. Salisbury's Labs., Brit. 711,564, July 7, 1954; CA $\underline{49}$, 14797.
611. Standard Oil Development Co., Brit. 579,980, Aug. 22, 1946; CA $\underline{41}$, 2570.

612. Abdon, A. H., J. Helminthol. <u>30</u>, 121-8 (1956).
613. Abramova, L. V., Sheverdina, N. I., and Kocheshkov, K. A., Doklady Akad. Nauk S.S.S.R. <u>123</u>, 681-4 (1958); CA <u>53</u>, 4928.
613a. Ainley, A. D., Elson, L. A., and Sexton, W. A., J. Chem. Soc. <u>1946</u>, 776-7; CA <u>41</u>, 392.
614. d'Ans, J., Zimmer, H., and Branchitsch, M. v., Ber. <u>88</u>, 1507-10 (1955); CA <u>51</u>, 1924.
614a. Anspon, H. D. and Pschorr, F. E., U.S. 2,683,705, July 13, 1954; CA <u>48</u>, 13275.
615. Anzilotti, W. F. and Tomsic, V. J., 5-th Symposium on combustion, Pittsburgh <u>1954</u>, 356-66; CA <u>50</u>, 1295.
616. Ballczo, H. and Schiffner, H., Z. anal. Chem. <u>152</u>, 3-18 (1956); CA <u>51</u>, 3371.
617. Arkhipov, A. S., Gigiena i Sanit. <u>22</u>, No. 2., 22-31 (1957); CA <u>51</u>, 12384.
618. Bambynek, W. and Freise, V., Z. physik. Chem. <u>7</u>, 317-31 (1956); CA <u>50</u>, 16231.
619. ----- and -----, Z. Naturforsch. <u>11a</u>, 168-9 (1956); CA <u>51</u>, 4768.
620. Barker, R. L., Brit. 761,568, Nov. 14, 1956; CA <u>51</u>, 10561.
621. Bartleson, J. D., U.S. 2,794,713-723, June 4, 1957; CA <u>51</u>, 13380.
622. Bartocha, B. and Gray, M. Y., Z. Naturforsch. <u>14b</u>, 350-1 (1959).
623. Bate, G. L., Miller, D. S. and Kulp, J. L., Anal. Chem. <u>29</u>, 84-8 (1957); CA <u>51</u>, 4840.
624. Bianchi, C., Ambanelli, U., and Salvi, G., Ateneo parmense <u>25</u>, 419-45 (1954); CA <u>49</u>, 8480.
625. Blum, M. S. and Bower, F. A., J. Econ. Entomol. <u>50</u>, 84-6 (1957); CA <u>51</u>, 10823.
626. Boyd, P. R., Walker, G., and Henderson, I. N., Lancet <u>272</u>, 181-5 (1957); CA <u>51</u>, 8302.
626a. Brinckman, F. G. and Stone, F. G. A., Paper presented at the 135-th ACS Meeting, Apr. 1959.
627. Brown, J. F. and Weir, R. J., J. Sci. Instr. <u>33</u>, 222-5 (1956); CA <u>50</u>, 13417.
628. Caldwell, J. R. and Wellman, J. W., U.S. 2,720,505, Oct. 11, 1955; CA <u>50</u>, 2205.
629. -----, U.S. 2,801,231, July 30, 1957; CA <u>51</u>, 18704.
629a. Calingaert, G. and Shapiro, H., U.S. 2,591,509, Apr. 1, 1952; CA <u>46</u>, 11229.
630. -----, U.S. 2,660,592-3-4-5-6, Nov. 24, 1953; CA <u>48</u>, 2085.
631. Cappuccio, V. and Cittadini, A., Ital. 519,728, Mar. 16, 1955; CA <u>51</u>, 16001.
632. Carrol, R. T., U.S. 2,597,920, May 27, 1952; Brit. 717,973, Nov. 3, 1954; CA <u>46</u>, 7824.
633. Caujolle, F., Lesbre, M., and Meynier, D., Compt. rend. <u>239</u>, 556-8 (1954); CA <u>49</u>, 2608.

634. Caujolle, F., Lesbre, M., and Meynier, D., Comp. rend. 239, 1091-3 (1954); CA 49, 3416.
635. -----, -----, Bru, M., Meynier, D., and Bru, Y., ibid., 240, 1829-30 (1955).
636. -----, -----, -----, -----, and -----, ibid., 241, 1420-1 (1955); CA 50, 5166.
637. -----, ----- and -----, Ann. farm. franç. 14, 88-97 (1956); CA 50, 13288.
638. -----, Dauty, R., Meynier, D., and Monnier, J., Compt. rend. soc. biol. 150, 431-3 (1956); CA 50, 171561
639. -----, Lesbre, M., Meynier, D., and Saqui-Sannes, G. de, Compt. rend. 243, 987-9 (1956); CA 51, 5277.
640. Chamberlain, G. H. W., Hoare, D. E., and Walsh, A. D., Discussions Faraday Soc. 1953, No. 14, 89-97; CA 48, 985.
641. Chatt, J. and Williams, A. A., Silicium, Schwefel, Phosphate, Colloq. Sek. Anorg. Chem. Intern. Union Reine u. Angew. Chem. Münster 1954, 60-3; CA 51, 13796.
641a. Churchill, J. W., U.S. 2,643,242, June 23, 1953; CA 47, 10904.
642. Comyns, D. E., Atomic Energy Research Estab. (Gt. Brit.) C/M-258, 7pp. (1957); CA 51, 17557.
643. Corke, A. T. K., J. Hort. Sci. 30, 197-200 (1955); CA 49, 13572.
644. Cornelius, W. and Caplan, J. D., S.A.E. Quart. Trans. 6, 488-510 (1952); CA 48, 11767.
645. Driel, H. van, Kooijman, P. L., and Reman, G. H., Proc. XI-th Intern. Congr. Pure and Appl. Chem. London 4, 559-89 (1947); CA 48, 9664.
646. Ducheylard, C., Lazard, B., and Roth, E., J. chim. phys. 50, 497-500 (1953); CA 48, 3786.
647. Duffy, R. and Holliday, A. K., Proc. Chem. Soc. 1959, 124-5.
648. Edgar, S. A., Poultry Sci. 35, 64-73 (1956); CA 50, 13272.
649. ----- and Teer, P. A., ibid., 36, 329-34 (1957); CA 51, 15772.
650. Edgell, W. F. and Ward, C. H., J. Am. Chem. Soc. 77, 6486-91 (1955); CA 50, 4641.
651. Eeman, G. E. M., Fr. 999,147, Jan. 28, 1952; CA 50, 16100.
652. Egerton, A. C. and Rudrakanchana, S., Fuel 33, 274-85 (1954); CA 48, 10327.
653. ----- and -----, Proc. Roy. Soc. (London) A225, 427-43 (1954); CA 49, 42.
654. Erhard, K. H. L. and Norrish, R. G. W., ibid., A234, 178-91 (1956); CA 50, 13417.
654a. Faragher, W. F., U.S. 2,634,257, Apr. 7, 1953; CA 47, 7250.
655. Flenner, A. L., U.S. 2,673,869, Mar. 30, 1954; CA 49, 5512.

655a. Freidlina, R. Kh. and Nesmeyanov, A. N., Compt. rend. acad. sci. U.R.S.S. 29, 567-70 (1940); CA 35, 3614.
656. Freimiller, L. R. and McKeever, C. H., U.S. 2,786,035 and U.S. 2,786,036, Mar. 19, 1957; CA 51, 9210.
657. Fritz, H. P., Chem. Ber. 92, 780-91 (1959).
658. Giraitis, A. P., Grandjean, W. B., and Neher, C. M., U.S. 2,777,867, Jan. 15, 1957; CA 51, 15551.
659. Gittens, T. W. and Mattison, E. L., U.S. 2,763,673, Sept. 18, 1956; CA 51, 4414.
660. Giuliani, V. and Belli, R., Folia Med. (Naples) 38, 1286-94 (1955); CA 50, 4394.
661. Gloskey, C. R., Brit. 751,499, June 27, 1956; CA 51, 1659.
661a. Good, W. D., Scott, D. W., Lacina, J. K., and McCullough, J. P., J. Phys. Chem. 63, 1139-42 (1959).
662. Gordy, W. and McCormick, C. G., J. Am. Chem. Soc. 78, 3243-6 (1956); CA 50, 16394.
662a. Gorsheleva, L. S., Zhur. Vysshei Nervnoi Deyatel'nosti im. I.P. Pavlova 1, 727-38 (1951); CA 47, 5549.
663. Griffiths, V. S. and Derwish, G. A. W., J. Mol. Spectroscopy 3, 165-76 (1959).
663a. Grim, S. O. and Seyferth, D., Chem. & Ind. 1959, 849-50.
664. Hartmann, H., Reiss, W., and Karbstein, B., Naturwissenschaften 46, 321 (1959).
665. ----- and Eschenbach, W., ibid.
666. -----, Niemöller, H., Reiss, W. and Karbstein, B., ibid.,
667. Hecker, A. C., U.S. 2,789,963, Apr. 23, 1957; CA 51, 17235.
668. Heiligmann, R. G. and Stickney, P. B., U.S. 2,765,297, Oct. 2, 1956; CA 51, 4051.
669. Hein, F. and Hecker, H., Z. Naturforsch. 11b, 667-8 (1956); CA 51, 6418.
670. Herold, P. and Asinger, F., Ger. 765,790, June 1, 1953; CA 49, 3238.
670a. Heron, S. D., U.S. 2,407,551, Sept. 10, 1946; CA 40, 7229.
671. Hill, E. F., U.S. 2,722,477, Nov. 1, 1955; CA 50, 3748.
672. ----- and DePree, D. O., U.S. 2,770,535, Nov. 13, 1956; CA 51, 3985.
673. Holder, C. H., Weeks, R. L., and Knox, Jr., W. T., U.S. 2,779,317, Jan. 29, 1957; CA 51, 6140.
674. Hobbs, Jr., C. L., U.S. 2,686,799, Aug. 17, 1954; CA 49, 3524.
675. Holliday, A. K. and Jeffers, W., J. Inorg. Nuclear Chem. 6, 134-7 (1958).
676. Hugher, M. B., Dissertation Abstr. 19, 1921 (1959).
677. Hügel, G., Kolloid-Z. 131, 4-10 (1953); CA 47, 9087.
678. Ireland, J., Brit. 713,727, Aug. 18, 1954; CA 49, 12530.
679. Jenkins, R. L., U.S. 2,646,403, July 21, 1953; CA 47, 10454.
680. Johnson, E. W., U.S. 2,801,258, July 30, 1957; CA 51, 18007.
680a. Juenge, E. C. and Cook, S. E., Paper presented at the 135-th ACS Meeting, Apr. 1959.

681. Kerk, G. J. M. van der and Noltes, J. G., J. Appl. Chem. 9, 176-9 (1959).
682. ----- and -----, ibid., 179-85.
683. Kerr, K. B. and Walde, A. W., Exptl. Parasitol. 5, 560-70 (1956); CA 51, 7566.
684. ----- and -----, U.S. 2,702,775-6-7-8, Feb. 22, 1955; CA 49, 7816.
685. Kien, L. K. van and Tuong, T., Compt. rend. soc. biol. 149, 2196-8 (1955); CA 50, 13275.
686. Komatsu, F., Honda, S. and Shimizu, Z., J. Sci. Labor (Japan) 29, 446-54 (1953); CA 48, 7287.
687. Komatsu, S. et al., Japan. 2317 ('56), Mar. 29; CA 51, 10561.
688. Korshak, V. V., Polyakova, A. M., Mironov, V. F., and Petrov, A. D., Izvest. Akad. Nauk S.S.S.R. Otdel Khim. Nauk 1959, 178-80; CA 53, 15959.
688a. -----, -----, and Tambovtseva, E. S., ibid., 742-44.
689. Koton, M. M., Zhur. Obshchei Khim. 22, 643-7 (1952); CA 47, 5376.
689a. -----, Moskvina, E. P., and Florinskii, F. S., ibid., 19, 1675-8 (1949); CA 44, 1436.
690. -----, Doklady Akad. Nauk S.S.S.R. 88, 991-3 (1956); CA 48, 8727.
691. Kraak, M. Dutch 68,578, Sept. 1951; CA 46, 5781.
692. Kuivila, H. G. and Beumel, O. F., Jr., J. Am. Chem. Soc. 80, 3798 (1958).
693. Lauderl, H. P. and Sturgis, B. M., Ind. Eng. Chem. 45, 1744-8 (1953); CA 47, 11709.
694. Laurie, V. W., J. Chem. Phys. 30, 1210-14 (1959).
695. Lecoq, R., Compt. rend. 239, 678-80 (1954); CA 49, 4180.
696. Leidi, G. and Girelli, A., Riv. combustibli 8, 668-86 (1954); CA 49, 11266.
697. Lucas, G. R., U.S. 2,598,402, May 27, 1952; CA 46, 9342.
698. Lyn, W. T., Inst. Mech. Ingrs., Jan. 1954, 17 pp.; CA 50, 9727.
699. Mack, G. P. and Parker, E., U.S. 2,592,926, Apr. 15, 1952; CA 46, 11767.
700. Magee, P. N., Stoner, H. B., and Barnes, J. M., J. Pathol. Bacteriol. 73, 107-34 (1957); CA 51, 9886.
701. LeGoff, P. and Letort, M., J. chim. phys. 53, 480-92 (1956); CA 51, 5511.
702. Leistner, W. E. and Setzler, W. E., U.S. 2,752,325, June 26, 1956; CA 51, 1264.
703. McDermott, J. P., U.S. 2,786,812, Mar. 26, 1957; CA 51, 10892.
704. -----, U.S. 2,786,813, Mar. 26, 1957; CA 51, 10892.
705. -----, U.S. 2,786,714, Mar. 26, 1957; CA 51, 10892.
706. Mack, G. P. and Parker, E., Brit. 735,030, Aug. 10, 1955; CA 50, 8736.

707. Mack, G. P., U.S. 2,745,820, May 15, 1956; CA 51, 6219.
708. -----, and Parker, E., U.S. 2,745,819, May 15, 1956; CA 51, 9214.
709. Madden, H. J., Brit. 701,708, Dec. 30, 1953; CA 49, 4012.
710. Maier, L., Angew. Chem. 71, 161 (1959).
711. -----, Tetrahedron Letters, 1959, No. 6, 1-6.
712. Miyanishi, M., J. Fuel Soc. Japan, 34, 373-9 (1955); CA 49, 13630.
713. Morelli, A., Folia med. (Naples) 36, 440-3 (1953); CA 48, 2253.
714. ----- and Preziosi, P., ibid., 526-37; 538-50; and 551-5 (1953); CA 48, 2916.
715. ----- and -----, ibid., 718-22 and 723-8 (1953); CA 48, 4697.
716. ----- and -----, ibid., 788-800 and 801-32 (1953); CA 48, 6571.
717. ----- and -----, Boll. soc. ital. biol. sper. 29, 1453-5 (1953); CA 48, 5354.
718. -----, Folia med. (Naples) 37, 541-61 and 562-73 (1954); CA 50, 13085.
719. ----- and Preziosi, P., ibid., 654-73 (1954); CA 49, 1965.
720. ----- and -----, Arch. ital. sci. farmacol. [3] 4, 302-4 (1954); CA 50, 3644.
721. -----, Rass. med. sper. 1, 43-7 (1954); CA 51, 8299.
722. Murin, A. N. and Nefedov, V. D., Primenenie Mechenykh Atomov v Anal. Khim., Akad. Nauk S.S.S.R., Inst. Geokhim. i Anal. Khim. 1955, 75-8; CA 50, 3915.
723. Nefedov, V. D. and Varshav, N. A., Zhur. Fiz. Khim. 28, 981-4 (1954); CA 49, 7354.
723a. Neiman, M. B. and Shushunov, V. A., J. Phys. Chem. (U.S.S.R.) 22, 161-6 (1948); CA 42, 5316.
724. ----- and -----, Doklady Akad. Nauk S.S.S.R. 60, 1347-50 (1950); CA 45, 425.
725. Nesmeyanov, A. N., Borisov, A. E., and Vel'chevskaya, V. D., Izvest. Akad. Nauk S.S.S.R., Otdel. Khim. Nauk 1949, 578-81; CA 44, 3381.
726. -----, -----, Savelyeva, I. S., and Golubeva, Ve. I., ibid., 1958, 1490-1; CA 53, 7973a.
727. -----, Lutsenko, I. F., and Ponomarev, S. V., Doklady Akad. Nauk S.S.S.R. 124, 1073-5 (1959); CA 53, 14984.
728. -----, Borisov, A. E., and Novikova, N. V., Izvest. Akad. Nauk S.S.S.R., Otdel. Khim. Nauk 1959, 259-262.
729. -----, -----, -----, and Osipova, M. A., ibid., 263-6.
730. -----, -----, and Novikova, N. V., ibid., 644-6.
731. Nitzsche, S. and Riedle, R., U.S. 2,770,611, Nov. 13, 1956; CA 51, 7761.
732. Noltes, J. G. and Kerk, G. J. M. van der, Chem. & Ind. (London) 1959, 294.
733. Osberghaus, O. and Taubert, R., Z. physik. Chem. (Frankfurt) [N.F.] 4, 264-85 (1955); CA 49, 12952.

734. Osthoff, R. C. and Rochow, E. G., J. Am. Chem. Soc. 74, 845 (1952); CA 47, 3639.
735. Padovani, C., Berti, V. and Leidi, G., Rev. combustibili 4, 411-32 (1950); CA 48, 975.
736. Pahnke, A. J., Cohen, P. M., and Sturgis, B. M., Ind. Eng. Chem. 46, 1024-9 (1954); CA 48, 8513.
737. Panov, E. M. and Kocheshkov, K. A., Doklady Akad. Nauk S.S.S.R. 123, 295-7 (1958); CA 53, 7133.
738. Passino, H. J and Lauer, G. G., U.S. 2,665,286, Jan. 5, 1954; CA 49, 367.
739. Pedinelli, M. and Stringari, M., Rass. med. ind. 19, 57-68 (1950); CA 45, 6773.
740. Pedley, J. B. and Skinner, H. A., Trans. Faraday Soc. 55, 544-7 (1959).
741. Polkinhorne, H. and Tapley, C. G., Brit. 761,357, Nov. 14, 1956; CA 51, 11382.
742. Pomerantseva, E. N., Klin. Med. (U.S.S.R.) 28, No. 3, 67-70 (1950); CA 47, 10122.
743. Prince, R. H., J. Chem. Soc. 1959, 1783-90.
744. Pulleybank, D. H. and Lovell, W. G., Petroleum Processing 5, 492-4 (1950); CA 44, 7521.
745. Ramsden, H. E., U.S. 2,744,876, May 8, 1956; CA 51, 459.
745a. -----, Weinberg, E. L., and Tomka, L. A., U.S. 2,789,104, Apr. 16, 1957; CA 51, 10939.
746. ----- and Banks, C. K., U.S. 2,789,994, Apr. 23, 1957; CA 51, 14786.
747. -----, Weinberg, E. L., and Tomka, L. A., U.S. 2,790,785, Apr. 30, 1957; CA 51, 10940.
748. -----, Brit. 742,975, Jan. 4, 1956; CA 51, 2033.
749. -----, Brit. 750,106, June 6, 1956; CA 51, 1247.
750. -----, Brit. 759,382, Oct. 17, 1956; CA 51, 13918.
751. Randaccio, C., Ital. 500,102, Nov. 17, 1954; CA 51, 10560.
752. Richard, W. R., U.S. 2,477,349, July 26, 1949; CA 44, 4718.
753. Rieche, A. and Dahlmann, J., Angew. Chem. 71, 194 (1959).
754. Rifkin, E. B. and Ewen, D. V., U.S. 2,580,243, Dec. 25, 1951; CA 46, 2790.
755. -----, Walcutt, C., and Betker, G. W., Jr., SAE Quart. Trans. 6, 472-87, 508-10 (1952); CA 48, 11767.
756. ----- and -----, Ind. Eng. Chem. 48, 1532-9 (1956); CA 51, 3976.
757. Roberti, G., Pipparelli, E., and Simonetti, A., Ricerca sci. 12, 347-52 (1951); CA 45, 6828.
758. Rose, A. and Rifkin, E. B., Ind. Eng. Chem. 48, 1528-32 (1956); CA 51, 3975.
759. Rossiter, W. T. and Currie, C.C., U.S. 2,742,368, Apr. 17, 1956; CA 50, 11713.
760. Rothman, L. A. and Becker, E. I., J. Org. Chem. 24, 294 (1959).
761. Rowland, G. P., Jr., and Reid, R. J., U.S. 2,445,730, July 20, 1948; CA 43, 440.

762. Rozhkov, I. V., Shimonaev, G. S., and Kornilova, E. N., Neftyanoe Khoz. 32, No. 5, 70-3 (1954); CA 48, 9663.
763. ----- and Kornilova, E. N., Khim. i Tekhnol. Topliva 1956, No. 12, 43-7; CA 51, 6132.
763a. ----- and -----, ibid., 1957, No. 4, 47-53; CA 51, 18565.
764. Sacco, A., Atti accad. nazl. Lincei, Rend. Classe sci. fiz., mat. e nat. 11, 101-105 (1951); CA 49, 158.
765. Saikina, M. K., Uchenye Zapiski Kazan. Gasudarst. Univ. im. V. I. Ul'yanova-Lenina Khim. 116, No. 2, 129-86 (1956); CA 51, 7191.
766. Salvini, M. and Massignan, G., Med. lavoro 45, 155-8 (1954); CA 48, 11651.
767. Sanders, N. D., Hensley, R. V. and Breitwieser, R., Natl. Advisory Comm. Aeronautics, Wartime Rept. A.R.R. No. E4I28 (1944).
768. Sathyamurthy, T. V., Swaminanthan, S. and Yeddanapalli, T. V., J. Indian. Chem. Soc. 27, 509-14 (1950).
769. Savidan, L., Bull. soc. chim. France 1953, 411-12; CA 47, 7372.
770. Scarinci, V., Studi Urbinati, Fac. farm. [c], 29, No. 4, 102-4 (1955); CA 51, 10736.
771. Seyferth, D., Naturwissenschaften 44, 34-5 (1957).
772. -----, J. Am. Chem. Soc. 81, 1844-7 (1959).
773. Shikhiyev, I. A., Shostakovski, M. F., Komarov, N. V., and Aslanov, I. A., Zhur. Obshchei Khim. 29, 1549-51 (1959).
774. Shimizu, K., J. Chem. Soc. Japan, Pure Chem. Sect. 77, 1284 (1956); CA 51, 2386.
775. Shimonaev, G. S. and Rozhkov, I. V., Zhur. Fiz. Khim. 29, 791-6 (1955); CA 51, 818.
776. ----- and -----, ibid., 31, 387-94 (1957); CA 51, 15232.
777. Shishido, K. and Kinukawa, J., Japan. 7889 ('54), Nov. 29; CA 50, 8250.
778. Shushunov, V. A. and Shlyaphikov, Yu. A., J. Phys. Chem. (U.S.S.R.) 22, 145-55 (1948); CA 42, 5315.
779. Serenson, R. A., U.S. 2,654,716, Oct. 6, 1953; CA 48, 2411.
780. Spano, H. and Kahn, M., J. Am. Chem. Soc. 74, 568-9 (1952); CA 47, 2054.
781. Stefl, E. P. and Best, C. E., U.S. 2,731,441, Jan. 17, 1956; CA 51, 1656.
782. ----- and -----, U.S. 2,731,482, Jan. 17, 1956; CA 51, 13918.
783. Stevens, C. D., Feldhake, C. J., and Kehoe, R. A., J. Pharmacol. Exptl. Therap. 117, 420-4 (1956); CA 51, 1454.
784. Street, J. C. and Thomas, A., Fuel 34, 4-36 (1955); CA 49, 2706.
785. Strohmeier, W., Hümpfner, K., Miltenberger, K., and Seifert, F., Z. Electrochem. 63, 537-39 (1959).
786. Talalaeva, T. V. and Kocheshkov, K. A., Doklady Akad. Nauk S.S.S.R. 77, 621-4 (1951); CA 45, 10191.

787. Talalaeva, T. V. and Kocheshkov, K. A., Izvest. Akad. Nauk S.S.S.R., Otdel. Khim. Nauk 1953, 126-34.
788. Thomas, A. B. and Rochow, E. G., J. Am. Chem. Soc. 79, 1843-8 (1957); CA 51, 10199.
789. ----- and -----, J. Inorg. & Nuclear Chem. 4, 205-15 (1957); CA 51, 13526.
789a. Tomka, L. A. and Weinberg, E. L., U.S. 2,789,105, Apr. 16, 1957; CA 51, 10939.
790. ----- and -----, U.S. 2,798,862, July 9, 1957; CA 51, 15166.
790a. ----- and -----, U.S. 2,789,106, Apr. 16, 1957; CA 51, 10939.
791. -----, U.S. 2,798,863, July 9, 1957; CA 51, 15166.
792. Thompson, R. B., U.S. 2,450,099, Sept. 28, 1948; CA 43, 1440.
793. Velling, E. I. and Preobrazhenskaya, A. A., Farmakol. i Toksikol. 9, No. 5, 48-9 (1946); CA 41, 7532.
794. Walcutt, C. and Rifkin, E. B., Ind. Eng. Chem. 43, 2844-9 (1951); CA 46, 2787.
795. -----, Mason, J. M., and Rifkin, E. G., ibid., 46, 1029-34 (1954); CA 48, 8525.
796. Walde, A. W., Van Essen, H. E., and Zbornik, T. W., U.S. 2,762,821, Sept. 11, 1956; CA 51, 4424.
797. Walker, G., and Boyd, P. R., Lancet 263, 467-9 (1952); CA 46, 10440.
798. Walker, A. T. and Fremon, G. V., U.S. 2,603,206, July 15, 1952; CA 47, 880.
799. Wallen, L. L., Iowa State Coll. J. Sci. 29, 526-8 (1955); CA 49, 10418.
800. Ward, C. H., Dissertation Abstr. 14, 456 (1954).
801. Weinberg, E. L., Brit. 718,471, Nov. 17, 1954; CA 49, 5028.
802. ----- and Ramsden, H. E., U.S. 2,746,946, May 22, 1956; CA 50, 12541.
803. -----, Brit. 749,722, May 30, 1956; CA 51, 2859.
804. -----, Brit. 753,998, Aug. 1, 1956; CA 51, 6683.
805. -----, U.S. 2,789,102, Apr. 16, 1957; CA 51, 10939.
806. ----- and Tomka, L. A., U.S. 2,789,103, Apr. 16, 1957; CA 51, 10939.
807. ----- and -----, U.S. 2,789,107, Apr. 16, 1957; CA 51, 10939.
808. -----, U.S. 2,796,412, June 18, 1957; CA 51, 16537.
809. Whitman, N., U.S. 2,657,225, Oct. 27, 1953; CA 48, 2358.
810. Whittingham, G., Nature 160, 671-2 (1947); CA 42, 1132.
811. Wyler, J. A., U.S. 2,486,773, Nov. 1, 1949; CA 44, 1709.
812. Yamakita, Y., Rev. Phys. Chem. Japan 10, 69-83, 85-96, and 98-104 (1945); CA 48, 2728.
813. Yamazaki, K. and Kato, Y., Rept. Inst. Sci. Technol. (Univ. Tokyo) 4, 167-71 (1950); CA 45, 8735.
814. Widmaier, O., Brennstoff-Chem. 34, 83-7 (1953); CA 47, 6641.

815. Amick, M. G., Parmerlee, A. E., and Stecher, J. L.,
U.S. 2,091,112, Aug. 24, 1937; CA 31, 7446.
816. Amiral, C., Lautie', R., and Misermont, R., Fr. 858,548,
Nov. 27, 1940; CA 42, 3168.
817. Baer, M., U.S. 2,561,044, July 17, 1951; CA 45, 8807.
818. Bardan, P., Alexiu, L., Schöbesch, O., and Ganea, I.,
Compt. rend. acad. sci. Roumanie 7, 325-31 (1943-5);
CA 42, 2675.
819. Bayan, S., Riv. ital. petrol. 15, No. 173, 11-15, No. 174,
10-14, 18 (1947); CA 43, 3793.
820. Bolea, G., Riv. med. aeronaut. 10, 324-35 (1947); CA 43,
2416.
821. Bonkaló, S., Orvosi Helilap 89, 10-13 (1948); CA 43,
8547.
822. Boulard, A. de, Tech. ind. pétrole (Science et industrie)
No. 284 bis, 35-40 (1939); CA 33, 7521.
823. Brauch, F., Z. klin. Med. 143, 194-218 and 38-87 (1943);
CA 40, 6682.
824. Breslow, D. S., U.S. 2,554,810, May 29, 1951; CA 45, 8809.
825. Breynan, T., Angew. Chem. 62A, 430 (1950); CA 45, 851.
826. Burt, S. L., U.S. 2,489,518, Nov. 29, 1949; CA 44, 5639.
827. Butkov, N. A. and Rabinovich, R. I., Neftyanoe Khoz. 1936,
No. 7, 67-70; CA 31, 2404.
828. Campbell, J. M., U.S. 2,400,915, May 28, 1946; CA 40,
5240.
829. Castellino, N., Folia med. (Naples) 29, 209-21 (1943);
CA 38, 6003.
830. Cattaneo, A. G. and Sanly, A. L., Ind. Eng. Chem. 33,
1370-3 (1948); CA 35, 8269.
831. Cavallazzi, D., Med. valoro 37, 367-73 (1946); CA 41,
3536.
832. Church, J. M. and Johnson, E. W., U.S. 2,580,730, Jan. 1,
1952; CA 46, 9575.
833. Clark, F. M., Chem. Eng. News 25, 2976-8 (1947); CA 42,
299.
834. Cleverdon, D. and Staudinger, J. J. P., U.S. 2,481,086,
Sept. 26, 1949; CA 44, 4718.
835. Draeger, A. A., Gwin, G. T., Leesemann, C. J. G., and
Morrow, M. R., Petroleum Refiner 30, No. 11, 139-41
(1951); CA 46, 2275.
836. Duckworth, J. B., Moore, G. T., and Domke, C. J., Oil Gas
J. 48, No. 52, 142-4, 148, 149, 151 (1950); CA 44,
7521.
837. Egerton, A. C. and Powling, J., Proc. Roy. Soc. (London)
A193, 172-90 (1948); CA 43, 3617.
838. ----- and Jain, B. D., Fuel 31, 62-74 (1952); CA 46, 2276.
839. Engel, Arbeitsschutz 1942, 301-8; CA 38, 3041.
840. Ershov, N. V., J. Applied Chem. (U.S.S.R.) 12, 871-4 (1939);
CA 34, 2578.

841. Evans, T. W., Vaughn, W. E., and Rust, F. F., U.S. 2,411,961, Dec. 3, 1946; CA 41, 2068.
842. Faragher, W. F., U.S. 2,502,444, Apr. 4, 1950; CA 44, 5632.
843. Flury, F., Festschr. Zangger. 2, 836-5 (1935); Ber. Ges. Physiol. exptl. Pharmacol. 86, 508; CA 31, 6711.
844. Frank, C. E. and Blackham, A. U., Natl. Advisory Comm. Aeronaut. Tech. Note 2549, 33 pp. (1952); CA 46, 5303.
845. Frommel, E., Herschberg, A. D., and Piquet, J., Compt. rend. soc. phys. hist. nat. Genève 60, 97-100 (1943).
846. Gaylor, P. J., Petroleum Engr. 18, No. 3, 43-8 (1946); CA 41, 1082.
847. Guthrie, J. E., Powick, W. C., and Bandel, D., North Am. Vet. 22, 22-4 (1940); CA 35, 2616.
848. Hanford, W. E. and Harmon, J., U.S. 2,418,832; CA 42, 581.
849. Hansom, T. K., and Coles, K. F., J. Inst. Petroleum 33, 589-97 (1947); CA 42, 3164.
850. Harding, J., U.S. 2,597,987, May 27, 1952; CA 46, 8893.
851. Hart, E. J., U.S. 2,476,661, July 19, 1949; CA 43, 9519.
852. Hochwalt, C. A., U.S. 2,552,278, May 8, 1951; CA 46, 139.
853. Hubner, W. H. and Egloff, G., Oil Gas J. 36, No. 36, 103-4, 106, 108, 112 (1938); CA 32, 8105.
854. Humperdinck, K., Deut. med. Wochschr. 68, 587-9 (1942); CA 36, 7134.
855. Jenkins, R. L., U.S. 2,578,359, Dec. 11, 1951; CA 46, 6143.
856. Johnson, E. W., U.S. 2,524,528, Oct. 3, 1950; CA 45, 2498.
857. Jonash, E. R., Meyer, C. L., and Branstetter, J. R., Natl. Advisory Comm. Aeronautics Rept. ARR No. E6C04, 20 pp. (1946); CA 42, 4734.
858. Jost, H., and Rögener, H., Z. Electrochem. 47, 307-9 (1941); CA 35, 5682.
859. Kaneko, Y., J. Chem. Soc. Japan 65, 616-20 (1944); CA 41, 3349.
860. King, R. O., Can. J. Research 25F, 326-41 (1947); CA 42, 1727.
861. -----, ibid., 26F, 125-50 (1948); CA 42, 5651.
862. Kooijman, P. L. and Ghijsen, W. L., Rec. trav. chim. 66, 673-9 (1947); CA 42, 1869.
863. Kraut, H. and Lehmann, G., Arch. Gewerbepath. Gewerbehyg. 11, 256-95 (1941); CA 37, 3591.
864. Lacrois, G., Rass. med. ind. 13, 474-95 (1942); CA 40, 2529.
865. Lang, E. P. and Kunze, F. M., J. Ind. Hyg. Toxicol. 30, 256-9; CA 42, 6943.
866. Laves, W., Deut. med. Wochschr. 65, 1746-8 (1939); CA 34, 2067.
867. Leininger, R. I., U.S. 2,476,422, July 19, 1949; CA 43, 8739.
868. Livingston, H. K., Hyde, J. L. and Campbell, M. H., Ind. Eng. Chem. 41, 2722-6 (1949); CA 44, 2215.

869. Livingston, H. K., Oil Gas J. <u>47</u>, No. 38, 67-77 (1949); CA <u>44</u>, 6612.
870. Lovell, R. G., Petroleum Refiner <u>22</u>, 321-8 (1943); CA <u>37</u>, 6862.
871. Mahler, E. R. W. de, Belg. 418,670, Dec. 31, 1936; CA <u>31</u>, 5816.
872. Marchenko, E. N. and Beilikis, G. A., Gigiena i Sanit. <u>11</u>, No. 9, 17-23 (1946); CA <u>41</u>, 4873.
873. Maxted, E. B. and Moon, K. L., J. Chem. Soc. <u>1949</u>, 2171-4; CA <u>44</u>, 3778.
874. Mazny, R. El , J. Roy. Egypt. Med. Assoc. <u>28</u>, 107-19 (1945); CA <u>41</u>, 6023.
875. Meaker, C. L., Wu, C. S., and Rainwater, L. J., Phys. Rev. <u>73</u>, 1240 (1948).
876. Meyer, C. L. and Branstetter, J. P., Natl. Advisory Comm. Aeronautics, Wartime Rept. ARR Nos. E4J05, E5A20, E5D16, E5D16a, and E6C05 (1946); CA <u>42</u>, 7515.
877. Miyanishi, M. and Murata, K., Sci. Papers Inst. Phys. Chem. Research (Tokyo) <u>41</u>, 99-107 (1943); CA <u>41</u>, 6696.
878. -----, ibid., <u>40</u>, 364-9 (1943); CA <u>41</u>, 6696.
879. Morita, N., Bull. Chem. Soc. Japan <u>21</u>, 25-30 (1948); CA <u>44</u>, 1399.
880. Mortensen, R. A., J. Ind. Hyg. Toxicol. <u>24</u>, 285-7 (1942); CA <u>37</u>, 944.
881. Obara, G., Tohoku J. Exptl. Med. <u>46</u>, 313-22 (1944); CA <u>42</u>, 7444.
882. Ramazzotti, P., Folia med. (Naples) <u>29</u>, 311-17 (1943); CA <u>38</u>, 6004.
883. Rice, F. O., U.S. 2,087,656, July 20, 1937; CA <u>31</u>, 6259.
884. Richard, A., Compt. rend. soc. biol. <u>123</u>, 959-61 (1936); CA <u>31</u>, 1881.
885. -----, J. physiol. path. gén. <u>35</u>, 735-45 (1937); CA <u>32</u>, 7566.
886. Rokhlin, S. and Starchevskaya, A., Farm. Zhur. <u>14</u>, No. 2, 29-30 (1941); CA <u>35</u>, 8102.
887. Romm, F. S., J. Gen. Chem. (U.S.S.R.) <u>10</u>, 1784-92 (1940); CA <u>35</u>, 3880.
888. Schuster, G., Ann. chim. anal. <u>25</u>, 55-6 (1943); CA <u>40</u>, 1116.
889. Shappirio, S., Can. 371,890, Feb. 8, 1938; CA <u>32</u>, 2545.
890. Shcholkin, K. and Sokolik, A., Acta physicochem. U.S.S.R. <u>7</u>, 589-96 (1937); CA <u>32</u>, 4319.
891. Smittenberg, J. and Kooijman, P. L., Rec. trav. chim. <u>59</u>, 593-600 (1940); CA <u>35</u>, 4943.
892. Sortman, C. W., Beatty, H. A., and Heron, S. D., Ind. Eng. Chem. <u>33</u>, 357-60 (1941); CA <u>35</u>, 2325.
893. Spencer, R. C., Natl. Advisory Comm. Aeronautics, Rept. No. 710, 6 pp. (1941); CA <u>40</u>, 2955.

894. Steiger, B., Petroleum Z. 33, No. 32, 1-7 (1937); CA 32, 3594.
895. Sullivan, F. W., Jr., U.S. 2,148,138, Feb. 21, 1939.
896. -----, U.S. 2,087,660, July 20, 1937; CA 31, 6454.
897. Taeger, H., Deut. med. Wochschr. 70, 186-8 (1944); CA 39, 1685.
898. Trimble, H. M. and Richardson, J. O., Oil Gas J. 36, No. 46, 130 (1938); CA 32, 8755.
899. Trusty, A. W., Refiner Natural Gasoline Mfr. 19, 93-6 (1940); CA 34, 4550.
900. Vaughn, W. E. and Rust, F. F., J. Org. Chem. 5, 449-71 (1940); CA 35, 372.
901. Widmaier, O., Automobiltech. Z. 43, 63-8 (1940); CA 36, 5340.
902. Yngve, V., U.S. 2,267,777, Dec. 30, 1941; CA 36, 2647.
903. -----, U.S. 2,267,778, Dec. 30, 1941; CA 36, 2647.
904. -----, U.S. 2,267,779, Dec. 30, 1941; CA 36, 2647.
905. N. V. de Bataafsche Petroleum Maatschappij, Dutch 60,768, Mar. 15, 1948; CA 41, 3949.
906. -----, Dutch 64,847, Dec. 15, 1949; CA 44, 5581.
907. -----, Brit. 628,014, Aug. 19, 1949; CA 44, 2556.
908. E. I. duPont de Nemours & Co., Brit. 469,518, July 26, 1937; CA 32, 592.
909. Esso Research & Engineering Co., Brit. 737,392, Sept. 28, 1955; CA 50, 9010.
910. Ethyl Gasoline Corporation, Fr. 841,534, May 22, 1939; CA 34, 4393.
911. -----, Fr. 841,535, May 22, 1939; CA 34, 4393.
912. Metal & Thermit Corp., Brit. 772,646, Apr. 17, 1957; CA 51, 15551.
913. Rohm & Haas Co., Brit. 771,746, Apr. 3, 1957; CA 51, 12542.
914. Solvay & Cie., Brit. 761,844, Nov. 21, 1956; CA 51, 10561.

IV. BIBLIOGRAPHY OF MONOGRAPHS AND REVIEW ARTICLES

Banks, F. R., "The Influence of Et_4Pb on Engine Design and Performance," J. Inst. Petroleum 35, 264-92 (1949). - A historical review with special attention to British conditions.

Edgar, G., "Tetraethyllead. Manufacture and Use," Ind. Eng. Chem. 31, 1439-46 (1939).

Elizarova, O. N., "Leaded Gasoline," Fel'dsher i Akusherka 1946, No. 3, 36-9. - A review of toxicology and antidotes.

Gastinger, E., "Trends in the Chemistry of Germanium," Fortschr. Chem. Forsch. 3, 603-56 (1955).

Gmelins Handbuch der anorganischen Chemie, "Germanium," System - Nummer 45, Ergänzungsband, 8th ed., Verlag Chemie, G. m. b. H., Weinheim/Berchtesgaden, 1958, pp. 545-557.

Hammerich, Th., "Behavior of Antiknock Agents I. Et_4Pb," Brennstoff-Chem. 30, 109-26 (1949). - A review of the literature and unpublished German results.

Hedges, E. S., "Organotin Compounds Find New Uses as Stabilizer, Anthelmintics, and Biocides," Mfg. Chemist. 28, 16-17 (1957).

Jones, R. G. and Gilman, H., "Methods of Preparation of Organometallic Compounds," Chem. Revs. 54, 835-881 (1954).

Johnson, O. H., "Organogermanium Compounds," in "Inorganic Syntheses," Th. Moeller, ed., McGraw-Hill Book Co., Inc., New York, 1957, Vol. V, pp. 64-70.

Kerk, G. J. M. van der, Luijten, J. G. A., and Noltes, J. G., "New Results of Organotin Research," Angew. Chem. 70, 298-306 (1958).

Korshak, V. V. and Kolesnikov, G. S., "Tetraethyllead," Uspekhi Khim. 15, 325-42 (1946). - A review of the applications, laboratory and commercial methods of preparation, and physical, chemical, and physiological properties of Et_4Pb.

Kraus, C. A., "The Chemistry of Organo-Metallic Compounds," J. Chem. Educ. 26, 45-9 (1949).

Leeper, R. W., Summers, L., and Gilman, H., "Organolead Compounds," Chem. Revs. 54, 101-67 (1954).

Luijten, J. G. A. and Kerk, G. J. M. van der, "Investigations in the Field of Organotin Chemistry," Tin Research Institute, Greenford, Middlesex, England, 1955. - A monograph.

Monserat, M. P., "Elevation of Octane Numbers of Liquid Combustibles by Means of Triethyllead Chloride," Combustibles (Zaragoza) 5, No. 25/6, 15-21, No. 27/8, 67-77 (1945).

Nesmeyanov, A. N., "Organic Compounds and the Periodic System," Uspekhi Khim. 14, 261-81 (1945). - The structure and reactions of organic compounds of Hg, Sn, As, Te, B, and Bi are reviewed.

Nickerson, S. P., "Tetraethyllead: A Product of American Research," J. Chem. Educ. 31, 560-71 (1954). - A review of the history of Et_4Pb manufacture and use.

Oikawa, H., "Germanium Compounds," J. Coal Research Inst. 6, 297-318, 426-7 (1955). - A review with 221 references.

Oikawa, H. and Shiota, M., "Synthesis of Organogemarium Compounds," Yûki Gôsei Kagaku Kyôkaishi 15, 411-18 (1957). - A review with 109 references.

Smith, H. V., "Organic Tin Stabilizers for Vinyl Plastics," Plastics (London) 17, 264 (1952). - A review with 24 references.

Smith, H. V., "Organo-Tin Stabilizers for Vinyl Polymers," Rubber Age & Synthetics 34, 206-7 (1953). - A review and discussion.

Smith, H. V., "Stabilizers for Vinyl Polymers. III. Organometallic Compounds," Brit. Plastics 27, 215-17 (1954). - A review with 75 references.

Tchakirian, A., "The Chemistry of Germanium," Ann. chim. 12, 415-99 (1939). - A review with 183 references.

Watt, G. W., "Reactions of Organic and Organometallic Compounds with Solution of Metals in Liquid Ammonia," Chem. Revs. 46, 317-79.

Zscharn, A., "The Decomposition of Tetraalkyllead Compounds," Chem. - Ztg. 64, 498 (1940).

V. LIST OF COMPOUNDS WITH PAGE INDEX

GERMANIUM

R_4Ge	Page
Me_4Ge	7
Et_4Ge	8
$(CH_2{:}CH)_4Ge$	8
Pr_4Ge	8
$(CH_2{:}CHCH_2)_4Ge$	8
Bu_4Ge	8
$i\text{-}Bu_4Ge$	9
$(n\text{-}C_5H_{11})_4Ge$	9
$(n\text{-}C_6H_{13})_4Ge$	9
$(n\text{-}C_7H_{15})_4Ge$	9
$(n\text{-}C_8H_{17})_4Ge$	9
$[MeCHEt(CH_2)_4]_4Ge$	9
$(n\text{-}C_{10}H_{21})_4Ge$	10
$(n\text{-}C_{12}H_{25})_4Ge$	10
$(n\text{-}C_{14}H_{29})_4Ge$	10
$(n\text{-}C_{16}H_{33})_4Ge$	10
$(n\text{-}C_{18}H_{37})_4Ge$	10
$(cyclo\text{-}C_6H_{11}CH_2CH_2)_4Ge$	10
Ph_4Ge	10
$(p\text{-}MeC_6H_4)_4Ge$	11
$(PhCH_2)_4Ge$	11
$(PhCH_2CH_2)_4Ge$	11
$[Ph(CH_2)_3]_4Ge$	11
$(1\text{-}C_{10}H_7)_4Ge$	11

$R_{4-n}GeR'_n$ and $R_2GeR'R''$	
Me_3GeEt	12
Me_2GeEt_2	12
Et_2GeBu_2	12
Et_2GePh_2	12
Et_3GeMe	12
Et_3GePr	12
Et_3GeBu	12
$Et_3Ge(n\text{-}C_5H_{11})$	12
Et_3GeCH_2Ph	12
$Et_3GeC_6H_4Ph\text{-}p$	12
$(n\text{-}C_5H_{11})_3GeEt$	13
$(n\text{-}C_6H_{13})_3GePh$	13
$(cyclo\text{-}C_6H_{11})_3GeMe$	13
$(cyclo\text{-}C_6H_{11})_3GeEt$	13
$(cyclo\text{-}C_6H_{11})_3GePr$	13
$(cyclo\text{-}C_6H_{11})_3GeBu$	13
$(cyclo\text{-}C_6H_{11})_3Ge(n\text{-}C_5H_{11})$	13
$(cyclo\text{-}C_6H_{11})_3GePh$	13
$(cyclo\text{-}C_6H_{11})_3GeCH_2Ph$	13
$Ph_3Ge(cyclo\text{-}C_6H_{11})$	13
$Ph_3Ge(n\text{-}C_6H_{13})$	13
$Ph_3Ge(n\text{-}C_8H_{17})$	13
$Ph_3Ge(n\text{-}C_{18}H_{37})$	13
$Ph_3GeC_6H_4Me\text{-}m$	14
$Ph_3GeCH_2CH_2Ph$	14
$Ph_3GeCH_2CHPh_2$	14
$Ph_3GeC_6H_4\text{-}p\text{-}CHPh_2$	14
Ph_3GeCPh_3	14
$Ph_2(PhCH_2CH_2)Ge(n\text{-}C_{18}H_{37})$	14
$Me_3GeCH_2CH{:}CH_2$	15
$Et_3GeCH_2CH{:}CH_2$	15
$Bu_3GeCH_2CH{:}CH_2$	15
$Ph_3GeCH_2CH{:}CH_2$	15
$Me_3GeCH_2CMe{:}CH_2$	16
$Me_2Ge(CH_2CH{:}CH_2)_2$	16
$(CH_2{:}CHCH_2)_3GeMe$	16
$Me_3GeCH{:}CH_2$	16
$Et_3GeCH{:}CH_2$	16
$Bu_3GeCH{:}CH_2$	16
$Et_2Ge(CH{:}CH_2)_2$	16
$(CH_2{:}CH)_3GeEt$	17
Me_3GeCH_2Cl	18
Me_3GeCH_2I	18
$Bu_3GeCH_2CH_2CCl_3$	18
$Bu_3GeCHBrCH_2Br$	18
$Me_3GeC_6H_4Br\text{-}p$	18
$Et_3GeC_6H_4Br\text{-}p$	18
$Ph_3GeC_6H_4Br\text{-}p$	18
Ph_3GeCH_2Cl	18
$Et_3GeCH_2CH_2CN$	18
$Bu_3GeCH_2CH_2CN$	18
$(n\text{-}C_5H_{11})_3GeCH_2CH_2CN$	18

	Page
$Et_3GeCH_2CH_2CHO$	19
$Bu_3GeCH_2CH_2CHO$	19
Et_3GeCH_2COMe	19
Et_3GeCH_2Bz	19
$Ph_3GeCHPhCH_2Bz$	20
Ph_3GeCO_2H	20
Ph_3GeCO_2Me	20
$Ph_3GeCO_2GePh_3$	20
$Me_3GeC_6H_4CO_2H\text{-}p$	20
$Et_3GeCH_2CH_2CO_2H$	21
$Et_3GeC_6H_4CO_2H\text{-}p$	21
$Pr_3GeCH_2CH_2CO_2Me$	21
$Bu_3GeCH_2CO_2Et$	21
$Bu_3GeCH_2CH_2CO_2Et$	21
$Ph_3GeC_6H_4CO_2H\text{-}m$	21
$Ph_3GeC_6H_4CO_2H\text{-}p$	21
$Et_3GeCH_2CHOHCH_2OH$	22
$Pr_3GeCH_2CH_2CH_2OH$	22
$Bu_3GeCH_2CH_2CH_2OH$	22
Ph_3GeCH_2OH	22
$Ph_3GeC(OH)Ph_2$	22
$Ph_3GeC_6H_4CH_2OH\text{-}p$	22
$m\text{-}Me_2NC_6H_4GePh_3$	23
$p\text{-}Me_2NC_6H_4GePh_3$	23
$Et_3Ge(CH_2)_3SCH_2CH_2OH$	23
$Et_3Ge(CH_2)_3SCH_2CO_2H$	23
$Et_3GeCH_2CH_2SCH_2CO_2H$	23
$Et_3GeCHBr(CH_2)_2CO_2Et$	23
$MeGe[C\colon CC(OH)Me_2]_3$	23
$Me_3GeC\colon CC(OH)Me_2$	24
$Me_2Ge[C\colon CC(OAC)Me_2]_2$	24
$Me_2Ge[C\colon CC(OH)Me_2]_2$	24
$(C_4H_3O)_4Ge$	24

MGeR₃

$LiGeEt_3$	24
$LiGePh_3$	25
$NaGePh_3$	25
$KGePh_3$	25

	Page
$R_{4-n}GeH_n$, $RR'GeH_2$, $R_2R'GeH$, $R_{4-n}GeD_n$, and R_2EtGeD	
$MeGeH_3$	26
$EtGeH_3$	26
$i\text{-}Pr_2GeH_2$	26
$Et(i\text{-}C_5H_{11})GeH_2$	26
Ph_2GeH_2	26
Et_3GeH	27
Pr_3GeH	27
Bu_3GeH	27
$(cyclo\text{-}C_6H_{11})_3GeH$	27
Ph_3GeH	28
$MeGeD_3$	29
$EtGeD_3$	29
$CH_2\colon CHGeH_3$	29
$PrGeH_3$	29
Me_2GeH_2	29
Me_2GeD_2	29
Et_2GeH_2	29
Me_3GeH	29
Me_3GeD	29
Me_2EtGeH	29
Me_2EtGeD	29
$(n\text{-}C_5H_{11})_3GeH$	30
$(i\text{-}C_5H_{11})_3GeH$	30
$Et_2(i\text{-}C_5H_{11})GeH$	30
$(n\text{-}C_6H_{13})_3GeH$	30
$(1\text{-}C_{10}H_7)_3GeH$	30

R_3GeX

Me_3GeF	31
Me_3GeCl	31
Me_3GeBr	32
Me_3GeI	32
$Me(CH_2Cl)_2GeCl$	32
$Me_2EtGeCl$	32
$Me_2EtGeBr$	32
Et_3GeF	32
Et_3GeCl	32
Et_3GeBr	33
Et_3GeI	33
$(CH_2\colon CH)_3GeBr$	33
$(CH_2\colon CH)_3GeI$	33
Pr_3GeF	33
Pr_3GeCl	33
Pr_3GeBr	33

	Page
Pr_3GeI	34
$i\text{-}Pr_3GeF$	34
$i\text{-}Pr_3GeCl$	34
$i\text{-}Pr_3GeBr$	34
$i\text{-}Pr_3GeI$	34
Bu_3GeF	34
Bu_3GeCl	35
Bu_3GeBr	35
Bu_3GeI	35
$i\text{-}Bu_3GeI$	35
$(n\text{-}C_5H_{11})_3GeI$	35
$(n\text{-}C_6H_{13})_3GeCl$	35
$(n\text{-}C_6H_{13})_3GeBr$	35
$(n\text{-}C_6H_{13})_3GeI$	36
$(cyclo\text{-}C_6H_{11})_3GeCl$	36
$(cyclo\text{-}C_6H_{11})_3GeBr$	36
$(cyclo\text{-}C_6H_{11})_3GeI$	36
Ph_3GeF	36
Ph_3GeCl	36
Ph_3GeBr	37
Ph_3GeI	37
$(p\text{-}MeC_6H_4)_3GeF$	37
$(1\text{-}C_{10}H_7)_3GeBr$	37

R_2GeX_2 and $RR'GeX_2$

	Page
Me_2GeF_2	38
Me_2GeCl_2	38
Me_2GeBr_2	39
$Me(CH_2Cl)GeCl_2$	39
$(ClCH_2)_2GeCl_2$	39
Et_2GeBr_2	39
Et_2GeCl_2	40
Et_2GeI_2	40
Pr_2GeF_2	40
Pr_2GeCl_2	40
Pr_2GeBr_2	40
Pr_2GeI_2	40
$i\text{-}Pr_2GeF_2$	41
$i\text{-}Pr_2GeCl_2$	41
$i\text{-}Pr_2GeBr_2$	41
$i\text{-}Pr_2GeI_2$	41
Ph_2GeCl_2	41
Ph_2GeBr_2	41
Ph_2GeI_2	41

$RGeX_3$

	Page
$MeGeF_3$	42
$MeGeCl_3$	42
$MeGeBr_3$	43
$MeGeI_3$	43
$ClCH_2GeCl_3$	43
Cl_3CGeCl_3	43
$EtGeCl_3$	43
$EtGeI_3$	44
$MeCHClGeCl_3$	44
$ClCH_2CH_2GeCl_3$	44
$CH_2\text{:}CHGeCl_3$	44
$PrGeCl_3$	44
$PrGeBr_3$	44
$i\text{-}PrGeCl_3$	45
$i\text{-}PrGeBr_3$	45
$CH_2\text{:}CHCH_2GeCl_3$	45
$CH_2\text{:}CMeCH_2GeCl_3$	45
$n\text{-}C_5H_{11}GeCl_3$	46
$MeCHCl(CH_2)_3GeCl_3$	46
$n\text{-}C_6H_{13}GeCl_3$	46
$n\text{-}C_6H_{12}ClGeCl_3$	46
$n\text{-}C_6H_{11}F_2GeCl_3$	46
$n\text{-}C_7H_{15}GeCl_3$	46
$Me_2CH(CH_2)_5GeCl_3$	46
$n\text{-}C_8H_{17}GeCl_3$	46
$n\text{-}C_{10}H_{21}GeCl_3$	46
$PhGeCl_3$	47
$CH_2(GeCl_3)_2$	47

$R_{4-n}Ge(CN)_n$, $R_{4-n}Ge(NCO)_n$, $R_{4-n}Ge(CNS)_n$ and $R_{4-n}Ge(NCS)_n$

	Page
$EtGe(CN)_3$ (?)	48
$(Me_2CH)_2Ge(CN)_2$	48
Et_3GeCN	48
Pr_3GeCN	48
$(Me_2CH)_3GeCN$	48
$EtGe(NCO)_3$	49
$Et_2Ge(NCO)_2$	49
$i\text{-}Pr_2Ge(NCO)_2$	49
$Et_3Ge(NCO)$	49
$Pr_3Ge(NCO)$	49
$i\text{-}Pr_3Ge(NCO)$	49
$Bu_3Ge(NCO)$	50

	Page		Page
$Me_2Ge(CNS)_2$	50	$Pr_2Ge(OOC_{10}H_{17})_2$	58
$Et_2Ge(NCS)_2$	51	$(CH_2)_5Ge(OOCMe_3)_2$	58
$(Me_2CH)_2Ge(NCS)_2$	51	$Pr_3GeOOCMe_3$	58
Et_3GeNCS	51	$Pr_3GeOOC_{10}H_{17}$	58
Pr_3GeNCS	51	$Ph_3GeOOCMe_3$	58
$(Me_2CH)_3GeNCS$	51	$Ph_3GeOOCPh_3$	58
Bu_3GeNCS	51	$Ph_3GeOOCMe_2Ph$	58
		$MeGe(OMe)_3$	59
$\underline{RGeO_2H \text{ and } (RGeO)_2O}$		$Me_2Ge(OMe)_2$	59
		Me_3GeOMe	59
$CH_2(GeO_2H)_2$	52	Ph_3GeOMe	59
$(PrGeO)_2O$	52		
$(PhGeO)_2O$	52	$\underline{R_{4-n}Ge(OOCR)_n}$	
$(p-MeOC_6H_4GeO)_2O$	52		
$(p-EtOC_6H_4GeO)_2O$	52	$Et_3GeSCH_2CO_2GeEt_3$	60
$(p-ClC_6H_4GeO)_2O$	52	$Ph_3GeCO_2GePh_3$	60
$(p-BrC_6H_4GeO)_2O$	52	For Other Organotin	
$(p-HO_2CC_6H_4GeO)_2O$	52	Carboxylates	
$(3-O_2N-4-Me_2NC_6H_3GeO)_2O$	52	See Table	61-65
$\underline{(RGeO)_n, (R_2GeO)_n, \text{ and } (R_3Ge)_2O}$		$\underline{R_2GeSO_4, (R_3Ge)_2SO_4, \text{ and}}$	
		$\underline{R_2GeCrO_4}$	
$(MeGeO)_n$	53		
$(Me_2GeO)_4$	53	Me_2GeSO_4	67
$(Et_2GeO)_n$	53	Et_2GeSO_4	67
$(Pr_2GeO)_n$	54	Pr_2GeSO_4	67
$(i-Pr_2GeO)_n$	54	$(Me_2CH)_2GeSO_4$	67
$(Ph_2GeO)_n$	54	$(Et_3Ge)_2SO_4$	67
$(Et_3Ge)_2O$	54	$(Pr_3Ge)_2SO_4$	67
$(Pr_3Ge)_2O$	54	$(i-Pr_3Ge)_2SO_4$	67
$(i-Pr_3Ge)_2O$	55	Me_2GeCrO_4	68
$(Bu_3Ge)_2O$	55		
$[(C_6H_{13})_3Ge]_2O$	55		
$(Ph_3Ge)_2O$	55	$\underline{(R_2GeS)_n, (R_3Ge)_2S, \text{ and } (R_3GeS)_2}$	
$[(p-MeC_6H_4)_3Ge]_2O$	56		
		$(Me_2GeS)_n$	68
$\underline{R'_3GeOH, (R_3GeO)_2, R_{4-n}Ge(OOR')_n}$		$(i-Pr_2GeS)_2$	68
$\underline{\text{and } R_{4-n}Ge(OR')_n}$		$(Et_3Ge)_2S$	69
		$(Ph_3Ge)_2S$	69
$i-Pr_3GeOH$	56	$[(PhCH_2)_3Ge]_2S$	69
$(cyclo-C_6H_{11})_3GeOH$	56	$[(p-MeC_6H_4)_3Ge]_2S$	69
Ph_3GeOH	57	$[(PhC_6H_4)_3Ge]_2S$	69
$(1-C_{10}H_7)_3GeOH$	57	$[(cyclo-C_6H_{11})_3GeS]_2$	70
$(Pr_3GeO)_2$	57		
$(Ph_3GeO)_2$	57		

	Page
Et_3GeSR	
See Table	71
$R_{4-n}Ge(NR'R'')_n$ and $(R_3Ge)_3N$	
$EtGe(NMe_2)_3$	72
$EtGe(NEt_2)_3$	72
Ph_3GeNH_2	72
$(Ph_3Ge)_3N$	72
$(Me_2N)_4Ge$	72
$(Et_2N)_4Ge$	72
$\equiv Ge-Si\equiv$, $\equiv Ge-R-Si\equiv$, and $\equiv Ge-Sn\equiv$	
$Ph_3SiGePh_3$	73
$(Ph_3Ge)_3SiH$	73
$(Ph_3Ge)_3SiBr$	73
$(Ph_3Ge)_3SiOH$	73
$(Ph_3Ge)_3SiCl$	73
$(Ph_3Ge)_3SiEt$	73
$(Ph_3Ge)_3SiNH_2$	74
$Ph_3GeSiEt_3$	74
$(Ph_3Ge)_2Si:Si(GePh_3)_2$ (?)	74
$[p-(Me_3Si)C_6H_4]_4Ge$	74
$Me_3GeCH_2SiMe(OMe)_2$	74
$[(Me_3GeCH_2)MeSiO-]_n$	74
$Ph_3GeSnPh_3$	74
$(R_3Ge)_2$, $(R_2XGe)_2$, $R_3Ge(CH_2)_xGeR_3$, and $R_3GeC:CGeR_3$	
$(Me_3Ge)_2$	75
$(Et_3Ge)_2$	75
$[(CH_2:CH)_3Ge]_2$	76
$[(cyclo-C_6H_{11})_3Ge]_2$	76
$(Ph_3Ge)_2$	76
$(Bu_2IGe)_2$ (?)	77
$(Et_3GeCH_2-)_2$	77
$(Et_3GeCH_2)_2CH_2$	77
$(Ph_3GeCH_2)_2CH_2$	77
$Ph_3Ge(CH_2)_5GePh_3$	77
$(Et_3GeC:)_2$	77
$[(cyclo-C_6H_{11})_3GeC:]_2$	77
$(Ph_3GeC:)_2$	77

	Page
$(RGe)_x$	
$(PhGe)_x$	78
$(i-PrGe)_x$	78
$(PrGe)_x$	78
R_2Ge	
Ph_2Ge (?)	78
TIN	
R_4Sn	
Me_4Sn	79
Et_4Sn	80
Pr_4Sn	81
$i-Pr_4Sn$	82
Bu_4Sn	82
$i-Bu_4Sn$	83
$(n-C_5H_{11})_4Sn$	83
$(i-C_5H_{11})_4Sn$	84
$(cyclo-C_5H_5)_4Sn$	84
$(cyclo-C_6H_9)_4Sn$	84
$(n-C_6H_{13})_4Sn$	84
$(n-C_7H_{15})_4Sn$	84
$(n-C_8H_{17})_4Sn$	85
$(BuCHEtCH_2)_4Sn$	85
$(Me_3CCH_2CHMeCH_2CH_2)_4Sn$	85
$(n-C_{12}H_{25})_4Sn$	85
$(n-C_{14}H_{29})_4Sn$	85
$(n-C_{16}H_{33})_4Sn$	85
$(n-C_{18}H_{37})_4Sn$	86
Ph_4Sn	86
$(m-MeC_6H_4)_4Sn$	88
$(o-MeC_6H_4)_4Sn$	88
$(p-MeC_6H_4)_4Sn$	88
$(Me_3C_6H_2)_4Sn$	88
$(PhCH_2)_4Sn$	89
$(C_9H_7)_4Sn$	89
$(1-C_{10}H_7)_4Sn$	89
$(o-PhC_6H_4)_4Sn$	89
$(m-PhC_6H_4)_4Sn$	89

	Page
$(p\text{-}PhC_6H_4)_4Sn$	89
$(C_{13}H_9)_4Sn$	89
$(C_{14}H_9)_4Sn$	90
$(ClCH_2)_4Sn$	91
$(BrCH_2)_4Sn$	91
$(MeCHCl)_4Sn$	91
$(p\text{-}ClC_6H_4)_4Sn$	91
$(m\text{-}HOC_6H_4)_4Sn$	91
$(p\text{-}MeOC_6H_4)_4Sn$	91
$(p\text{-}EtOC_6H_4)_4Sn$	91
$(p\text{-}Me_2NC_6H_4)_4Sn$	91
$(p\text{-}Me_3SiC_6H_4)_4Sn$	91
$(CH_2{:}CH)_4Sn$	91, 92
$(CH_2{:}CMe)_4Sn$	92
$(CH_2{:}CHCH_2)_4Sn$	92
$(PhC{:}C)_4Sn$	92
$(SCH{:}CHCH{:}CH)_4Sn$	93

$\underline{R_2SnR'_2}$

See Table	94-96

$\underline{R_3SnR'}$

See Table	97-100

$\underline{R_{4-n}Sn\,(\text{organo-halide})_n}$

See Table	102-103

$\underline{R_{4-n}Sn(\text{organo-cyanide or -thiocyanide})_n}$

See Table	104

$\underline{R_{4-n}Sn(R'CO_2H)_n}$ (R' = divalent radical)

See Table	106

$\underline{R_{4-n}Sn(R'CO_2R'')_n}$ (R' = divalent radical)

See Table	107-108

$\underline{R_3Sn(CH_2)_xCONH_2}$ and $\underline{R_3SnR'COOCO}$ (R' = divalent radical)

$Pr_3SnCH_2CH_2CONH_2$	109
$Ph_3SnCH_2CH_2CONH_2$	109
$Ph_3Sn\overline{CHCHCH{:}CHCHCHCOOCOCH}$	109

$\underline{R_{4-n}Sn[CR'_xR''_y(COR''')_{3-(x+y)}]_n}$

See Table	110-111

$\underline{R_{4-n}Sn(\text{hydroxyhydrocarbyl})_n}$

See Table	113

$\underline{R_{4-n}Sn(-R'OR'')_n}$ and $\underline{R_{4-n}Sn(-R'OOCR'')_n}$ (R' = divalent radical)

See Table	114

$\underline{R_{4-n}Sn(-R'NR''_2)_n}$ and $\underline{R_{4-n}Sn(-R'NHCOR'')_n}$ (R' = divalent radical)

See Table	116

$\underline{R_{4-n}Sn(\text{alkenyl})_n}$

See Table	118-121

	Page
$R_{4-n}Sn(heterocyclyl)_n$	
See Table	123
$R_3SnC\vdots CR'$	
See Table	124
Polyfunctional $R_{4-n}SnR'_n$	
$m\text{-}(EtO_2CCH_2O)C_6H_4SnPh_3$	124
$Ph_3Sn(CH_2)_3OCH_2CH_2CN$	124
$Ph_3SnC\vdots CCH(OEt)_2$	124
$Ph_3SnCH_2CH(OH)CH_2Cl$	124
$Bu_2Sn[CH(CN)CO_2Me]_2$	125
$Bu_2Sn(CHAcCO_2Et)_2$	125
$Bu_2Sn(CHAcCO_2Bu)_2$	125
$Bu_2Sn(CHBzCO_2Et)_2$	125
$Bu_2Sn(CAc_2CO_2Et)_2$	125
$Bu_2Sn(CBuAcCO_2Et)_2$	125
$Bu_3SnCHAcCO_2CHMeCH_2CHMe_2$	125
$3\text{-}(p\text{-}O_2NC_6H_4N_2)\text{-}4\text{-}Me_2NC_6H_3SnPh_3$	126
$3\text{-}(p\text{-}BrC_6H_4N_2)\text{-}4\text{-}Me_2NC_6H_3SnPh_3$	126
$3\text{-}(p\text{-}ClC_6H_4N_2)\text{-}4\text{-}Me_2NC_6H_3SnPh_3$	126
$6\text{-}(p\text{-}O_2NC_6H_4N_2)\text{-}3\text{-}Me_2NC_6H_3SnPh_3$	126
$6\text{-}(p\text{-}BrC_6H_4N_2)\text{-}3\text{-}Me_2NC_6H_3SnPh_3$	126
$6\text{-}(p\text{-}HO_2CC_6H_4N_2)\text{-}3\text{-}Me_2NC_6H_3SnPh_3$	126
M_nSnR_{4-n} (M=alkali metal)	
$LiSnBu_3$	127
$LiSnPh_3$	127
$LiSn(C_6H_4Me)_3$	127
$NaSnEt_3$	127
$NaSnPh_3$	128
$NaSn(CPh_3)Ph_2$	128
Na_2SnEt_2	128
$(Et_2SnNa)_2$	128

	Page
$R_{4-n}SnH_n$	
$MeSnH_3$	128
$EtSnH_3$	128
$CH_2\text{:}CHSnH_3$	128
$PrSnH_3$	128
$BuSnH_3$	129
Me_2SnH_2	129
Et_2SnH_2	129
Pr_2SnH_2	129
Bu_2SnH_2	129
Ph_2SnH_2	129
Me_3SnH	129
Et_3SnH	130
Pr_3SnH	130
$i\text{-}Pr_3SnH$	130
Bu_3SnH	130
$(n\text{-}C_6H_{13})_3SnH$	131
$(n\text{-}C_8H_{17})_3SnH$	131
Ph_3SnH	131
$R_{4-n}SnX_n$ (X = halogen)	
Me_3SnF	132
Me_3SnCl	133
Me_3SnBr	133
Me_3SnI	133
Et_3SnF	134
Et_3SnCl	134
Et_3SnBr	135
Et_3SnI	136
Pr_3SnF	136
Pr_3SnCl	136
Pr_3SnBr	137
Pr_3SnI	137
$i\text{-}Pr_3SnCl$	137
$i\text{-}Pr_3SnBr$	137
Bu_3SnCl	137
Bu_3SnBr	138
Bu_3SnI	138
$i\text{-}Bu_3SnF$	139
$t\text{-}Bu_3SnCl$	139
$(n\text{-}C_5H_{11})_3SnCl$	139
$(i\text{-}C_5H_{11})_3SnF$	139
$(i\text{-}C_5H_{11})_3SnI$	139

	Page
$(cyclo\text{-}C_6H_{11})_3SnCl$	139
$(n\text{-}C_6H_{13})_3SnF$	139
$(n\text{-}C_7H_{15})_3SnI$	139
$(n\text{-}C_8H_{17})_3SnCl$	139
$(n\text{-}C_8H_{17})_3SnI$	139
$(n\text{-}C_{12}H_{25})_3SnCl$	140
$(n\text{-}C_{14}H_{29})_3SnCl$	140
$(n\text{-}C_{18}H_{37})_3SnCl$	140
Ph_3SnF	140
Ph_3SnCl	140
Ph_3SnBr	141
Ph_3SnI	142
$(MeC_6H_4)_3SnCl$	142
$(MeC_6H_4)_3SnBr$	142
$(PhCH_2)_3SnF$	142
$(PhCH_2)_3SnCl$	143
$(p\text{-}MeC_6H_4CH_2)_3SnCl$	143
$(o\text{-}PhC_6H_4)_3SnBr$	143
$(ClCH_2)_3SnCl$	144
$(BrCH_2)_3SnBr$	144
$(MeCHCl)_3SnCl$	144
$(ClCH{:}CH)_3SnCl$	144
$(p\text{-}ClC_6H_4)_3SnBr$	144
$(p\text{-}ClC_6H_4)_3SnI$	145
$(o\text{-}ClC_6H_4CH_2)_3SnF$	145
$(CH_2{:}CH)_3SnF$	145
$(CH_2{:}CH)_3SnCl$	145
$(CH_2{:}CH)_3SnBr$	145
$(CH_2{:}CH)_3SnI$	145
$(MeO_2CCH_2CH_2)_3SnBr$	145
$(p\text{-}EtO_2CC_6H_4)_3SnCl$	146

$R_2R'SnX$

See Table 147-148

$R_2(\text{Halo-}R')SnX$ (R' = divalent radical)

See Table 150

$R_2(\text{Cyano-}R')SnX$ (R' = divalent radical)

See Table 151

$R_2(\text{Alkenyl})SnX$

See Table 152-153

$R_{3-x}R'_xSnX$ (R' = Carboxyaryl, Carbalkoxy-alkyl or -aryl, or Hydroxyalkyl)

	Page
$Pr(MeO_2CCH_2CH_2)_2SnBr$	153
$Pr_2(EtO_2CCH_2CHMe)SnBr$	153
$Bu_2(MeO_2CCH_2CH_2)SnBr$	153
$Ph_2(MeO_2CCH_2CH_2)SnI$	153
$Ph_2(o\text{-}HO_2CC_6H_4)SnCl$	154
$Ph_2(o\text{-}MeO_2CC_6H_4)SnCl$	154
$Ph_2(HOCH_2CH_2CH_2)SnBr$	154

$R_{3-x}R'_xSnX$ (R' = $Me_3SiCH_2\text{-}$)

See Table 154

R_2SnX_2, $RR'SnX_2$, and R_2SnXX'

	Page
Me_2SnCl_2	156
Me_2SnBr_2	157
Me_2SnI_2	157
$EtMeSnCl_2$	157
Et_2SnF_2	157
Et_2SnCl_2	158
Et_2SnBr_2	159
Et_2SnI_2	159
Pr_2SnCl_2	159
Pr_2SnBr_2	160
Pr_2SnI_2	160
$i\text{-}Pr_2SnCl_2$	160
$i\text{-}Pr_2SnI_2$	160
$BuPrSnCl_2$	160
$Bu_2SnBrCl$	160
Bu_2SnCl_2	161
Bu_2SnBr_2	162
$i\text{-}Bu_2SnI_2$	162
$(i\text{-}C_5H_{11})_2SnI_2$	162
$(n\text{-}C_6H_{13})_2SnCl_2$	162
$(n\text{-}C_8H_{17})_2SnCl_2$	162
$(BuCHEtCH_2)_2SnCl_2$	162
$(Me_3CCH_2CHMeCH_2CH_2)_2\text{-}SnCl_2$	163

	Page
$(n\text{-}C_{12}H_{25})_2SnCl_2$	163
$(cyclo\text{-}C_6H_{11})_2SnCl_2$	163
Ph_2SnCl_2	163
Ph_2SnBr_2	164
$(MeC_6H_4)_2SnCl_2$	164
$(MeC_6H_4)_2SnBr_2$	164
$(PhCH_2)_2SnCl_2$	165
$(1\text{-}C_{10}H_7)_2SnCl_2$	165
$(1\text{-}C_{10}H_7)_2SnBr_2$	165
$(1\text{-}C_{10}H_7)_2SnI_2$	165
$(o\text{-}PhC_6H_4)_2SnCl_2$	165
$(p\text{-}PhC_6H_4)_2SnCl_2$	166
$(p\text{-}PhC_6H_4)_2SnBr_2$	166
$(9\text{-Phenanthryl})_2SnCl_2$	166
$(CH\!:\!CH)_2SnCl_2$	166
$(CH_2\!:\!CH)_2SnBr_2$	166
$(CH_2\!:\!CHCH_2)_2SnBr_2$	166
$(MeCH\!:\!CH_2)_2SnCl_2$	167
$(MeCH\!:\!CH_2)_2SnBr_2$	167
$(CH_2\!:\!CMe)_2SnBr_2$	167
$(CH_2\!:\!CH)BuSnCl_2$	168
$(ClCH\!:\!CH)_2SnCl_2$	168
$(ClCH_2)_2SnCl_2$	168
$(BrCH_2)_2SnBr_2$	168
$(MeCHCl)_2SnCl_2$	169
$(ClCH_2CH_2)_2SnI_2$ (?)	169
$(PrCHCl)_2SnCl_2$	169
$(p\text{-}ClC_6H_4)_2SnCl_2$	169
$(p\text{-}IC_6H_4)_2SnCl_2$	169
$(p\text{-}BrC_6H_4)_2SnCl_2$	169
$(o\text{-}MeC_6H_4)_2SnCl_2$	169
$(p\text{-}MeOC_6H_4)_2SnCl_2$	170
$(p\text{-}MeOC_6H_4)_2SnBr_2$	170
$(p\text{-}EtOC_6H_4)_2SnCl_2$	170
$(EtO_2CCH_2)_2SnBr_2$	170
$(EtO_2CCHMe)_2SnBr_2$	170
$(MeO_2CCH_2CH_2)_2SnBr_2$	170
$(p\text{-}EtO_2CC_6H_4)_2SnCl_2$	170
$(EtO_2CC_6H_4)_2SnBr_2$	171
$(EtO_2CC_6H_4)_2SnI_2$	171
$Ph(MeOCCH_2CH_2)SnBr_2$	171
$Ph(NCCH_2CH_2CH_2)SnBr_2$	171
$(Me_3SiCH_2)_2SnBr_2$	171
$(Me_3SiCH_2)_2SnI_2$	171
$(SCH\!:\!CHCH\!:\!C)_2SnCl_2$	171

$RSnX_3$

$MeSnCl_3$	172
$MeSnBr_3$	172
$MeSnI_3$	172
$EtSnCl_3$	173
$EtSnI_3$	173
$PrSnI_3$	173
$i\text{-}PrSnCl_3$	173
$BuSnCl_3$	173
$BuSnI_3$	174
$PhSnCl_3$	174
$PhSnI_3$	174
$o\text{-}MeC_6H_4SnCl_3$	174
$p\text{-}MeC_6H_4SnCl_3$	175
$1\text{-}C_{10}H_7SnCl_3$	175
$ClCH_2SnCl_3$	175
$BrCH_2SnBr_3$	175
$(MeCHCl)SnCl_3$	175
$p\text{-}BrC_6H_4SnBr_3$	175
$p\text{-}IC_6H_4SnBr_3$	176
$ClCH\!:\!CHSnCl_3$	176
$CH_2\!:\!CHSnCl_3$	176
$(CH_2\!:\!CHSnBr_3)_3$	176
$CH_2\!:\!CHCH_2SnBr_3 \cdot 2HBr$	176
$(CH_2\!:\!CHCH_2SnBr_3)_3$	176
$PhCH\!:\!CPhSnCl_3$	176
$o\text{-}MeO_2CC_6H_4SnCl_3$	176

$RSnO_2H$ and $RSnOX$

$MeSnO_2H$	177
$EtSnO_2H$	177
$CH_2\!:\!CHSnO_2H$	177
$CH_2\!:\!CHCH_2SnO_2H$	177
$BuSnO_2H$	178
$n\text{-}C_{12}H_{25}SnO_2H$	178
$PhSnO_2H$	178
$MeC_6H_4SnO_2H$	178
$p\text{-}ClC_6H_4SnO_2H$	178
$p\text{-}BrC_6H_4SnO_2H$	178
$o\text{-}MeO_2CC_6H_4SnO_2H$	179
$PhCH\!:\!CPhSnO_2H$	179
$1\text{-}C_{10}H_7SnO_2H$	179

	Page
$(-CH_2SnO_2H)_2$	179
$(-CH_2SnOCl)_2$	179
$(-CH_2SnONHCH_2CH_2NH_2)_2$	179
$CH_2(CH_2SnO_2H)_2$	179
$(-CH_2CH_2SnO_2H)_2$	179

$RSn(OCOR')_3$

$i\text{-}PrSn(OCOCHMeCH_2SCH_2Ph)_3$	180

$(R_2SnO)_n$

$(Me_2SnO)_n$	180
$(Et_2SnO)_n$	181
$[(CH_2{:}CH)_2SnO]_n$	182
$[(MeCHCl)_2SnO]_n$	182
$(Pr_2SnO)_n$	182
$(Bu_2SnO)_n$	181
$[(n\text{-}C_6H_{13})_2SnO]_n$	182
$[(n\text{-}C_8H_{17})_2SnO]_n$	182
$[(BuEtCHCH_2)_2SnO]_n$	182
$[(Me_3CCH_2CHMeCH_2CH_2)_2SnO]_n$	182
$(n\text{-}C_{12}H_{25})_2SnO$ (?)	182
$(Ph_2SnO)_n$	183
$[(MeC_6H_4)_2SnO]_n$	184
$[(p\text{-}ClC_6H_4)_2SnO]_n$	184
$[(p\text{-}BrC_6H_4)_2SnO]_n$	184
$[(p\text{-}IC_6H_4)_2SnO]_n$	184
$[(p\text{-}O_2NC_6H_4)_2SnO]_n$	184
$[(o\text{-}MeO_2CC_6H_4)_2SnO]_n$	184
$[(p\text{-}EtO_2CC_6H_4)_2SnO]_n$	184
$[(p\text{-}MeOC_6H_4)_2SnO]_n$	184
$[(p\text{-}EtOC_6H_4)_2SnO]_n$	184
$[(C_{10}H_7)_2SnO]_n$	184
$[(p\text{-}PhC_6H_4)_2SnO]_n$	184
$[(PhCH{:}CPh)_2SnO]_n$	184

$(R_2SnO)_n \cdot R_2SnX_2$

See Table	185

$R_2SnO \cdot R_2Sn(OH)X$

$Me_2SnO \cdot Me_2Sn(OH)I$	186
$Et_2SnO \cdot Et_2Sn(OH)I$	186
$i\text{-}Pr_2SnO \cdot i\text{-}Pr_2Sn(OH)I$	186
$i\text{-}Bu_2SnO \cdot i\text{-}Bu_2Sn(OH)I$	186
$(i\text{-}C_5H_{11})_2SnO \cdot (i\text{-}C_5H_{11})_2Sn(OH)I$	186

$R_2Sn(OR')_2$

See Table	187-188

$R_2Sn(OR')(OCOR'')$

See Table	189

$R_2Sn(OCOR')_2$, $R_2Sn(O_3SR')_2$, $R_2Sn[O_2As(O)R']$, and $R_2Sn[OAs(O)(OH)R']_2$

See Table	191-193

$R'O[R_2SnO]_xR'$, $R'CO_2[R_2SnO]_x\text{-}OCR'$, and $R'O[R_2SnO]_xR' \cdot R''_2SnX_2$

See Table	195-196
$HO[(PhCH_2)_2SnO]_3Ac$	197
$AcO(Bu_2SnO)_xAc$	197

Miscellaneous R_2Sn Salts

Me_2SnSO_3	197
$Me_2SnCrO_4 \cdot Me_2SnO$	197
$Me_2SnO_2As(O)OH$	197
Me_2SnMoO_4	197
Me_2SnWO_4	197
$Me_2Sn(IO_3)_2$	197
$Me_2Sn(VO_3)_2$ (?)	197

	Page
R_3SnOH	
Me_3SnOH	198
Et_3SnOH	198
$(CH_2{:}CH)_3SnOH$	199
Pr_3SnOH	199
Bu_3SnOH	199
Ph_3SnOH	199
$(p\text{-}MeC_6H_4)_3SnOH$	199
$(PhCH_2)_3SnOH$	200
$(p\text{-}ClC_6H_4)_3SnOH$	200
$(p\text{-}BrC_6H_4)_3SnOH$	200
$(o\text{-}MeO_2CC_6H_4)_3SnOH$	200
$(R_3Sn)_2O$ and $(R_2R'Sn)_2O$	
$(Me_3Sn)_2O$	200
$(Et_3Sn)_2O$	200
$(Pr_3Sn)_2O$	201
$(i\text{-}Pr_3Sn)_2O$	201
$(Bu_3Sn)_2O$	201
$(i\text{-}Bu_3Sn)_2O$	201
$[(n\text{-}C_6H_{13})_3Sn]_2O$	202
$[(n\text{-}C_8H_{13})_3Sn]_2O$	202
$(Ph_3Sn)_2O$	202
$[Bu_2(CH_2{:}CH)Sn]_2O$	202
$[(PhCH_2)_3Sn]_2O$	202
$[(o\text{-}BrC_6H_4CH_2)_3Sn]_2O$	202
$[(o\text{-}ClC_6H_4CH_2)_3Sn]_2O$	202
$[(o\text{-}FC_6H_4CH_2)_3Sn]_2O$	202
R_3SnOR'	
Et_3SnOEt	203
Et_3SnOPh	203
$Et_3SnOC_6H_4Me\text{-}p$	203
$Et_3SnOC_6H_4NO_2\text{-}p$	203
Pr_3SnOPh	203
Bu_3SnOBu	204
Bu_3SnOMe	204
$Bu_3SnOCMe_3$	204
$Bu_3SnOC_6H_{11}$	204
Bu_3SnOCH_2Ph	204
$Bu_3SnOCMe{:}CHAc$	204
$(n\text{-}C_8H_{17})_3SnOMe$	204

	Page
$Bu_3SnOC_{12}H_{25}$	204
$Bu_3SnOCMe_2CH_2Ac$	204
$Bu_2(C_9H_7)SnOMe$	204
$Bu_2(BuEtCHCH_2OCOCHPh)\text{-}SnOMe$	204
$(Et_3SnOCH_2CH_2\text{-})_2$	205
$Me_3SnO(CH_2)_3SiMe_2Et$	205
R_3SnOOR'	
$Me_3SnOOCMe_3$	205
$Et_3SnOOCMe_3$	205
$Et_3SnOOC_6H_4CHMe_2$	205
$Bu_3SnOOCMe_3$	205
$R_3SnOCOR'$	
See Table	206-209
$R_2R'SnOCOR''$	
See Table	210
$R_2\overline{SnR'COO}$ and $R_2\overline{SnSR'COO}$	
(R = divalent radical)	
$Pr_2\overline{SnCH_2CH_2COO}$	211
$Bu_2\overline{SnCH_2CH_2COO}$	211
$Ph_2\overline{SnCH_2CH_2COO}$	211
$Ph_2\overline{SnC_6H_4COO}$	211
$Bu_2\overline{SnSCH_2COO}$	211
$Bu_2\overline{SnSCH_2CH(NH_2)COO}$	211
$R_2SnO_2P\equiv$ and $R_{4-n}Sn(OP\equiv)_n$	
See Table	212

	Page		Page
$R_{4-n}Sn[OP(O)(X)R']_n$ and $R_{4-n}Sn[OP(O)R'R'']_n$ (X = R'' or OR''')		$R_2Sn(SCN)_2$, $R_{4-n}Sn[SC(S)OR']_n$, $R_2Sn[SC(S)NR'_2]_2$, and $R_2Sn[SP(S)(OR')_2]_2$	
See Table	214-215	See Table	226
$(R_3Sn)_2SO_4$ and $R_3SnOAs(O)(OH)R'$		$(R_3Sn)_2S$	
$(Me_3Sn)_2SO_4$	216	See Table	227
$(Et_3Sn)_2SO_4$	216		
$Ph_3SnOAs(O)(OH)C_6H_4NO_2$-p	216	R_3SnSR'	
$(RSnS)_2S$ and $(RSnSe)_2Se$		See Table	228-230
$(o\text{-}MeC_6H_4SnS)_2S$	216	R_3SnCH_2MgCl	231
$(MeSnSe)_2Se$	216		
$(EtSnSe)_2Se$	216	Organo-Sn derivs. of B and Al	
$RSn(SR')_3$		$(Me_3SnCH_2)_3B$	231
See Table	218-219	$Bu_2Sn[OB(OCH_2CHEtBu)_2]_2$	231
		$Ph_3SnHAl(OEt)_3$	231
$(R_2SnS)_n$		$Ph_2Sn[HAl(OEt)_3]_2$	231
See Table	220	$PhSn[HAl(OEt)_3]_3$	231
$[(Me_3SiCH_2)_2SnS]_n$	220	$(cyclo\text{-}C_6H_{11})_3SnHAl(OEt)_3$	231
$R_2Sn(SR')_2$		Organo-Sn Derivs. of Si and Ge	
See Table	222-225	$(Me_3SiCH_2)Me_2SnBr$	154
		$(Me_3SiCH_2)Bu_2SnBr$	154
$R_2Sn[SR'OB(OR'')_2]_2$ and $(R_2Sn)_3[(SR'O)_3B]_2$ (R = divalent radical)		$(Me_3SiCH_2)_2MeSnBr$	154
		$(Me_3SiCH_2)_2MeSnI$	154
		$(Me_3SiCH_2)_2BuSnI$	154
		$(Me_3SiCH_2)_2SnBr_2$	171
See Table	225	$(Me_3SiCH_2)_2SnI_2$	171
		$Me_2Sn(OSiMe_3)_2$	187
		$Me_3SnO(CH_2)_3SiMe_2Et$	205
		$[(Me_3SiCH_2)_2SnS]_n$	220
		$(Me_3SiCH_2)_3SnI$	232
		$Me_2Sn(CH_2SiMe_3)_2$	232
		$Me_2Sn(CH_2SiHMe_2)_2$	232
		$Me_2Sn[(CH_2)_3SiHMe_2]_2$	233

	Page
$(CH_2:CH)_2Sn(CH_2SiMe_3)_2$	233
$Bu_2Sn(CH_2SiHMe_2)_2$	233
$Bu_2Sn[CH_2Si(OEt)Me_2]_2$	233
$Bu_2Sn(CH_2SiMe_3)_2$	233
$Ph_2Sn(CH_2SiMe_3)_2$	233
$Bu_2(Me_3SiCH_2)SnOCOCF_3$	234
$Me_3SnCH_2SiMe_3$	234
$Me_3SnCH_2SiMeCl_2$	234
$Me_3SnCH_2SiMe(OMe)_2$	234
$Et_3SnCH_2CH_2SiCl_3$	234
$Et_3SnCH_2CH_2Si(OMe)_3$	234
$Bu_3SnCH_2SiHMe_2$	234
$Bu_3SnCH_2Si(OEt)Me_2$	235
$Bu_3SnCH_2SiMe_2CH_2CH_2SiMe_2Cl$	235
$(Bu_3SnCH_2SiMe_2)_2O$	235
$(Bu_3SnCH_2SiMe_2CH_2CH_2SiMe_2)_2O$	235
$Ph_3SnSiPh_4$	235
$(Me_3SiCH_2)_4Sn$	235
$(Me_3SiC_6H_4)_4Sn$	91, 235
$\overline{Me_2SiCH_2SnMe_2CH_2SiMe_2O}$	236
$\overline{Me_2Si(CH_2)_3SnMe_2(CH_2)_3SiMe_2O}$	236
$\overline{Me_2SiCH_2SnBu_2CH_2SiMe_2O}$	236
$[(Me_3SnCH_2)MeSiO]_x$	236
$[(Me_3SiCH_2)_2SnO]_x$	236
$[(-OR_2Si-)_n=OSnEt_2-]_x$	236
$\overline{Me_2SiCH_2SnMe_2CH_2SiMe_2O}$	242
$\overline{Me_2SiCH_2SnBu_2CH_2SiMe_2O}$	242
$\overline{Me_2Si(CH_2)_3SnMe_2(CH_2)_3SiMe_2O}$	242
$Ph_3SnGePh_3$	237

$R_{4-n}Sn(N=)_n$

	Page
$Et_3SnN(CO)_2C_6H_4$	237
$Bu_2Sn(\overline{NCOCH_2CH_2CO})_2$	237
$Bu_2Sn(\overline{NCH(O)CH_2Et})_2$ or $Bu_2Sn(ON:CHPr)_2$ (?)	237

	Page
$R_{4-n}Sn(NR'O_2SR'')_n$ and $R_2Sn(NR'O_2SR)(OR)$	
See Table	238

Organostannyl-Phosphonium Complexes

	Page
$[Me_2Sn(CH_2PPh_3)_2][Me_2SnBr_4]$	238
$[Me_2Sn(CH_2PPh_3)_2][Cr(NH_3)_2(SCN)_4]_2$	238
$[Me_2Sn(CH_2PPh_3)_2][BPh_4]_2$	238
$[Me_2Sn(CH_2PPh_3)_2][HgBr_2Cl]_2$	238
$[Me_3SnCH_2PPh_3][Me_3SnBr_2]$	239

$R_2Sn[Fe(CO)_4]$, $R_2Sn[Co(CO)_4]_2$, $(R_3Sn)_2[Fe(CO)_4]$, and $R_3Sn[Co(CO)_4]$

See Table 240

Organotin Heterocycles

	Page
$\overline{C_6H_4CH_2CH_2C_6H_4SnCl_2}$	241
$\overline{C_6H_4CH_2CH_2C_6H_4SnMe_2}$	241
$\overline{C_6H_4CH_2CH_2C_6H_4SnPh_4}$	241
$\overline{C_6H_4CH_2CH_2C_6H_4SnO}$	241
$\overline{C_6H_4CH_2CH_2C_6H_4SnS}$	242
$\overline{Bu_2SnSCH_2CH_2O}$	242
$\overline{SnCl_2CH_2SnCl_2CH_2SnCl_2O} \cdot Et_2O$	242
$\overline{Bu_2SnSCH_2CO_2CH_2CHMeOCOCH_2S}$	242
$\overline{Bu_2SnSBu_2SnSBu_2SnO}$ or $\overline{Bu_2SnOBu_2SnOBu_2SnS}$	242
$\overline{Me_2SiCH_2SnMe_2CH_2SiMe_2O}$	242
$\overline{Me_2SiCH_2SnBu_2CH_2SiMe_2O}$	242
$\overline{Me_2Si(CH_2)_3SnMe_2(CH_2)_3SiMe_2O}$	243

	Page
$(R_3Sn)_2$, and $(R_2SnX)_2$	
$(Me_3Sn)_2$	243
$(Et_3Sn)_2$	243
$(Pr_3Sn)_2$	244
$(Bu_3Sn)_2$	244
$(Ph_3Sn)_2$	244
$[(p-ClC_6H_4)_3Sn]_2$	245
$[(o-MeC_6H_4)_3Sn]_2$	245
$[(p-MeC_6H_4)_3Sn]_2$	245
$[(2-PhC_6H_4)_3Sn]_2$	245
$(p-MeC_6H_4)_3SnSnMe_3$	245
$(Et_2SnCl)_2$	246
$(Pr_2SnCl)_2$	246
$(Bu_2SnBr)_2$	246
$(Bu_2SnCl)_2$	246
$[(n-C_5H_{11})_2SnCl]_2$	246
$(Ph_2SnCl)_2$	246
$(Bu_2SnNO_3)_2$	246

$(R_3SnC?)_2$

See Table	247

$R_3SnR'SnR_3$ (R = divalent radical)

See Table	248

Miscellaneous Organotin Compounds

Et_3SnCN	249
Ph_3SnCN	249
$(Me_2Sn)_2[Fe(CN)_6]\cdot Me_2SnO$	249
$(Me_2Sn)_3[Fe(CN)_6]_2$	249
$Et_3Sn(NCO)$	249
$Et_3Sn(NCS)$	250
$EtOSnEt_2F$	250
$(Me_3Sn)_3OBr$	250
$(Me_3Sn)_3OI$	250
$ClSnR_2[OSnR_2]_xCl$	250
$-SnCl_2(SnCl_2CH_2)_n-$	250

	Page
$R'O[SnR_2O]_nSnR_2OR'$	250
$[-O-CR(OR)-O-SnR_2''-O-SnR_2'']_n$	251

R_2Sn

Et_2Sn	252
$(cyclo-C_5H_5)_2Sn$	252
Ph_2Sn	252
$(C_{14}H_9)_2Sn$	252
$(\overline{SCH:CHCH:C})_2Sn$	252
$\overline{C_6H_4CH_2CH_2C_6H_4}Sn$	253

LEAD

R_4Pb

Me_4Pb	255
Et_4Pb	257
Pr_4Pb	262
$i-Pr_4Pb$	263
Bu_4Pb	263
$i-Bu_4Pb$	263
$(n-C_5H_{11})_4Pb$	263
$(cyclo-C_6H_{11})_4Pb$	264
$(n-C_7H_{15})_4Pb$	264
$(n-C_{14}H_{29})_4Pb$	264
$(n-C_{16}H_{33})_4Pb$	264
Ph_4Pb	264
$(m-MeC_6H_4)_4Pb$	265
$(o-\text{and }p-MeC_6H_4)_4Pb$	266
$(PhCH_2)_4Pb$	266
$(CH_2:CH)_4Pb$	267
$(CH_2:CHCH_2)_4Pb$	267
$(o-FC_6H_4CH_2)_4Pb$	267
$(o-ClC_6H_4CH_2)_4Pb$	267
$(o-BrC_6H_4CH_2)_4Pb$	267
$(MeOC_6H_4)_4Pb$	268
$(EtOC_6H_4)_4Pb$	268
$(Me_2NC_6H_4)_4Pb$	268

	Page
R_3PbR'	
See Table	270-271
$R_2PbR'_2$	
See Table	272
$R_2R'R''Pb$	
See Table	273

Functionally Substituted $R_{4-n}SnR'_n$

$R_{4-n}SnR'_n$ Contg. Olefinic Groups

See Table	275

Halogen-Substd. $R_{4-n}SnR'_n$

See Table	275

$R_{4-n}SnR'_n$ Contg. Amino Groups

See Table	276

$R_{4-n}SnR'_n$ Contg. -COO- Groups

See Table	277

$R_{4-n}SnR'_n$ Contg. Hydroxy or Alkoxy Group

See Table	278

$R_{4-n}SnR'_n$ Contg. Mixed Functional Groups

See Table	280

	Page
$(\overline{SCH:CHCH:C})_4Pb$	281

$MPbR_3$ (M = alkali metal)

$NaPbEt_3$	281
$NaPb(cyclo-C_6H_{11})_3$	281
$NaPbPh_3$	281
$LiPbPh_3$	281
Me_3PbH	282

R_3PbX

Me_3PbCl	282
Me_3PbI	283
Et_3PbCl	283
Et_3PbBr	283
Bu_3PbCl	284
$(cyclo-C_6H_{11})_3PbI$	284
Ph_3PbCl	284
Ph_3PbBr	285
Ph_3PbI	285
For Other R_3SnX See Table	286

$R_2R'SnX$

See Table	287

R_2PbX_2

Et_2PbCl_2	288
$(ClCH:CH)_2PbCl_2$	289
Ph_2PbCl_2	289
For Other R_2SnX_2 See Table	290

$RPbX_3$

$EtPbI_3$	291
$PrPbI_3$	291
$BuPbI_3$	291

	Page		Page
$RPbO_2H$		$R_{4-n}Pb(OH)_n$	
$PhPbO_2H$	292	$Et_2Pb(OH)_2$	300
$p\text{-}MeC_6H_4PbO_2H$	292	$(H_2NC_6H_4)_2Pb(OH)_2$	300
$1\text{-}C_{10}H_7PbO_2H$	292	Me_3PbOH	300
		Et_3PbOH	300
$RPb(OCOR')_3$		Pr_3PbOH	301
		Bu_3PbOH	301
		$(cyclo\text{-}C_6H_{11})_3PbOH$	301
$PhPb(OAc)_3$	293	Ph_3PbOH	301
$PhPb(OCOCHMe_2)_3$	293		
$p\text{-}MeC_6H_4Pb(OAc)_3$	293	$R_{4-n}Pb(OR')_n$	
$PhPb(OBz)_3$	294		
$p\text{-}MeC_6H_4Pb(OCOCMe:CH_2)_3$	294	$Et_2Pb(OC_6H_4NO_2\text{-}p)_2$	302
$1\text{-}C_{10}H_7Pb(OAc)_3$	294	$Et_3PbOC_6H_2(NO_2)_3$	302
$1\text{-}C_{10}H_7Pb(OCOCHMe_2)_3$	294	Et_3PbOEt	302
$1\text{-}C_{10}H_7Pb(OBz)_3$	294		
		$R_2Pb(O_3SR')_2$	
R_2PbO			
		See Table	302
Me_2PbO	295		
Et_2PbO	295		
Ph_2PbO	295	$R_3Pb(O_3SR')$ and $R_3Pb(O_2SR')$	
$(p\text{-}MeC_6H_4)_2PbO$	295		
$(O_2NC_6H_4)_2PbO$	295	See Table	303
		$(Et_3Pb)_2CO_3$	304
$R_2Pb(OCOR')_2$			
		$R_3PbOCOR'$	
Et_2PbCO_3	296		
$Et_2Pb(OAc)_2$	296	See Table	305-308
$Et_2Pb(OCOCH_2Cl)_2$	296		
$(ClCH:CH)_2Pb(OAc)_2$	296	R_2PbS, $(R_3Pb)_2S$, R_3PbSR', and	
$(ClCH:CH)_2Pb(OBz)_2$	297	$R_{4-n}Pb(SCOR')_n$	
$(SCH:CHCH:C)_2Pb(OCOCHMe_2)_2$	297		
$Ph_2Pb(OAc)_2$	297	Ph_2PbS	309
$Ph_2Pb(OCOCHMe_2)_2$	297	$(Et_3Pb)_2S$	309
$Ph_2Pb(OBz)_2$	297	Et_3PbSEt	309
For Other $R_2Sn(OCOR')_2$		Et_3PbSPh	310
See Table	298-299	$Et_2Pb(SAc)_2$	310
		Et_3PbSAc	310

	Page
$R_3PbNCOR'CO$ (R = divalent radical)	
See Table	310
$R_3PbNR'SO_2R''$ and $R_3PbNCOR'''SO_2$ (R = divalent radical)	
See Table	311
$R_{4-n}PbX_n$ (X = CN, CNO, CNS, CNSe, NO_3, SO_3, SeO_3, SO_4 and $-O_2P(O)F$)	
See Table	312–314

$(R_3Pb)_2$

$(Me_3Pb)_2$	315
$(Et_3Pb)_2$	315
$(Pr_3Pb)_2$	316
$(Bu_3Pb)_2$	316
$[(cyclo-C_6H_{11})_3Pb]_2$	316
$(Ph_3Pb)_2$	316
$[(m-MeC_6H_4)_3Pb]_2$	317
$[(o-MeC_6H_4)_3Pb]_2$	317
$[(p-MeC_6H_4)_3Pb]_2$	317
$[(o-MeOC_6H_4)_3Pb]_2$	317
$[(p-MeOC_6H_4)_3Pb]_2$	317
$[(o-EtOC_6H_4)_3Pb]_2$	317
$[(p-EtOC_6H_4)_3Pb]_2$	318
$[(2,4,6-Me_3C_6H_2)_3Pb]_2$	318
$[(o-ClC_6H_4CH_2)_3Pb]_2$	318
$[(o-BrC_6H_4CH_2)_3Pb]_2$	318
$[(1-C_{10}H_7)_3Pb]_2$	318

$(R_3PbC!)_2$

See Table	319

	Page
$R_3PbR'ObR_3$ (R' = divalent radical)	
See Table	320
$[R_2PbFe(CO)_4]_2$, $(R_3Pb)_2Fe(CO)_4$, and $R_3PbCo(CO)_4$	
See Table	321

Miscellaneous Organolead Compounds

$Ph_2Pb(C_6H_4CO_2H-o)Cl$	322
$Ph_2Pb(C_6H_4CO_2H-o)OH$	322
$Ph_3SiOPbPh_3$	322
$Ph_2PbC_6H_4COO$	322

R_2Pb

Et_2Pb	323
$[(EtO_2C)_2CH]_2Pb$	323
$(C_5H_5)_2Pb$	323
$(C_{13}H_9)_2Pb$	323
$(Me_2NC_6H_4)_2Pb$	323
$[m-O_2NC_6H_4C(NO_2)_2]_2Pb$	323

MIX
Papier aus verantwortungsvollen Quellen
Paper from responsible sources
FSC® C105338

If you have any concerns about our products,
you can contact us on
ProductSafety@springernature.com

In case Publisher is established outside the EU,
the EU authorized representative is:
**Springer Nature Customer Service Center GmbH
Europaplatz 3, 69115 Heidelberg, Germany**

Printed by Libri Plureos GmbH
in Hamburg, Germany